Quadratic Formula

If $ax^2 + bx + c = 0$, for $a \neq 0$, then

$$x = \frac{-b \pm \sqrt{b^2 - 4ac}}{2a}$$

Cartesian Coordinates

Let $P_1 = (x_1, y_1)$ and $P_2 = (x_2, y_2)$; then:

1 The distance d between P_1 and P_2 is

$$d = \sqrt{(x_2 - x_1)^2 + (y_2 - y_1)^2}$$

2 The slope m of the line segment $\overline{P_1 P_2}$ is

$$m = \frac{y_2 - y_1}{x_2 - x_1}$$

3 The slope intercept form of a line is

$$y = mx + b$$

4 The point slope form of a line is

$$y - y_1 = m(x - x_1)$$

Logarithm Properties

1 $\log_b MN = \log_b M + \log_b N$

2 $\log \dfrac{M}{N} = \log_b M - \log_b N$

3 $\log_b M^k = k \log_b M$

Conic Sections

1 Circle $x^2 + y^2 = r^2$

2 Parabola $y^2 = 4cx$ or $x^2 = 4cy$

3 Ellipse $\dfrac{x^2}{a^2} + \dfrac{y^2}{b^2} = 1$

4 Hyperbola $\dfrac{x^2}{a^2} - \dfrac{y^2}{b^2} = 1$ or $\dfrac{y^2}{b^2} - \dfrac{x^2}{a^2} = 1$

Intermediate Algebra

Intermediate Algebra

FOURTH EDITION

M. A. MUNEM

W. TSCHIRHART

MACOMB COMMUNITY COLLEGE

WORTH PUBLISHERS, INC.

INTERMEDIATE ALGEBRA, FOURTH EDITION

PRINTED IN THE UNITED STATES OF AMERICA .

LIBRARY OF CONGRESS CATALOG CARD NUMBER: 87-51217

ISBN: 0-87901-377-X

FIRST PRINTING, JANUARY 1988

PRODUCTION: BARBARA ANNE SEIXAS

DESIGN: MALCOLM GREAR DESIGNERS

TYPOGRAPHER: SYNTAX INTERNATIONAL PTE LTD

PRINTING AND BINDING: VON HOFFMANN PRESS, INC.

COVER PHOTOGRAPHY: PAUL CLANCY

PHOTO CREDITS: **p. 62** NASA; **p. 141** Donald Dietz/Stock, Boston; **p. 170** Peter Menzel/Stock, Boston; **p. 183** Owen Franken/ Stock, Boston; **p. 241** The Bettmann Archive; **p. 327** Ira Kirschenbaum/Stock, Boston; **p. 301** Owen Franken/Stock, Boston; **p. 331** The Bettmann Archive; **p. 356** Swiss National Tourist Office; **p. 425** UPI/Bettmann Newsphoto; **p. 482** Marty Stouffer/ Animals Animals

WORTH PUBLISHERS, INC.

33 IRVING PLACE

NEW YORK, NEW YORK 10003

Preface

In this fourth edition of *Intermediate Algebra*, we have made many improvements intended to give students a better preparation for courses in college algebra and precalculus mathematics. We have also added many new examples, problems, and applications that should help students to put their new skills to use.

Prerequisites

One year of high-school algebra or an equivalent course in beginning algebra is sufficient background to do well in this course. Conscientious students with less preparation should be able to master the material in this book if time is devoted to the review topics in Chapters 1–3. The *Study Guide* will also help these and other students who need additional guidance or practice.

Important Features

We have taken care to ensure that the average student will be able to read and follow the development of each chapter. Topics are presented in brief sections that progress logically. We frequently remind students of what they know before we introduce new material, always starting with familiar ideas. Several features of the book contribute to its effectiveness:

1. *Examples*   Each concept and procedure is clearly explained and illustrated with worked-out examples, showing all substitutions. We have re-examined all of the examples in the book to be sure that they are clear and concrete, and we have added many new examples.

2. *Procedural Guides*   Concise, step-by-step guides explain the basic algebraic operations and show students "how to do it."

3. *Problem Sets*   A set of problems follows each brief section. Each problem set is graded in difficulty, building from very simple drill problems that parallel the examples to those that require greater skill.

4. *Review Problem Sets*   Review problems cover the entire chapter, giving students an opportunity to check their mastery of all topics before moving on. There are a total of 5,054 problems in the section and review problem sets. The answers to all odd-numbered problems are provided at the end of the book.

5. *Chapter Tests*   New to this edition is a brief test at the end of each chapter that students can take for practice and to check their own readiness for a class test. All test answers are provided.

6. *Word Problems and Applications*   We give a great deal more attention in this edition to helping students translate verbal expressions into algebraic expressions. There are many more applications, both in examples and in problem sets, to show students how algebra can be used in their personal lives and in their other studies. There are sections devoted entirely to applications in Chapters 4, 6, 8, and 9, and additional applications throughout the book.

7. *Use of Calculators*   Problems designed to be solved with a calculator are labeled with the symbol $\boxed{c}$

8. *Common Student Errors*   These are clearly identified at appropriate places in the book. We warn students about common errors in such a way that there will be no confusion about what is correct and what is not.

The Major Improvements

We have rewritten and reorganized many topics and added new material. Some of the major changes:

1. A new section on the *operations of fractions and their applications* has been added to Chapter 1.

2. *Scientific notation*, previously in Chapter 8, is now covered in Chapters 1 and 5.

3. A new Section 4.4 on *translating verbal expressions into algebraic form* prepares students for translating word problems from different fields into equations in Section 4.5.

4. Section 4.6 now includes the *union and intersection of two sets*.

5. The section on *equations and inequalities involving absolute values* has been expanded to Sections 4.8 and 4.9.

6. Section 5.1 now shows how to calculate *car and mortgage payments*. The section on *operations involving radicals* has been expanded to two sections (5.4 and 5.5), and additional applications are now included.

7. Chapters 7 and 8 have been reorganized, with many added examples and applications.

8. Basic concepts from *geometry* are now introduced in Chapter 12.

Student Aids

The *Study Guide* has been thoroughly revised with this edition. The Guide now includes worked-out solutions to every other odd-numbered problem in the textbook, as well as study objectives, semiprogrammed problems for the most difficult material in each chapter, and additional word problems. There are two practice tests for each chapter—a multiple-choice test and a problem-solving test. A cumulative review problem set for each chapter, which includes representative problems from all prior chapters, has been added to prevent the dulling of skills acquired earlier in the course. The answers to the cumulative reviews and the chapter tests are included in the *Study Guide*.

Computer tutorial programs, for use in a computer lab or for individual student purchase, have been developed for this new edition. Key topics from the chapters, for use on an IBM P.C. or the Apple II family, give students additional practice in mastering the material.

Instructor Aids

The *Instructor's Resource Manual* begins with a placement test, which has been used successfully with many thousands of students. This test efficiently assesses each student's ability in order to designate placement in the appropriate course. The manual also includes chapter overviews and suggestions for covering the material in one-quarter and one-semester courses. The testing resources include a bank of questions for the whole book and two exams for each chapter. Finally, the *Instructor's Resource Manual* contains the *answers for all even-numbered problems* in the textbook and all answers for the chapter tests and the final examination in the manual.

Acknowledgments

By the time a book reaches its fourth edition, a great many people have contributed to its development. We wish to thank the many users and reviewers of previous editions for their helpful suggestions, especially William Coppage, *Wright State University*; Linda Exley, *DeKalb Community College, Clarkston*; Merle Friel, *Humboldt State University*; Gene Hall, *DeKalb Community College, Clarkston*; Gus Pekara, *Oklahoma City Community College*; Howard E. Taylor, *West Georgia College* (Emeritus); Stuart Thomas, *Oregon State University*; Roger Willig, *Montgomery County Community College*; and Jim Wolfe, *Portland Community College*. Thanks again to Murray Blose and Gerald Goff at *Oklahoma State University*, who developed the placement test.

We are also grateful for the many helpful suggestions from these reviewers of the fourth edition: John E. Butcher, *Kean College of New Jersey*; Kathleen Seagraves Higdon, *Oregon State University*; Julia R. Monte, *Daytona Beach Community College*; and Laurie Pieracci, *Sierra Community College*.

Thanks are due to our many colleagues at *Macomb Community College* who have taught from previous editions of the book and shared their experiences with us. We especially wish to thank Steve Fasbinder of *Oakland University* and Wayne Hille of *Wayne State University*, who assisted in the proofreading and solved all the problems in the book. We are also grateful, as always, to the staff of Worth Publishers.

M. A. Munem
W. Tschirhart

Contents

1 Numbers and Their Properties

Algebra is a generalization of arithmetic in which letter symbols, such as x or y, are used to represent numbers. An understanding of arithmetic is therefore an essential foundation for the study of algebra. This will become clear as you review the basic concepts in this chapter.

1.1 Symbols and Notation of Algebra

Algebra begins with a systematic study of the operations and rules of arithmetic. The operations of *addition, subtraction, multiplication* and *division* serve as a basis for all arithmetic calculations. The symbols used for these operations are

$$+, \; -, \; \times \text{ or } \cdot, \text{ and } \div$$

In mathematics, the words that describe the four operations are the *sum, difference, product* and *quotient*. In algebra, letters such as $a, b, c, x,$ or y are used to represent particular numbers. Thus, if a and b are two numbers, then the *operations symbols* are described in the following table:

Operation	Symbols	Words
Addition	$a + b$	The **sum** of a and b
Subtraction	$a - b$	The **difference** of a and b
Multiplication	$ab, a \cdot b, a(b), (a)b, (a)(b)$	The **product** of a and b (or a times b)
Division	$a \div b, a/b, \text{ or } \dfrac{a}{b}$	The **quotient** of a and b

For example, we say that the sum of 3 and 5 is 8, which means that both $3 + 5$ and 8 represent the sum of 3 and 5. We also say that the product of 2 and 7 is 14, which means that both $2 \cdot 7$ and 14 represent the product of 2 and 7.

Additional symbols are used for comparing numbers and expressions. Thus, if a and b are numbers, the *comparison symbols* are described in the following table:

Symbols	Words
$a = b$	a is equal to b
$a \neq b$	a is not equal to b
$a < b$	a is less than b
$a \leq b$	a is less than or equal to b
$a > b$	a is greater than b
$a \geq b$	a is greater than or equal to b

EXAMPLE 1   Translate each statement into symbols.

(a) Nine equals four plus five.

(b) The difference of x and 3 is less than 2.

(c) The sum of twice x and 4 is greater than 5.

(d) The quotient of y and 2 is 7.

SOLUTION   The statements are translated into symbols as follows:

(a) $9 = 4 + 5$ (b) $x - 3 < 2$

(c) $2x + 4 > 5$ (d) $\dfrac{y}{2} = 7$

EXAMPLE 2   Write equivalent word statements for each statement represented by the symbols.

(a) $3x - 5 \leq 8$ (b) $\dfrac{u}{5} = 3$

(c) $7y + 3 > 2$ (d) $2(b - 9) \geq 8$

SOLUTION   The equivalent word statements are as follows:

(a) The difference of three times x and 5 is less than or equal to 8.

(b) The quotient of u and 5 is 3.

(c) The sum of seven times y and 3 is greater than 2.

(d) Twice the difference of b and 9 is greater than or equal to 8.

Another notation is designed to clarify ideas and simplify calculations by permitting us to write some expressions compactly and efficiently. The use of *exponents* allows us to write repeated multiplication in a more compact form. For example, $3 \cdot 3$ can be written simply as 3^2, and $3 \cdot 3 \cdot 3 \cdot 3$ as 3^4. The expression 3^4 is called an **exponential form,** while $3 \cdot 3 \cdot 3 \cdot 3$ is called the **expanded form** of 3^4. In 3^4, 3 is called the **base** and 4 is called the **exponent** or the **power** to which the base is raised. The exponent 4 tells us the number of times the base 3 appears in the product. That is,

$$3^4 = 3 \cdot 3 \cdot 3 \cdot 3 = 81.$$

We say that the **value** of 3^4 is 81. Another way of saying "find the value of" is "simplify." For instance, we simplified 3^4 as 81. In Examples 3–5, identify the base and the exponent, then write each expression in expanded form and find the value of each expression.

EXAMPLE 3 6^2

SOLUTION   The base is 6 and the exponent is 2.

$$6^2 = 6 \cdot 6 = 36.$$

EXAMPLE 4 5^3

SOLUTION   The base is 5 and the exponent is 3.

$$5^3 = 5 \cdot 5 \cdot 5 = 125.$$

EXAMPLE 5 2^6

SOLUTION   The base is 2 and the exponent is 6.

$$2^6 = 2 \cdot 2 \cdot 2 \cdot 2 \cdot 2 \cdot 2 = 64.$$

Order of Operation

In evaluating mathematical expressions involving more than one operation, we often use **grouping symbols** such as

parentheses   (   )

brackets   [   ]

braces   {   }

to indicate which operations are to be performed first. For example, to evaluate the expression

$$4 \cdot (5 + 7),$$

we do the addition within the parentheses first, then multiply. Thus,

$$4 \cdot (5 + 7) = 4 \cdot 12 = 48.$$

We often indicate multiplication without the multiplication symbol. For example, the expression $4 \cdot (5 + 7)$ may be written as $4(5 + 7)$.

Now we consider the problem of evaluating a mathematical expression that involves more than one operation but does not contain grouping symbols. For instance, suppose that we wish to evaluate the mathematical expression

$$4 \cdot 5 + 7.$$

In this case, it is not obvious which operation is to be performed first, the addition or the multiplication. If we add 5 and 7 first and then multiply by 4, we get 48 as an answer. That is

$$4 \cdot 5 + 7 = 4 \cdot 12 = 48.$$

On the other hand, if we multiply 4 and 5 first and then add 7, we get 27 as an answer. That is,

$$4 \cdot 5 + 7 = 20 + 7 = 27.$$

To decide which answer is correct, we must agree on which operations are to be done first. The following rules tell us the order in which operations are to be performed.

RULE 1   **Order of Operations**

1. Work from within the innermost grouping symbols to the outermost.
2. Find any powers indicated by exponents.
3. Perform all multiplications and divisions working from left to right.
4. Perform all additions and subtractions working from left to right.

In Examples 6–10, find the value of each expression using the rule for the order of operations.

EXAMPLE 6 $37 + 8 \cdot 2$

SOLUTION First, we perform the multiplication, then the addition.

$$37 + 8 \cdot 2 = 37 + 16 \qquad \text{(We multiplied } 8 \cdot 2\text{)}$$
$$= 53. \qquad \text{(We added 37 and 16)}$$

EXAMPLE 7 $3 \cdot 5 + 2(7 + 3)$

SOLUTION
$$3 \cdot 5 + 2(7 + 3) = 3 \cdot 5 + 2 \cdot 10 \qquad \text{(We added the numbers inside the parentheses)}$$
$$= 15 + 20 \qquad \text{(We multiplied)}$$
$$= 35. \qquad \text{(We added)}$$

EXAMPLE 8 $9 + 12 \div 4 - 2 \cdot 3$

SOLUTION
$$9 + 12 \div 4 - 2 \cdot 3 = 9 + 3 - 6 \qquad \text{(We divided and multiplied from left to right)}$$
$$= 12 - 6 \qquad \text{(We added)}$$
$$= 6. \qquad \text{(We subtracted)}$$

EXAMPLE 9 $3 \cdot 2^3 + 10 \div 5 - 3^2$

SOLUTION
$$3 \cdot 2^3 + 10 \div 5 - 3^2 = 3 \cdot 8 + 10 \div 5 - 9 \qquad \text{(We found the value of each number with an exponent)}$$
$$= 24 + 2 - 9 \qquad \text{(We multiplied and divided)}$$
$$= 26 - 9 \qquad \text{(We added)}$$
$$= 17. \qquad \text{(We subtracted)}$$

EXAMPLE 10 $5[4^3 + 3(6^2 - 3 \cdot 2)]$

SOLUTION
$$5[4^3 + 3(6^2 - 3 \cdot 2)] = 5[64 + 3(36 - 6)] \qquad \text{(We found the value of each number with an exponent and multiplied)}$$
$$= 5[64 + 3 \cdot 30] \qquad \text{(We subtracted numbers inside parentheses)}$$
$$= 5[64 + 90] \qquad \text{(We multiplied inside the bracket)}$$
$$= 5[154] \qquad \text{(We added inside the bracket)}$$
$$= 770. \qquad \text{(We multiplied)}$$

To simplify a mathematical expression of the form

$$\frac{9 + 5}{4 + 3}$$

we divide the simplified value of the **numerator** by the simplified value of the **denominator.** Thus

$$\frac{9+5}{4+3} = \frac{14}{7} = 2.$$

The horizontal bar symbol implies that grouping symbols are around the numerator and around the denominator. For instance,

$$\frac{9+5}{4+3} \text{ is the same as } (9+5) \div (4+3).$$

EXAMPLE 11   Simplify

$$\frac{5^2 + 2 \cdot 5}{3 + 10 \div 5}$$

SOLUTION   We find the values of the numerator and the denominator separately. The value of the numerator is

$$5^2 + 2 \cdot 5 = 25 + 10 = 35.$$

The value of the denominator is

$$3 + 10 \div 5 = 3 + 2 = 5.$$

Therefore,

$$\frac{5^2 + 2 \cdot 5}{3 + 10 \div 5} = \frac{35}{5} = 7.$$

PROBLEM SET 1.1

In problems 1–14, translate each statement into symbols.

1. The sum of y and 3 is 15.
2. Five times the product of 7 and x is 10.
3. The difference of t and 3 is 7.
4. The sum of three-eighths of a number x and five-twelfths of a number x is 48.
5. 2 less than a number x is greater than 5.
6. The quotient of 7 and a number x is 4.
7. The product of a number x and 7 decreased by 4 is less than or equal to 35.
8. A number n added to the product of 5 and the sum of n and 4 is greater than 12.
9. The quotient of y and 3 is equal to the difference of y and 5.
10. Five times the sum of x and 4 is equal to 21.

11. Twice a number x plus the product of 2 more than the number x and 3 is less than 5.

12. A number n less one-fourth of the number n is greater than or equal to 5 plus the number n.

13. Three times the product of y and 8 is less than or equal to seven times the sum of y and 8.

14. The difference of x and 8 is less than the product of x and 8.

In problems 15–28, write an equivalent word statement for each given statement represented by the symbols.

15. $x + 2 < 10$

16. $3y + 5 \neq 14$

17. $3x \leq 15$

18. $7y > 21$

19. $3t + 2 > 8$

20. $6x + 3 > 9$

21. $10x - 2 \neq 15$

22. $7x + 8 \leq 15$

23. $5y + 1 \geq 6$

24. $3u - 1 < 5$

25. $\dfrac{u}{4} + 3 = 7$

26. $\dfrac{y}{2} - 1 = 8$

27. $\dfrac{x}{3} - \dfrac{5}{8} = \dfrac{3}{8}$

28. $-5x + \dfrac{2x}{3} \geq 5$

In problems 29–40, identify the base and the exponent, then write each expression in expanded form and find the value of each expression.

29. 5^2

30. 4^3

31. 10^3

32. 7^2

33. 2^7

34. 5^4

35. 3^5

36. 6^3

37. 11^2

38. 7^3

39. 9^3

40. 13^2

In problems 41–84, find the value of each expression by using the rule for the order of operations.

41. $5 \cdot 7 - 4$

42. $7 \cdot 3 + 5 \cdot 4$

43. $9 \cdot 6 + 11 - 2$

44. $17 - 3 \cdot 5 + 7$

45. $4 + 4 \cdot 4 - 4 \div 4$

46. $(6 + 4) \div 5 + 3$

47. $8 - (4 \div 2) + 7$

48. $(6 - 4) \div 2 + 5$

49. $6 + 2 - 7 + 9 \cdot 2 \div 3$

50. $3[2 + 3(7 - 4)]$

51. $8[3 + (9 - 2)] + 6$

52. $(12 - 3) \div (16 - 9)$

53. $8 - (12 - 2) \div (4 + 1)$

54. $3[(16 \div 8) + 2] + 4$

55. $8[(4 - 2) \cdot 5 - 3] + 1$

56. $15 \cdot (6 - 4) \div 3 + 3$

57. $5 \cdot 6 + 11$

58. $5(6 + 11)$

59. $2^2 + 3^2$

60. $2 \cdot 7^2 - 3 \cdot 2^2$

61. $(9 + 6)^2 \div 5$

62. $(9 + 7)^2 \div 2^3$

63. $7 + (12 - 9)^2 \cdot 2 + 3$

64. $12 + 3 \cdot 4^3 \div 4^2 - 5$

65. $3^2 + 4^2$

66. $(3 + 4)^2 - 2 \cdot 3 \cdot 4$

67. $(7 - 3)^2$

68. $7^2 + 3^2 - 2 \cdot 3 \cdot 7$

69. $2^2 + 3^2 + 4^2$

70. $(2 + 3 + 4)^2$

71. $3^4 - 2^4$

72. $(3 - 2)^4 + 7^3$

73. $9 \cdot 2^2 + 10 \div 5 + 4^3$

74. $100 - 20 \div 2 + 4 \cdot 2 + 2^5$

75. $2^3[16^2 - (4^2 + 3^2)]$

76. $3[3^2 + (6 - 3)^3] + 2^2$

77. $15^2 - [11 \cdot 5 + 4(2 \cdot 4 - 3)]$

78. $15^2 + [7^2 - (5 \cdot 2 - 3)]$

79. $3^3 + 2^3 + 3 \cdot 3^2 \cdot 2 - 3 \cdot 3 \cdot 2^2$

80. $5^3 + 3^3 + 3 \cdot 5^2 \cdot 3 - 3 \cdot 5 \cdot 3^2 + 3^3$

81. $\dfrac{24 + 6^2 \div 3}{7 + 6 \cdot 3}$

82. $\dfrac{8^2 + 12 \div 3 - 2^2}{10 \div 2}$

83. $\dfrac{25^2 + 18 \div 2 \cdot 3 + 8 \cdot 5}{2 \cdot 2}$

84. $\dfrac{3^2(3 + 2) + 3(3 + 2)^2}{5 \cdot 2 - 2(3 - 1)^2}$

85. Suppose you are given a number: add 3 to the number, multiply the sum by 6, divide the product by 2, subtract 9, then divide the difference by 3. Now write an expression, first using the number 5, and then 7, to describe this process. Also find the value of each expression.

86. Suppose that Debbie's age is 26: subtract 5 from her age, multiply the difference by 16, divide the product by 8, add 10, then divide the result by 2. Now write a mathematical expression to describe this process, and find the value of the expression. Repeat the procedure for Ilene's age, which is 35, and for Lisa's age, which is 16. How about doing your age while you are at it?

1.2 Sets of Numbers

The idea of a *set* allows us to classify numbers. A set may be thought of as a collection of objects. Any one of the objects in a set is called an **element** or a **member** of the set. Capital letters, such as A, B, C, and D, or braces $\{\ \}$ enclosing the elements in a set are often used to denote sets. Thus, if we write $A = \{1, 2, 3, 4, 5\}$, we mean the set A whose elements are the numbers 1, 2, 3, 4, and 5. The symbol used to show that an element belongs to a set is $\in$. In our example,

$1 \in A$ is read "1 is an element (member) of set A"

$2 \in A$ is read "2 is an element of set A"

$3 \in A$ is read "3 is an element of set A"

$4 \in A$ is read "4 is an element of set A"

$5 \in A$ is read "5 is an element of set A."

The symbol $\notin$ is used to indicate that an element does not belong to a set. The expression $6 \notin A$, for example, is read "6 is *not* an element of A." If a set is so defined that it does not have any element, we call it an **empty set** or **null set**. The symbol for an empty set is $\varnothing$ or $\{\ \}$. For example, the set P of all presidents of the United States who died before their thirty-fifth birthday is an empty set, since one must be at least 35 years old to qualify for the presidency.

Sometimes it is inconvenient or impossible to list the elements of a set. In such cases, we often use **set-builder notation.** For example, suppose that P is the set of all presidents of the United States. Using set-builder notation, we write

$$P = \{x \mid x \text{ is a president of the United States}\},$$

which is read "the set of all x, such that x is a president of the United States."

Two sets are said to be **equal** if they contain exactly the same elements.

For example,

$$\{1, 2, 3\} = \{3, 1, 2\},$$

because both sets contain the same elements. Set A is called a **subset** of a set B, and is written $A \subseteq B$, if every element in A is also an element of B. That is,

$A \subseteq B$ if and only if A is contained in B.

For example, the set $A = \{1, 2, 3\}$ is a subset of a set $B = \{1, 2, 3, 5\}$, since every element in set A is also an element of set B.

EXAMPLE 1   List all possible subsets of $C = \{1, 2, 3\}$.

SOLUTION $\varnothing$, $\{1\}$, $\{2\}$, $\{3\}$, $\{1, 2\}$, $\{1, 3\}$, $\{2, 3\}$, and $\{1, 2, 3\}$.

A set A is said to be a **finite set** if the elements in A can be counted, in the usual way, using counting numbers $1, 2, 3, 4, \ldots$, and if this counting process eventually terminates, resulting in a specific number n. A set that is neither finite nor empty is called an **infinite set.** For example, the set of odd counting numbers $A = \{1, 3, 5, 7, \ldots\}$, where the dots mean the set continues in the same manner, is an infinite set. Whereas, the set $B = \{2, 4, 6, \ldots, 500\}$ is a finite set.

The Real Numbers

Sets of numbers can be visualized by using a **number line** or a **coordinate axis.** To construct a number line, we draw a horizontal line L, with arrowheads at both ends to show that the line extends endlessly in both directions. Then we choose a point O on this line, and we associate it with the number 0. This point is called the **origin.** Next, we select another point U on the number line, called the **unit point.** We associate the number 1 with the point U. The distance between the origin O and the unit point U is called the **unit distance.** This distance may be 1 inch, 1 centimeter, or one unit of whatever measure you choose (Figure 1a).

Figure 1

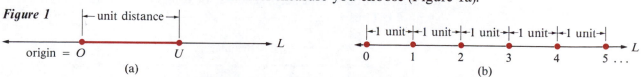

(a)

(b)

By measuring one unit distance to the right of the point U, we find a point that we associate with the number 2. Repeating this process, we can find points to associate with $3, 4, 5, \ldots$, (the three dots shown here and in Figure 1b mean "and so on").

If the line *L* is drawn horizontally, the positive numbers are usually represented to the right of the origin *O* and the negative numbers to the left. An arbitrary unit length is chosen, and points corresponding to the positive and negative numbers are marked off from *O* (Figure 2).

Figure 2

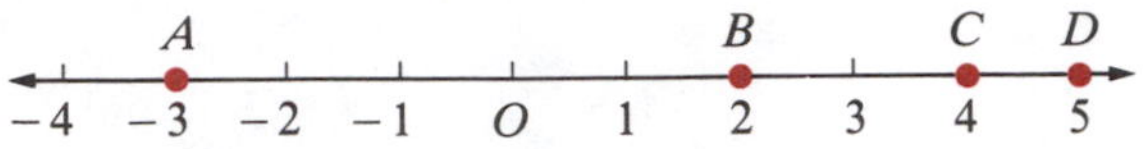

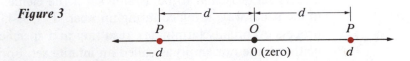

The point associated with a number on a number line is called the **graph** of that number, and the number is called the **coordinate** of that point. For example, the coordinates of points *A*, *B*, *C*, and *D* on the number line in Figure 2 are −3, 2, 4, and 5, in that order.

The origin *O* is assigned the coordinate 0 (zero). If *d* is the distance between the origin and a point *P*, then the coordinate of *P* is *d* or −*d*, depending on whether *P* is to the right or to the left of the origin (Figure 3).

Figure 3

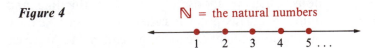

In algebra, we encounter many different sets of numbers. Following is a summary of the common sets (and their symbols) used throughout this textbook. Note that we illustrate a set of numbers on a number line by *coloring* the points whose coordinates are members of the set.

1. The **natural numbers,** also called **counting numbers** or **positive integers,** are the numbers 1, 2, 3, 4, 5, . . . The set of natural numbers {1, 2, 3, 4, 5, . . .}, represented by the symbol $\mathbb{N}$, is illustrated in Figure 4.

Figure 4

$\mathbb{N}$ = the natural numbers

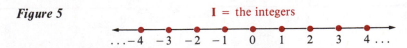

2. The **integers** consist of all the natural numbers, the negatives of the natural numbers, and zero. The set of integers is represented by the symbol **I**; Thus **I** = {. . . , −4, −3, −2, −1, 0, 1, 2, 3, 4, . . .} (Figure 5).

Figure 5

I = the integers

3. The **rational numbers** are the numbers that can be written in the form *a*/*b*, in which *a* and *b* are integers and *b* ≠ 0. Because *b* may equal 1, every integer is a rational number. The set of all rational numbers

in which a and b are integers and $b \neq 0$ is often represented by $\mathbb{Q}$. Some numbers in the set $\mathbb{Q}$ are

$$\frac{1}{2}, \frac{4}{3}, \frac{30}{7}, \frac{20}{4}, -\frac{3}{4}, -\frac{11}{5}, -\frac{15}{3}, -\frac{2}{3}, \text{ and } -\frac{5}{8}$$

(Figure 6). In the expression a/b, a is called the numerator and b is called the denominator.

Figure 6

4. The **irrational numbers** are the numbers that cannot be expressed as a quotient of two integers. Some examples of these numbers are $\sqrt{2}$, $\sqrt{3}$, and π.

5. The **positive real numbers** correspond to points to the right of the origin (Figure 7a), and the **negative real numbers** correspond to points to the left of the origin (Figure 7b). The **real numbers** consist of the rational numbers and the irrational numbers. It can be shown that the graph of the set of real numbers is the entire number line and that every point on the number line corresponds to exactly one real number (Figure 7c). The set of all real numbers is represented by the symbol $\mathbb{R}$.

Figure 7

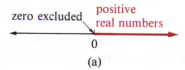

(a)

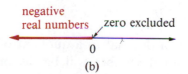

(b)

$\mathbb{R}$ = the real numbers

(c)

Decimals

We often find it useful to change rational numbers into decimal form. This makes it easier to compare two or more rational numbers. To express a rational number as a **decimal,** we divide the numerator by the denominator. The decimal form of a rational number may be **terminating,** as in the following examples:

$$\frac{2}{5} = 2 \div 5 = 0.4 \qquad \text{and} \qquad -\frac{3}{4} = -3 \div 4 = -0.75.$$

The decimal form of a rational number may also be **nonterminating** and **repeating,** as in the examples below:

$$\frac{2}{3} = 0.6666\ldots, \qquad \frac{7}{9} = 0.7777\ldots, \qquad \text{and} \qquad \frac{1}{7} = 0.1428571428571\ldots.$$

A repeating decimal such as 0.6666... is often written as $0.\bar{6}$, where the overbar indicates the block of digits that repeats:

$$\frac{2}{3} = 0.\bar{6}, \qquad 0.7777\ldots = 0.\bar{7}, \qquad \text{and}$$

$$0.142857142857\ldots = 0.\overline{142857}.$$

EXAMPLE 2   Express each rational number as a decimal.

(a) $\frac{3}{8}$ (b) $\frac{10}{3}$ (c) $-\frac{5}{6}$ (d) $\frac{73}{99}$

SOLUTION   (a) The decimal form of $\frac{3}{8}$ is obtained by dividing 3 by 8; that is, $\frac{3}{8} = 0.375$.

(b) $\frac{10}{3} = 3.3333\ldots = 3.\bar{3}$

(c) $-\frac{5}{6} = -0.83333\ldots = -0.8\bar{3}$

(d) $\frac{73}{99} = 0.7373\ldots = 0.\overline{73}$

Example 2 shows that every rational number can be expressed as a decimal that is either terminating or repeating. It is also true that every terminating or repeating decimal represents a rational number.

EXAMPLE 3   Express each terminating decimal as a quotient of integers.

(a) 0.7   (b) −0.53   (c) 1.025

SOLUTION   We change a decimal to a quotient of integers as follows: The numerator of the rational number will be the original number without the decimal point, and the denominator will be some multiple of 10, with as many zeros after the one as there are digits after the decimal point of the original form.

(a) $0.7 = \frac{7}{10}$ (b) $-0.53 = -\frac{53}{100}$ (c) $1.025 = \frac{1.025}{1.000} = \frac{41}{40}$

In Section 4.1 of Chapter 4, you will see how to rewrite a nonterminating repeating decimal as a quotient of two integers.

We may obtain decimal representations of irrational numbers. However, such decimals are nonterminating and nonrepeating. For instance, decimal representations of $\sqrt{2}$, $\sqrt{3}$, and π are: $\sqrt{2} = 1.4142135\ldots$, $\sqrt{3} = 1.7320508\ldots$, and $\pi = 3.1415926\ldots$.

The process of finding decimal representations of irrational numbers is usually difficult. Often, some method of approximation is employed. For example, we may approximate $\sqrt{2}$, $\sqrt{3}$, and π as follows: $\sqrt{2} \approx 1.41$, $\sqrt{3} \approx 1.73$, and $\pi \approx 3.14$. The symbol "$\approx$" denotes that the numerical value is only an *approximation*.

EXAMPLE 4   Identify each number as being rational or irrational.

(a) $-\frac{4}{7}$ (b) $\sqrt{6}$

(c) $0.\overline{35}$ (d) $0.12112111211112\ldots$

(e) $\sqrt{36}$

SOLUTION   (a) $-\frac{4}{7}$ is a quotient of two integers; thus it is rational.

(b) $\sqrt{6}$ is a square root of a positive integer that is not a perfect square; thus it is irrational.

(c) $0.\overline{35}$ is a repeating decimal; thus it is rational.

(d) $0.12112111211112\ldots$ is a nonterminating, nonrepeating decimal; thus it is irrational.

(e) $\sqrt{36} = 6$, which is an integer; therefore $\sqrt{36}$ is a rational number.

PROBLEM SET 1.2

1. Let $A = \{4, 5, 6, 7, 8, 9, 10\}$ and $B = \{3, 6, 8, 9, 10\}$. Insert in the following blanks the correct symbol, $\in$ or $\notin$.

   5 ___ A 5 ___ B 10 ___ A 4 ___ B

   10 ___ B 12 ___ A 8 ___ B 9 ___ A

2. Use set notation to describe the set A of the counting numbers greater than 1 but less than 9.

3. Use set notation to describe the set A of all months of the year whose names begin with the letter M.

4. Use set notation to describe the set of all days of the week whose names begin with the letter T.

5. Use set-builder notation to describe the set of counting numbers C greater than 4 and less than 17. Is the set C finite or infinite?

6. Use set-builder notation to describe the set of the first six letters in the English alphabet.

In problems 7–10, list all the subsets of each set.

7. $\{2, 3\}$ 8. $\{a, b, c\}$ 9. $\{5, 6, 7, 8\}$ 10. $\{a, b, c, d\}$

In problems 11–20, represent each set on a number line.

11. $\{-4, -3, -2, -1, 2, 4\}$ 12. $\{-5, -3, 0, 3, 5\}$

13. $\{-\frac{5}{3}, -\frac{2}{3}, -\frac{1}{3}, \frac{1}{3}, \frac{2}{3}, \frac{5}{3}\}$ 14. $\{\frac{1}{5}, \frac{2}{5}, \frac{3}{5}, \frac{4}{5}\}$

15. $\{-2, -1, 0, \frac{1}{2}, \frac{3}{2}, \frac{7}{2}\}$ 16. $\{-\frac{5}{9}, -\frac{4}{9}, -\frac{3}{9}, -\frac{2}{9}, -\frac{1}{9}\}$

17. $\{0, 2, 4, 6, \ldots\}$

18. $\{-\frac{3}{2}, -\frac{1}{2}, \frac{1}{2}, \frac{3}{2}, \frac{5}{2}, \ldots\}$

19. $\{\ldots, -5, -3, -1, 0\}$

20. $\{0, \frac{1}{7}, \frac{8}{7}, \frac{15}{7}, \frac{22}{7}, \frac{29}{7}, \ldots\}$

In problems 21–30, express each rational number as a decimal.

21. $\frac{3}{5}$ **22.** $-\frac{7}{4}$ **23.** $\frac{3}{2}$ **24.** $\frac{7}{2}$ **25.** $\frac{4}{5}$

26. $\frac{9}{100}$ **27.** $-\frac{5}{4}$ **28.** $\frac{5}{9}$ **29.** $-\frac{7}{3}$ **30.** $\frac{6}{7}$

In problems 31–40, express each decimal as a quotient of integers.

31. 0.27   **32.** 1.72   **33.** 2.64   **34.** 7.155   **35.** −0.125

36. −0.008   **37.** 0.0527   **38.** 0.0098   **39.** −0.00329   **40.** −0.00051

In problems 41–54, identify each number as being rational or irrational.

41. $-\frac{5}{9}$ **42.** $-\frac{3}{7}$ **43.** $\sqrt{14}$ **44.** $\sqrt{13}$

45. $0.\overline{27}$ **46.** $0.\overline{37}$ **47.** 3.464464446 …   **48.** 4.575575557 …

49. $\sqrt{16}$ **50.** $\sqrt{25}$ **51.** 0.374   **52.** 0.671

53. $-\sqrt{81}$ **54.** $-\sqrt{49}$

1.3 Properties of Real Numbers

In Section 1.2 we outlined the procedure for locating real numbers on a number line. In this section, we examine the basic properties of real numbers. These properties serve as a foundation for the algebraic steps we will use in later chapters. You may want to review how to add, subtract, multiply, and divide positive real numbers before you proceed. [The rules for signed (negative and positive) numbers are presented in Section 1.4.]

The basic properties of real numbers can be expressed in terms of the operations of addition and multiplication. If a and b are real numbers, there is a unique real number $a + b$, called their **sum.** The sum is formed by the process of **addition.** There is also a real number $a \cdot b$, called the **product** of a and b. The product is formed by the process of **multiplication.** The notation $a \times b$ for the product is not often used. The preferred notation is $a \cdot b$ or ab. Assume that the letters a, b, and c represent any real numbers.

1 The Commutative Properties

(i) For addition: **(ii) For multiplication:**

$$a + b = b + a \qquad\qquad a \cdot b = b \cdot a$$

For example,

$$13 + 20 = 20 + 13,$$

because

$$13 + 20 = 33 \quad \text{and} \quad 20 + 13 = 33.$$

Also,

$$3 \cdot 2 = 2 \cdot 3,$$

since

$$3 \cdot 2 = 6 \quad \text{and} \quad 2 \cdot 3 = 6.$$

2 The Associative Properties

(i) **For addition:**   (ii) **For multiplication:**

$$a + (b + c) = (a + b) + c \qquad a \cdot (b \cdot c) = (a \cdot b) \cdot c$$

For example,

$$7 + (3 + 9) = (7 + 3) + 9,$$

because

$$7 + (3 + 9) = 7 + 12 = 19 \quad \text{and} \quad (7 + 3) + 9 = 10 + 9 = 19.$$

Also,

$$3 \cdot (5 \cdot 2) = (3 \cdot 5) \cdot 2,$$

since

$$3 \cdot (5 \cdot 2) = 3 \cdot 10 = 30 \quad \text{and} \quad (3 \cdot 5) \cdot 2 = 15 \cdot 2 = 30.$$

3 The Distributive Properties

(i) $a \cdot (b + c) = a \cdot b + a \cdot c$ (ii) $(b + c) \cdot a = b \cdot a + c \cdot a$

For example,

$$6 \cdot (5 + 7) = (6 \cdot 5) + (6 \cdot 7),$$

because

$$6 \cdot (5 + 7) = 6 \cdot 12 = 72 \quad \text{and} \quad (6 \cdot 5) + (6 \cdot 7) = 30 + 42 = 72.$$

Also,

$$(5 + 7) \cdot 2 = (5 \cdot 2) + (7 \cdot 2).$$

because

$$(5 + 7) \cdot 2 = 12 \cdot 2 = 24 \quad \text{and} \quad (5 \cdot 2) + (7 \cdot 2) = 10 + 14 = 24.$$

4 The Identity Properties

(i) For addition:	(ii) For multiplication:
$a + 0 = 0 + a = a$	$a \cdot 1 = 1 \cdot a = a$

Zero is called the **additive identity** and one is called the **multiplicative identity.** For example, $5 + 0 = 0 + 5 = 5$ and $6 \cdot 1 = 1 \cdot 6 = 6$.

5 The Inverse Properties

(i) For each real number a, there is a real number $-a$, called the **additive inverse** or **negative** of a, such that

$$a + (-a) = (-a) + a = 0.$$

(ii) For each real number $a \neq 0$, there is a real number $1/a$, called the **multiplicative inverse** or **reciprocal** of a, such that

$$a \cdot \frac{1}{a} = \frac{1}{a} \cdot a = 1.$$

For example, the additive inverse of 3 is -3, that is,

$$3 + (-3) = (-3) + 3 = 0$$

and the multiplicative inverse of 7 is $\frac{1}{7}$; that is,

$$7 \cdot \frac{1}{7} = \frac{1}{7} \cdot 7 = 1.$$

EXAMPLE 1   State the properties that justify each of the following equalities.

(a) $3 \cdot (-4) = (-4) \cdot 3$

(b) $5 \cdot (6 \cdot 3) = (5 \cdot 6) \cdot 3$

(c) $14 \cdot (2 + \sqrt{3}) = 28 + 14 \cdot \sqrt{3}$

(d) $11 \cdot \frac{1}{11} = 1$

(e) $\frac{2}{3} \cdot (5 + \frac{3}{8}) = (5 + \frac{3}{8}) \cdot \frac{2}{3}$

(f) $7 + (-7) = 0$

SOLUTION

(a) Commutative property for multiplication

(b) Associative property for multiplication

(c) Distributive property

(d) Multiplicative inverse

(e) Commutative property for multiplication

(f) Additive inverse

EXAMPLE 2

State the property for each step.

(a) $y + [x + (-y)] = y + [(-y) + x]$

(b) $\qquad\qquad = [y + (-y)] + x$

(c) $\qquad\qquad = [0] + x$

(d) $\qquad\qquad = x$

SOLUTION

In writing steps (a)–(d), we use the following properties.

(a) $y + [x + (-y)] = y + [(-y) + x]$ (Commutative property of addition)

(b) $\qquad\qquad = [y + (-y)] + x$ (Associative property of addition)

(c) $\qquad\qquad = [0] + x$ (Additive inverse)

(d) $\qquad\qquad = x$ (Additive identity)

These five properties will be used as reasons or justifications for much of what you do in algebra. Following is a list of other properties that can be *derived* from these properties.

6 The Cancellation Properties

(i) For addition:

$$\text{if} \quad a + c = b + c, \qquad \text{then} \qquad a = b$$

(ii) For multiplication:

$$\text{if} \quad ac = bc \quad \text{and} \quad c \neq 0, \quad \text{then} \quad a = b$$

For example,

$$\text{if} \quad x + 3 = y + 3, \qquad \text{then} \qquad x = y.$$

Also,

$$\text{if} \quad 7x = 7y, \qquad \text{then} \qquad x = y.$$

7 The Negative (or Opposite) Properties

> (i) $-(-a) = a$ (ii) $(-a)b = a(-b) = -(ab)$
>
> (iii) $(-a)(-b) = ab$

For example,

$$-(-8) = 8, \qquad (-2)(4) = 2(-4) = -(2 \cdot 4) = -8,$$

and

$$(-3)(-7) = 3 \cdot 7 = 21.$$

8 The Zero-Factor Properties

> (i) $a \cdot 0 = 0 \cdot a = 0$
>
> (ii) If $a \cdot b = 0$, then $a = 0$ or $b = 0$ (or both)

For example,

$$\text{if} \quad a = 1{,}000{,}000, \qquad \text{then} \qquad a \cdot 0 = (1{,}000{,}000) \cdot 0 = 0,$$

and if $a = 7$ and we know that $a \cdot b = 0$, then $b = 0$.

The properties and definitions introduced so far have all involved the use of "equality," which is denoted by the symbol "$=$." The statement $a = b$ means that a and b are two names for the same thing. We shall list the following addition properties for elements a, b, and c of a set of real numbers $\mathbb{R}$.

9 The Reflexive Property

> $$a = a$$

10 The Symmetric Property

> If $a = b$, then $b = a$.

11 The Transitive Property

> If $a = b$ and $b = c$, then $a = c$.

Another important and frequently used property is the substitution property.

12 The Substitution Property

> If $a = b$, then a can be substituted for b in any statement involving b without affecting the truthfulness of the statement.

For example, if $a = b$ and $b + 3 = 10$, then $a + 3 = 10$.

PROBLEM SET 1.3

In problems 1–8, verify the commutative properties by performing actual computations.

1. $15 + 17$ and $17 + 15$

2. $21 + 9$ and $9 + 21$

3. $8 \cdot 9$ and $9 \cdot 8$

4. $14 \cdot 11$ and $11 \cdot 14$

5. $0.6 + 0.5$ and $0.5 + 0.6$

6. $2.25 + 3.50$ and $3.50 + 2.25$

7. $0.8 \cdot 0.41$ and $0.41 \cdot 0.8$

8. $0.76 \cdot 5.2$ and $5.2 \cdot 0.76$

In problems 9–16, verify the associative properties by performing actual computations.

9. $2 + (7 + 6)$ and $(2 + 7) + 6$

10. $10 + (9 + 3)$ and $(10 + 9) + 3$

11. $3 \cdot (8 \cdot 5)$ and $(3 \cdot 8) \cdot 5$

12. $6 \cdot (5 \cdot 2)$ and $(6 \cdot 5) \cdot 2$

13. $4.1 + (6.8 + 3.3)$ and $(4.1 + 6.8) + 3.3$

14. $2.25 + (3.5 + 1.75)$ and $(2.25 + 3.5) + 1.75$

15. $0.67 \cdot (0.5 \cdot 0.4)$ and $(0.67 \cdot 0.5) \cdot 0.4$

16. $4.2 \cdot (5.1 \cdot 3.6)$ and $(4.2 \cdot 5.1) \cdot 3.6$

In problems 17–22, verify the distributive properties by performing actual computations.

17. $3 \cdot (5 + 8)$ and $3 \cdot 5 + 3 \cdot 8$

18. $6.1 \cdot (7.6 + 4.8)$ and $6.1 \cdot 7.6 + 6.1 \cdot 4.8$

19. $(7 + 9) \cdot 3$ and $7 \cdot 3 + 9 \cdot 3$

20. $(15 + 12) \cdot 6$ and $15 \cdot 6 + 12 \cdot 6$

21. $2.1 \cdot (1.1 + 3.4)$ and $2.1 \cdot 1.1 + 2.1 \cdot 3.4$

22. $(3.25 + 2.33) \cdot 1.85$ and $3.25 \cdot 1.85 + 2.33 \cdot 1.85$

In problems 23–42, state the properties that justify each of the following statements.

23. $3 + (-8) = (-8) + 3$

24. $\frac{1}{2} \cdot 5 = 5 \cdot \frac{1}{2}$

25. $8 \cdot (\sqrt{5} \cdot 4) = (8 \cdot \sqrt{5}) \cdot 4$

26. $1 \cdot \sqrt{11} = \sqrt{11}$

27. $(7 + \sqrt{2}) \cdot 3 = 21 + 3\sqrt{2}$

28. $\sqrt{7} + (-\sqrt{7}) = 0$

29. $\left(\dfrac{1}{\sqrt{3}}\right) \cdot \sqrt{3} = 1$

30. $1 \cdot \frac{7}{9} = \frac{7}{9}$

31. $(-2) \cdot (-3) = 2 \cdot 3$

32. $-2 \cdot 5 = -10$

33. If $5a = 0$, then $a = 0$.

34. If $-3x = -3y$, then $x = y$.

35. If $9x = 9y$, then $x = y$.

36. $4 \cdot (u + v) = 4u + 4v$

37. $15 \cdot 0 = 0$

38. $\frac{3}{4} + (1 + \frac{1}{2}) = (\frac{3}{4} + 1) + \frac{1}{2}$

39. $3 + \sqrt{11} = \sqrt{11} + 3$

40. If $t + 3 = 5 + 3$, then $t = 5$.

41. $-8 + 0 = -8$

42. If $x \neq 0$, then $x \cdot \dfrac{1}{x} = 1$.

In problems 43 and 44, state the property that justifies each step.

43. (a) $x + (-x) = x(1) + x(-1)$
    (b) $\qquad\quad = x[1 + (-1)]$
    (c) $\qquad\quad = x[0]$
    (d) $\qquad\quad = 0$

44. (a) $a(b - c) = a[b + (-c)]$
    (b) $\qquad\quad = ab + a(-c)$
    (c) $\qquad\quad = ab + [a(-1)c]$
    (d) $\qquad\quad = ab + (-1)(ac)$
    (e) $\qquad\quad = ab + (-ac)$
    (f) $\qquad\quad = ab - ac$

In problems 45–48, fill in the blanks to make the statement an illustration of the given property of equality.

45. The Symmetric Property: If $-3 = x$, then _____.

46. The Reflexive Property: $3 + x =$ _____.

47. The Transitive Property: If $3t + 2t = (3 + 2)t$ and $(3 + 2)t = 5t$, then _____.

48. The Substitution Property: If $3x + 2y = 7$ and $y = x + 2$, then
$3x + 2(\underline{\qquad}) = 7$.

1.4 Operations with Signed Numbers

We now discuss the concepts and terminology commonly used with integers and then review the basic operations with integers.

    Throughout this book, the negative integers are written with a negative sign affixed, and the positive integers are indicated by the absence of a sign. Positive and negative numbers together are called **signed numbers.** Hence, if a nonzero number has no sign in front of it, it is understood to be positive.

For example, the coordinate of a point that corresponds to the number $+5$ is written as 5.*

You learned the rules for adding, subtracting, multiplying, and dividing positive numbers in your study of arithmetic. The concept of absolute value is helpful in extending these rules to include negative numbers.

Consider the distance between the origin and each of the points labeled 5 and -5 on a number line. Both of these points are located five units from the origin. The distance between the origin and the point with coordinate 5 is the same as the distance between the origin and the point with coordinate -5. This common distance is 5 (Figure 1). Such distances can be interpreted by using the absolute value concept. The *absolute value* of a number indicates how far the number is from the origin, regardless of whether it is to the left or to the right of the origin.

Figure 1

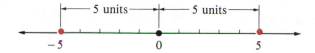

Since the numbers 5 and -5 are each 5 units from the origin 0, they have the same absolute value 5. We denote the absolute value of a number by writing vertical lines on either side of the number. Thus,

$$|5| = 5, \text{ reads as "the absolute value of 5 is 5"}$$

and

$$|-5| = 5, \text{ reads as "the absolute value of } -5 \text{ is 5."}$$

Notice that the absolute value of a positive number (or zero) is that number. For instance, $|3| = 3$, $|8| = 8$, $|47| = 47$, and $|0| = 0$, $|0.75| = 0.75$, and $|\frac{2}{3}| = \frac{2}{3}$. On the other hand, the absolute value of a negative number is the additive inverse (or the **opposite**) of the number. For example,

$$|-2| = -(-2) = 2, \qquad |-9| = -(-9) = 9, \qquad \text{and} \qquad |-85| = -(-85) = 85.$$

More formally, we have the following definition:

DEFINITION   **Absolute Value**

> If x is a real number, then $|x|$, the **absolute value** of x, is defined by
>
> $$|x| = \begin{cases} x, \text{ if } x \text{ is positive or } x \text{ is zero} \\ -x, \text{ if } x \text{ is negative.} \end{cases}$$

* Note: Some authors write the numbers preceded by the raised symbol $+$ for the points to the right of zero and the raised symbol $-$ for the points to the left of zero. For example, a point corresponding to the number "positive five" is denoted by $^{+}5$, and a point corresponding to the number "negative five" is denoted by $^{-}5$. However, in this book, we denote these numbers by 5 and -5.

For example,

$$|6| = 6, \text{ because 6 is positive,}$$

$$|-6| = -(-6) = 6, \text{ because } -6 \text{ is negative,}$$

and

$$|0| = 0.$$

Note that if x is negative, then $-x$ is a positive number.

EXAMPLE 1   Find the value of each expression.

(a) $|8|$ (b) $\left|-\frac{2}{3}\right|$ (c) $\left|\frac{7}{5}\right|$ (d) $-|-17|$

SOLUTION   (a) $|8| = 8$, because 8 is positive.

(b) $\left|-\frac{2}{3}\right| = -\left(-\frac{2}{3}\right) = \frac{2}{3}$, because $-\frac{2}{3}$ is negative.

(c) $\left|\frac{7}{5}\right| = \frac{7}{5}$, since $\frac{7}{5}$ is positive.

(d) $|-17| = 17$, so $-|-17| = -17$.

Addition of Signed Numbers

The addition of real numbers can be illustrated on the number line. Using the properties of real numbers, we can represent addition by directed moves (changes of position) on the line—a move to the right if the number is positive and to the left if the number is negative. For example, to add 2 and 3 on the number line (Figure 2), we start at the origin and move two units to the right. Thus, the number 2 is represented by an arrow from 0 to 2. Next, start at the number 2 and move three units to the right, so that the number 3 is represented by an arrow between 2 and 5. Together, the sum of the two directed moves is $2 + 3 = 5$ (Figure 2a). It can also be seen that $3 + 2 = 5$ (Figure 2b). This illustrates the commutative property of addition: $3 + 2 = 2 + 3$.

Figure 2

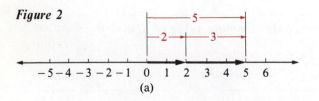
(a)

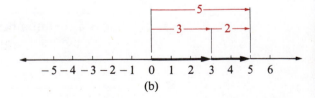
(b)

Now consider the sum $(-2) + (-3)$. Visualize the addition of a negative number as a movement to the left on the number line. The sum can be found by moving to the left of the origin: $(-2) + (-3) = -5$ (Figure 3).

Figure 3

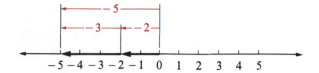

Note that in both of these cases the sums can be found by adding the absolute values of the numbers and retaining the common sign of the numbers. These results illustrate the following rule:

> To **add** two numbers with like signs, add their absolute values and keep their common sign.

For example,

$$5 + 7 = |5| + |7| = 12$$

and

$$(-5) + (-8) = -(|-5| + |-8|) = -(5 + 8) = -13.$$

The addition of numbers with different signs can also be easily illustrated on a number line. For example, the sum $(-3) + 8$ can be interpreted as a movement of three units to the left and then eight units to the right (Figure 4a). The sum $(-3) + 8$ can also be interpreted as a movement of eight units to the right and then three units to the left (Figure 4b). In both cases, the same sum is obtained; that is, $(-3) + 8 = 5$.

Figure 4

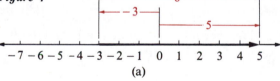

(a)

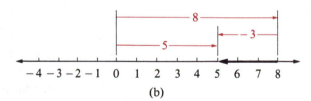

(b)

Note that, in the example $-3 + 8 = 5$, the sum can be obtained by finding the difference of the absolute values of the numbers and retaining the sign of the number with the larger absolute value. This result is an example of the following rule:

> To **add** two numbers with unlike signs, find the difference of their absolute values by subtracting the smaller absolute value from the larger. Retain the sign of the number with the larger absolute value.

For example,

$$6 + (-2) = (|6| - |-2|) = (6 - 2) = 4$$

and

$$2 + (-5) = (-5) + 2 = -(|-5| - |2|) = -(5 - 2) = -3.$$

EXAMPLE 2   Find the following sums.

(a) $9 + 7$ (b) $(-5) + (-7)$

(c) $7 + (-23)$ (d) $(-4) + 7 + 5$

SOLUTION   (a) $9 + 7 = 16$

(b) $(-5) + (-7) = -(5 + 7) = -12$

(c) $7 + (-23) = -(23 - 7) = -16$

(d) $(-4) + 7 + 5 = (7 - 4) + 5$
$\qquad\qquad\qquad = 3 + 5 = 8$

Subtraction of Signed Numbers

To subtract b from a, an operation denoted by $a - b$, we use the following definition:

DEFINITION   **Subtraction**

> If a and b are real numbers, the **difference** $a - b$ is defined to be $a + (-b)$, where $-b$ is the additive inverse of b. In other words, to subtract a number b, change its sign and add.

This definition, together with the rules for adding signed numbers, provides a method for subtracting signed numbers. For example, the difference $(-7) - 3$ can be found as follows:

$$(-7) - 3 = (-7) + (-3) = -10.$$

We can illustrate this subtraction on a number line by interpreting the subtraction of 3 from -7 as a movement of seven units to the left and then three more units to the left; that is, $(-7) - 3 = -10$ (Figure 5).

Figure 5

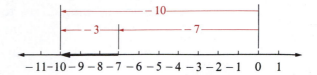

The rule for subtracting signed numbers can be stated as follows:

To **subtract** one signed number from another, change the sign of the number to be subtracted, and then add by following the rules for adding signed numbers.

For example,

$$7 - 4 = 7 + (-4) = 3 \quad \text{and} \quad (-9) - 6 = (-9) + (-6) = -15.$$

EXAMPLE 3   Find the following differences.

(a) $5 - 2$ (b) $(-15) - (-8)$ (c) $(-9) - 4$

SOLUTION   (a) $5 - 2 = 5 + (-2) = 3$

(b) $(-15) - (-8) = (-15) + 8 = -7$

(c) $(-9) - 4 = (-9) + (-4) = -13$

Multiplication of Signed Numbers

The multiplication of two positive integers is often described as repeated addition. For example, $3 \cdot 2$ can be interpreted as $3 + 3 = 6$ or $2 + 2 + 2 = 6$. The same approach can be used to describe the multiplication of a negative number by a positive number. Thus, $(-6) \cdot 3$ can be interpreted as $(-6) + (-6) + (-6) = -18$ or $(-3) + (-3) + (-3) + (-3) + (-3) + (-3) = -18$. Also, we can write $(-5) \cdot (2)$ as $(-5) + (-5) = -10$ or as $(-2) + (-2) + (-2) + (-2) + (-2) = -10$.

Because multiplication is commutative, it must also be true that $(-6) \cdot 3 = 3 \cdot (-6) = -18$ and $(-5) \cdot 2 = 2 \cdot (-5) = -10$.

These examples suggest that:

The product of a negative number and a positive number is a negative number. That is, if a and b are positive numbers, then

$$a(-b) = -(ab)$$

or

$$(-a)b = -(ab).$$

In order to understand how to find the product of two negative numbers, it might be helpful to examine the following pattern:

$$(-4) \cdot 3 = -12$$
$$(-4) \cdot 2 = -8$$
$$(-4) \cdot 1 = -4$$
$$(-4) \cdot 0 = 0.$$

Notice that as the second number decreases by 1, the product increases by 4. Thus, you may expect the following:

$$(-4) \cdot (-1) = 4$$
$$(-4) \cdot (-2) = 8$$
$$(-4) \cdot (-3) = 12.$$

In fact, as these two examples suggest, it is true that:

> The product of two negative numbers is a positive number. That is, if both a and b are positive numbers, then
> $$(-a) \cdot (-b) = +(a \cdot b).$$

EXAMPLE 4 Find each product.

(a) $(-9) \cdot 6$ (b) $5 \cdot (-8)$ (c) $(-5) \cdot (-3)$

SOLUTION (a) $(-9) \cdot 6 = -(9 \cdot 6) = -54$ (b) $5 \cdot (-8) = -(5 \cdot 8) = -40$

(c) $(-5) \cdot (-3) = 5 \cdot 3 = 15$

When multiplying more than two signed numbers, the numbers can be paired to determine the sign of the product. For example, the product $(-3) \cdot 2 \cdot (-5) \cdot 6$ can be written

$$[(-3) \cdot (-5)] \cdot [2 \cdot 6] = 15 \cdot 12 = 180.$$

In this example, there are two negative factors—that is, an even number of negative factors—and so the product is positive. In the case of the product $(-4) \cdot (-7) \cdot (-2) \cdot 3$, we have

$$[(-4) \cdot (-7)] \cdot [(-2) \cdot 3] = 28 \cdot (-6) = -168.$$

In this example, there are three negative factors—an odd number of negative factors—and so the product is negative. We can state a more general rule for finding the product of signed numbers:

> The product of signed numbers is positive if there is an even number of negative factors and negative if there is an odd number of negative factors.

EXAMPLE 5   Determine the following products.

(a) $3 \cdot (-2) \cdot 6 \cdot (-7) \cdot (-8) \cdot 2 \cdot (-4)$

(b) $(-0.5) \cdot 7 \cdot (-0.75) \cdot (-8) \cdot (-4) \cdot (-7)$

SOLUTION   (a) There is an even number of negative factors, so the product is positive:

$$3 \cdot (-2) \cdot 6 \cdot (-7) \cdot (-8) \cdot 2 \cdot (-4) = 16{,}128.$$

(b) There is an odd number of negative factors, so the product is negative:

$$(-0.5) \cdot 7 \cdot (-0.75) \cdot (-8) \cdot (-4) \cdot (-7) = -588.$$

Division of Signed Numbers

To divide one signed number by another, we use the following definition: The **quotient** $a \div b$, in which $b \neq 0$, is defined to be $a \cdot (1/b)$ in which $1/b$ is the multiplicative inverse of b. That is,

$$a \div b = a \cdot \frac{1}{b}.$$

More formally, we have:

DEFINITION   **Division**

> $a \div b$ is the unique number c for which $c \cdot b = a$; that is, if $b \neq 0$,
>
> $$a \div b = c \qquad \text{if and only if} \quad c \cdot b = a.$$

For instance,

$12 \div 4 = 3$	because	$3 \cdot 4 = 12$
$12 \div (-4) = -3$	because	$(-3) \cdot (-4) = 12$
$(-12) \div 4 = -3$	because	$(-3) \cdot 4 = -12$
$(-12) \div (-4) = 3$	because	$3 \cdot (-4) = -12$

These examples suggest the following rules:

(i) If two numbers have the same signs, their quotient is positive.

(ii) If two numbers have different signs, their quotient is negative.

EXAMPLE 6   Perform each division.

(a) $45 \div (-9)$ (b) $(-42) \div 3$ (c) $(-6.5) \div (-1.3)$

SOLUTION   (a) $45 \div (-9) = -(45 \div 9) = -5$

(b) $(-42) \div 3 = -(42 \div 3) = -14$

(c) $(-6.5) \div (-1.3) = 6.5 \div 1.3 = 5$

It should be noted that *division by zero is not possible*. This can be explained as follows: Suppose there is a number c, such that $5 \div 0 = c$. For this to be true, $c \cdot 0$ must equal 5. We ask, what number c multiplied by 0 will give 5? There is *no such number*. No matter what number we substitute for c, when we multiply c by 0, we cannot get 5, because $c \cdot 0 = 0$. Therefore we say, in general, if $a \neq 0$, then

$$\frac{a}{0} \text{ is undefined.}$$

On the other hand, if $b \neq 0$, then

$$\frac{0}{b} = 0, \quad \text{since} \quad 0 \cdot b = 0.$$

If we try to divide 0 by 0, we run into problems for a different reason. Suppose there is a number c such that $0 \div 0 = c$. This means that $0 \cdot c = 0$. But this is true for all c. That is, $0 \div 0$ is not *unique*. Therefore, we say that

$$\frac{0}{0} \text{ is } \textit{indeterminate.}$$

In Section 1.1, we used the order of operations to evaluate numerical expressions involving mixed operations. We now extend this evaluation to expressions containing signed numbers.

In Examples 7–9, evaluate each expression.

EXAMPLE 7 $(-3)[(-21) \div (-7)]$

SOLUTION $\qquad (-3)[(-21) \div (-7)] = (-3)3$ (We divided within the bracket)
$$= -9.$$ (We multiplied)

EXAMPLE 8 $5 + 4 \cdot 6 \div (-3) - (-7)$

SOLUTION $\qquad 5 + 4 \cdot 6 \div (-3) - (-7) = 5 + 24 \div (-3) - (-7)$ (We multiplied)
$$= 5 + (-8) - (-7)$$ (We divided)
$$= (-3) - (-7)$$ (We added)
$$= (-3) + 7$$ (We subtracted)
$$= 4.$$ (We added)

EXAMPLE 9 $(-3)^2 + 7 \cdot [(-6) + 4] \div (-2)$

SOLUTION $\qquad (-3)^2 + 7 \cdot [(-6) + 4] \div (-2)$
$$= (-3)^2 + 7 \cdot [(-2)] \div (-2)$$ (We added within parentheses)
$$= 9 + 7 \cdot (-2) \div (-2)$$ (We raised (-3) to the power 2)
$$= 9 + (-14) \div (-2)$$ (We multiplied)
$$= 9 + 7$$ (We divided)
$$= 16.$$ (We added)

PROBLEM SET 1.4

In problems 1–10, find the value of each expression.

1. $|3|$ **2.** $|2.7|$ **3.** $|-11|$ **4.** $|-5\frac{1}{2}|$ **5.** $-|16|$

6. $-|-7.31|$ **7.** $-|-21|$ **8.** $-|0|$ **9.** $-|-\frac{5}{7}|$ **10.** $|-11.8|$

In problems 11–34, find the following sums.

11. $7 + (-3)$ **12.** $(-23) + 14$ **13.** $(-6) + (-8)$ **14.** $(-11) + (-7)$

15. $27 + (-39)$ **16.** $(-100) + 43$ **17.** $(-8) + 8$ **18.** $7 + (-7)$

19. $10 + (-17)$ **20.** $(-19) + 27$ **21.** $(-21) + 9$ **22.** $(-20) + (-17)$

23. $(-35) + (-18)$ **24.** $(-75) + 25$ **25.** $15.2 + (-13.7)$ **26.** $(-21.3) + (-14.6)$

27. $(-9.7) + (-8.5)$ **28.** $7.28 + (-2.71)$ **29.** $3 + (-5) + (-7)$ **30.** $(-8) + 12 + (-7)$

31. $(-3) + 7 + (-11)$ **32.** $18 + (-13) + (-17)$ **33.** $6 + (-9) + (-1)$ **34.** $(-1) + (-2) + 8$

In problems 35–52, find the following differences.

35. $18 - (-19)$ **36.** $(-25) - 8$ **37.** $(-25) - (-45)$

38. $(-97) - (-39)$ **39.** $(-22) - (-7)$ **40.** $16 - (-30)$

41. $(-8) - 8$ **42.** $(-8) - (-5)$ **43.** $11.1 - (-0.9)$

44. $0 - (-2)$ **45.** $0.052 - (-0.007)$ **46.** $(-5.03) - (-4.83)$

47. $0 - (-8)$ **48.** $(-8.7) - 5.8$ **49.** $(-111) - (-17)$

50. $4.8 - (-3.71)$ **51.** $(-327) - (-481)$ **52.** $(-3,125) - (-1,241)$

In problems 53–68, find the following products.

53. $3 \cdot (-5)$
54. $(-6) \cdot 4$
55. $(-2) \cdot (-8)$
56. $(-12) \cdot 0.5$
57. $(-9) \cdot 0.3$
58. $0.4 \cdot 10$
59. $(-8) \cdot 0 \cdot (-4)$
60. $0.4 \cdot (-0.8)$
61. $(-3) \cdot (-2) \cdot (-4)$
62. $(-4) \cdot (-6) \cdot (-2)$
63. $(-6) \cdot (-6) \cdot 7$
64. $(-11) \cdot 0 \cdot (-7) \cdot (-6)$
65. $(-1) \cdot (-2) \cdot (-3) \cdot (-5)$
66. $(-2) \cdot (-3) \cdot 4 \cdot (-5)$
67. $5 \cdot (-7) \cdot 3 \cdot (-2)$
68. $3 \cdot (-2) \cdot (-10) \cdot (-4)$

In problems 69–84, find the following quotients.

69. $(-10) \div 5$
70. $27 \div (-9)$
71. $18 \div (-9)$
72. $(-10) \div (-2)$
73. $(-57) \div (-19)$
74. $(-32) \div 8$
75. $(-3.9) \div 0.3$
76. $(-\frac{5}{4}) \div (-2)$
77. $(-75) \div (-5)$
78. $12.5 \div (-1.25)$
79. $(22.5) \div (-0.015)$
80. $(-0.49) \div (-0.7)$
81. $(-350) \div (-70)$
82. $(-2,400) \div (-20)$
83. $(-4,400) \div (-1,100)$
84. $46.72 \div (-6.4)$

In problems 85–110 evaluate each expression.

85. $(-2)[9 + (-11)]$
86. $(-8)[(-7) - (-3)]$
87. $6[(-8) - (-5)]$
88. $7[(-4) + (-3)]$
89. $(-5)[(-15) \div 3]$
90. $(-4)[(-35) \div (-7)]$
91. $[(-28) - (-7)] \div (-2)$
92. $[(-48) - (-12)] \div (-2)$
93. $(-36) \div [(-8) \div (-4)]$
94. $(-18) \div [(-24) \div (-4)]$
95. $(-15) \div 3 + 2 \cdot (-5)$
96. $(-11) + (-16) \div (-4) - (-3)$
97. $17 - 5 \cdot 2 \div 2 - 5 + 4$
98. $24 \div 3 - 5 + 4[(-3) \cdot 2 - 1]$
99. $(-42) \div (-6) - 7 + 8[(-4) \cdot 2 - 1]$
100. $[8 - (-3) \cdot 2][8 + (-3) \cdot 2]$
101. $5 \cdot (-2)^3 \cdot (-3)^2 \div (-8)$
102. $5^2 \cdot (-3)^2 \cdot (-2)^3 \div (-9)$
103. $(-2)[(-3) - (-2)^3] + (4 - 6)^2$
104. $3 + [(-4) + 2]^3 + 2^3 \cdot (-8)$
105. $(-5)^2 + 4 \cdot [(-5) + 3] \div (-2)$
106. $[(-3) - (-2)^5]^2 + [9 - (-3)]^2$
107. $[6 + (-8)^2] \div (-2) + [8 + (-3)]^2$
108. $(-4)(-3)^3 - 5 + (7 - 2)^2$
109. $5 + 20 \div (-2)^2 - 3^4$
110. $(5 + 3)^2 + 5(-2)^3 - 3(-5)^2 + 2$

111. The expression

$$\frac{(x + y) + |x - y|}{2}$$

always equals the larger of the numbers x and y. Verify this fact for $x = 9$ and $y = -2$.

112. The expression

$$\frac{(x + y) - |x - y|}{2}$$

always equals the smaller of the numbers x and y. Verify this fact for $x = 9$ and $y = -2$.

113. Find the *mistake*, then give the correct answer.

$$|(-8) - 3|^2 = |(-8)|^2 - |3|^2 = 64 - 9 = 55.$$

In problems 114–117, use signed numbers to solve the word problems.

114. Before going to bed, Carlos read the thermometer and saw that the temperature was 18°. When he awoke, he found the temperature was $-2°$. How many degrees did the temperature fall during the night?

115. If the temperature is 10° below 0°C and then increases by 25°, what is the new temperature?

116. David has a savings account with a balance of $270. If he makes a deposit of $100 and then two withdrawals of $40 and $65, what is his new balance?

117. In one series of three plays, a football team loses 4 yards, gains 7 yards, and then loses 5 yards. What is the team's position now in relation to the original line of scrimmage?

1.5 Operations on Rational Numbers

Recall from Section 1.2 that a rational number can be represented in the form of a/b, where a and b are integers, and $b \neq 0$.

This form of a rational number is usually called a **fraction,** where a is the *numerator* of the fraction and b is the *denominator* of the fraction. In this section we review the operations on rational numbers that are written as fractions.

Two fractions are said to be **equivalent** if they have the same value but different forms.

For example, the fractions

$$\frac{3}{5}, \quad \frac{-6}{-10}, \quad \frac{12}{20}, \quad \text{and} \quad \frac{15}{25}$$

are equivalent.

We denote two equivalent fractions by writing an *equal sign* between them. In the preceding example, we write

$$\frac{3}{5} = \frac{-6}{-10} = \frac{12}{20} = \frac{15}{25}.$$

Two fractions are equivalent, if their **cross products** are equal. For example, we can write

$$\frac{2}{3} = \frac{10}{15} \quad \text{as} \quad \frac{2}{3} \overset{\longleftarrow}{\underset{\longrightarrow}{=}} \frac{10}{15}.$$

The diagonal lines with arrowheads show which numbers form the cross products. Notice that $2(15) = 3(10)$. This pattern is generalized as follows:

$$\frac{a}{b} = \frac{c}{d}$$

whenever $ad = bc$, and $b \neq 0$ and $d \neq 0$

For example,

$$\frac{5}{7} = \frac{15}{21}, \quad \text{since } 5 \cdot 21 = 7 \cdot 15.$$

If the numerator and denominator of a fraction are integers, and the denominator is not zero, the fraction is said to be in **lowest terms,** or **reduced,** or **simplified,** when the numerator and the denominator have no common factor other than 1. To reduce a fraction to lowest terms, we use the following **fundamental principle of fractions:**

$$\frac{ak}{bk} = \frac{a}{b}$$

Where a and b have no common factor, and $b \neq 0$ and $k \neq 0$.

EXAMPLE 1　Reduce each fraction to lowest terms.

(a) $\dfrac{35}{55}$　(b) $\dfrac{45}{-63}$　(c) $\dfrac{-26}{91}$　(d) $\dfrac{13}{27}$

SOLUTION　(a) $\dfrac{35}{55} = \dfrac{7 \cdot \cancel{5}}{11 \cdot \cancel{5}} = \dfrac{7}{11}$

> The slanted lines through the common factors in the numerator and the denominator indicate that we have divided out, or cancelled, them.

(b) $\dfrac{45}{-63} = \dfrac{5 \cdot \cancel{9}}{-7 \cdot \cancel{9}} = \dfrac{5}{-7}$

(c) $\dfrac{-26}{91} = \dfrac{-2 \cdot \cancel{13}}{7 \cdot \cancel{13}} = \dfrac{-2}{7}$

(d) Fraction is already in lowest terms.

We now turn our attention to the following rules of signs for fractions:

$$-\frac{a}{b} = \frac{a}{-b} = \frac{-a}{b}$$

and

$$\frac{a}{b} = \frac{-a}{-b} = -\frac{-a}{b} = -\frac{a}{-b}$$

EXAMPLE 2   Use the preceding rules of signs to simplify each fraction.

(a) $\dfrac{-6}{-15}$ (b) $\dfrac{25}{-10}$ (c) $-\dfrac{-35}{14}$

SOLUTION   (a) $\dfrac{-6}{-15} = \dfrac{6}{15} = \dfrac{2 \cdot \cancel{3}}{5 \cdot \cancel{3}} = \dfrac{2}{5}$ (b) $\dfrac{25}{-10} = -\dfrac{25}{10} = -\dfrac{5 \cdot \cancel{5}}{2 \cdot \cancel{5}} = -\dfrac{5}{2}$

(c) $-\dfrac{-35}{14} = \dfrac{35}{14} = \dfrac{5 \cdot \cancel{7}}{2 \cdot \cancel{7}} = \dfrac{5}{2}$

Addition and Subtraction of Fractions

Before two fractions can be added or subtracted, they must have the same denominator. Fractions with the same denominator are called **like fractions.** For example, $\frac{5}{9}$ and $\frac{2}{9}$ are like fractions.

The addition or subtraction of two like fractions is accomplished by a direct application of the distributive property. For example, to add $\frac{5}{9}$ and $\frac{2}{9}$, we have

$$\frac{5}{9} + \frac{2}{9} = 5 \cdot \frac{1}{9} + 2 \cdot \frac{1}{9} = (5 + 2) \cdot \frac{1}{9} = 7 \cdot \frac{1}{9} = \frac{7}{9}.$$

In general,

$$\frac{a}{b} + \frac{c}{b} = \frac{a + c}{b}, \qquad b \neq 0$$

Since subtraction can be defined in terms of addition, we have

$$\frac{a}{b} - \frac{c}{b} = \frac{a}{b} + \left(-\frac{c}{b} \right) = \frac{a}{b} + \frac{-c}{b} = \frac{a + (-c)}{b} = \frac{a - c}{b}.$$

That is,

$$\frac{a}{b} - \frac{c}{b} = \frac{a - c}{b}, \qquad b \neq 0$$

EXAMPLE 3   Perform each operation.

(a) $\dfrac{2}{7} + \dfrac{3}{7}$ (b) $\dfrac{8}{11} - \dfrac{5}{11}$

SOLUTION   (a) $\dfrac{2}{7} + \dfrac{3}{7} = \dfrac{2+3}{7} = \dfrac{5}{7}$ (b) $\dfrac{8}{11} - \dfrac{5}{11} = \dfrac{8-5}{11} = \dfrac{3}{11}$

EXAMPLE 4   Perform the operations $\dfrac{11}{17} + \dfrac{5}{17} - \dfrac{7}{17}$.

SOLUTION

$$\dfrac{11}{17} + \dfrac{5}{17} - \dfrac{7}{17} = \dfrac{11+5-7}{17} = \dfrac{9}{17}.$$

If the denominators of fractions are not the same, we say that the fractions are **unlike fractions.** To add or subtract unlike fractions, we must first find a common denominator, then use the fundamental principle of fractions to change each fraction into an equivalent fraction that has the common denominator. For example, to add the fractions $\frac{5}{7}$ and $\frac{3}{5}$, we have

$$\dfrac{5}{7} + \dfrac{3}{5} = \dfrac{5 \cdot 5}{7 \cdot 5} + \dfrac{3 \cdot 7}{5 \cdot 7} = \dfrac{25}{35} + \dfrac{21}{35} = \dfrac{25+21}{35} = \dfrac{46}{35}.$$

In general,

$$\dfrac{a}{b} + \dfrac{c}{d} = \dfrac{ad}{bd} + \dfrac{bc}{bd} = \dfrac{ad+bc}{bd}.$$

That is,

$$\boxed{\dfrac{a}{b} + \dfrac{c}{d} = \dfrac{ad+bc}{bd}}$$

Subtraction of unlike fractions is performed according to the rule:

$$\boxed{\dfrac{a}{b} - \dfrac{c}{d} = \dfrac{ad-bc}{bd}}$$

EXAMPLE 5   Perform each operation.

(a) $\dfrac{2}{3} + \dfrac{3}{4}$ (b) $\dfrac{2}{3} - \dfrac{3}{7}$

SOLUTION   (a) $\dfrac{2}{3} + \dfrac{3}{4} = \dfrac{2 \cdot 4 + 3 \cdot 3}{3 \cdot 4} = \dfrac{8 + 9}{12} = \dfrac{17}{12}$

(b) $\dfrac{2}{3} - \dfrac{3}{7} = \dfrac{2 \cdot 7 - 3 \cdot 3}{3 \cdot 7} = \dfrac{14 - 9}{21} = \dfrac{5}{21}$

There are occasions when a direct application of the addition or subtraction rules is not the most convenient approach. Take, for example, the following addition of fractions:

$$\dfrac{3}{8} + \dfrac{5}{6} = \dfrac{3 \cdot 6 + 8 \cdot 5}{8 \cdot 6} = \dfrac{18 + 40}{48} = \dfrac{58}{48} = \dfrac{29 \cdot 2}{24 \cdot 2} = \dfrac{29}{24}.$$

Although this process is always correct, it is unnecessarily complicated if the denominators have a common factor, such as the preceding example. The same addition can be performed as follows:

$$\dfrac{3}{8} + \dfrac{5}{6} = \dfrac{3 \cdot 3}{8 \cdot 3} + \dfrac{5 \cdot 4}{6 \cdot 4} = \dfrac{9}{24} + \dfrac{20}{24} = \dfrac{9 + 20}{24} = \dfrac{29}{24}.$$

Note that in the second solution we found the smallest positive integer that both denominators divide into, which is called the **least common denominator,** or **LCD**. To find the **LCD** of a set of fractions, we follow these steps:

> 1. Factor each denominator into a product of prime numbers (positive integers greater than 1, whose only factors are themselves and 1) or into a product of prime numbers and -1.
>
> 2. List each prime factor with the largest exponent it has in any factored denominator; then find the product of the factors so listed.

EXAMPLE 6   Perform each operation.

(a) $\dfrac{11}{35} + \dfrac{3}{14}$

(b) $\dfrac{11}{12} - \dfrac{7}{16}$

SOLUTION   (a) The denominators are factored as follows:

$$35 = 5 \cdot 7 \quad \text{and} \quad 14 = 2 \cdot 7.$$

The prime factors are 2, 5, and 7, and therefore the LCD of the fractions is $2 \cdot 5 \cdot 7 = 70$. Rewriting the fractions as equivalent fractions and adding, we have:

$$\frac{11}{35} + \frac{3}{14} = \frac{11 \cdot 2}{35 \cdot 2} + \frac{3 \cdot 5}{14 \cdot 5}$$

$$= \frac{22}{70} + \frac{15}{70} = \frac{22 + 15}{70} = \frac{37}{70}.$$

(b) Writing the denominators in factored form, we have:

$$12 = 2 \cdot 2 \cdot 3 = 2^2 \cdot 3 \qquad \text{and} \qquad 16 = 2 \cdot 2 \cdot 2 \cdot 2 = 2^4,$$

so that, the LCD of the fractions is $2^4 \cdot 3 = 48$. Rewriting the fractions as equivalent fractions and subtracting, we have:

$$\frac{11}{12} - \frac{7}{16} = \frac{11 \cdot 4}{12 \cdot 4} - \frac{7 \cdot 3}{16 \cdot 3}$$

$$= \frac{44}{48} - \frac{21}{48} = \frac{44 - 21}{48} = \frac{23}{48}.$$

EXAMPLE 7   Perform the operations $\dfrac{7}{12} + \dfrac{1}{10} - \dfrac{2}{45}$ and simplify.

SOLUTION   Writing the denominators in factored form, we have:

$$12 = 2 \cdot 2 \cdot 3 = 2^2 \cdot 3, \qquad 10 = 2 \cdot 5, \qquad \text{and} \qquad 45 = 3 \cdot 3 \cdot 5 = 3^2 \cdot 5$$

so that, the LCD of the fractions are $2^2 \cdot 3^2 \cdot 5 = 180$. Thus,

$$\frac{7}{12} + \frac{1}{10} - \frac{2}{45} = \frac{7 \cdot 15}{12 \cdot 15} + \frac{1 \cdot 18}{10 \cdot 18} - \frac{2 \cdot 4}{45 \cdot 4}$$

$$= \frac{105}{180} + \frac{18}{180} - \frac{8}{180}$$

$$= \frac{105 + 18 - 8}{180} = \frac{115}{180} = \frac{23}{36}.$$

Multiplication and Division of Fractions

To multiply two fractions, multiply their numerators and their denominators. That is, if a/b and c/d are fractions, and b and $d \neq 0$, then

$$\boxed{\frac{a}{b} \cdot \frac{c}{d} = \frac{a \cdot c}{b \cdot d}}$$

It is important to remember that the resulting fraction should always be simplified.

EXAMPLE 8   Multiply and simplify the following fractions.

(a) $\dfrac{4}{8} \cdot \dfrac{5}{6}$ (b) $\dfrac{5}{11} \cdot \left(-\dfrac{22}{65}\right)$

SOLUTION   (a) $\dfrac{4}{8} \cdot \dfrac{5}{6} = \dfrac{4 \cdot 5}{8 \cdot 6} = \dfrac{20}{48} = \dfrac{5 \cdot \cancel{4}}{12 \cdot \cancel{4}} = \dfrac{5}{12}$

(b) $\dfrac{5}{11} \cdot \left(\dfrac{-22}{65}\right) = \dfrac{5 \cdot (-22)}{11 \cdot 65} = \dfrac{-110}{715} = \dfrac{\cancel{55} \cdot (-2)}{\cancel{55} \cdot 13} = \dfrac{-2}{13} = -\dfrac{2}{13}$

Division of fractions follows from the definition of division; that is,

$$A \div B = A \cdot \dfrac{1}{B}.$$

Thus, if a/b and c/d are fractions, and $(c/d) \neq 0$, then

$$\dfrac{a}{b} \div \dfrac{c}{d} = \dfrac{a}{b} \cdot \dfrac{1}{c/d} = \dfrac{a}{b} \cdot \dfrac{d}{c} = \dfrac{a \cdot d}{b \cdot c}$$

or

$$\boxed{\dfrac{a}{b} \div \dfrac{c}{d} = \dfrac{a}{b} \cdot \dfrac{d}{c} = \dfrac{a \cdot d}{b \cdot c}}$$

In other words, to divide two fractions, invert the second fraction and multiply.

EXAMPLE 9   Divide and simplify the following fractions.

(a) $\dfrac{5}{8} \div \dfrac{3}{4}$ (b) $\dfrac{91}{95} \div \left(\dfrac{-39}{57}\right)$

SOLUTION   (a) $\dfrac{5}{8} \div \dfrac{3}{4} = \dfrac{5}{8} \cdot \dfrac{4}{3} = \dfrac{5}{2 \cdot 2 \cdot 2} \cdot \dfrac{2 \cdot 2}{3} = \dfrac{5}{2 \cdot 3} = \dfrac{5}{6}$

(b) $\dfrac{91}{95} \div \left(\dfrac{-39}{57}\right) = -\dfrac{91}{95} \cdot \dfrac{57}{39} = -\dfrac{7 \cdot \cancel{13}}{5 \cdot \cancel{19}} \cdot \dfrac{3 \cdot \cancel{19}}{3 \cdot \cancel{13}} = -\dfrac{7}{5}$

The order of operations discussed in Section 1.1 can now be applied to fractions.

In Examples 10 and 11, perform the indicated operations and simplify.

EXAMPLE 10

$$\frac{10}{8} \div \frac{25}{16} \cdot \frac{7}{15}$$

SOLUTION Using the order of operations by performing divisions and multiplications from left to right, we have

$$\frac{10}{8} \div \frac{25}{16} \cdot \frac{7}{15} = \left(\frac{10}{8} \cdot \frac{16}{25}\right) \cdot \frac{7}{15}$$

$$= \frac{4}{5} \cdot \frac{7}{15} = \frac{28}{75}.$$

EXAMPLE 11

$$\frac{2}{5} - \frac{1}{4} \div \frac{2}{3} + \frac{3}{4}$$

SOLUTION Using the order of operations, we have

$$\frac{2}{5} - \frac{1}{4} \div \frac{2}{3} + \frac{3}{4} = \frac{2}{5} - \frac{1}{4} \cdot \frac{3}{2} + \frac{3}{4} \qquad \text{(We performed division from left to right)}$$

$$= \frac{2}{5} - \frac{3}{8} + \frac{3}{4} \qquad \text{(We performed multiplication)}$$

$$= \frac{2 \cdot 8 - 5 \cdot 3}{40} + \frac{3}{4} \qquad \text{(We performed subtraction)}$$

$$= \frac{16 - 15}{40} + \frac{3}{4} = \frac{1}{40} + \frac{3}{4}$$

$$= \frac{1 + 3 \cdot 10}{40} = \frac{1 + 30}{40} = \frac{31}{40}. \qquad \text{(We added)}$$

PROBLEM SET 1.5

In problems 1–12, reduce each fraction to lowest terms.

1. $\dfrac{15}{45}$

2. $\dfrac{28}{42}$

3. $\dfrac{77}{121}$

4. $\dfrac{-56}{96}$

5. $\dfrac{-65}{91}$

6. $\dfrac{98}{168}$

7. $\dfrac{-250}{350}$

8. $\dfrac{-156}{-520}$

9. $\dfrac{-45}{-72}$

10. $-\dfrac{65}{-39}$

11. $-\dfrac{-6}{8}$

12. $-\dfrac{-15}{-35}$

In problems 13–40, combine the fractions and simplify.

13. $\dfrac{2}{5} + \dfrac{1}{5}$

14. $\dfrac{2}{3} + \dfrac{5}{3}$

15. $\dfrac{5}{11} - \dfrac{3}{11}$

16. $\dfrac{7}{13} - \dfrac{5}{13}$

17. $\dfrac{7}{12} + \dfrac{3}{12}$

18. $\dfrac{7}{15} - \dfrac{3}{15}$

19. $\dfrac{21}{25} - \dfrac{16}{25}$

20. $\dfrac{9}{49} + \dfrac{2}{49}$

21. $\dfrac{1}{2} + \dfrac{3}{8}$

22. $\dfrac{3}{4} + \dfrac{1}{12}$

23. $\dfrac{1}{3} - \dfrac{1}{6}$

24. $\dfrac{7}{8} - \dfrac{1}{2}$

25. $\dfrac{11}{12} + \dfrac{7}{16}$

26. $\dfrac{17}{24} + \dfrac{11}{18}$

27. $\dfrac{5}{6} - \dfrac{3}{10}$

28. $\dfrac{35}{63} - \dfrac{5}{18}$

29. $\dfrac{7}{12} + \dfrac{4}{15}$

30. $\dfrac{13}{9} - \dfrac{14}{15}$

31. $\dfrac{5}{9} - \dfrac{7}{12}$

32. $\dfrac{11}{12} + \dfrac{3}{16}$

33. $\dfrac{2}{3} + \dfrac{7}{6} - \dfrac{3}{4}$

34. $\dfrac{5}{8} + \dfrac{3}{4} - \dfrac{5}{6}$

35. $\dfrac{6}{7} - \dfrac{2}{3} + \dfrac{5}{14}$

36. $\dfrac{11}{18} - \dfrac{5}{8} - \dfrac{5}{9}$

37. $\dfrac{7}{12} - \dfrac{3}{2} + \dfrac{4}{9}$

38. $\dfrac{8}{9} - \dfrac{5}{12} - \dfrac{1}{3}$

39. $\dfrac{7}{8} + \dfrac{5}{3} - \dfrac{7}{4}$

40. $\dfrac{4}{7} + \dfrac{3}{14} - \dfrac{2}{21}$

In problems 41–82, perform the indicated operations and simplify.

41. $\dfrac{4}{5} \cdot \dfrac{10}{12}$

42. $\dfrac{11}{52} \cdot \left(\dfrac{-30}{80} \right)$

43. $\dfrac{15}{14} \cdot \left(\dfrac{-7}{24} \right)$

44. $\dfrac{14}{36} \cdot \left(\dfrac{-24}{35} \right)$

45. $\left(\dfrac{-57}{34} \right) \cdot \left(\dfrac{51}{19} \right)$

46. $\dfrac{19}{31} \cdot \left(\dfrac{-62}{-38} \right)$

47. $\left(\dfrac{9}{5} \right) \cdot \left(\dfrac{-5}{27} \right)$

48. $\dfrac{4}{5} \cdot \left(\dfrac{-55}{72} \right)$

49. $\left(\dfrac{-7}{5} \right) \cdot \left(\dfrac{15}{21} \right)$

50. $\left(\dfrac{-32}{121} \right) \cdot \left(\dfrac{33}{-144} \right)$

51. $\left(\dfrac{-11}{9} \right) \cdot \left(\dfrac{-18}{55} \right)$

52. $\left(\dfrac{-21}{25} \right) \cdot \left(\dfrac{-15}{28} \right)$

53. $\left(\dfrac{-18}{35} \right) \cdot \left(\dfrac{-14}{27} \right)$

54. $\left(\dfrac{-6}{16} \right) \cdot \left(\dfrac{-12}{18} \right)$

55. $\dfrac{8}{9} \div \dfrac{24}{27}$

56. $\left(\dfrac{-9}{16} \right) \div \dfrac{21}{30}$

57. $\left(\dfrac{-25}{24} \right) \div \dfrac{15}{16}$

58. $\dfrac{35}{24} \div \left(\dfrac{-15}{84} \right)$

59. $\left(\dfrac{-51}{21} \right) \div \left(\dfrac{35}{84} \right)$

60. $\left(\dfrac{-16}{106} \right) \div \left(\dfrac{-93}{155} \right)$

61. $\dfrac{20}{21} \div \left(\dfrac{-15}{14} \right)$

62. $\left(\dfrac{-21}{45} \right) \div \left(\dfrac{-30}{14} \right)$

63. $\left(\dfrac{-5}{18} \right) \div \left(\dfrac{-25}{12} \right)$

64. $\dfrac{336}{55} \div \left(\dfrac{-168}{125} \right)$

65. $\left(\dfrac{-18}{35} \right) \div \dfrac{18}{42}$

66. $\dfrac{12}{33} \div \left(\dfrac{-8}{21} \right)$

67. $\left(\dfrac{-68}{21} \right) \div \left(\dfrac{-51}{28} \right)$

68. $\left(\dfrac{-32}{27} \right) \div \left(\dfrac{-57}{27} \right)$

69. $\left(\dfrac{-39}{51} \right) \div \left(\dfrac{26}{34} \right)$

70. $\left(\dfrac{-21}{40} \right) \div \dfrac{9}{15}$

71. $\dfrac{8}{5} \cdot \dfrac{15}{14} \div \dfrac{24}{7}$

72. $\dfrac{4}{3} \cdot \dfrac{9}{8} \div \left(\dfrac{-4}{5} \right)$

73. $\dfrac{9}{16} \cdot \left(\dfrac{-2}{3} \right) \div \dfrac{5}{21}$

74. $\dfrac{2}{5} \cdot \dfrac{3}{7} \div \left(\dfrac{-35}{4} \right)$

75. $\dfrac{2}{3} \div \dfrac{4}{5} \cdot \dfrac{3}{7} \cdot \dfrac{14}{9}$

76. $\dfrac{10}{9} \cdot \dfrac{12}{15} \div \left(\dfrac{-10}{21} \right)$

77. $\dfrac{12}{33} \div \dfrac{21}{16} \cdot \left(\dfrac{-33}{40} \right)$

78. $\dfrac{4}{9} \cdot \left(\dfrac{-5}{10} \right) + \dfrac{5}{18}$

79. $\dfrac{7}{9} - \dfrac{21}{12} \div \dfrac{5}{4} + \dfrac{5}{12}$

80. $\dfrac{4}{3} + \dfrac{21}{8} \div \dfrac{6}{15} - \dfrac{3}{4}$

81. $\dfrac{3}{4} + \dfrac{3}{5} \div \dfrac{7}{6} - \dfrac{1}{2}$

82. $\dfrac{2}{3} \div \dfrac{4}{5} + \dfrac{3}{7} - \dfrac{3}{4}$

83. An airplane is carrying 160 passengers. If this load is $\frac{8}{11}$ of its capacity, what is the capacity of the airplane?

84. Kathy won $3500 in a state lottery. If $\frac{1}{7}$ of her prize money was withheld for taxes, how much money did she actually receive?

85. A mathematics teacher graded two sets of multiple-choice examinations. The first examination took $\frac{3}{4}$ hour to grade, and the second test took $\frac{3}{5}$ hour. How much longer did it take the teacher to grade the first examination?

86. A board is 10 meters long. A carpenter cuts two pieces from the board. One piece is $3\frac{1}{4}$ meters long and the other piece is $4\frac{1}{2}$ meters long. Find the length of the piece that remains.

1.6 Calculators and Approximations

Today many students own or have access to an electronic calculator.* A scientific calculator with special keys will expedite many of the calculations required in this book. The symbol ⓒ marks problems and examples, or groups of problems and examples, for which the use of a calculator is recommended. If you don't have access to a calculator, you can work most of these problems by using the tables in the appendixes.

Learn to use your calculator properly by studying the instruction booklet furnished with it. In particular, practice performing chain calculations so you can do them as efficiently as possible, using whatever memory features your calculator may possess to store intermediate results. After you learn *how* to use a calculator, take care to learn *when* to use it and especially when *not* to use it.

EXAMPLE 1 ⓒ Change each rational number to a decimal.

(a) $\frac{7}{11}$ (b) $\frac{5}{17}$ (c) $6\frac{8}{21}$ (d) $\frac{11}{19}$

SOLUTION If we use an eight-digit calculator, we find:

(a) $\frac{7}{11} = 0.63636364$ (b) $\frac{5}{17} = 0.29411765$

(c) $6\frac{8}{21} = 6.3809524$ (d) $\frac{11}{19} = 0.57894737$

---

* There are two types of calculators available, those using *algebraic notation* (AN) and those using *reverse Polish notation* (RPN). Advocates of AN claim that it is more "natural," while supporters of RPN say that RPN is just as "natural" but avoids the parentheses required when sequential calculations are made in AN. Before purchasing a scientific calculator, you should familiarize yourself with both AN and RPN so that you can make an intelligent decision based on your own preferences.

EXAMPLE 2   C Change each square root to a decimal.

(a) $\sqrt{13}$ (b) $\sqrt{19}$

(c) $\sqrt{11.2}$ (d) $\sqrt{28.7}$

SOLUTION   If we use an eight-digit calculator, we find:

(a) $\sqrt{13} = 3.6055513$ (b) $\sqrt{19} = 4.3588989$

(c) $\sqrt{11.2} = 3.3466401$ (d) $\sqrt{28.7} = 5.3572381$

Approximations

Numbers produced by a calculator are often approximate, because the calculator can work only with a finite number of decimal places. For instance, a ten-digit calculator gives

$$\tfrac{2}{3} = 0.6666666667 \quad \text{and} \quad \sqrt{2} = 1.414213562,$$

both of which are **approximations** of the true values. Therefore, unless we explicitly ask for numerical approximations or indicate that a calculator is recommended, it's usually best to leave the answer in fractional form or in radical form.

Most numbers obtained from measurements of real-world quantities are subject to error and must also be regarded as approximations.

DEFINITION   **Significant Digits**

> The digits of a number beginning with the first nonzero digit to the left of the decimal point (or with the first nonzero digit after the decimal point if there is no nonzero digit to the left of the decimal point) and ending with the last digit to the right of the decimal point are known as **significant digits.**

Number	Significant Digits
0.0037	3, 7
3.8	3, 8
4.60	4, 6, 0

For instance, in the number 1.304, the digits 1, 3, 0, and 4 are significant, and we say that the number is accurate to four significant digits. In this case, the string of significant digits begins with the first nonzero digit and ends with the last digit specified.

Other examples of significant digits are illustrated in the table on the left.

When it is impractical, undesirable, or impossible to write all of the available decimal places for a number, the number must be **rounded off** to, or approximated by, a number with fewer significant digits. Some scientific calculators can be set to round off all displayed numbers to a particular number of decimal places or significant digits. However, it's easy enough to round off numbers without a calculator: Simply drop all unwanted digits to the right of the digits to be retained, and increase the last retained digit by 1 if the first dropped digit is 5 or greater. It may be necessary to replace dropped digits by zero in order to hold the decimal point. For instance, we round off 3,187.3 to the nearest hundred as 3,200.

Rounding off should be done in one step rather than digit by digit. Digit-by-digit rounding off may produce an incorrect result. For example, if 4.4248 is rounded off to four significant digits as 4.425, and then rounded off to three significant digits, the result is 4.43. However, 4.4248 is correctly rounded off in one step to three significant digits as 4.42.

EXAMPLE 3   Round off the given number as indicated.

(a) 6.438 to the nearest tenth

(b) 28.9825 to the nearest thousandth

(c) 3,782 to the nearest hundred.

SOLUTION   (a) To the nearest tenth, $6.438 \approx 6.4$.

(b) To the nearest thousandth, $28.9825 \approx 28.983$.

(c) To the nearest hundred, $3,782 \approx 3,800$.

EXAMPLE 4   Round off each number to three significant digits.

(a) 3.14159   (b) 2.831   (c) 3.796

(d) 7.075   (e) 0.027468

SOLUTION   (a) $3.14159 \approx 3.14$ (b) $2.831 \approx 2.83$ (c) $3.796 \approx 3.80$

(d) $7.075 \approx 7.08$ (e) $0.027468 \approx 0.0275$

EXAMPLE 5   © Use a calculator to change each number to a decimal, then round off the answer to three significant digits.

(a) $\frac{5}{6}$ (b) $3\frac{11}{13}$ (c) $\sqrt{7}$ (d) $\sqrt{31.7}$

SOLUTION   Using an eight-digit calculator, we find:

(a) $\frac{5}{6} = 0.83333333 \approx 0.833$ (b) $3\frac{11}{13} = 3.8461538 \approx 3.85$

(c) $\sqrt{17} = 4.1231056 \approx 4.12$ (d) $\sqrt{31.7} = 5.6302753 \approx 5.63$

Scientific Notation

Many problems in science and business require the use of very large numbers. For example, the speed C of light is approximately

$$C = 300,000,000 \text{ meters per second.}$$

The number N of heartbeats in a person's normal lifetime is given by

$$N = 2,200,000,000.$$

A convenient way of expressing such numbers is by *scientific notation*. A positive number x is written in **scientific notation** if x has the form

$$x = s \times 10^n$$

where s is a number between 1 and 10 $(1 \leq s < 10)$ and n is a positive integer.

    To change a large number from ordinary decimal form to scientific notation, we move the decimal point n places to the left to obtain a number between 1 and 10. Then, we multiply this number by 10^n. For example, to change the number 468,000 to scientific notation, we move the decimal point 5 places to the left. Then we multiply this number by 10^5. That is,

$$468,000 = 4.68 \times 10^5.$$

5 places

Also,

$$300,000,000 = 3 \times 10^8 \quad \text{and} \quad 2,200,000,000 = 2.2 \times 10^9.$$

EXAMPLE 6   Rewrite 487,000,000 in scientific notation.

SOLUTION   We move the decimal point 8 places to the left. Then, we multiply this number by 10^8. The number is written in scientific notation as

$$487,000,000 = 4.87 \times 10^8.$$

The preceding procedure can be reversed whenever a number is given in scientific notation and we wish to write it in ordinary decimal form.

EXAMPLE 7   Rewrite 7.85×10^4 in ordinary decimal form.

SOLUTION   We first note that $10^4 = 10,000$, then we multiply 7.85 by 10,000. That is,

$$7.85 \times 10^4 = 7.85 \times 10,000 = 78,500.$$

Multiplying 7.85 by 10^4 simply moves the decimal point 4 places to the right. Therefore, we can omit the middle step and write

$$7.85 \times 10^4 = 78,500.$$

4 places

Many calculators automatically switch to scientific notation whenever the number is too large to be displayed in ordinary decimal form. When a number such as 4.675×10^{13} is displayed, the multiplication sign and the base do not appear and the display shows simply

4.675   13.

PROBLEM SET 1.6

[C] In problems 1–8, use a calculator to change each number to a decimal.

1. $\frac{4}{29}$ **2.** $\frac{8}{37}$ **3.** $\frac{15}{17}$ **4.** $6\frac{8}{23}$ **5.** $\sqrt{26}$

6. $\sqrt{27.8}$ **7.** $\sqrt{71.3}$ **8.** $\sqrt{93.8}$

In problems 9–18, round off the given number to the nearest tenth.

9. 0.27   **10.** 0.553   **11.** 5.32   **12.** 6.07   **13.** 7.998

14. 11.349   **15.** 15.016   **16.** 22.991   **17.** 24.052   **18.** 4.96

In problems 19–28, round off the given number to the nearest hundredth.

19. 3.1872   **20.** 11.147   **21.** 14.3649   **22.** 223.5949   **23.** 21.0038

24. 42.703   **25.** 16.507   **26.** 156.155   **27.** 23.697   **28.** 17.006

In problems 29–38, round off each number to three significant digits.

29. 1.732   **30.** 4.809   **31.** 14.276   **32.** 121.5   **33.** 368.1

34. 97.04   **35.** 5,139   **36.** 5.1372   **37.** 27.98   **38.** 738,199

[C] In problems 39–48, use a calculator to change the following rational numbers to decimals and then round off the results to three significant digits.

39. $\frac{4}{3}$ **40.** $\frac{6}{17}$ **41.** $\frac{1}{7}$ **42.** $\frac{10}{19}$ **43.** $2\frac{5}{13}$

44. $7\frac{4}{7}$ **45.** $\frac{11}{6}$ **46.** $\frac{19}{21}$ **47.** $5\frac{13}{27}$ **48.** $6\frac{3}{23}$

[C] In problems 49–58, use a calculator to change the following square roots to decimals and then round off the results to four significant digits.

49. $\sqrt{3}$ **50.** $\sqrt{5}$ **51.** $\sqrt{7}$ **52.** $\sqrt{11}$ **53.** $\sqrt{23}$

54. $\sqrt{37}$ **55.** $\sqrt{47}$ **56.** $\sqrt{93}$ **57.** $\sqrt{111}$ **58.** $\sqrt{253}$

In problems 59–64, rewrite each number in scientific notation.

59. 3,782   **60.** 385,000   **61.** 384,000

62. 5,800   **63.** 780,000,000   **64.** 431,000,000

In problems 65–70, rewrite each number in expanded form.

65. 2.1×10^2 **66.** 8.6×10^3 **67.** 7.5×10^5

68. 5.41×10^4 **69.** 3.12×10^7 **70.** 1.87×10^8

In problems 71–74, rewrite each statement so that all numbers are expressed in scientific notation.

71. The diameter of the sun is approximately 870,000 miles.

72. The sun contains about 1,489,000,000,000,000,000,000,000,000 tons of hydrogen.

73. The mass of the earth is approximately 5,980,000,000,000,000,000,000,000 tons.

74. One gram of hydrogen contains 602,300,000,000,000,000,000,000 atoms.

1.7 The Language of Algebra

Perhaps the most important characteristic of algebra is the use of symbols other than the usual numerals to stand for numbers. The symbols commonly employed are letters. By combining symbols, or symbols and numbers, in certain ways, we are able to make a few algebraic symbols take the place of many words. For instance, to represent the total cost of an unknown number of suits that cost $200 each, we may use some letter, say x, to designate the number of suits purchased. Then the total cost (in dollars) of the suits is given by the expression $200x$. Thus,

if $x = 3$,   the total cost is $\$200(3) = \$600$;

and

if $x = 10$,   the total cost is $\$200(10) = \$2,000$.

In the expression $200x$, the 200 is referred to as a constant, and x is called a variable. In general, a **variable** is a symbol that may represent different numbers. (These numbers may be chosen either at random or according to some law.) A **constant** represents one number, usually an explicit number.

Algebraic expressions are formed by using constants, variables, mathematical operations (such as addition, subtraction, multiplication, and division), other operations (such as raising to powers and taking roots), and grouping symbols.

For example,

$$8, \qquad 3 + 2x, \qquad 3(y + 5), \qquad \frac{4x + 1}{3y + 2}, \qquad \text{and} \qquad 2\{t - 3[t + 2(t + 4)]\}$$

are all algebraic expressions.

The **value** of an algebraic expression is the number the expression represents when specific numbers are substituted for the letters in the expression. The process we use to find the value of the expression is called **evaluating** the expression.

For example, if x is replaced by -3 and y by 7, then the value of the algebraic expression $8x + 5y$ is given by

$$8x + 5y = 8(-3) + 5(7) = -24 + 35 = 11.$$

EXAMPLE 1   Find the value of the expression $2x + 3y - xy$, when $x = 7$ and $y = 4$.

SOLUTION   Substituting 7 for x and 4 for y, we have,

$$2x + 3y - xy = 2(7) + 3(4) - (7)(4)$$
$$= 14 + 12 - 28 = -2.$$

If an algebraic expression is formed by multiplying two or more expressions, each of these expressions is called a **factor** of the product. For instance, in the product

$$3uv$$

the factors are 3, u, and v. The factors of the product

$$5x(2y + 7)$$

are 5, x, and $2y + 7$.

Recall from Section 1.1 on page 3 that a multiplication involving the same factor two times is indicated by $x \cdot x$, three times by $x \cdot x \cdot x$, and n times by

$$\overbrace{x \cdot x \cdot x \cdots x}^{n \text{ times}}.$$

We may also use exponents to provide an alternative notation for these products. For example, $x \cdot x$ can be written as x^2 and $x \cdot x \cdot x$ as x^3. In general, if n is a positive integer,

$$\overbrace{x \cdot x \cdot x \cdots x}^{n \text{ factors}} = x^n.$$

In using the **exponential notation** x^n, x is the **base** and n is the **exponent** or **power** to which the base is raised. The value of x^n is often referred to as the nth *power* of x.

EXAMPLE 2   Rewrite each expression using exponential notation.

(a) $3 \cdot 3 \cdot 3 \cdot 3 \cdot 3 \cdot 3$ (b) $b \cdot b \cdot b \cdot b \cdot c \cdot c \cdot c$

(c) $(-y) \cdot (-y) \cdot (-y) \cdot (-y) \cdot (-y)$

SOLUTION   (a) $3 \cdot 3 \cdot 3 \cdot 3 \cdot 3 \cdot 3 = 3^6$ (b) $b \cdot b \cdot b \cdot b \cdot c \cdot c \cdot c = b^4 c^3$

(c) $(-y) \cdot (-y) \cdot (-y) \cdot (-y) \cdot (-y) = (-y)^5$

EXAMPLE 3   Rewrite each expression in expanded form.

(a) 2^4 (b) w^5 (c) $(-t)^7$ (d) $2^4 x^5 y^3$

SOLUTION   (a) $2^4 = 2 \cdot 2 \cdot 2 \cdot 2$

(b) $w^5 = w \cdot w \cdot w \cdot w \cdot w$

(c) $(-t)^7 = (-t) \cdot (-t) \cdot (-t) \cdot (-t) \cdot (-t) \cdot (-t) \cdot (-t)$

(d) $2^4 x^5 y^3 = 2 \cdot 2 \cdot 2 \cdot 2 \cdot x \cdot x \cdot x \cdot x \cdot x \cdot y \cdot y \cdot y$

Formulas

An *equation* is a mathematical statement that two expressions are equal. Thus,

$$3x + 2 = 5 \quad \text{and} \quad 7y - 1 = 4 + 3y$$

are equations. One important type of equation is a *formula*. A **formula** is an equation that expresses one variable in terms of one or more other variables. Formulas have many important applications in different fields. For example, if C represents the temperature on the Celsius scale, and F represents the temperature on the Fahrenheit scale, then a formula for C in terms of F is

$$C = \frac{5}{9}(F - 32).$$

If $F = 212°$ (the boiling point of water), then

$$C = \frac{5}{9}(212 - 32) = \frac{5}{9}(180) = 100°.$$

If $F = 32°$ (the freezing point of water), then

$$C = \frac{5}{9}(32 - 32) = 0°,$$

and if $F = 0°$, then

$$C = \frac{5}{9}(0 - 32) = -\frac{160°}{9}.$$

Other examples of using formulas follow.

EXAMPLE 4   The formula for the perimeter P of a rectangle of length ℓ and width w (Figure 1) is

$$P = 2\ell + 2w.$$

Find the perimeter P of a rectangle whose length is 5 centimeters and whose width is 3 centimeters.

Figure 1

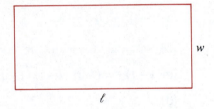

SOLUTION   Here $\ell = 5$ and $w = 3$. Thus,

$$P = 2\ell + 2w = 2(5) + 2(3) = 10 + 6 = 16.$$

Therefore, the perimeter is 16 centimeters.

EXAMPLE 5   In business, the formula

$$S = P(1 + rt)$$

expresses the total amount S (accumulated principal and interest) if a principal amount P is loaned at an annual simple interest rate r for t years. Find the total amount S if:

(a) $P = \$4,000$, $r = 7\%$, and $t = 5$ years.

© (b) $P = \$5,732$, $r = 6.8\%$, and $t = 4$ years.

SOLUTION   (a) Substituting 4,000 for P, $7\% = 0.07$ for r, and 5 for t in the formula for S, we have

$$S = P(1 + rt) = 4{,}000[1 + (0.07)5]$$
$$= 4{,}000[1 + 0.35]$$
$$= 4{,}000[1.35] = 5{,}400.$$

Therefore, the total amount accumulated is $5,400.

(b) Here, $P = 5{,}732$, $r = 0.068$, and $t = 4$, so that

$$S = P(1 + rt) = 5{,}732[1 + (0.068)4]$$
$$= 7{,}291.10.$$

Therefore, the total amount accumulated is $7,291.10.

EXAMPLE 6   In physics, the formula

$$S = V_0 t - \tfrac{1}{2}gt^2$$

expresses the altitude S of an object fired vertically upward from the ground, with an initial speed V_0, at time t, with g a constant due to gravity. Suppose that a ball is thrown vertically upward with an initial speed of 64 feet per second. Find the altitude of the ball after 3 seconds if $g = 32$ feet per second per second.

SOLUTION   Substituting 64 for V_0, 3 for t, and 32 for g, we have,

$$S = V_0 t - \tfrac{1}{2}gt^2 = 64(3) - \tfrac{1}{2}(32)(3^2)$$
$$= 192 - 16(9) = 192 - 144$$
$$= 48.$$

Therefore, the ball was 48 feet high.

EXAMPLE 7   Ⓒ The formula for the volume V for a right circular cylinder of base radius r and height h (Figure 2) is given by

Figure 2

$$V = \pi r^2 h.$$

Its total surface area S, including the top and the bottom, is given by the formula

$$S = 2\pi rh + 2\pi r^2.$$

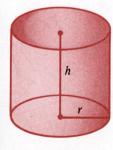

Find the volume and the total surface area of a standard cylindrical oil drum of radius 27.95 centimeters and of height 86.35 centimeters (use $\pi = 3.14$).

SOLUTION   Substituting 27.95 for r and 86.35 for h in the formulas for V and S, we have

$$V = \pi r^2 h = 3.14(27.95)^2(86.35)$$
$$= 211{,}814.46$$

and

$$S = 2\pi rh + 2\pi r^2 = 2(3.14)(27.95)(86.35) + 2(3.14)(27.95)^2$$
$$= 20{,}062.62.$$

Therefore the volume is 211,814.46 cubic centimeters and the total surface area is 20,062.62 square centimeters.

It should be noted that some of these formulas will be used in the applications of Chapters 4, 6, and 8.

PROBLEM SET 1.7

In problems 1–8, find the value of each algebraic expression.

1. $x(y - 3)$; $x = 2$, $y = 5$

2. $7(x - y)$; $x = 8$, $y = 2$

3. $\dfrac{x + y}{5}$; $x = 10$, $y = 15$

4. $\dfrac{8}{x - y}$; $x = -5$, $y = -2$

5. $\dfrac{5}{y^2} + 8x$; $x = -2$, $y = -10$

6. $\dfrac{4}{3}r^3$; $r = 6$

ⓒ **7.** $\left(\dfrac{2x}{y^4}\right)^3$; $x = 5.7$, $y = 3.2$

ⓒ **8.** $\left(\dfrac{-6a^2}{b}\right)^4$; $a = 7.2$, $b = 3.5$

In problems 9–18, rewrite each expression using exponential notation.

9. $5 \cdot 5 \cdot 5$

10. $7 \cdot 7 \cdot 7 \cdot 7 \cdot 7$

11. $x \cdot x \cdot x \cdot x$

12. $z \cdot z \cdot z \cdot z \cdot z \cdot z$

13. $(-t)(-t)(-t)(-t)(-t)$

14. $x \cdot x \cdot y \cdot y \cdot y$

15. $u \cdot u \cdot u \cdot v \cdot v \cdot v \cdot v \cdot v$

16. $(-a)(-a)(-a)(-b)(-b)$

17. $3 \cdot 3 \cdot x \cdot x \cdot x \cdot x \cdot y \cdot y$

18. $2 \cdot 2 \cdot 2 \cdot (-x)(-x)(-x)(-x)$

In problems 19–28, rewrite each expression in expanded form.

19. 8^4

20. $3 \cdot 9^4$

21. y^5

22. $5t^5$

23. $5^3 t^4$

24. $9^3 x^4 y$

25. $(-x)^4 y^3$

26. $(-t)^3(-s)^4$

27. $4^3 u^4 v$

28. $5^3 u^4 v^7$

In problems 29–34, use the formula $F = \frac{9}{5}C + 32$ to find the temperature F in degrees Fahrenheit when the temperature C in degrees Celsius is the given number.

29. $C = 0°$

30. $C = 5°$

31. $C = 10°$

32. $C = 15°$

33. $C = 20°$

34. $C = 25°$

In problems 35–40, use the formula

$$A = s^2$$

for the area A of a square of side S (Figure 3) and the formula

$$P = 4s$$

for the perimeter P of a square. Find (a) the area of a square, and (b) the perimeter of a square.

Figure 3

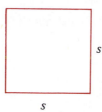

s

s

35. $s = 9$ inches

36. $s = 3.4$ meters

37. $s = 4$ feet

38. $s = 0.7$ yard

39. $s = 1.7$ inches

40. $s = 5.3$ centimeters

C In problems 41–46, use the formula

$$A = \pi r^2$$

for the area A of a circle of radius r (Figure 4), and the formula

$$C = 2\pi r$$

for the circumference C. Find (a) the area of a circle, and (b) the circumference of a circle. (Use $\pi = 3.14$.) Approximate the answer to two decimal places.

Figure 4

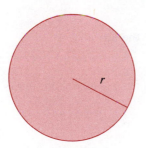

r

41. $r = 10$ inches

42. $r = 15$ feet

43. $r = 5$ feet

44. $r = 7$ yards

45. $r = 6.2$ meters

46. $r = 11.3$ centimeters

In problems 47–52, use the formula

$$A = \ell w$$

for the area A of a rectangle with length ℓ and width w. Also use the formula for the perimeter P on page 48. Find (a) the area of a rectangle, and (b) the perimeter of a rectangle.

47. $\ell = 7$ inches, $w = 5$ inches

48. $\ell = 7.3$ centimeters, $w = 6.1$ centimeters

49. $\ell = 8$ meters, $w = 3$ meters

50. $\ell = 14$ feet, $w = 10.3$ feet

51. $\ell = 15$ feet, $w = 10$ feet

52. $\ell = 9.3$ yards, $w = 8.6$ yards

In problems 53–58, use the formula

$$V = s^3$$

for the volume V of the cube of side s (Figure 5) and the formula

$$A = 6s^2$$

for the surface area A of the cube. Find (a) the volume of the cube, and (b) the surface area of the cube.

Figure 5

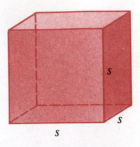

53. $s = 3$ inches   C 54. $s = 2.3$ meters   55. $s = 6$ centimeters
C 56. $s = 5.2$ feet   57. $s = 10$ centimeters   C 58. $s = 5.7$ meters

C In problems 59–64, use the formula

$$V = \frac{4\pi r^3}{3}$$

for the volume V of the sphere of radius r (Figure 6), and the formula

$$A = 4\pi r^2$$

for the surface area A of the sphere. Find (a) the volume of the sphere, and (b) the surface area of the sphere. (Use $\pi = 3.14$.)

Figure 6

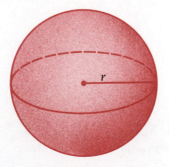

59. $r = 5$ centimeters   60. $r = 3$ meters   61. $r = 4.1$ inches
62. $r = 7.9$ feet   63. $r = 8.2$ meters   64. $r = 13.2$ centimeters

In problems 65–70, use the formula from *physics*

$$d = rt$$

to find the distance d an object travels when moving at a uniform rate r during time t.

65. $r = 50$ miles per hour, $t = 3$ hours

66. $r = 70$ miles per hour, $t = 2\frac{1}{2}$ hours

67. $r = 66$ feet per second, $t = 15$ seconds

68. $r = 44$ feet per second, $t = 21.5$ seconds

69. $r = 600$ centimeters per second, $t = 13$ seconds

70. $r = 0.7$ meter per second, $t = 25$ seconds

C In problems 71–74, use the formula on page 48 to find the total amount of money S accumulated if a principal amount P is invested at an annual simple interest rate r for t years.

71. $P = \$8,320$, $r = 7.8\%$, $t = 5$ years

72. $P = \$17,317$, $r = 6.7\%$, $t = 4$ years

73. $P = \$25,125$, $r = 9.25\%$, $t = 7$ years

74. $P = \$75,000$, $r = 11.25\%$, $t = 8$ years

75. People who want to improve their level of cardiovascular fitness through exercise sometimes determine their maximum desirable heart rate. Medical researchers have found that the desirable maximum heart rate R (in beats per minute) of a person exercising is given by the formula

$$R = 220 - A$$

where A is the person's age. What is the desirable maximum heart rate of a 45-year-old while exercising?

C **76.** In statistics, the following formula

$$v = \frac{(x_1 - \bar{x})^2 + (x_2 - \bar{x})^2 + (x_3 - \bar{x})^2}{3}$$

is used, where x_1, x_2, and x_3 represent real numbers, and

$$\bar{x} = \frac{x_1 + x_2 + x_3}{3},$$

where $\bar{x}$ is the average of the three numbers. Compute v if $x_1 = 1.8$, $x_2 = 1.4$, and $x_3 = 3.3$.

REVIEW PROBLEM SET

In problems 1–4, translate each statement into symbols.

1. The difference of x and 11 is 21.

2. The product of 6 and x is greater than x plus 5.

3. Twice the sum of y and 3 equals 27.

4. The difference of z and 7 is less than the product of z and 2.

In problems 5–8, write an equivalent word statement for each given statement represented by the symbols.

5. $y - 2 \leq 11$ **6.** $2x + 3 = x - 5$ **7.** $3 + 4x = 18$ **8.** $3(8 - z) \neq 2z$

In problems 9–12, identify the base and the exponent, then write each expression in expanded form and find the value of each expression.

9. 3^4 **10.** 5^3 **11.** 4^5 **12.** 17^2

In problems 13–16, find the value of each expression by using the rule for the order of operations.

13. $6 + (5 - 3)^4 \cdot 2 - 5$ **14.** $10 + 4 \cdot 3^4 \div 2^2 - 7$
15. $3^2 - 4^2 \div 2^3 + 15$ **16.** $(3^2 - 2^3) + 4^2 \div 8 \cdot 3$

17. Use set notation to describe the set A of the counting numbers greater than 3 but less than 12.

18. Use set-builder notation to describe the set of counting numbers B greater than 5 but less than 15.

In problems 19–22, represent each set on a number line.

19. $\{-5, -2, 3, 5\}$ **20.** $\{-\frac{3}{2}, \frac{1}{2}, 0, 2, 3\}$ **21.** $\{0, 3, 6, 9, \ldots\}$ **22.** $\{\ldots, -\frac{5}{2}, -\frac{3}{2}, -\frac{1}{2}\}$

In problems 23–26, express each rational number as a decimal.

23. $\frac{7}{5}$ **24.** $-\frac{5}{3}$ **25.** $3\frac{1}{8}$ **26.** $-\frac{11}{4}$

In problems 27–32, express each decimal as a quotient of integers.

27. 0.34 **28.** 5.22 **29.** 0.184 **30.** -4.7 **31.** -6.814 **32.** 0.226

In problems 33–44, identify each number as being rational or irrational.

33. $-\frac{2}{31}$ **34.** $\sqrt{11}$ **35.** $4.142242224\ldots$ **36.** $0.\overline{56}$
37. $-0.\overline{74}$ **38.** $\sqrt{81}$ **39.** $\sqrt{24.1}$ **40.** $5.374474447\ldots$
41. -5.7 **42.** 0.378 **43.** $8.6\overline{14}$ **44.** $\frac{11}{16}$

In problems 45–60, justify each statement by giving the appropriate property. Assume that all letters represent real numbers.

45. $x + 5 = 5 + x$ **46.** $a + (b + 3) = (a + b) + 3$ **47.** $3t = t3$
48. $u(v + w) = uv + uw$ **49.** $2(xy) = (2x)y$ **50.** $1 \cdot 4 = 4$
51. $5 + 0 = 5$ **52.** If $2x = 2y$, then $x = y$.   **53.** $0 \cdot k = 0$
54. $(-2) \cdot (-3) = 2 \cdot 3$ **55.** If $x + 9 = y + 9$, then $x = y$.   **56.** $uv = vu$
57. $3(x + 2) = 3x + 6$ **58.** $7(mn) = (7m)n$ **59.** $4t + 0 = 4t$
60. $1 \cdot y = y$

In problems 61–88, evaluate each expression.

61. $|-13|$ **62.** $-|14|$ **63.** $-|-21|$
64. $\left|-\frac{7}{3}\right|$ **65.** $(-5) + 2$ **66.** $(-13) + (-21)$

67. $(-6) + (-8)$ **68.** $(-15) + 9$ **69.** $5 - (-9)$
70. $(-5) + 13$ **71.** $17 - (-13)$ **72.** $0 - (-3)$
73. $(-7) + (-3) + (-2)$ **74.** $(-12) - 8$ **75.** $(-4) \cdot (-12)$
76. $(-11) \cdot (-10)$ **77.** $8 \cdot (-5)$ **78.** $(-6) \cdot 7$
79. $(-16) \cdot (-1) \cdot (-2)$ **80.** $(-7) \cdot 2 \cdot (-1) \cdot 6$ **81.** $(-3) \cdot (-7) \cdot 5$
82. $(-1) \cdot (-2) \cdot (-3) \cdot (-4)$ **83.** $(-63) \div 7$ **84.** $25 \div (-5)$
85. $(-40) \div (-5)$ **86.** $(-36) \div (-6)$ **87.** $21 \div (-7)$
88. $(-56) \div 7$

In problems 89–92, reduce each fraction to lowest terms.

89. $\dfrac{12}{30}$ **90.** $\dfrac{-52}{65}$ **91.** $\dfrac{21}{-28}$ **92.** $-\dfrac{-15}{35}$

In problems 93–104, perform the indicated operations and simplify.

93. $\dfrac{3}{7} + \dfrac{2}{7}$ **94.** $\dfrac{7}{11} - \dfrac{3}{11}$ **95.** $\dfrac{3}{4} - \dfrac{5}{12}$
96. $\dfrac{5}{8} + \dfrac{7}{12}$ **97.** $\dfrac{2}{3} + \dfrac{3}{4} - \dfrac{1}{2}$ **98.** $\dfrac{7}{9} - \dfrac{1}{3} - \dfrac{1}{6}$
99. $\dfrac{5}{12} \cdot \dfrac{8}{15}$ **100.** $\dfrac{3}{7} \cdot \left(\dfrac{-21}{5}\right)$ **101.** $\dfrac{7}{12} \div \dfrac{21}{4}$
102. $\left(\dfrac{-8}{11}\right) \div \left(\dfrac{5}{-22}\right)$ **103.** $\dfrac{3}{5} \cdot \dfrac{7}{8} \div \left(\dfrac{21}{-4}\right)$ **104.** $\left(\dfrac{-6}{13}\right) \div \left(\dfrac{5}{26}\right) \cdot \left(\dfrac{15}{-2}\right)$

In problems 105–110, round off the given number to the nearest hundredth.

105. 1.078 **106.** 145.395 **107.** 3.816
108. 26.1632 **109.** 0.0371 **110.** 0.0482

In problems 111–116, round off each number to four significant digits.

111. 1.03468 **112.** 0.00137215 **113.** 1.0127415
114. 278.516 **115.** 29.3019 **116.** 3.001287

C In problems 117–128, use a calculator to change each number to a decimal, and then round off the result to three significant digits.

117. $\frac{5}{7}$ **118.** $\frac{11}{14}$ **119.** $\frac{16}{37}$ **120.** $\sqrt{32}$
121. $\sqrt{53}$ **122.** $4\frac{7}{13}$ **123.** $\sqrt{71.2}$ **124.** $\sqrt{63.7}$
125. $\frac{5}{113}$ **126.** $\frac{-3}{764}$ **127.** $\frac{8}{451}$ **128.** $\sqrt{118.4}$

In problems 129–132, rewrite each number in scientific notation.

129. 5,872 **130.** 16,380 **131.** 123,400 **132.** 2,800,000

In problems 133–136, rewrite each number in expanded form.

133. 6.872×10^4 **134.** 5.72×10^6 **135.** 2.91×10^5 **136.** 3.232×10^8

In problems 137–140, find the value of each algebraic expression.

137. $x(2y - x)$; $x = 3$, $y = 5$

138. $9(x + y)$; $x = 5$, $y = 7$

139. $\dfrac{x - y}{3}$; $x = -12$, $y = 6$

140. $\dfrac{12}{x - y}$; $x = 2$, $y = -4$

In problems 141–146, rewrite each expression using exponential notation.

141. $2 \cdot 2 \cdot x \cdot x \cdot x$

142. $9 \cdot 9 \cdot w \cdot w \cdot w \cdot w$

143. $5 \cdot u \cdot u \cdot v \cdot v \cdot v$

144. $3 \cdot 3 \cdot m \cdot m \cdot m \cdot n \cdot n \cdot n \cdot n$

145. $(-2)(-2)(-t)(-t)(-s)(-s)$

146. $(-p)(-p)(-p)(-q)(-q)(-q)(-q)(-q)$

In problems 147–152, rewrite each expression in expanded form.

147. $x^2 y^4$

148. $5t^3 s^2$

149. $3^3 u^2 v^4$

150. $-3x^4 y^5$

151. $m^3 n^4 p$

152. $p^3 q t^6$

In problems 153–160, evaluate the formula for the given values.

153. $F = \frac{9}{5}C + 32$, for $C = 30°$

154. $F = \frac{9}{5}C + 32$, for $C = 35°$

155. $A = \ell w$, for $\ell = 6$ inches, $w = 5$ inches

156. $A = \ell w$, for $\ell = 4.7$ meters, $w = 3.2$ meters

157. $P = 2\ell + 2w$, for $\ell = 4$ feet, $w = 3$ feet

158. $P = 2\ell + 2w$, for $\ell = 5.3$ centimeters, $w = 4.8$ centimeters

C **159.** $A = \pi r^2$, for $r = 7$ inches (use $\pi \approx 3.14$)

C **160.** $A = \pi r^2$, for $r = 9$ feet

CHAPTER 1 TEST

1. Translate each statement into symbols.

   (a) The sum of twice x and 3 is 15.

   (b) The product of 7 and y is less than the difference of y and 5.

2. Identify the base and exponent of 4^3, then evaluate 4^3 by first writing it in expanded form.

3. Evaluate the expression $7 + (8 - 5)^3 \div 3^2 - 5 \cdot 2$.

4. Use set notation to describe the set A of the counting numbers greater than or equal to 3 but less than 11.

5. Express $\frac{8}{5}$ as a decimal.

6. Express -3.46 as a quotient of integers.

7. Evaluate each expression.

   (a) $|-7|$ (b) $(-7) + (-8)$ (c) $(-27) - (-18)$

   (d) $(-4)(3)$ (e) $(-12) \div (-6)$ (f) $(-1) \cdot (-3) \cdot (4) \cdot (-2)$

8. Reduce each fraction to lowest terms.

   (a) $\dfrac{-28}{52}$ (b) $-\dfrac{-51}{63}$

9. Perform the indicated operations and simplify.

(a) $\dfrac{5}{16} + \dfrac{3}{16}$ (b) $\dfrac{7}{8} - \dfrac{3}{5}$ (c) $\dfrac{3}{28} \cdot \dfrac{7}{15}$

(d) $\dfrac{14}{15} \div \dfrac{7}{10}$ (e) $\dfrac{4}{5} \cdot \dfrac{3}{8} \div \left(-\dfrac{2}{5}\right)$

10. Round off each number as indicated.

(a) 3.238 to the nearest hundredth

(b) 5.00239 to four significant digits

11. Rewrite 3,196,000 in scientific notation.

12. Rewrite 5.27×10^4 in expanded form.

13. Find the value of $3y(4x - y)$ for $x = 3$, $y = 7$.

14. Rewrite $2 \cdot 2 \cdot (-x) \cdot (-x) \cdot (-x) \cdot y \cdot y$ using exponential notation.

15. Rewrite $4u^3 v^4$ in expanded form.

16. Evaluate the formula $P = 2\ell + 2w$ for $\ell = 3.5$ centimeters and $w = 2.25$ centimeters.

2 The Algebra of Polynomials

We continue our study of algebra by applying the properties of real numbers to algebraic expressions known as *polynomials*. In this chapter, we learn to add, subtract, multiply, divide, and factor polynomials. Later in the book, we will see that many of the most practical applications of algebra require us to translate statements using words into statements using polynomials.

2.1 Polynomials

The following algebraic expressions are examples of polynomials:

$$4x, \qquad t - 5, \qquad y^2 + 4y + 7, \qquad \text{and} \qquad -z^5 + 3wz^4 + 2w^2 + 4.$$

More formally, a **polynomial** is an algebraic expression in which the variables appear only with nonnegative integer exponents. That is, all exponents of the variables of a polynomial are nonnegative integers, and there can be no division by a variable (no variable will appear in the denominator of a fraction).

The algebraic expressions

$$2x^4 + 3x - 5, \qquad -\frac{7}{2}t^3 + 5t^2 - 2t + 1, \qquad \frac{3}{2}y,$$

and

$$-w^3 + 5^{1/2}w^2 - w + 7$$

are polynomials.

The expressions

$$\frac{3}{x^2}, \qquad 4 - \frac{1}{y}, \qquad \text{and} \qquad \frac{2t - 1}{t + 7}$$

are *not* polynomials, because the variables x, y, and t appear in the denominators. Also, the expression $\sqrt{x}$ is not a polynomial, because, as is shown in Chapter 5, $\sqrt{x} = x^{1/2}$, and the exponent $\frac{1}{2}$ is not a nonnegative integer.

If a polynomial consists of parts connected by plus or minus signs, then each of the parts, together with the sign preceding it, is called a **term** of the polynomial. For instance, in the polynomial

$$2x^4 + 3x - 5,$$

the terms are $2x^4$, $3x$, and -5.

The variables involved in a term of a polynomial are called the **literal factors,** and the numerical factor is called the **numerical coefficient.** For example, in the term $-5xy$, the x and y are literal factors and -5 is the numerical coefficient. The **coefficients** of a polynomial are the numerical coefficients of the terms of the polynomial. For example, the coefficients of the polynomial

$$-2x^5 + 3x^4 + 5x^3 - 4x^2 + 7x + 8$$

are -2, 3, 5, -4, 7, and 8.

When there is no specific number in a term, as in x, the numerical coefficient is 1; also, the coefficient of the constant term 9 is 9.

Three special types of polynomials are defined as follows:

(i) A polynomial containing only one term is called a **monomial.**

(ii) A polynomial containing exactly two terms is called a **binomial.**

(iii) A polynomial containing exactly three terms is called a **trinomial.**

For example,

$4x$ is a monomial,

$3y - 5$ is a binomial, and

$3z^2 + 4z + 7$ is a trinomial.

Each nonzero polynomial has a degree, which we define formally as follows:

DEFINITION   **Degree of a Polynomial**

The **degree of a term** in a polynomial is the sum of all the exponents of the variable factors in the term. The **degree of the polynomial** is the highest degree among all the terms in the polynomial that have nonzero coefficients.

Note that, when adding exponents to determine degree, a variable with no exponent is regarded as having exponent 1. A nonzero real number (such as 8) is a polynomial of degree zero. The real number zero is a polynomial with no degree assigned to it. For example, $2x = 2x^1$ and $8 = 8x^0$.

EXAMPLE 1   In each case, identify the polynomial as a monomial, binomial, or trinomial. Give the degree and the numerical coefficients of the polynomial.

(a) $-7y$ (b) $3w^4 - 2$ (c) $-2x^2 + 3x + 4$

SOLUTION   (a) Monomial of degree 1 with coefficient -7

(b) Binomial of degree 4 with coefficients 3 and -2

(c) Trinomial of degree 2 with coefficients -2, 3, and 4

EXAMPLE 2   Find the degree of each polynomial.

(a) $3x^3y - 2x^2y + 7xy^2 - y^3$ (b) $a^2b^3c + a^3b^2c^4 - 5$

SOLUTION   (a) The sums of the exponents in each term of $3x^3y - 2x^2y + 7xy^2 - y^3$ are

first term: $3 + 1 = 4$

second term: $2 + 1 = 3$

third term: $1 + 2 = 3$

fourth term: 3.

Because the highest sum is 4, the degree of the polynomial is 4.

(b) The degree of the polynomial $a^2b^3c + a^3b^2c^4 - 5$ is 9, the sum of the exponents of the second term, because that is the term with the highest degree.

Evaluation of a Polynomial

Polynomials such as $2x + 3$ can be evaluated for specific values of x. For example, if x is replaced by 4, we obtain

$$2x + 3 = 2 \cdot 4 + 3.$$

Notice that the order in which we complete the preceding operation will determine the final results. For example, the expression

$$2 \cdot 4 + 3$$

will be evaluated as $(2 \cdot 4) + 3 = 8 + 3 = 11$, since by our rule for order of operations we multiply first and then perform the additions or subtractions. The rules for order of operations extends to polynomials, so, when evaluating polynomials, we perform operations in the following order:

Order of Operations for Evaluating Polynomials

(i) Evaluate any expression involving exponents.

(ii) Perform all multiplications or divisions from left to right as they occur.

(iii) Perform all remaining additions or subtractions from left to right.

In Examples 3 and 4, evaluate the polynomials for the indicated values of the variables.

EXAMPLE 3   (a) $40 - 9x^2$, for $x = 3$ (b) $96t - 16t^2$, for $t = -2$

SOLUTION   (a) We replace x by 3:

$$
\begin{aligned}
40 - 9x^2 &= 40 - 9(3)^2 \\
&= 40 - 9(9) &&\text{[We evaluated } (3)^2.\text{]} \\
&= 40 - 81 &&\text{(We multiplied.)} \\
&= -41. &&\text{(We subtracted.)}
\end{aligned}
$$

(b) We replace t by -2:

$$
\begin{aligned}
96t - 16t^2 &= 96(-2) - 16(-2)^2 \\
&= 96(-2) - 16 \cdot (4) &&\text{[We evaluated } (-2)^2.\text{]} \\
&= -192 - 64 &&\text{(We multiplied.)} \\
&= -256. &&\text{(We subtracted.)}
\end{aligned}
$$

EXAMPLE 4   (a) $3x^2 - xy + y - 3$, for $x = 2$ and $y = 3$

(b) $4uv^3 - 2u^2v + 7$, for $u = -1$ and $v = -2$

SOLUTION   (a) We replace x by 2 and y by 3:

$$
\begin{aligned}
3x^2 - xy + y - 3 &= 3(2)^2 - 2(3) + 3 - 3 \\
&= 3(4) - 2(3) + 3 - 3 \\
&= 12 - 6 + 3 - 3 = 6.
\end{aligned}
$$

(b) We replace u by -1 and v by -2:

$$
\begin{aligned}
4uv^3 - 2u^2v + 7 &= 4(-1)(-2)^3 - 2(-1)^2(-2) + 7 \\
&= 4(-1)(-8) - 2(1)(-2) + 7 \\
&= 32 + 4 + 7 = 43.
\end{aligned}
$$

EXAMPLE 5   A rocket is fired vertically upward, and it is h meters above the ground t seconds after being fired, where

$$h = 560t - 16t^2$$

and the positive direction is upward. Find the height of the rocket after (a) 2 seconds (b) 5 seconds.

SOLUTION   We begin by substituting the numbers 2 and 5 in the formula

$$h = 560t - 16t^2.$$

(a) If $t = 2$, then

$$h = 560(2) - 16(2^2)$$
$$= 1{,}120 - 64 = 1{,}056.$$

Therefore the height of the rocket after 2 seconds is 1,056 meters.

(b) If $t = 5$, then

$$h = 560(5) - 16(5^2)$$
$$= 2{,}800 - 400 = 2{,}400.$$

Therefore the height of the rocket after 5 seconds is 2,400 meters.

PROBLEM SET 2.1

In problems 1–10, identify the polynomial as a monomial, binomial, or trinomial. Give the degree and the coefficients.

1. $3x - 2$ **2.** -3 **3.** $4y^2$ **4.** $5x^3 - 5$

5. $t^2 - 5t + 6$ **6.** $w^3 - w - 1$ **7.** $2u^7 - 13$ **8.** $4 - 7y$

9. $-x^4 - x^2 + 13$ **10.** $\frac{1}{2}p^4 + \frac{3}{2}p^2 + 8$

In problems 11–20, find the degree of each polynomial.

11. $5x^2y + 16x$ **12.** $3m^2 + 5mnp - 1$ **13.** $t^2 - 5ts + 2s^2$

14. $13x^4 - 7xy + 5xz^4$ **15.** $7u^2 + 13u^2v^3 + 9v^4$ **16.** $4w^2 - 3wz^2 + 5wz^3$

17. $2x^3y - 8xy^5 + 5x^2y^3$ **18.** $-7u^5v^3 + 11u^7v - 4$ **19.** $11 - 4pq^6 - 6p^4q^5$

20. $14y^3z^4 - 16y^4z^3$

In problems 21–30, evaluate the polynomial for the given value of the variable.

21. $4x - 1$, for $x = 2$ **22.** $3z + 2$, for $z = -2$

23. $2u^2 - u + 4$, for $u = 3$ **24.** $3y^3 - 5y + 3$, for $y = 4$

25. $y^3 - 2y^2 - y + 1$, for $y = -1$ **26.** $-2x^4 + 3x^2 + 15$, for $x = -3$

27. $4 - t - t^2$, for $t = -2$ **28.** $3w^4 + 3w^2 + 15$, for $w = -1$

29. $x^5 - x^4 + x^3 - x^2 + x - 2$, for $x = 2$ **30.** $m^4 - 3m^2 + 17m + 23$, for $m = 3$

In problems 31–36, evaluate the polynomial for $x = 1$, $y = -1$, and $z = 2$.

31. $2xy^2 - yz$ **32.** $3x^2 + xy - z^2y$ **33.** $x^3y^2 - 2yz^2 + xy^2$

34. $4zy^2 + x^2z - 3yz^3$ **35.** $3xy^2 - 2y^2z - xy$ **36.** $2xyz + x^2z^3$

In problems 37–40, indicate whether or not the given algebraic expression is a polynomial.

37. $\dfrac{13}{x} + 75$ **38.** $7u^5 - 3u^4 - 3u^3 + 11$

39. $-\frac{1}{7}t^8 + \frac{1}{5}t^5 + 1$ **40.** $\dfrac{5}{y^2} + 9y + 2$

41. A ball is thrown vertically upward from the ground with an initial speed of 64 feet per second. If the positive direction of the distance from the ground is up, then its height h above the ground t seconds later is given by the formula

$$h = -16t^2 + 64t.$$

Find the height after (a) 2 seconds (b) 3 seconds.

42. The total gross earnings T (in millions of dollars) of a particular corporation t years from January 1, 1988, is given by the formula

$$T = \frac{2}{3}t^2 + 2t + 10.$$

Find the total earnings after (a) 2 years (b) 5 years.

2.2 Addition and Subtraction of Polynomials

In Chapter 1, we solved many problems involving addition and subtraction of real numbers, and we used the commutative, associative, and distributive properties of real numbers. These properties can be extended to polynomials because polynomials are expressions that represent real numbers. In fact, if P, Q, and R represent polynomials, then:

> (i) $P + Q = Q + P$
>
> (ii) $P + (Q + R) = (P + Q) + R$
>
> (iii) (a) $P(Q + R) = PQ + PR$
>   (b) $(P + Q)R = PR + QR$

Suppose that you want to add the monomials $3x^2$ and $4x^2$. By the distributive property, you have

$$3x^2 + 4x^2 = (3 + 4)x^2 = 7x^2.$$

The terms $3x^2$ and $4x^2$ are called *like* or *similar terms*. These can be defined more precisely as follows:

DEFINITION   **Like Terms**

> In an algebraic expression, two or more terms that have the same variable part, although they may have different numerical coefficients, are called **like,** or **similar,** terms.

For example, $3x^2$ and $-5x^2/2$ are like terms, and so are $4y^7$ and $2y^7/3$. However, $4x^2$ and $3x$ are not like terms, since the exponents of the variables are not the same.

To add or subtract monomials, only like terms can be combined.

EXAMPLE 1   Perform each operation by combining like terms.

(a) $3x + 5x$ (b) $4t^2 - 2t^2$

(c) $2y^3 + 4y^3 - 3y^3$ (d) $-4w^4 - (-6w^4)$

(e) $3xy^2 + 7xy^2 + 2x^3y^2$ (f) $3a + 2b + 7c$

SOLUTION   (a) $3x + 5x = (3 + 5)x = 8x$

(b) Here, we use the fact $a - b = a + (-b)$, so that

$$4t^2 - 2t^2 = 4t^2 + (-2t^2) = [4 + (-2)]t^2 = 2t^2.$$

(c) $2y^3 + 4y^3 - 3y^3 = [2 + 4 + (-3)]y^3 = 3y^3$

(d) $-4w^4 - (-6w^4) = [-4 - (-6)]w^4 = 2w^4$

(e) The only like terms in the expression are the first two. Combining like terms, we have

$$3xy^2 + 7xy^2 + 2x^3y^2 = (3 + 7)xy^2 + 2x^3y^2 = 10xy^2 + 2x^3y^2.$$

(f) Because there are no like terms in the expression, no combining can be done.

Addition and subtraction of polynomials with more than one term can be performed by combining like terms. To do this, we use some of the basic properties of real numbers and the rules for addition and subtraction of real numbers. For example, to perform the addition

$$(3x^3 + 2x^2 + 4) + (7x^3 + 5x^2 + 8),$$

we combine like terms, by regrouping terms, so that

$$(3x^3 + 2x^2 + 4) + (7x^3 + 5x^2 + 8) = 3x^3 + 2x^2 + 4 + 7x^3 + 5x^2 + 8$$
$$= 3x^3 + 7x^3 + 2x^2 + 5x^2 + 4 + 8$$
$$= (3 + 7)x^3 + (2 + 5)x^2 + (4 + 8)$$
$$= 10x^3 + 7x^2 + 12.$$

With enough practice, writing the regrouping of like terms becomes unnecessary.

EXAMPLE 2   Perform the addition $(4y^3 + 7y - 3) + (y^3 - y + 2)$.

SOLUTION   We remove the parentheses and combine like terms:

$$(4y^3 + 7y - 3) + (y^3 - y + 2) = 4y^3 + 7y - 3 + y^3 - y + 2$$
$$= (4 + 1)y^3 + (7 - 1)y + (-3 + 2)$$
$$= 5y^3 + 6y - 1.$$

To subtract $3x^2 + 7x - 11$ from $7x^2 - 9x + 13$ we extend the definition of subtraction to polynomials:

$$P - Q = P + (-Q).$$

Thus,

$$(7x^2 - 9x + 13) - (3x^2 + 7x - 11) = (7x^2 - 9x + 13) + [-(3x^2 + 7x - 11)].$$

Now, we form the opposite of $3x^2 + 7x - 11$ by taking the opposite of each term. That is,

$$-(3x^2 + 7x - 11) = -3x^2 - 7x + 11,$$

so that

$$(7x^2 - 9x + 13) - (3x^2 + 7x - 11)$$
$$= (7x^2 - 9x + 13) + (-3x^2 - 7x + 11)$$
$$= 7x^2 - 9x + 13 - 3x^2 - 7x + 11$$
$$= (7 - 3)x^2 + (-9 - 7)x + (13 + 11)$$
$$= 4x^2 - 16x + 24.$$

EXAMPLE 3   Perform the subtraction $(2w^3 - w^2 - 8w) - (-4w^3 + 5w^2 - 7w + 3)$.

SOLUTION

$$(2w^3 - w^2 - 8w) - (-4w^3 + 5w^2 - 7w + 3)$$
$$= (2w^3 - w^2 - 8w) + [-(-4w^3 + 5w^2 - 7w + 3)]$$
$$= (2w^3 - w^2 - 8w) + (4w^3 - 5w^2 + 7w - 3)$$
$$= 2w^3 - w^2 - 8w + 4w^3 - 5w^2 + 7w - 3$$
$$= (2 + 4)w^3 + (-1 - 5)w^2 + (7 - 8)w - 3$$
$$= 6w^3 - 6w^2 - w - 3.$$

When we add and subtract polynomials, it is often useful to rearrange like terms by using a "vertical scheme." To do this, we line up like terms beneath each other and then add or subtract numerical coefficients. The vertical scheme is most useful when we have several polynomials to add or subtract. For example, the polynomials $5x^2 + 3x + 2$ and $3x^2 + 2x + 7$ can be added as follows:

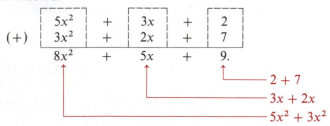

On the other hand, the polynomial $4x^2 - 5x + 7$ can be subtracted from $-3x^2 + 7x - 11$ simply by changing all signs in $4x^2 - 5x + 7$, and then proceeding as in addition. That is, subtract each term in the bottom row from the like term in the top row, as shown in the following.

$$
\begin{array}{r}
-3x^2 + 7x - 11 \\
(-) \quad 4x^2 - 5x + 7 \\
\hline
\end{array}
\qquad
\begin{array}{r}
-3x^2 + 7x - 11 \\
(+) -4x^2 + 5x - 7 \\
\hline
-7x^2 + 12x - 18.
\end{array}
$$

$(-11) + (-7)$

$7x + 5x$

$(-3x^2) + (-4x^2)$

In Examples 4–8, perform each addition and subtraction by using the vertical scheme.

EXAMPLE 4 $(2x - 7x^2 + 8) + (-x^3 - x^2 - 1)$

SOLUTION   We line up like terms and add coefficients:

$$
\begin{array}{r}
-7x^2 + 2x + 8 \\
(+) \quad -x^3 - x^2 - 1 \\
\hline
-x^3 - 8x^2 + 2x + 7.
\end{array}
$$

EXAMPLE 5 $(7 - 3t^2 + t) + (2t^2 + 8 + 13t) + (3t - 7 + 2t^2)$

SOLUTION   We arrange like terms in columns and combine:

$$
\begin{array}{r}
-3t^2 + t + 7 \\
2t^2 + 13t + 8 \\
(+) \quad 2t^2 + 3t - 7 \\
\hline
t^2 + 17t + 8.
\end{array}
$$

EXAMPLE 6   Subtract $9t^2 - 13t + 21$ from $2t^2 + 5t - 43$.

SOLUTION   We arrange like terms in columns:

$$
\begin{array}{r}
2t^2 + 5t - 43 \\
(-)\ 9t^2 - 13t + 21.
\end{array}
$$

We form the opposite of the bottom polynomial [that is, $-(9t^2 - 13t + 21) = -9t^2 + 13t - 21$] and add:

$$
\begin{array}{r}
2t^2 + 5t - 43 \\
(+)\ -9t^2 + 13t - 21 \\
\hline
-7t^2 + 18t - 64.
\end{array}
$$

EXAMPLE 7 $(z^2 - 3z + 4) + (2z^2 - z + 2) - (z^2 - 5z + 8)$

SOLUTION   First we perform the addition of $z^2 - 3z + 4$ and $2z^2 - z + 2$:

$$
\begin{array}{r}
z^2 - 3z + 4 \\
(+)\ 2z^2 - z + 2 \\
\hline
3z^2 - 4z + 6.
\end{array}
$$

Then we perform the subtraction by adding the opposite of the last polynomial

$$
\begin{array}{r}
3z^2 - 4z + 6 \\
(-)\ z^2 - 5z + 8 \\
\hline
\end{array}
$$

$$
\text{or}\quad
\begin{array}{r}
3z^2 - 4z + 6 \\
(+)\ -z^2 + 5z - 8 \\
\hline
2z^2 + z - 2
\end{array}
$$

If polynomials contain more than one variable, addition and subtraction are still performed by combining like terms.

EXAMPLE 8   Subtract $3ab^2 - 5a^2b + 11$ from $-7a^2b + 8ab^2 - 18$.

SOLUTION   We arrange the like terms in columns. We have

$$
\begin{array}{r}
-7a^2b + 8ab^2 - 18 \\
(-)\ -5a^2b + 3ab^2 + 11.
\end{array}
$$

Now we form the opposite of the bottom polynomial and add:

$$
\begin{array}{r}
-7a^2b + 8ab^2 - 18 \\
(+)\ 5a^2b - 3ab^2 - 11 \\
\hline
-2a^2b + 5ab^2 - 29.
\end{array}
$$

Polynomials may be used to represent quantities that occur frequently in applications in business and economics. For example, in economics, suppose that C is the **total cost** of producing x units of a commodity. If R dollars is the **total revenue** obtained when x units of a commodity are sold, then the **total profit** P obtained by producing and selling x units of a commodity is given by the formula

$$P = R - C.$$

EXAMPLE 9   A firm can sell x units of a particular commodity it produces. The total cost C (in dollars) of each day's production when x units are produced is given by $C = x^2 + 20x + 700$.

(a) Find the daily profit P if the total revenue R obtained from selling all x items produced is given by $R = 100x$.

(b) How much profit will the firm make if it produces and sells 40 items per day?

SOLUTION   Using the formula $P = R - C$, we have

(a) $P = 100x - (x^2 + 20x + 700)$
$\quad = -x^2 + 80x - 700.$

(b) When $x = 40$, we have

$$\begin{aligned} P &= -40^2 + 80(40) - 700 \\ &= -1{,}600 + 3{,}200 - 700 \\ &= 900. \end{aligned}$$

Therefore, the daily profit is $900.

PROBLEM SET 2.2

In problems 1–14, perform the indicated additions of the polynomials.

1. $5x^2 + 7x^2$

2. $3y^3 + 8y^3$

3. $2v^3 + 9v^3$

4. $4x^2 + 5x^2$

5. $-3t^2 + 7t^2$

6. $(-8z) + (-5z)$

7. $(3x + 4) + (5x + 3)$

8. $(-5u - 3) + (7u + 1)$

9. $(2z^2 + 3z + 1) + (5z^2 + 2z + 4)$

10. $(3x + 8x^3 + 5) + (2x^2 + 7x + 5)$

11. $(3 - 4x + 7x^2) + (2x - 3x^2 - 5)$

12. $(7w - 2w^2 - 3) + (5 + 3w^2 - 9w)$

13. $(5c^2 - 3c^3 - c + 2c^4) + (4c^3 + 3c^4 - c^2 + 2c)$

14. $(1 + 2x - 3x^2 + 4x^3) + (5x^3 - x^2 + 3x - 7)$

In problems 15–28, perform the indicated subtractions of the polynomials.

15. $7u - 3u$

16. $10y - 3y$

17. $3x^2 - x^2$

18. $4w^3 - w^3$

19. $4v^3 - (-2v^3)$

20. $5t^4 - (-2t^4)$

21. $(3t^3 + 2) - (-t^3 + 4)$

22. $(-11z^3 + 10z) - (9z - 12z^3)$

23. $(3x^2 + 5x + 8) - (x^2 + 3x + 4)$

24. $(4w^3 - 7w - 8) - (-2w^3 + 4w - 2)$

25. $(3s^4 - 4s^3 + 6s^2 + s - 1) - (4 - s + 2s^2 - 3s^3 - s^4)$

26. $(5y^3 - 3y^2 + 2y - 8) - (y - y^3 + 3y^2 - 1)$

27. $(-4t^3 + 8t^2 - 7) - (3t^2 - 4t + 11)$

28. $(x^6 - 2x^4 - 3x^2) - (x^5 - 2x^3 - 3x - 4)$

In problems 29–42, perform the indicated operations by combining like terms.

29. $xy + 3xy + 8xy$

30. $5xyz + xyz + 3xyz$

31. $10uv - 7uv$

32. $12t^2s - 9t^2s$

33. $5mn^2 + (-3mn^2) + 2m^2n$

34. $14x^3y - 4x^3y - 4xy^3 + 9xy^3$

35. $-5x^2y - (-3x^2y)$

36. $-11u^3v^2 - (-12u^3v^2)$

37. $(7ts - 4) + (-3ts - 2)$

38. $(-4w^2z + 6) + (5w^2z - 8)$

39. $(-8x^2y + 6xy - 7xy^2) + (3xy^2 - 4xy + 2x^2y)$

40. $(11mn - 9m^2n^2 + 13m^3n^3) + (3m^3n^3 + 5mn + 7m^2n^2 + 13)$

41. $(3w^2z + 4wz - 7wz^2) - (-2w^2z + 3wz^2 + wz)$

42. $(7 - 8xy - 5x^3y^3) - (2xy - 2x^3y^3)$

In problems 43–50, perform the indicated operations and combine like terms.

43. $(8x^2 + 3x - 7) + (-5x^2 + 2x + 1) + (2x^2 - 3x + 4)$

44. $(-2p^2 + 5p + 2) + (3p^2 - 2p + 3) + (p^2 - 3p - 7)$

45. $(t^3 - 2t^2 + 3t + 1) + (2t^3 + t^2 - 2t + 2) + (-t^3 + 3t^2 - 2t - 1)$

46. $(3y^4 - 4y^3 + y^2 - 2y + 3) + (7y^4 + 5y^3 + 2y^2 - y - 7) + (6y^3 - 6y + 5)$

47. $(2w^2 - 3w + 4) + (5w - 1 + w^2) - (w + 2w^2 - 6)$

48. $(-x^2 + 8x - 11) - (x^2 - 3x - 2) + (3x^2 - 10x + 3)$

49. $(7u^3v^2 - 3u^2v + 2w) - (4w + u^2v - u^3v^2) - (-2u^3v^2 + 5u^2v + 3w)$

50. $(2xy - 3xz + 4yz) - (5xz - 3yz + 4xy) - (-2yz + 3xy - xz)$

51. A firm manufactures and sells portable radios. If x radios are manufactured each day, then C (in dollars) is the daily cost of production where C is given by $C = 0.15x^2 + 25x + 100$.

  (a) Find the daily profit P if the total revenue R obtained from selling x radios is given by $R = 75x$.

  (b) How much profit will the firm make if it manufactures and sells 200 radios per day?

52. A furniture company estimates that the weekly total cost C (in dollars) of manufacturing x desks is given by $C = x^3 - 3x^2 - 80x + 500$.

  (a) Find the weekly profit P if the total revenue obtained from selling x desks is given by $R = 2{,}800x$.

  (b) What is the total profit P per week if 32 desks were manufactured and sold per week?

53. A newspaper stand contains dimes and quarters only. There are 6 fewer quarters than dimes. If x represents the number of dimes, write an algebraic expression that represents the total value of all coins in cents. Also, simplify the expression.

54. Rose jogs for some time at a rate of 6 kilometers per hour, and then rides her bicycle for twice as much time at a rate of 10 kilometers per hour. Let t be the time in hours that Rose took to jog. Write an algebraic expression for the total distance that Rose jogs and rides the bicycle. (Use distance = rate × time.)

2.3 Properties of Positive Integral Exponents

Recall from Chapter 1, Section 1.7, that if n is a positive integer and x is a real number, then

$$x^n = \overbrace{x \cdot x \cdot x \cdots x}^{n \text{ factors}},$$

where x is the **base** and n is the **power** or the **exponent.**

If $n = 1$ in the above definition, we have

$$x^1 = x.$$

Thus, when no exponent is shown, we assume it to be the number 1. For example,

$$3 = 3^1,$$
$$a = a^1, \quad \text{and}$$
$$x^2 y = x^2 y^1.$$

Suppose that we wish to find the product of x^2 and x^3. We know that $x^2 = x \cdot x$ and $x^3 = x \cdot x \cdot x$. Therefore,

$$x^2 \cdot x^3 = (x \cdot x)(x \cdot x \cdot x)$$

and since this expression contains five factors of x, we have

$$x^2 \cdot x^3 = x^5.$$

Notice that the exponent of the product x^5 is the sum of the exponents of the factors x^2 and x^3, that is,

$$x^2 \cdot x^3 = x^{2+3}$$
$$= x^5.$$

Similarly, the expression $(x^2)^3$ can be written as a single power of x by applying the definition of positive exponents as follows:

$$
\begin{aligned}
(x^2)^3 &= x^2 \cdot x^2 \cdot x^2 && \text{(Three factors of } x^2) \\
&= (x \cdot x)(x \cdot x)(x \cdot x) && \text{(Replacing } x^2 \text{ by } x \cdot x) \\
&= x^6.
\end{aligned}
$$

The exponent in the result is the product of exponents 2 and 3, that is,

$$(x^2)^3 = x^{2 \cdot 3}$$
$$= x^6.$$

We may now generalize to obtain the following properties of exponents:

Properties of Exponents

If a and b are real numbers and m and n are positive integers, then:

(i) **Multiplication property:** $a^m \cdot a^n = a^{m+n}$

(ii) **Power-of-a-power property:** $(a^m)^n = a^{mn}$

(iii) **Power-of-a-product property:** $(ab)^n = a^n b^n$

(iv) **Power-of-a-quotient property:** $\left(\dfrac{a}{b}\right)^n = \dfrac{a^n}{b^n}$ if $b \neq 0$

(v) **Division property:** $\dfrac{a^m}{a^n} = a^{m-n}$ if $a \neq 0$ and m is greater than n

Notice that in Property (v), we assume that m is greater than n, so that $m - n$ represents a positive exponent. For instance,

$$\frac{u^8}{u^3} = u^{8-3} = u^5 \qquad \text{for} \qquad u \neq 0.$$

However, if $m = n$, we have

$$\frac{a^m}{a^n} = \frac{a^m}{a^m} = 1.$$

For example,

$$\frac{x^3}{x^3} = 1 \qquad \text{for} \qquad x \neq 0.$$

Also,

$$\frac{(-y)^5}{(-y)^5} = 1 \qquad \text{for} \qquad y \neq 0.$$

If m is less than n, we have

$$\frac{a^m}{a^n} = \frac{1}{a^{n-m}} \qquad \text{for} \qquad a \neq 0$$

where $n - m$ represents a positive exponent. For instance,

$$\frac{x^2}{x^5} = \frac{1}{x^{5-2}} = \frac{1}{x^3} \qquad \text{for} \qquad x \neq 0.$$

Also,

$$\frac{(-t)^7}{(-t)^{10}} = \frac{1}{(-t)^3} \qquad \text{for} \qquad t \neq 0.$$

In working with expressions like those in Properties (iv) and (v), always remember that the *denominator of a fraction cannot be zero*.

These five properties can be verified by applying the definition of positive integral exponents. We shall verify Properties (i), (iii), and (v) here and leave the verifications of Properties (ii) and (iv) to the reader. (See Problems 70 and 72 of Problem Set 2.3.)

PROOF OF
PROPERTY (i)

$$a^m a^n = a^{m+n}$$

$$a^m a^n = \underbrace{(a \cdot a \cdots a)}_{m \text{ factors}}\underbrace{(a \cdot a \cdot a \cdots a)}_{n \text{ factors}} \qquad \text{(Definition)}$$

$$= \underbrace{a \cdot a \cdot a \cdot a \cdot a \cdots a \cdot a}_{m + n \text{ factors}}$$

$$= a^{m+n}. \qquad \text{(Definition)}$$

PROOF OF
PROPERTY (iii)

$$(ab)^n = a^n b^n$$

$$(ab)^n = \underbrace{(ab)(ab) \cdots (ab)}_{n \text{ factors}} \qquad \text{(Definition)}$$

$$= \underbrace{(a \cdot a \cdots a)}_{n \text{ factors}}\underbrace{(b \cdot b \cdots b)}_{n \text{ factors}} \qquad \text{(Why?)}$$

$$= a^n b^n \qquad \text{(Definition)}$$

PROOF OF
PROPERTY (v)

$$\frac{a^m}{a^n} = a^{m-n} \qquad \text{if } m \text{ is greater than } n \text{ and } a \neq 0$$

$$\frac{a^m}{a^n} = \frac{\overbrace{a \cdot a \cdot a \cdot a \cdot a \cdots a}^{m \text{ factors}}}{\underbrace{a \cdot a \cdots a}_{n \text{ factors}}} = \frac{\overbrace{\underbrace{(a \cdot a \cdots a)}_{n \text{ factors}}\underbrace{(a \cdot a \cdot a \cdots a)}_{m - n \text{ factors}}}^{m \text{ factors}}}{\underbrace{a \cdot a \cdots a}_{n \text{ factors}}}$$

$$= \underbrace{a \cdot a \cdots a}_{m - n \text{ factors}}$$

$$= a^{m-n}.$$

We can use the five properties to rewrite algebraic expressions containing exponents as compactly as possible. When we do this we say that the expression has been **simplified.** Although the word "simplify" has no precise mathematical definition, the meaning it conveys is usually clear from the context.

EXAMPLE 1   Use Property (i) to simplify the following expressions.

(a) $2^3 \cdot 2^4$ (b) $x^4 \cdot x^6$ (c) $(-a)^2 \cdot (-a)^4$ (d) $-y^2 \cdot y^4$

SOLUTION   To apply Property (i), we keep the common base and add the exponents. Thus:

(a) $2^3 \cdot 2^4 = 2^{3+4} = 2^7 = 128$

(b) $x^4 \cdot x^6 = x^{4+6} = x^{10}$

(c) $(-a)^2 \cdot (-a)^4 = (-a)^{2+4}$
$= (-a)^6$

Because the negative sign within the parentheses is part of the base, we have

$$(-a)^6 = (-a)(-a)(-a)(-a)(-a)(-a) = a^6.$$

(d) $-y^2 \cdot y^4 = -y^{2+4} = -y^6$

Warning: It is a common error to try to find a product like $4^3 \cdot 2^5$ by multiplying the bases ($4 \cdot 2 = 8$) and then adding the exponents ($3 + 5 = 8$). You would then arrive at the solution $4^3 \cdot 2^5 \overset{?}{=} 8^8$. This is wrong, of course, since $4^3 \cdot 2^5 = 64 \cdot 32 \neq 8^8$.

EXAMPLE 2   Use Property (ii) to remove the parentheses in the following expressions.

(a) $(3^2)^3$ (b) $(x^4)^3$

(c) $(y^7)^n$ n is any positive integer

SOLUTION   Applying Property (ii), we have:

(a) $(3^2)^3 = 3^{2 \cdot 3} = 3^6 = 729$ (b) $(x^4)^3 = x^{4 \cdot 3} = x^{12}$

(c) $(y^7)^n = y^{7 \cdot n} = y^{7n}$

EXAMPLE 3   Use Property (iii) to remove the parentheses in the following expressions.

(a) $(3x)^3$ (b) $(-x)^5$ (c) $(xyz)^4$

SOLUTION

Applying Property (iii), we have:

(a) $(3x)^3 = 3^3 \cdot x^3 = 27x^3$

(b) $(-x)^5 = [(-1)x]^5 = (-1)^5 x^5 = (-1)x^5 = -x^5$

(c) $(xyz)^4 = [x(yz)]^4$ (Associative Property)
    $= x^4(yz)^4 = x^4 y^4 z^4$

EXAMPLE 4

Use Property (iv) to remove the parentheses in the following expressions.

(a) $\left(\dfrac{2}{3}\right)^3$ (b) $\left(\dfrac{x}{y}\right)^5$ (c) $\left(\dfrac{-a}{3}\right)^2$

SOLUTION

Applying Property (iv), we raise both the numerator and the denominator of the quotient to the indicated power. Thus:

(a) $\left(\dfrac{2}{3}\right)^3 = \dfrac{2^3}{3^3} = \dfrac{8}{27}$ (b) $\left(\dfrac{x}{y}\right)^5 = \dfrac{x^5}{y^5}$ (c) $\left(\dfrac{-a}{3}\right)^2 = \dfrac{(-a)^2}{3^2} = \dfrac{a^2}{9}$

EXAMPLE 5

Use Property (v) to simplify the following expressions.

(a) $\dfrac{2^5}{2^2}$ (b) $\dfrac{x^7}{x^2}$ (c) $\dfrac{(-a)^5}{(-a)^2}$ (d) $\dfrac{-a^5}{-a^2}$

SOLUTION

Applying Property (v), that is, $a^m/a^n = a^{m-n}$, where m is larger than n, we keep the common base and subtract exponents:

(a) $\dfrac{2^5}{2^2} = 2^{5-2} = 2^3 = 8$

(b) $\dfrac{x^7}{x^2} = x^{7-2} = x^5$

(c) $\dfrac{(-a)^5}{(-a)^2} = (-a)^{5-2} = (-a)^3 = -a^3$

(d) $\dfrac{-a^5}{-a^2} = \dfrac{a^5}{a^2}$ (Why?); and $\dfrac{a^5}{a^2} = a^{5-2} = a^3$, so $\dfrac{-a^5}{-a^2} = a^3$

EXAMPLE 6

Use Properties (i)–(v) to simplify the given expressions.

(a) $(3x^2)^3$ (b) $\left(\dfrac{x^3}{y^2}\right)^4$ (c) $\dfrac{x^7}{x^3 \cdot x^2}$ (d) $\left(\dfrac{x^3 y^5}{2xy^2}\right)^7$

SOLUTION

(a) $(3x^2)^3 = 3^3(x^2)^3$ [Property (iii)]

$= 27x^{2 \cdot 3}$ [Property (ii)]

$= 27x^6$

(b) $\left(\dfrac{x^3}{y^2}\right)^4 = \dfrac{(x^3)^4}{(y^2)^4}$ [Property (iv)]

$= \dfrac{x^{3 \cdot 4}}{y^{2 \cdot 4}}$ [Property (ii)]

$= \dfrac{x^{12}}{y^8}$

(c) $\dfrac{x^7}{x^3 \cdot x^2} = \dfrac{x^7}{x^{3+2}} = \dfrac{x^7}{x^5}$ [Property (i)]

$= x^{7-5}$ [Property (v)]

$= x^2$

(d) $\left(\dfrac{x^3 y^5}{2xy^2}\right)^7 = \left(\dfrac{x^{3-1}y^{5-2}}{2}\right)^7$ [Property (v) applied to x and y within the parentheses]

$= \left(\dfrac{x^2 y^3}{2}\right)^7$

$= \dfrac{(x^2)^7(y^3)^7}{2^7}$ [Properties (iii) and (iv)]

$= \dfrac{x^{2 \cdot 7}y^{3 \cdot 7}}{128}$ [Property (ii)]

$= \dfrac{x^{14}y^{21}}{128}$

EXAMPLE 7   Ⓒ The formula

$$A = P(1 + r)^t$$

is used to find the amount A that an investment of P dollars grows to when invested in an account paying compound interest at an annual interest rate r for t years. Find the amount A if \$2,000 is invested at 6.5% compounded annually for 4 years.

SOLUTION

Substituting $P = 2,000$, $r = 6.5\% = 0.065$ and $t = 4$ in the formula, we have

$$A = P(1 + r)^t$$
$$= 2{,}000(1 + 0.065)^4$$
$$= 2{,}000(1.065)^4$$
$$= 2{,}572.93.$$

Therefore, the amount is \$2,572.93.

EXAMPLE 8   Use the properties of exponents to simplify each expression and write each answer in scientific notation.

$$\text{(a) } (4 \times 10^7)(2 \times 10^5) \qquad \text{(b) } \frac{35 \times 10^7}{5 \times 10^3}$$

SOLUTION

$$\text{(a) } (4 \times 10^7)(2 \times 10^5) = (4)(2)(10^7 \times 10^5)$$
$$= 8 \times 10^{7+5}$$
$$= 8 \times 10^{12}$$

$$\text{(b) } \frac{35 \times 10^7}{5 \times 10^3} = \left(\frac{35}{5}\right)\left(\frac{10^7}{10^3}\right) = 7 \times 10^{7-3} = 7 \times 10^4$$

PROBLEM SET 2.3

In problems 1–10, use Property (i) to simplify each expression.

1. $3^2 \cdot 3^3$ **2.** $2^4 \cdot 2^2$ **3.** $(-2)^3 \cdot (-2)^2$

4. $(-3) \cdot (-3)^2(-3)^3$ **5.** $-x^5 \cdot x^7$ **6.** $-y^6 \cdot y^4$

7. $t^3 \cdot t^4 \cdot t^5$ **8.** $(-u)^9 \cdot (-u)^3 \cdot (-u)$ **9.** $(-v)^3 \cdot (-v)^5$

10. $(-x)^7 \cdot (-x)^3 \cdot x^2$

In problems 11–20, use Property (ii) to remove the parentheses and brackets in each expression.

11. $(2^2)^3$ **12.** $(3^2)^4$ **13.** $[(-2)^3]^2$ **14.** $[(-3)^2]^2$ **15.** $(x^7)^5$

16. $(y^3)^{12}$ **17.** $(t^2)^{11}$ **18.** $(z^4)^5$ **19.** $[(-w)^3]^4$ **20.** $[(c^2)^3]^5$

In problems 21–30, use Property (iii) to remove the parentheses and brackets in each expression.

21. $(2x)^4$ **22.** $(5y)^2$ **23.** $(uv)^5$ **24.** $(ab)^4$

25. $(xyz)^7$ **26.** $(3mn)^5$ **27.** $(-2w)^3$ **28.** $(-2xyz)^4$

29. $[-3(-x)y]^3$ **30.** $[-3(-x)(-y)]^2$

In problems 31–40, use Property (iv) to remove the parentheses in each expression.

31. $\left(\dfrac{3}{4}\right)^2$ **32.** $\left(\dfrac{1}{2}\right)^5$ **33.** $\left(\dfrac{-2}{3}\right)^3$ **34.** $\left(-\dfrac{3}{2}\right)^3$ **35.** $\left(\dfrac{x}{y}\right)^4$

36. $\left(\dfrac{y}{z}\right)^7$ **37.** $\left(\dfrac{a}{-b}\right)^5$ **38.** $\left(-\dfrac{m}{n}\right)^8$ **39.** $\left(\dfrac{-t}{s}\right)^6$ **40.** $\left(-\dfrac{w}{c}\right)^9$

In problems 41–50, use Property (v) to simplify each expression.

41. $\dfrac{3^5}{3^2}$ **42.** $\dfrac{(-2)^6}{(-2)^3}$ **43.** $\dfrac{4^9}{4^6}$ **44.** $\dfrac{(-3)^4}{(-3)^6}$

45. $\dfrac{-x^8}{-x^3}$ **46.** $\dfrac{-m^{25}}{-m^{11}}$ **47.** $\dfrac{y^{25}}{y^{20}}$ **48.** $\dfrac{(-t)^5}{(-t)^5}$

49. $\dfrac{w^{13}}{w^{10}}$ **50.** $\dfrac{x^n}{x^2}$ (n is a positive integer greater than 2)

In problems 51–60, use Properties (i)–(v) to simplify each expression.

51. $(3x^2y^3)^4$ **52.** $(-2z^4)^3$ **53.** $\dfrac{u^4v^7}{uv^3}$ **54.** $\dfrac{x^{16}y^5}{x^5y^9}$ **55.** $\left(\dfrac{w^2}{2z}\right)^3$

56. $\left(\dfrac{3m^2}{4n}\right)^2$ **57.** $\left(\dfrac{3a^5b^3}{2a^2b^6}\right)^3$ **58.** $\left(\dfrac{4t^2s^3}{8t^3s}\right)^5$ **59.** $\dfrac{(-4xy^4z^5)^3}{(2x^2yz^3)^2}$ **60.** $\dfrac{(8u^4v^7)^3}{(4u^2v^3)^2}$

In problems 61–66, use the properties of exponents to simplify each expression. Write each answer in scientific notation.

61. $(2 \times 10^5)(4.8 \times 10^9)$ **62.** $(3.4 \times 10^7)(2 \times 10^4)$ **63.** $(4.5 \times 10^7)(2 \times 10^4)$

64. $(3.6 \times 10^7)(2 \times 10^7)$ **65.** $\dfrac{7.8 \times 10^9}{3.9 \times 10^5}$ **66.** $\dfrac{8.2 \times 10^{11}}{4.1 \times 10^4}$

© **67.** Find the amount that an investment of \$2,500 grows to when compounded annually at a rate of $5\frac{1}{2}\%$ for 10 years.

© **68.** Find the amount that an investment of \$5,275 grows to when compounded annually at a rate of $6\frac{3}{4}\%$ for 7 years.

69. Find the value of $(-1)^n$ if n is an even positive integer.

70. Verify Property (ii), page 71.

71. Find the value of $(-1)^n$ if n is an odd positive integer.

72. Verify Property (iv), page 71.

2.4 Multiplication of Polynomials

To multiply polynomials, we extend the commutative and associative properties for the multiplication of real numbers to polynomials. That is, if P, Q, and R are polynomial expressions, then:

(i) $PQ = QP$	(ii) $P(QR) = (PQ)R$

For example, to multiply the monomials $8x^3$ and $2x^2$, we apply the associative and commutative properties:

$$(8x^3)(2x^2) = (8 \cdot 2)(x^3 \cdot x^2)$$
$$= 16x^5.$$

Note that we found the product of the two monomials essentially by regrouping the factors, multiplying the coefficients, and applying the rule for multiplying like bases by adding exponents.

EXAMPLE 1   Perform each multiplication and simplify the results.

(a) $(4x^2)(7x)$ (b) $(-3y)(2y^3)$ (c) $(-\frac{2}{3}a^2b^3)(-6a^2b)$

SOLUTION   We regroup the coefficients together and the variables together, and then multiply:

(a) $(4x^2)(7x) = (4 \cdot 7)(x^2 \cdot x) = 28x^3$

(b) $(-3y)(2y^3) = (-3 \cdot 2)(y \cdot y^3) = -6y^4$

(c) $(-\frac{2}{3}a^2b^3)(-6a^2b) = [(-\frac{2}{3}) \cdot (-6)](a^2 \cdot a^2 \cdot b^3 \cdot b) = 4a^4b^4$

To multiply a monomial by a polynomial, we apply the distributive property to polynomials, that is, if P, Q, and R are polynomials, then:

(i) $P(Q + R) = PQ + PR$	(ii) $(P + Q)R = PR + QR$

For example, to find the product of $3x^2$ and $2x^3 + 4x$, we apply the preceding distributive property, so that

$$3x^2(2x^3 + 4x) = (3x^2)(2x^3) + (3x^2)(4x)$$
$$= 6x^5 + 12x^3.$$

EXAMPLE 2   Find the following products.

(a) $4x(3x - 2)$ (b) $2y^3(y^2 - 3y + 4)$ (c) $-2t^4(-3t^2 - 7t + 4)$

SOLUTION   Using the distributive property, we have

(a) $4x(3x - 2) = 4x[3x + (-2)]$
$$= (4x)(3x) + (4x)(-2)$$
$$= 12x^2 - 8x$$

(b) Multiplying the expression $2y^3$ (outside the parentheses) by each term inside the parentheses, we have

$$2y^3(y^2 - 3y + 4) = (2y^3)(y^2) + (2y^3)(-3y) + (2y^3)(4)$$
$$= 2y^5 - 6y^4 + 8y^3.$$

(c) $-2t^4(-3t^2 - 7t + 4) = (-2t^4)(-3t^2) + (-2t^4)(-7t) + (-2t^4)(4)$
$$= 6t^6 + 14t^5 - 8t^4$$

To multiply one polynomial by another, use the distributive property to reduce the given multiplication to a multiplication of monomials.

For example, to multiply $x + 2$ by $x + 4$, we can think of $x + 4$ as one *term*—call it u for the time being—and then apply the distributive property, so that

$$
\begin{aligned}
(x + 2)(x + 4) &= (x + 2)u \\
&= xu + 2u && \text{(We used the distributive property.)} \\
&= x(x + 4) + 2(x + 4) && \text{(We replaced } u \text{ by } x + 4.) \\
&= x^2 + 4x + 2x + 8 && \text{(We used the distributive property.)} \\
&= x^2 + 6x + 8.
\end{aligned}
$$

Although it is important to remember that the distributive property is what allows us to multiply polynomials, at times it takes too long to write out all the steps involved. A shorter procedure follows:

> To **multiply** two polynomials, multiply each term of one polynomial by each term of the other and then simplify the result by combining like terms.

EXAMPLE 3 Find each product.

(a) $(x + 2)(2x - 3)$ (b) $(3t - 1)(2t^2 + 7t + 4)$

SOLUTION (a) Multiply each term of $x + 2$ by each term of $2x - 3$:

$$
\begin{aligned}
(x + 2)(2x - 3) &= (x)(2x) + (x)(-3) + (2)(2x) + (2)(-3) \\
&= 2x^2 - 3x + 4x - 6 \\
&= 2x^2 + x - 6.
\end{aligned}
$$

$$
\begin{aligned}
\text{(b) } (3t - 1)(2t^2 + 7t + 4) &= (3t)(2t^2) + (3t)(7t) + (3t)(4) + (-1)(2t^2) \\
&\quad + (-1)(7t) + (-1)(4) \\
&= 6t^3 + 21t^2 + 12t - 2t^2 - 7t - 4 \\
&= 6t^3 + 19t^2 + 5t - 4
\end{aligned}
$$

The actual computations can become quite tedious in many multiplications involving polynomials (as we have seen in the previous examples). Therefore, any device to help perform the computations and to simplify the work is desirable. One such device, the vertical scheme, is illustrated by the following example. To find the product of $2x - 1$ and $x^3 + x^2 - 2x - 1$, we can arrange the polynomials in a vertical scheme:

$$
\begin{array}{l}
x^3 + x^2 - 2x - 1 \\
2x - 1 \qquad\qquad \text{multiply} \\
\hline
2x^4 + 2x^3 - 4x^2 - 2x \qquad\qquad [(2x)(x^3 + x^2 - 2x - 1)]\text{(partial product)} \\
 - x^3 - x^2 + 2x + 1 \quad \text{add} \quad [(-1)(x^3 + x^2 - 2x - 1)]\text{(partial product)} \\
\hline
2x^4 + x^3 - 5x^2 + 1. \qquad\qquad \text{(product)}
\end{array}
$$

This shortcut involves arranging the partial products so that like terms are in the same column, ready for the final step of addition. Note that the "mechanics" of this method are based on properties you have already learned. This was also true for the vertical scheme for addition.

In Examples 4 and 5, use the vertical scheme to multiply each of the following.

EXAMPLE 4 $(x + 4)(2x - 3)$

SOLUTION   Using the vertical scheme, we have

$$
\begin{array}{r}
x + 4 \\
2x - 3 \quad \text{multiply} \\
\hline
2x^2 + 8x \\
-3x - 12 \quad \text{add} \\
\hline
2x^2 + 5x - 12.
\end{array}
$$

EXAMPLE 5 $(x^2 - 2x + 1)(x^2 + x + 2)$

SOLUTION   Using the vertical scheme, we have

$$
\begin{array}{r}
x^2 - 2x + 1 \\
x^2 + x + 2 \quad \text{multiply} \\
\hline
x^4 - 2x^3 + x^2 \\
x^3 - 2x^2 + x \\
2x^2 - 4x + 2 \quad \text{add} \\
\hline
x^4 - x^3 + x^2 - 3x + 2.
\end{array}
$$

We can also multiply polynomials with more than one variable. In doing so, it is good practice to express the mixed-letter parts of terms in alphabetical order. This makes it easier to spot like terms.

In Examples 6 and 7 find each product.

EXAMPLE 6 $(x - y)(3x^3 - 5xy + 7y^2)$

SOLUTION   Using the vertical scheme, we have

$$
\begin{array}{r}
3x^3 - 5xy + 7y^2 \\
x - y \\
\hline
3x^4 - 5x^2y + 7xy^2 \\
5xy^2 - 3x^3y - 7y^3 \quad \text{add} \\
\hline
3x^4 - 5x^2y + 12xy^2 - 3x^3y - 7y^3.
\end{array}
$$

EXAMPLE 7 $(u^2 + 2uv - v^2)(u + uv + v)$

SOLUTION   Applying the vertical scheme, we have

$$u^2 + 2uv\ - v^2$$

$$u\ +\ uv\ + v$$

$$\overline{u^3 + 2u^2v - \ uv^2}$$

$$u^3v + 2u^2v^2 - uv^3$$

$$u^2v + 2uv^2 \qquad\qquad\qquad - v^3 \quad \text{add}$$

$$\overline{u^3 + 3u^2v + \ uv^2 + u^3v + 2u^2v^2 - uv^3 - v^3.}$$

We can also apply a device for simplifying the computations required when multiplying two binomials in which the first and second terms are like terms. For example, consider the product $(x + 2)(x + 3)$. If we apply distributive properties, we have

$$(x + 2)(x + 3) = x(x + 3) + 2(x + 3)$$

$$= x^2 + 3x + 2x + 6$$

$$= x^2 + 5x + 6.$$

The result of the preceding multiplication is a trinomial whose terms are determined as follows:

First term: $(x + 2)(x + 3)$ $=$ $x^2 + 5x + 6$

$(x)(x)$

Middle term: $(x + 2)(x + 3)$ $=$ $x^2 + 5x + 6$

$\oplus$ $(3)(x) + (2)(x)$

Last term: $(x + 2)(x + 3)$ $=$ $x^2 + 5x + 6$

$(2)(3)$

This method enables us to multiply two binomials and write the product directly, without having to show any intermediate steps.

In Examples 8–10, find the following products by using the method just illustrated.

EXAMPLE 8 $(x + 4)(2x + 1)$

SOLUTION   First term: $(x + 4)(2x + 1) \qquad = \qquad 2x^2 + \underline{\quad} + \underline{\quad}$

$(x)(2x)$

Middle term: $(x + 4)(2x + 1) \qquad = \qquad 2x^2 + 9x + \underline{\quad}$

$\oplus \qquad (1)(x) + (4)(2x)$

Last term: $(x + 4)(2x + 1) \qquad = \qquad 2x^2 + 9x + 4$

$(4)(1)$

Therefore,

$$(x + 4)(2x + 1) = 2x^2 + 9x + 4.$$

EXAMPLE 9 $(3x - 4)(2x + 3)$

SOLUTION   First term: $(3x - 4)(2x + 3) \qquad = \qquad 6x^2 + \underline{\quad} + \underline{\quad}$

$(3x)(2x)$

Middle term: $(3x - 4)(2x + 3) \qquad = \qquad 6x^2 + \quad x + \underline{\quad}$

$\oplus \qquad (3x)(3) + (-4)(2x)$

Last term: $(3x - 4)(2x + 3) \qquad = \qquad 6x^2 + \quad x \quad + (-12)$

$(-4)(3)$

Therefore,

$$(3x - 4)(2x + 3) = 6x^2 + x - 12.$$

EXAMPLE 10 $(4x - 3y)(x - y)$

SOLUTION   We can obtain the product by using the following scheme:

$$(4x - 3y)(x - y) \qquad = \qquad 4x^2 - 7xy + 3y^2$$

First term: $(4x)(x) = 4x^2$

Middle term: $(4x)(-y) + (-3y)(x) = -7xy$

Last term: $(-3y)(-y) = 3y^2$

At times it is essential that you learn to multiply first-degree polynomials of the types in Examples 8, 9, and 10 mentally. For instance, consider the product of the binomials in Example 8: $(x + 4)(2x + 1)$. We start by multiplying each term in the first binomial by each term in the second binomial, and naming these products as follows:

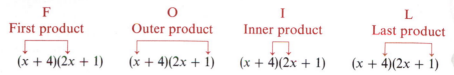

F	O	I	L
First product	Outer product	Inner product	Last product

$$(x + 4)(2x + 1) \quad (x + 4)(2x + 1) \quad (x + 4)(2x + 1) \quad (x + 4)(2x + 1)$$

Performing these operations on one line, we obtain

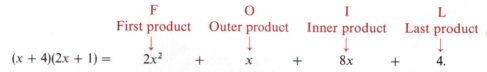

	F	O	I	L
	First product	Outer product	Inner product	Last product

$$(x + 4)(2x + 1) = \quad 2x^2 \quad + \quad x \quad + \quad 8x \quad + \quad 4.$$

The inner and outer products are like terms and hence can be combined into one term, called the *middle term*. Thus

$$(x + 4)(2x + 1) = 2x^2 + 9x + 4.$$

To speed up the process we can combine the inner and outer products mentally. The method just described is known as the **FOIL method.** For example, we can rework Example 10 mentally as follows:

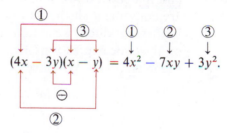

$$(4x - 3y)(x - y) = 4x^2 - 7xy + 3y^2.$$

A child's sandbox is to be made by cutting equal squares from the corners of a rectangular sheet of galvanized iron 3 meters long and 2 meters wide, and turning up the sides.

Figure 1

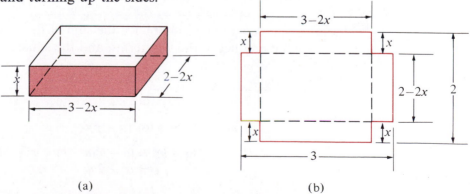

(a) (b)

If x represents the number of meters in a side of each of the squares, write a single polynomial to represent the volume of the resulting box. Evaluate this polynomial for the volume when $x = 0.3$ meters.

SOLUTION   The volume V of the sandbox (Figure 1a) is given by

$$\begin{aligned} V &= \ell wh \\ &= (3 - 2x)(2 - 2x)x \\ &= (6 - 10x + 4x^2)x \\ &= 6x - 10x^2 + 4x^3. \end{aligned}$$

When $x = 0.3$ meters,

$$\begin{aligned} V &= 6(0.3) - 10(0.3)^2 + 4(0.3)^3 \\ &= 1.8 - 0.9 + 0.108 = 1.008. \end{aligned}$$

Therefore the volume is 1.008 cubic meters.

Special Products

The multiplication of certain polynomials occurs often enough in algebra and in applications of algebra to be worthy of special consideration. There are convenient shortcuts to determining the products of these polynomials. In these products, we assume that the symbols a and b represent real numbers or algebraic expressions. Consider, for example, the problem of squaring the binomial $a + b$. We have

$$\begin{aligned} (a + b)^2 = (a + b)(a + b) &= a^2 + ab + ba + b^2 \\ &= a^2 + 2ab + b^2. \end{aligned}$$

We state the result as follows:

Special Product (i): Squaring a Binomial

The square of a binomial is the sum of the square of the first term, twice the product of the two terms, and the square of the last term. That is,

$$(a + b)^2 = a^2 + 2ab + b^2.$$

A similar result occurs for $(a - b)^2$:

$$\begin{aligned} (a - b)^2 = (a - b)(a - b) &= a^2 - ab - ba + b^2 \\ &= a^2 - 2ab + b^2 \end{aligned}$$

Therefore, we have the following:

Special Product (ii): Squaring a Binomial

$$(a - b)^2 = a^2 - 2ab + b^2.$$

Special Products (i) and (ii) are called **perfect squares.**

EXAMPLE 12   Use Special Product (i) or (ii) to carry out the multiplications.

(a) $(x + 3)^2$ (b) $(c + 2d)^2$ (c) $(3u - 2)^2$ (d) $(2m^2n^3 - \frac{1}{2})^2$

SOLUTION   (a) Substituting x for a and 3 for b in Special Product (i), we have

$$(x + 3)^2 = (x)^2 + 2(x)(3) + (3)^2$$
$$= x^2 + 6x + 9.$$

(b) Substituting c for a and $2d$ for b in Special Product (i), we have

$$(c + 2d)^2 = (c)^2 + 2(c)(2d) + (2d)^2$$
$$= c^2 + 4cd + 4d^2.$$

(c) Substituting $3u$ for a and 2 for b in Special Product (ii), we have

$$(3u - 2)^2 = (3u)^2 - 2(3u)(2) + (2)^2$$
$$= 9u^2 - 12u + 4.$$

(d) Substituting $2m^2n^3$ for a and $\frac{1}{2}$ for b in Special Product (ii), we have

$$\left(2m^2n^3 - \frac{1}{2}\right)^2 = (2m^2n^3)^2 - 2(2m^2n^3)\left(\frac{1}{2}\right) + \left(\frac{1}{2}\right)^2$$
$$= 4m^4n^6 - 2m^2n^3 + \frac{1}{4}.$$

Another special product that occurs frequently in mathematics and the sciences has the form $(a - b)(a + b)$. You will notice that the middle term of this product is zero.
Thus

$$(a - b)(a + b) = a^2 + ab - ba - b^2$$
$$= a^2 - b^2.$$

Hence, we have the following:

Special Product (iii)

$$(a - b)(a + b) = a^2 - b^2.$$

Notice that the binomial on the right side of the equation in Special Product (iii) is the **difference of two squares.**

EXAMPLE 13   Use Special Product (iii) to carry out the multiplications.

(a) $(c + 5)(c - 5)$ (b) $(2u + 3v)(2u - 3v)$

SOLUTION   (a) We substitute c for a and 5 for b in Special Product (iii):

$$(c + 5)(c - 5) = c^2 - 5^2$$
$$= c^2 - 25.$$

(b) $(2u + 3v)(2u - 3v) = (2u)^2 - (3v)^2$
$$= 4u^2 - 9v^2.$$

Four more special products occur frequently enough to be mentioned here. They are as follows:

Special Products (iv)–(vii)

> (iv) $(a + b)^3 = a^3 + 3a^2b + 3ab^2 + b^3$
>
> (v) $(a - b)^3 = a^3 - 3a^2b + 3ab^2 - b^3$
>
> (vi) $(a + b)(a^2 - ab + b^2) = a^3 + b^3$
>
> (vii) $(a - b)(a^2 + ab + b^2) = a^3 - b^3$

These special products can be verified by direct multiplication and are left as exercises (see Problems 71–74 in Problem Set 2.4).

EXAMPLE 14   Use Special Products (iv)–(vii) to carry out the multiplications.

(a) $(x + 2)^3$ (b) $(2x - y)^3$

(c) $(2u + v)(4u^2 - 2uv + v^2)$ (d) $(3y - 2z)(9y^2 + 6yz + 4z^2)$

SOLUTION   (a) We substitute x for a and 2 for b in Special Product (iv):

$$(x + 2)^3 = (x)^3 + 3(x)^2(2) + 3(x)(2)^2 + (2)^3$$
$$= x^3 + 6x^2 + 12x + 8.$$

(b) We substitute $2x$ for a and y for b in Special Product (v):

$$(2x - y)^3 = (2x)^3 - 3(2x)^2(y) + 3(2x)(y)^2 - (y)^3$$
$$= 8x^3 - 12x^2y + 6xy^2 - y^3.$$

(c) We substitute $2u$ for a and v for b in Special Product (vi). Note that $4u^2 = (2u)^2$.

$$(2u + v)(4u^2 - 2uv + v^2) = (2u)^3 + (v)^3$$
$$= 8u^3 + v^3.$$

(d) Substitute $3y$ for a and $2z$ for b in Special Product (vii). Note that $9y^2 = (3y)^2$ and $4z^2 = (2z)^2$.

$$(3y - 2z)(9y^2 + 6yz + 4z^2) = (3y)^3 - (2z)^3$$
$$= 27y^3 - 8z^3.$$

PROBLEM SET 2.4

In problems 1–10, find the products of the given monomials.

1. $(2x^2)(3x^4)$ **2.** $(3y^3)(5y^2)$ **3.** $(-5t^3)(6t^4)$

4. $(7m^2)(-8m^5)$ **5.** $(7u^2v^3)(-4u^3v^4)$ **6.** $(-3xy^7)(6x^5y)$

7. $(-3x^2yz^3)(-4xy^2z)$ **8.** $(-10p^3q^4r^5)(-2pq^2r^3)$ **9.** $(2ab)(3a^2c)(-4b^2c^3)$

10. $(3x^2yz^3)(5xyz^2)(-2yz)$

In problems 11–20, use the distributive property to find each product.

11. $x(x + 1)$ **12.** $-y(2y - 3)$ **13.** $t^2(t + 2)$

14. $x^3(2x - 4)$ **15.** $3w(2w^2 - 4)$ **16.** $5u(3u^4 + 7u^2)$

17. $-2xy^2(2x^2 - 3xy + 5y^2)$ **18.** $m^2n^3(4m^2n - 4mn + 5mn^2)$

19. $4c^2d(3c^3d - 2c^2d^2 + cd^3)$ **20.** $-3x^4z^3(-2xz^2 - 4x^2y + 5x^3 - 3)$

In problems 21–30, use the vertical scheme to find each product.

21. $(x + y)(x + y - 1)$ **22.** $(u - v)(u - v + 1)$ **23.** $(2t + s)(t^2 + 2ts + s^2)$

24. $(x - 2y)(x^2 - 3xy - y^2)$ **25.** $(m^2 + 3)(m^3 + 2m^2 - 3m + 4)$ **26.** $(4 - x^2)(x^4 - 2x^2 - 3x + 4)$

27. $(y^2 - 5y + 6)(y^2 + 4y + 4)$ **28.** $(w^2 - 2w + 1)(w^2 + 2w + 1)$

29. $(x^2 + 2xy + y^2)(x^3 - 3x^2y + 3xy^2 - y^3)$ **30.** $(2u^2 - uv + v^2)(u^3 - u^2v + uv^2 - v^3)$

In problems 31–50, use the method illustrated in Examples 8, 9, and 10, page 82, to find each product.

31. $(x + 1)(x + 2)$ **32.** $(w + 3)(w + 4)$ **33.** $(u - 4)(u - 5)$ **34.** $(z - 6)(z - 8)$

35. $(t + 5)(t - 2)$ **36.** $(x + 3y)(3x + y)$ **37.** $(y + 3)(y - 6)$ **38.** $(4t - 7)(t + 4)$

39. $(3x - 1)(2x + 1)$ **40.** $(9c + 2)(c - 3)$ **41.** $(5w + 4)(w - 1)$ **42.** $(11x - 3y)(2x + y)$

43. $(2x + y)(x + 3y)$ **44.** $(3m - 2p)(2m + 3p)$ **45.** $(6m - 5n)(4m + 3n)$ **46.** $(2y - 1)(y - 3)$

47. $(7x + 3y)(4x - 5y)$ **48.** $(12x - 5y)(3x - 2y)$ **49.** $(10v - 7)(5v - 8)$ **50.** $(9u - 7v)(5u - 4v)$

In problems 51–60, use Special Product (i) or (ii) to find each product.

51. $(x + 1)^2$ **52.** $(w + 5)^2$ **53.** $(2s + t)^2$ **54.** $(x + 3y)^2$ **55.** $(u - 3v)^2$

56. $(m - 2n)^2$ **57.** $(3x - 5)^2$ **58.** $(7z - 4)^2$ **59.** $(4y + 5z)^2$ **60.** $(9y + 8z)^2$

In problems 61–70, use Special Product (iii) to find each product.

61. $(x + y)(x - y)$ **62.** $(t - s)(t + s)$ **63.** $(w - 7)(w + 7)$ **64.** $(x + 11)(x - 11)$

65. $(2m + 9)(2m - 9)$ **66.** $(10 - 3y)(10 + 3y)$ **67.** $(8x - y)(8x + y)$ **68.** $(6w - 7z)(6w + 7z)$

69. $(5u + 6v)(5u - 6v)$ **70.** $(9t + 2s)(9t - 2s)$ **71.** Verify Special Product (iv).

72. Verify Special Product (v).   **73.** Verify Special Product (vi).   **74.** Verify Special Product (vii).

In problems 75–90, use Special Products (iv)–(vii) to find each product.

75. $(x + 1)^3$ **76.** $(u + v)^3$ **77.** $(x + 1)(x^2 - x + 1)$

78. $(m + n)(m^2 - mn + n^2)$ **79.** $(c - 2d)^3$ **80.** $(2x - 5y)^3$

81. $(u - 3v)(u^2 + 3uv + 9v^2)$ **82.** $(3x - 4y)(9x^2 + 12xy + 16y^2)$ **83.** $(5t + 2s)^3$

84. $(w + 7)(w^2 - 7w + 49)$ **85.** $(2x + 3y)(4x^2 - 6xy + 9y^2)$ **86.** $(4t - 3s)^3$

87. $(x^3 - 2)^3$ **88.** $(u^2 + v^2)^3$ **89.** $(u^2 - v^2)(u^4 + u^2v^2 + v^4)$

90. $(2x^2 + y^2)(4x^4 - 2x^2y^2 + y^4)$

91. An open cardboard box is to be made from a rectangular sheet of cardboard 48 centimeters long and 36 centimeters wide by cutting squares from each corner and folding up the sides. If x represents the number of centimeters in a side of each of the squares, write a single polynomial to represent the volume of the resulting box. Evaluate the volume for $x = 6$ centimeters.

92. Advertising fliers are to be made from rectangular sheets of paper 15 centimeters wide and 24 centimeters long. If the margins at the top and bottom are each x centimeters, and if the margins at the sides are each x centimeters (Figure 2), write a single polynomial to represent the area of the printed part of the rectangle. Evaluate the area for $x = 2$ centimeters.

Figure 2

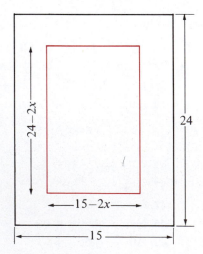

2.5 Factoring Polynomials

In mathematics, it is often useful to represent a polynomial as a product of two or more polynomials. Each polynomial that is multiplied to form the product is called a **factor** of the product. For example, consider the product

$$a(2a - b)(a + b).$$

The factors are a, $2a - b$, and $a + b$.

We often start with a product (in its expanded form) and wish to find its factors. This is done by a process called **factoring;** and we say that the polynomial is **factorable.** A polynomial that has no factors other than itself and 1, or its negative and -1, is said to be **prime.** When a polynomial is written as a product of prime factors, we say that the polynomial is **factored completely.**

Common Factors

The distributive property,

$$P(Q + R) = PQ + PR,$$

provides a bridge between factors and products. If you write the distributive property in the "reverse" order,

$$PQ + PR = P(Q + R),$$

you obtain a general principle of common factoring. In the example above, P is a common factor of PQ and PR on the left side of the equation. The factors on the right side of the equation are P and $Q + R$.

The polynomial expression $x^2 + x$ can be written as $x \cdot x + x \cdot 1$. Thus, by applying the distributive property, we can factor out the common factor x. Therefore, the polynomial is factored completely as follows:

$$x^2 + x = x(x + 1).$$

When factoring a polynomial expression, we begin by determining the **greatest common factor (GCF)** of the terms of the polynomial. *When the terms have no common variable factors, the GCF is the largest integer that is a factor of all of the coefficients.* For example, to factor $15x + 35$, we see that 5 is the GCF of the terms $15x$ and 35, because 5 is the largest factor of both $15x$ and 35. Using the distributive property we can factor out the GCF as follows:

$$15x + 35 = 5(3x) + 5(7)$$
$$= 5(3x + 7)$$

When the terms have common variable factors, the GCF is the monomial with the largest integer exponent that divides (is a factor of) each term of the polynomial. For example, to factor the polynomial $28x^7 - 21x^4 + 35x^2$, we see that the largest number that divides each of the coefficients 28, -21, and 35 is 7. The highest power of the variable x that is a factor of x^7, x^4, and x^2 is x^2. Therefore, the greatest common factor for $28x^7$, $-21x^4$, and $35x^2$ is $7x^2$. The polynomial expression $28x^7 - 21x^4 + 35x^2$ can be written as

$$7x^2(4x^5) - 7x^2(3x^2) + 7x^2(5).$$

Now, we apply the distributive property:

$$28x^7 - 21x^4 + 35x^2 = 7x^2(4x^5) - 7x^2(3x^2) + 7x^2(5)$$
$$= 7x^2(4x^5 - 3x^2 + 5).$$

This procedure for factoring is further illustrated by the following examples.

In Examples 1 and 2, factor each polynomial by factoring out the common factor.

EXAMPLE 1   (a) $2y^3 + 4y^2$ (b) $4u^5 + 8u^3$

(c) $3x^3y^4 - 9x^2y^3 - 6xy^2$ (d) $-15a^5b^4 - 25a^2b^5 - 10a^2b^3$

SOLUTION   (a) The greatest common factor of the coefficients is 2. The variable y is in all terms and y^2 is the greatest power that is common. Therefore, the GCF $= 2y^2$. Thus, rewriting the expression and applying the distributive property, we have

$$2y^3 + 4y^2 = (2y^2)(y) + (2y^2)(2)$$
$$= 2y^2(y + 2).$$

(b) The greatest common factor is $4u^3$, since

$$4u^5 = (4u^3)(u^2) \qquad \text{and} \qquad 8u^3 = (4u^3)(2).$$

Using the distributive property to factor $4u^3$ from each term, we have

$$4u^5 + 8u^3 = (4u^3)(u^2) + (4u^3)(2)$$
$$= 4u^3(u^2 + 2).$$

(c) The greatest common factor is $3xy^2$, since

$$3x^3y^4 = (3xy^2)(x^2y^2) \qquad 9x^2y^3 = (3xy^2)(3xy) \qquad \text{and} \qquad 6xy^2 = (3xy^2)(2).$$

Using the distributive property to factor $3xy^2$ from each term, we have

$$3x^3y^4 - 9x^2y^3 - 6xy^2 = (3xy^2)(x^2y^2) - (3xy^2)(3xy) - (3xy^2)(2)$$
$$= (3xy^2)(x^2y^2 - 3xy - 2).$$

(d) The greatest common factor is $-5a^2b^3$, since

$$-15a^5b^4 = (-5a^2b^3)(3a^3b) \qquad -25a^2b^5 = (-5a^2b^3)(5b^2) \qquad \text{and}$$
$$-10a^2b^3 = (-5a^2b^3)(2)$$

Using the distributive property to factor $-5a^2b^3$ from each term, we have

$$-15a^5b^4 - 25a^2b^5 - 10a^2b^3 = -5a^2b^3(3a^3b) - 5a^2b^3(5b^2) - 5a^2b^3(2)$$
$$= -5a^2b^3(3a^3b + 5b^2 + 2).$$

Factoring by Grouping

Some algebraic expressions are written so that all the terms have a common binomial factor that can be factored out by the methods just shown. The following example illustrates this.

EXAMPLE 2   (a) $5x(y + z) + 2(y + z)$ (b) $a(c - d) + 2(d - c)$

SOLUTION   (a) Because the binomial $y + z$ is common factor of each term, we have

$$5x(y + z) + 2(y + z) = (5x + 2)(y + z).$$

(b) Because $d - c = -(c - d)$, we have

$$a(c - d) + 2(d - c) = a(c - d) - 2(c - d) = (a - 2)(c - d).$$

It is sometimes possible to factor polynomials that do not contain factors common to every term by using the method of common factors. To do this, we first group the terms of the polynomials, and then we look for common polynomial factors in each group. For example, consider the polynomial $3xm + 3ym - 2x - 2y$. We note that two terms of the expression contain a factor of x and that the other two terms contain a factor of y. Grouping these terms accordingly, we obtain

$$3xm + 3ym - 2x - 2y = (3xm - 2x) + (3ym - 2y).$$

Factoring out the common monomial x from the first group and the common monomial y from the second group, we have

$$(3xm - 2x) + (3ym - 2y) = (3m - 2)x + (3m - 2)y.$$

We factor out the common binomial $3m - 2$:

$$(3m - 2)x + (3m - 2)y = (3m - 2)(x + y).$$

We could also have grouped the terms in another way, with the same result:

$$3xm + 3ym - 2x - 2y = 3m(x + y) - 2(x + y)$$
$$= (3m - 2)(x + y).$$

Therefore, in either case, we have

$$3xm + 3ym - 2x - 2y = (3m - 2)(x + y).$$

EXAMPLE 3   Factor each expression by grouping the terms in a suitable way.

(a) $ac - d - c + ad$

(b) $3u + uv + 3u^2 + v$

(c) $2ax^2 + 2ay^2 - bx^2 - by^2$

(d) $3ax + 3ay + 3az - 2bx - 2by - 2bz$

(e) $pq - pr - sr + sq$

SOLUTION   (a) $ac - d - c + ad = (ac - c) + (ad - d)$
$$= c(a - 1) + d(a - 1)$$
$$= (c + d)(a - 1)$$

(b) $3u + uv + 3u^2 + v = (3u + 3u^2) + (uv + v)$
$$= 3u(1 + u) + v(u + 1)$$
$$= (3u + v)(1 + u)$$

(c) $2ax^2 + 2ay^2 - bx^2 - by^2 = (2ax^2 + 2ay^2) + (-bx^2 - by^2)$
$$= 2a(x^2 + y^2) - b(x^2 + y^2)$$
$$= (2a - b)(x^2 + y^2)$$

(d) $3ax + 3ay + 3az - 2bx - 2by - 2bz$
$$= (3ax + 3ay + 3az) + (-2bx - 2by - 2bz)$$
$$= 3a(x + y + z) - 2b(x + y + z)$$
$$= (3a - 2b)(x + y + z)$$

(e) $pq - pr - sr + sq = (pq - pr) + (-sr + sq)$
$$= p(q - r) + s(-r + q)$$
$$= p(q - r) + s(q - r)$$
$$= (p + s)(q - r)$$

PROBLEM SET 2.5

In problems 1–20, factor each polynomial by factoring out the common factor.

1. $x^2 - x$

2. $4x^2 + 2x$

3. $9x^2 + 3x$

4. $10x^2 - 5x$

5. $4x^2 + 7xy$

6. $9x^3 + 3x^4$

7. $a^2b - ab^2$

8. $17x^3y^2 - 34x^2y$

9. $6p^2q + 24pq^2$

10. $12a^3b^2 + 36a^2b^3$

11. $6ab^2 + 30a^2b$

12. $5abc + 20abc^2$

13. $12x^3y - 48x^2y^2$

14. $4x^3 - 2x^2 + x$

15. $2a^3b - 8a^2b^2 - 6ab^3$

16. $4xy^2z + x^2y^2z^2 - x^3y^3$

17. $x^3y^2 + x^2y^3 + 2xy^4$

18. $3x^2y^2 + 6x^2z^2 - 9x^2$

19. $9m^2n + 18mn^2 - 27mn$

20. $8xy^2 + 24x^2y^3 + 4xy^3$

In problems 21–30, factor out the common binomial factor in each expression.

21. $3x(2a + b) + 5y(2a + b)$ **22.** $(2m + 3)x - (2m + 3)y$ **23.** $5x(a + b) + 9y(a + b)^2$

24. $x(y - z) - (z - y)$ **25.** $m(x - y) + (y - x)$ **26.** $(x - y) + 5(x - y)^2$

27. $7x(2a + 7b) + 14(2a + 7b)^2 + (2a + 7b)^3$ **28.** $(x + y)^3 + x(x + y)^2 + 5y(x + y)$

29. $y(xy + 2)^3 - 5x(xy + 2)^2 + 7(xy + 2)$ **30.** $x^2(a + b)^3 - 4xy(a + b)^2 + 3(a + b)$

In problems 31–45, factor each expression by grouping.

31. $ax + ay + bx + by$ **32.** $x^2a + x^2b + a + b$ **33.** $x^5 + 3x^4 + x + 3$

34. $ax^5 + b - bx^5 - a$ **35.** $yz + 2y - z - 2$ **36.** $a^2x - 1 - a^2 + x$

37. $ab^2 - b^2c - ad + cd$ **38.** $2x^2 - yz^2 - x^2y + 2z^2$ **39.** $2ax + by - 2ay - bx$

40. $x^3 + x^2 - 5x - 5$ **41.** $x^2 - ax + bx - ab$ **42.** $x^2 + ax + bx + ab$

43. $ax + bx + ay + by + a + b$ **44.** $2ax - b + 2bx - c + 2cx - a$

45. $2x^3 + x^2y - x^2 + 2xy + y^2 - y$

2.6 Factoring by Recognizing Special Products

When Special Products (i)–(vii) in Section 2.4 are read from right to left, they reveal patterns useful for factoring. We restate some of these products below for reference.

> **(i) Difference of two squares**
>
> $$a^2 - b^2 = (a - b)(a + b).$$

EXAMPLE 1   Factor each expression completely.

(a) $x^2 - 25$ (b) $16s^2 - 25t^2$

(c) $16u^2 - (v - 2w)^2$ (d) $16x^4 - y^4$

SOLUTION   (a) We write $x^2 - 25 = x^2 - 5^2$, which is the difference between two squares. Then we use the special product $a^2 - b^2 = (a - b)(a + b)$ with $a = x$ and $b = 5$. We have

$$x^2 - 25 = (x - 5)(x + 5).$$

(b) Notice that $16s^2 = (4s)^2$ and $25t^2 = (5t)^2$. Thus, the given expression is the difference between two squares. Use the special product $a^2 - b^2 = (a - b)(a + b)$ with $a = 4s$ and $b = 5t$. We have

$$16s^2 - 25t^2 = (4s)^2 - (5t)^2 = (4s - 5t)(4s + 5t).$$

(c) The expression is the difference between two squares. We have

$$16u^2 - (v - 2w)^2 = (4u)^2 - (v - 2w)^2$$
$$= [4u - (v - 2w)][4u + (v - 2w)]$$
$$= (4u - v + 2w)(4u + v - 2w).$$

(d) The expression is the difference between two squares, so that

$$16x^4 - y^4 = (4x^2)^2 - (y^2)^2$$
$$= (4x^2 - y^2)(4x^2 + y^2).$$

However, $4x^2 - y^2$ is also the difference between two squares, so that

$$4x^2 - y^2 = (2x - y)(2x + y).$$

Thus, the original expression is completely factored as follows:

$$16x^4 - y^4 = (2x - y)(2x + y)(4x^2 + y^2).$$

Note that $4x^2 + y^2$ is a prime polynomial and cannot be factored.

Although it is usually difficult to factor third-degree polynomials, polynomials representing the *sum* or the *difference* of two cubes can be factored directly. Rewriting Special Products (vi) and (vii) from Section 2.4, we have:

(ii) Sum of two cubes

$$a^3 + b^3 = (a + b)(a^2 - ab + b^2).$$

(iii) Difference of two cubes

$$a^3 - b^3 = (a - b)(a^2 + ab + b^2).$$

EXAMPLE 2   Factor each of the following expressions.

(a) $x^3 + 27$ (b) $8u^3 - v^3$ (c) $(x + y)^3 - (z - w)^3$

SOLUTION   (a) Because $27 = 3^3$, the expression $x^3 + 27$ is a sum of two cubes. We use the special product $a^3 + b^3 = (a + b)(a^2 - ab + b^2)$ with $a = x$ and $b = 3$.

$$x^3 + 27 = x^3 + 3^3 = (x + 3)(x^2 - 3x + 9).$$

(b) Because $8u^3 = (2u)^3$, the expression $8u^3 - v^3$ is a difference of two cubes. We use the special product $a^3 - b^3 = (a - b)(a^2 + ab + b^2)$:

$$8u^3 - v^3 = (2u)^3 - v^3 = (2u - v)[(2u)^2 + (2u)v + v^2]$$
$$= (2u - v)(4u^2 + 2uv + v^2).$$

(c) $(x + y)^3 - (z - w)^3$
$$= [(x + y) - (z - w)][(x + y)^2 + (x + y)(z - w) + (z - w)^2]$$
$$= (x + y - z + w)[(x + y)^2 + (x + y)(z - w) + (z - w)^2].$$

Factoring by Combining Methods

We have discussed factoring by determining common factors, by grouping, and by recognizing special products. Factoring some polynomials may require the application of more than one of these methods. To accomplish this, we suggest the following steps:

Step 1. Factor out common factors (if there are any).

Step 2. Examine the remaining polynomial factors to see if each is prime (that is, continue to factor until you reach prime factored form).

Step 3. Determine if factoring by grouping can be applied.

In Examples 3–8, factor each expression until you reach prime factored form.

EXAMPLE 3 $75b^2 - 243$

SOLUTION   First we factor out 3 so that
$$75b^2 - 243 = 3(25b^2 - 81).$$
The expression $25b^2 - 81$ is a difference between two squares:
$$25b^2 - 81 = (5b)^2 - 9^2$$
$$= (5b - 9)(5b + 9).$$

The original expression is factored completely as follows:
$$75b^2 - 243 = 3(25b^2 - 81)$$
$$= 3(5b - 9)(5b + 9).$$

EXAMPLE 4 $5t^6 + 40$

SOLUTION   First we factor out 5 so that
$$5t^6 + 40 = 5(t^6 + 8).$$
We see that $t^6 + 8 = (t^2)^3 + 2^3$. Thus, the expression is the sum of two cubes. Therefore,
$$t^6 + 8 = (t^2 + 2)[(t^2)^2 - 2t^2 + 2^2]$$
$$= (t^2 + 2)(t^4 - 2t^2 + 4).$$

The original expression is factored completely as follows:

$$5t^6 + 40 = 5(t^6 + 8)$$
$$= 5(t^2 + 2)(t^4 - 2t^2 + 4).$$

EXAMPLE 5 $64u^6 - 1$

SOLUTION   The expression $64u^6 - 1$ can be factored as the difference of two squares:

$$64u^6 - 1 = (8u^3)^2 - 1^2$$
$$= (8u^3 - 1)(8u^3 + 1).$$

However,

$$8u^3 - 1 = (2u)^3 - 1^3$$
$$= (2u - 1)(4u^2 + 2u + 1)$$

and

$$8u^3 + 1 = (2u)^3 + 1^3$$
$$= (2u + 1)(4u^2 - 2u + 1).$$

The complete factorization is

$$64u^6 - 1 = (2u - 1)(2u + 1)(4u^2 + 2u + 1)(4u^2 - 2u + 1).$$

EXAMPLE 6 $x^2 + 4x + 4$

SOLUTION   We can write the perfect square trinomial in Special Product (i), Section 2.4, page 84, in reverse:

$$a^2 + 2ab + b^2 = (a + b)^2.$$

If we set $a = x$ and $b = 2$, we have

$$x^2 + 4x + 4 = x^2 + 2(x)(2) + 2^2 = (x + 2)^2.$$

EXAMPLE 7 $x^2 - 2xy + y^2 - a^2 - 2ab - b^2$

SOLUTION   Using step 3, we group the terms that contain x or y, and those that contain a or b. In this way we are able to recognize special products.

$$x^2 - 2xy + y^2 - a^2 - 2ab - b^2 = (x^2 - 2xy + y^2) + (-a^2 - 2ab - b^2)$$
$$= (x^2 - 2xy + y^2) - (a^2 + 2ab + b^2).$$

We apply Special Products (i) and (ii) of Section 2.4:

$$(x^2 - 2xy + y^2) - (a^2 + 2ab + b^2) = (x - y)^2 - (a + b)^2.$$

Because this latter form is the difference of two squares, the complete factorization is

$$x^2 - 2xy + y^2 - a^2 - 2ab - b^2 = (x - y)^2 - (a + b)^2$$
$$= [(x - y) - (a + b)][(x - y) + (a + b)]$$
$$= (x - y - a - b)(x - y + a + b).$$

EXAMPLE 8 $\quad x^4 + 2x^2y^2 + 9y^4$

SOLUTION $\quad$ If the middle term were $6x^2y^2$ rather than $2x^2y^2$, we could factor the expression as

$$x^4 + 6x^2y^2 + 9y^4 = (x^2 + 3y^2)^2. \qquad \text{(Special Product (i), Section 2.4)}$$

But the middle term can be changed to $6x^2y^2$ by adding $4x^2y^2$ to $2x^2y^2$ (already there) and then subtracting $4x^2y^2$ at the end of the expression. (This will *not* change the value of the expression.) Thus,

$$x^4 + 2x^2y^2 + 9y^4 = (x^4 + 6x^2y^2 + 9y^4) - 4x^2y^2$$
$$= (x^2 + 3y^2)^2 - 4x^2y^2$$
$$= (x^2 + 3y^2)^2 - (2xy)^2$$
$$= [(x^2 + 3y^2) - 2xy][(x^2 + 3y^2) + 2xy]$$
$$= (x^2 - 2xy + 3y^2)(x^2 + 2xy + 3y^2).$$

PROBLEM SET 2.6

In problems 1–20, use the difference of two squares to factor each expression completely.

1. $x^2 - 4$
2. $100 - x^2$
3. $1 - 9y^2$
4. $25 - 4a^2$

5. $36 - 25t^2$
6. $16x^2y^2 - 9$
7. $16u^2 - 25v^2$
8. $25m^2 - 49n^2$

9. $a^2b^2 - c^2$
10. $49w^2 - 81z^2$
11. $(a - b)^2 - 100c^2$
12. $144p^2 - (q - 3)^2$

13. $81x^4 - 1$
14. $256x^4 - y^4$
15. $u^8 - v^8$
16. $625w^4 - 81z^4$

17. $(x + y)^2 - (a - b)^2$
18. $(3x + 2y)^2 - 25z^2$

19. $t^4 - 81(r + s)^4$
20. $16(x + y)^4 - 81(w - z)^4$

In problems 21–38, use the sum or difference of two cubes to factor each expression completely.

21. $x^3 + 1$
22. $y^3 + 125$
23. $64 - t^3$
24. $27m^3 - n^3$

25. $27w^3 + z^3$
26. $x^3y^3 - 64$
27. $8x^3 - 27y^3$
28. $125t^3 - 216$

29. $w^3 - 8y^3z^3$
30. $64a^3 + 27b^3$
31. $(x + 2)^3 - y^3$
32. $(c + 3)^3 + d^3$

33. $(y + 1)^3 + (w + 2)^3$
34. $64 - (t - 1)^3$
35. $w^6 + 8z^6$
36. $u^6 - 27$

37. $x^9 - 1$
38. $v^9 + 512$

In problems 39–62, use common factors, special products, or grouping to factor each expression until you reach prime factored form.

39. $8x^3 - 2xy^2$ **40.** $3u^3v - 27uv^3$ **41.** $64y - 4y^3$ **42.** $36s^2t^2 - 4s^4$

43. $3u^4v - 24uv$ **44.** $216x^4y^3 + 27xy^3$ **45.** $7x^7y + 7xy^7$ **46.** $3t^8 + 81t^2$

47. $t^6 - 1$ **48.** $64 - x^6$ **49.** $2u^7 - 128u$ **50.** $64w^6 - 729$

51. $y^2 + 8y + 16$ **52.** $t^2 - 12t + 36$ **53.** $9u^2 - 42uv + 49v^2$ **54.** $25x^2 - 40xy + 16y^2$

55. $x^2 + y^2 - z^2 - 9 + 2xy - 6z$ **56.** $u^2 + 4uv + 4v^2 - 4a^2 + 4ab - b^2$

57. $w^2 - y^2 + 2w - 2yz + 1 - z^2$ **58.** $t^2 - 10t - 16s^2 + 24sr + 25 - 9r^2$

59. $x^4 + x^2y^2 + y^4$ **60.** $9w^4 + 2w^2z^2 + z^4$

61. $4m^4 + n^4$ **62.** $25x^4 + 4x^2y^2 + 4y^4$

2.7 Factoring Trinomials of the Form $ax^2 + bx + c$

Some trinomial expressions of the form

$$ax^2 + bx + c,$$

in which a, b, and c are integers and $a \neq 0$, can be factored into a product of two binomials.

Recall from Section 2.4 that the product of two binomials can be a trinomial. Review the process of multiplying two binomials. This should suggest an idea for factoring. For instance,

$$
\begin{aligned}
(3x + 4)(2x + 5) &= 6x^2 + 15x + 8x + 20 \\
&= 6x^2 + (15 + 8)x + 20 \\
&= 6x^2 + 23x + 20.
\end{aligned}
$$

Notice that the first term in the trinomial is the product of the first terms in each binomial. The last term in the trinomial is the product of the last terms in each binomial. The middle term in the trinomial is found by adding the product of the outside terms to the product of the inside terms.

The relationship between the coefficients of the trinomial and the coefficients of the factors is illustrated in the following diagrams. The coefficients 6 and 20 are obtained as follows:

times

$$6x^2 + 23x + 20 = (3x + 4)(2x + 5).$$

times

The coefficient of the middle term in the trinomial, 23, is obtained as follows:

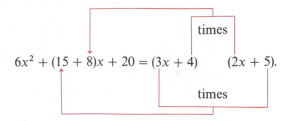

$$6x^2 + (15 + 8)x + 20 = (3x + 4) \qquad (2x + 5).$$

These diagrams illustrate how to factor a trinomial of the form $ax^2 + bx + c$.

We begin by writing

$$ax^2 + bx + c = (\underline{}x + \underline{})(\underline{}x + \underline{}).$$

Next, we fill in the blanks with the numbers so that:

1. The product of the first term of each binomial is ax^2.

2. The sum of the product of the outside terms and the product of the inside terms is bx.

3. The product of the last terms is c.

We try all possible choices of factors of a and c until we find a combination that gives us the desired middle term. If all the coefficients of the trinomial are positive, it is only necessary to try combinations of positive integers. If some of the coefficients of the trinomial are negative, we will have to try combinations that include negative integers. For example, to factor the trinomial $x^2 + 7x + 10$, we only try combinations of positive integers. Since the product of the first terms of the binomials is x^2, we have 1 as the coefficient of each first term, so that

$$x^2 + 7x + 10 = (x + \underline{})(x + \underline{}).$$

Using only positive integer factors of 10, we have two possibilities:

$$10 = 10 \cdot 1 \qquad \text{or} \qquad 10 = 2 \cdot 5.$$

That is, there are only two possible ways to fill in the remaining blanks:

$$(x + 10)(x + 1) \qquad \text{or} \qquad (x + 5)(x + 2).$$

We test the two factorizations to see whether either one produces the correct middle term, $7x$:

Possible Combinations	Products
$(x + 10)(x + 1)$	$x^2 + 11x + 10$
$(x + 5)(x + 2)$	$x^2 + 7x + 10$

Note that the combination $(x + 5)(x + 2)$ gives $7x$ as the middle term. Therefore, the correct factoring is

$$x^2 + 7x + 10 = (x + 5)(x + 2).$$

To factor $3x^2 + x - 2$, first factor the first term, $3x^2$, as $3x \cdot x$. Because the sign of the last term is negative, we know that the signs of the second terms in the binomials must be opposites of one another. (If both were positive or both were negative, the sign of the last term in the trinomial would be positive.) Therefore, we have

$$3x^2 + x - 2 = (3x + \underline{\quad})(x - \underline{\quad}) \qquad \text{or} \qquad (3x - \underline{\quad})(x + \underline{\quad}).$$

Next, factor the last term, 2. (Ignore the negative sign now because we've already accounted for it by setting up the signs in the binomials.) Thus,

$$2 = 1 \cdot 2 \qquad \text{or} \qquad 2 = 2 \cdot 1.$$

Note that we might have to try both $1 \cdot 2$ and $2 \cdot 1$, since the order of the factors makes a difference when we substitute the factors in the blanks. For example, $(3x + 1)(x - 2)$ has a different product than $(3x + 2)(x - 1)$. Try the two different factorizations of 2 in the blanks. We will see that the possible combinations of binomial factors are:

Possible Combinations	Products
$(3x + 1)(x - 2)$	$3x^2 - 5x - 2$
$(3x - 1)(x + 2)$	$3x^2 + 5x - 2$
$(3x + 2)(x - 1)$	$3x^2 - x - 2$
$(3x - 2)(x + 1)$	$3x^2 + x - 2$

The combination that gives the correct middle term, x, is $(3x - 2)(x + 1)$. Therefore, the correct factorization is

$$3x^2 + x - 2 = (3x - 2)(x + 1).$$

In Examples 1–6, factor each trinomial.

EXAMPLE 1 $x^2 + 6x + 8$

SOLUTION   Since all the coefficients are positive, we only consider combinations of positive integers. The first term, x^2, is factored as $x \cdot x$, so the trinomial can be expressed in the form

$$x^2 + 6x + 8 = (x + \underline{\quad})(x + \underline{\quad}).$$

The last term, 8, is factored as

$$8 = 8 \cdot 1 \qquad \text{or} \qquad 8 = 2 \cdot 4.$$

Trying different possible combinations of the factors of 8 in the blanks, we see that $(x + 4)(x + 2)$ produces the correct middle term, $6x$:

$(x + 4)(x + 2)$ $\qquad\qquad\qquad\qquad$ $6x$ $\qquad$ (Middle Term)

$$x \cdot 2 + 4 \cdot x = 2x + 4x$$

Therefore, the factors of the trinomial are $x + 4$ and $x + 2$, and

$$x^2 + 6x + 8 = (x + 4)(x + 2).$$

EXAMPLE 2 $y^2 - 9y + 20$

SOLUTION   Factoring the first term, we have $y^2 = y \cdot y$. Since the sign of the last term is positive, the signs of the second terms in the binomials must either both be positive or both be negative. The sign of the middle term of the polynomial is negative, which tells us that the signs in the binomials must be negative. So we have

$$y^2 - 9y + 20 = (y - \underline{\quad})(y - \underline{\quad}).$$

Next we factor the last term, 20, in the following ways:

$$20 = 20 \cdot 1, \qquad 20 = 10 \cdot 2, \qquad \text{or} \qquad 20 = 5 \cdot 4.$$

Substituting different possible combinations of the factors of 20, we see that $(y - 5)(y - 4)$ produces the correct middle term, $-9y$:

$(y - 5)(y - 4)$ $\qquad\qquad\qquad\qquad$ $-9y$ $\qquad$ (Middle Term)

$$y \cdot (-4) + (-5) \cdot y = -4y - 5y$$

Therefore, $y - 5$ and $y - 4$ are the factors of the trinomial, and

$$y^2 - 9y + 20 = (y - 5)(y - 4).$$

EXAMPLE 3 $2r^2 + 5r - 3$

SOLUTION   We begin by factoring the first term, $2r^2$, as $2r \cdot r$. Since the sign of the last term is negative, the signs of the second terms in the binomials are opposites of one another, so we have

$$2r^2 + 5r - 3 = (2r + \underline{\quad})(r - \underline{\quad}) \qquad \text{or} \qquad (2r - \underline{\quad})(r + \underline{\quad}).$$

Next we factor the last term, 3. (Again, ignore the negative sign because we've already accounted for it by setting up the signs in the binomial.) We have

$$3 = 3 \cdot 1.$$

Substituting the possible combinations of the factors of 3 in the blanks, we see that $(2r - 1)(r + 3)$ gives the correct middle term, $5r$:

$(2r - 1)(r + 3)$ $5r$ (middle term)

$$2r \cdot 3 + (-1) \cdot r = 6r - r$$

Therefore,

$$2r^2 + 5r - 3 = (2r - 1)(r + 3).$$

EXAMPLE 4 $6m^2 - 11m - 10$

SOLUTION   Factoring the first term, $6m^2$, we have

$$6m^2 = 6m \cdot m \qquad \text{or} \qquad 6m^2 = 3m \cdot 2m.$$

Because the sign of the last term is negative, the signs of the second terms of the binomials are opposites of one another, so that

$$6m^2 - 11m - 10 = (6m + \underline{\quad})(m - \underline{\quad}) \qquad \text{or} \qquad (6m - \underline{\quad})(m + \underline{\quad}),$$

or

$$6m^2 - 11m - 10 = (3m + \underline{\quad})(2m - \underline{\quad}) \qquad \text{or} \qquad (3m - \underline{\quad})(2m + \underline{\quad}).$$

Next we factor the last term, 10, as

$$10 = 10 \cdot 1 \qquad \text{or} \qquad 10 = 2 \cdot 5.$$

We try different combinations of the factors of 10, and we find that $(3m + 2)(2m - 5)$ gives the desired middle term, $-11m$:

$(3m + 2)(2m - 5)$ $-11m$ (middle term)

$$3m \cdot (-5) + 2 \cdot 2m = -15m + 4m$$

Therefore,

$$6m^2 - 11m - 10 = (3m + 2)(2m - 5).$$

EXAMPLE 5 $7u^2 - 37uv + 10v^2$

SOLUTION   We start by factoring the first term, $7u^2$, as $7u \cdot u$. Because the sign of the last term is positive, the signs of the second terms in the binomials must either both be positive or both be negative. The sign of the middle term in the polynomial is negative. Therefore, the signs in the binomials must be negative. Thus, we have

$$7u^2 - 37uv + 10v^2 = (7u - \underline{\hspace{1cm}})(u - \underline{\hspace{1cm}}).$$

Note that the middle term of the trinomial includes the factors uv and that the last term includes v^2. If these variables appear in the trinomial, the binomials must both have a v in their second terms. Thus, we can factor the last term, $10v^2$, in the following ways:

$$10v^2 = 10v \cdot v \qquad \text{or} \qquad 10v^2 = 2v \cdot 5v.$$

We try different possible combinations of the factors of $10v^2$ in the blanks, and we find that $(7u - 2v)(u - 5v)$ gives the desired middle term, $-37uv$:

$$(7u - 2v)(u - 5v) \qquad\qquad\qquad -37uv \qquad \text{(middle term)}$$
$$7u \cdot (-5v) + (-2v) \cdot u = -35uv - 2uv$$

Therefore,

$$7u^2 - 37uv + 10v^2 = (7u - 2v)(u - 5v).$$

The trinomials we are now factoring are more complicated than the trinomials we started with. However, it is still important that we begin by looking for common factors in the terms. The following example makes this clear.

EXAMPLE 6   Factor the trinomial $60a^3b + 25a^2b^2 - 15ab^3$.

SOLUTION   First we take out the common factor $5ab$:

$$60a^3b + 25a^2b^2 - 15ab^3 = 5ab(12a^2 + 5ab - 3b^2).$$

Next, we try to factor the trinomial $12a^2 + 5ab - 3b^2$. We start by factoring the first term, $12a^2$, as $12a \cdot a$, $6a \cdot 2a$, or $4a \cdot 3a$. Because the sign of the last term is negative, the signs of the second terms in the binomials are opposites of one another, so that

$$12a^2 + 5ab - 3b^2 = (12a + \underline{\hspace{0.7cm}})(a - \underline{\hspace{0.7cm}}) \qquad \text{or} \qquad (12a - \underline{\hspace{0.7cm}})(a + \underline{\hspace{0.7cm}}),$$

or

$$12a^2 + 5ab - 3b^2 = (6a + \underline{\hspace{0.7cm}})(2a - \underline{\hspace{0.7cm}}) \qquad \text{or} \qquad (6a - \underline{\hspace{0.7cm}})(2a + \underline{\hspace{0.7cm}}),$$

or

$$12a^2 + 5ab - 3b^2 = (4a + \underline{\hspace{0.7cm}})(3a - \underline{\hspace{0.7cm}}) \qquad \text{or} \qquad (4a - \underline{\hspace{0.7cm}})(3a + \underline{\hspace{0.7cm}}).$$

Factoring the last term, $3b^2$, we have

$$3b^2 = 3b \cdot b.$$

Trying the different possible combinations of the factors of $3b^2$, we find that $(4a + 3b)(3a - b)$ gives the correct middle term, $5ab$:

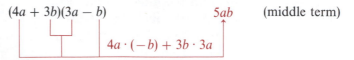

Therefore,

$$12a^2 + 5ab - 3b^2 = (4a + 3b)(3a - b),$$

and the complete factorization of the original trinomial is

$$60a^3b + 25a^2b^2 - 15ab^3 = 5ab(4a + 3b)(3a - b).$$

In working with some mathematical applications, we may encounter problems in which factoring is needed, but no direction is given as to what method should be used. In this case, we have to decide which methods are appropriate. It is therefore worthwhile to review basic procedures that have been used throughout the chapter.

1. If a polynomial expression has a greatest common factor (other than 1), then we factor out the greatest common factor. For example,

$$4x^2y^3 - 6xy^4 = 2xy^3(2x - 3y).$$

2. If a polynomial expression has two terms (a binomial), we check to see whether it is
  (i) the difference of two squares, or
  (ii) the sum of two cubes, or
(iii) the difference of two cubes,
and then factor.
For example,

$$\text{(a)}\ 4x^2 - y^2 = (2x - y)(2x + y)$$

$$\text{(b)}\ x^3 + 8y^3 = (x + 2y)(x^2 - 2xy + 4y^2)$$

$$\text{(c)}\ 27x^3 - y^3 = (3x - y)(9x^2 + 3xy + y^2)$$

3. If a polynomial expression has three terms (a trinomial), we check to see whether it is
  (i) a perfect square trinomial, which will factor into the square of a
      binomial, or

(ii) a trinomial that is not a perfect square, which will factor by trial and error.

For example,

$$\text{(a)} \quad x^2 - 6x + 9 = (x - 3)^2$$

$$\text{(b)} \quad 4y^2 - 5y - 6 = (4y + 3)(y - 2)$$

4. If a polynomial expression has more than three terms, then we try to factor it by grouping. For example,

$$y^3 - 5y^2 + 3y - 15 = (y^2 + 3)(y - 5).$$

PROBLEM SET 2.7

In problems 1–40, factor each trinomial completely.

1. $x^2 + 4x + 3$
2. $y^2 + 5y + 6$
3. $t^2 - 3t + 2$
4. $x^2 - 3x - 4$
5. $y^2 + 15y + 36$
6. $z^2 + 3z - 10$
7. $x^2 - 2x - 15$
8. $a^2 - 3ab - 28b^2$
9. $u^2 - 16u + 63$
10. $x^2 - 3x - 40$
11. $z^2 + 11zw + 30w^2$
12. $w^2 - 9wz - 10z^2$
13. $x^2 - 7x - 18$
14. $x^2 - 17x + 30$
15. $m^2 + 2mn - 120n^2$
16. $m^2 - 13mn - 30n^2$
17. $12 - x^2 - 4x$
18. $40 - 3x - x^2$
19. $-5t + 36 - t^2$
20. $16 - y^2 - 6y$
21. $2w^2 + 7w + 3$
22. $2w^2 + w - 6$
23. $3x^2 + 5x - 2$
24. $2y^2 + 9y - 5$
25. $5y^2 - 11y + 2$
26. $4x^2 - 35xy - 9y^2$
27. $3c^2 + 7cd + 2d^2$
28. $10t^2 - 19t + 6$
29. $6x^2 + 13x + 6$
30. $6y^2 - y - 7$
31. $6z^2 + 5zy - 6y^2$
32. $6x^2 - 7xy - 5y^2$
33. $12v^2 + 17v - 5$
34. $6c^2 - 7cd - 3d^2$
35. $56x^2 - 83x + 30$
36. $42z^2 + z - 30$
37. $12 - 2w^2 - 5w$
38. $18x^2 + 101x + 90$
39. $6rs + 5r^2 - 8s^2$
40. $24x^2 - 67xy + 8y^2$

In problems 41–50, use common factors and the factoring of trinomials to factor the polynomials.

41. $5x^3 - 55x^2 + 140x$
42. $x^2yz^2 - xyz^2 - 12yz^2$
43. $128st^3 - 32s^2t^2 + 2s^3t$
44. $bx^2c + 7bcx + 12bc$
45. $x^2y^2 + 10xy^2 + 21y^2$
46. $a^2x^2z^2 + 5a^2xz^2 - 14a^2z^2$
47. $4m^2n^2 + 24m^2n - 28m^2$
48. $7hkx^2 + 21hkx + 14hk$
49. $wx^2y - 9wxy + 14wy$
50. $pq^2x^2y - 2pq^2xy - 15pq^2y$

2.8 Division of Polynomials

In Section 2.3, we introduced the property of exponents,

$$\frac{a^m}{a^n} = a^{m-n}.$$

This property is used in the division of monomials. For example, to divide $4x^5y^3$ by $2x^2y$, we have

$$\frac{4x^5y^3}{2x^2y} = 2x^{5-2}y^{3-1} = 2x^3y^2.$$

Similarly, we divide x^3yz^2 by xyz^5 as follows:

$$\frac{x^3yz^2}{xyz^5} = (x^{3-1})(1)\left(\frac{1}{z^{5-2}}\right) = \frac{x^2}{z^3}.$$

> To divide a polynomial by a monomial, we divide each term of the polynomial by the monomial.

For instance, to divide $x^3 + 2x^2 + x$ by x, we write

$$\frac{x^3 + 2x^2 + x}{x} = \frac{x^3}{x} + \frac{2x^2}{x} + \frac{x}{x}$$
$$= x^2 + 2x + 1.$$

The following examples illustrate this procedure.

EXAMPLE 1   Divide $5x^3 + 3x^2 + 7x$ by x and simplify.

SOLUTION   We divide each term of the polynomial by the monomial

$$\frac{5x^3 + 3x^2 + 7x}{x} = \frac{5x^3}{x} + \frac{3x^2}{x} + \frac{7x}{x}$$
$$= 5x^2 + 3x + 7.$$

EXAMPLE 2   Divide $x^3 + 2x^2 - 3x + 7$ by x^2 and simplify.

SOLUTION   We divide each term of the polynomial by the monomial

$$\frac{x^3 + 2x^2 - 3x + 7}{x^2} = \frac{x^3}{x^2} + \frac{2x^2}{x^2} - \frac{3x}{x^2} + \frac{7}{x^2}$$
$$= x + 2 - \frac{3}{x} + \frac{7}{x^2}.$$

EXAMPLE 3   Divide $-4u^5v^6 - 12u^4v^5 + 8u^3v^4$ by $2uv^3$ and simplify.

SOLUTION $$\frac{-4u^5v^6 - 12u^4v^5 + 8u^3v^4}{2uv^3} = \frac{-4u^5v^6}{2uv^3} - \frac{12u^4v^5}{2uv^3} + \frac{8u^3v^4}{2uv^3}$$
$$= -2u^4v^3 - 6u^3v^2 + 4u^2v.$$

Warning: An *error* often made when dividing a polynomial by a monomial is to divide only one term of the polynomial by the monomial. Consider the following example:

$$\frac{8x^3 + 2x}{x} \overset{?}{=} 8x^2 + 2x.$$

This is *not* correct. The correct method is

$$\frac{8x^3 + 2x}{x} = \frac{8x^3}{x} + \frac{2x}{x}$$

$$= 8x^2 + 2.$$

To divide one polynomial by another polynomial, we use a method similar to the "long-division" method used in arithmetic; for example, $6,741 \div 21$. We usually do this long division in the following way:

$$
\begin{array}{r}
321 \leftarrow \text{quotient} \\
\text{divisor} \longrightarrow 21 \overline{)6741} \leftarrow \text{dividend} \\
63 \quad (= 21 \cdot 3) \\
\hline
44 \\
42 \quad (= 21 \cdot 2) \\
\hline
21 \\
21 \quad (= 21 \cdot 1) \\
\hline
0 \leftarrow \text{remainder}
\end{array}
$$

The result of this calculation can be expressed as $6,741 = (21)(321) + 0$, that is,

---

dividend = (divisor)(quotient) + remainder.

We can also perform this division by changing the dividend 6,741 and the divisor 21 to their expanded forms first:

$$21 = 2 \cdot 10 + 1$$

and

$$6,741 = 6 \cdot 10^3 + 7 \cdot 10^2 + 4 \cdot 10 + 1$$

and then dividing:

$$
\begin{array}{r}
3 \cdot 10^2 + 2 \cdot 10 + 1 \quad\quad \leftarrow \text{quotient} \\
\text{divisor} \longrightarrow 2 \cdot 10 + 1 \overline{)6 \cdot 10^3 + 7 \cdot 10^2 + 4 \cdot 10 + 1} \leftarrow \text{dividend} \\
6 \cdot 10^3 + 3 \cdot 10^2 \quad\quad\quad [= (2 \cdot 10 + 1)(3 \cdot 10^2)] \\
\hline
4 \cdot 10^2 + 4 \cdot 10 \\
4 \cdot 10^2 + 2 \cdot 10 \quad\quad [= (2 \cdot 10 + 1)(2 \cdot 10)] \\
\hline
2 \cdot 10 + 1 \\
2 \cdot 10 + 1 \quad\quad [= (2 \cdot 10 + 1)(1)] \\
\hline
0 \leftarrow \text{remainder}
\end{array}
$$

Note that the expanded forms of the numbers in this example are polynomials. Each term of the polynomials has a known base of 10.

Let us change this problem slightly by changing base 10 to base x. We now have the long division of $6x^3 + 7x^2 + 4x + 1$ by $2x + 1$:

$$\boxed{\dfrac{6x^3}{2x} = 3x^2} \qquad \boxed{\dfrac{4x^2}{2x} = 2x} \qquad \boxed{\dfrac{2x}{2x} = 1}$$

$$
\begin{array}{r}
3x^2 + 2x\ + 1 \quad \longleftarrow \text{quotient} \\
\text{divisor} \longrightarrow 2x + 1 \,\overline{)\, 6x^3 + 7x^2 + 4x + 1} \quad \longleftarrow \text{dividend} \\
\text{subtract} \longrightarrow \underline{6x^3 + 3x^2} \qquad\qquad\qquad [= (2x+1)(3x^2)] \\
4x^2 + 4x \\
\text{subtract} \longrightarrow \underline{4x^2 + 2x} \qquad\qquad [= (2x+1)(2x)] \\
2x + 1 \\
\text{subtract} \longrightarrow \underline{2x + 1} \qquad [= (2x+1)(1)] \\
0 \quad \longleftarrow \text{remainder}
\end{array}
$$

This example illustrates the following systematic step-by-step procedure for dividing one polynomial by another:

Procedure for Long Division

> Step 1. Arrange both polynomials in descending powers of one variable, and write the missing terms of the dividend with zero coefficients.
>
> Step 2. Find the first term of the quotient by dividing the first term of the dividend by the first term of the divisor.
>
> Step 3. Multiply the quotient term obtained in step 2 by the entire divisor.
>
> Step 4. Subtract the product obtained in step 3 from the dividend; and bring down the next term of the original dividend to form the new dividend.
>
> Step 5. Repeat the procedure in steps 2, 3, and 4 for the new dividend; keep repeating the steps until the degree of the remainder is less than the degree of the divisor.
>
> Step 6. Check the calculation: does
>
> $$(\text{divisor})(\text{quotient}) + \text{remainder} = \text{dividend}?$$

EXAMPLE 4   Divide $3x^2 + 3x^3 + x^4 - 2x - 3$ by $x + 1$.

SOLUTION   Follow the procedure given:

Step 1. Arrange both polynomials in descending powers of x, so that

$$\text{divisor} \longrightarrow x + 1 \,\overline{)\, x^4 + 3x^3 + 3x^2 - 2x - 3} \quad \longleftarrow \text{dividend}$$

Step 2. Divide x^4, the first term of the dividend, by x, the first term of the divisor, to obtain

$$\frac{x^4}{x} = x^3.$$

Step 3. Multiply x^3, the first quotient term, by $x + 1$, the divisor, to get

$$x^3(x + 1) = x^4 + x^3.$$

Step 4. Put the product $x^4 + x^3$ under the dividend. Subtract and bring down the next term to obtain a new dividend, so that

$$
\begin{array}{r}
x^3 \qquad\qquad\qquad\qquad \\
x + 1 \,\overline{)\, x^4 + 3x^3 + 3x^2 - 2x - 3} \\
\text{subtract} \longrightarrow x^4 + \ x^3 \qquad\qquad\qquad \\
\hline
2x^3 + 3x^2 \longleftarrow \text{new dividend}
\end{array}
$$

Step 5. Repeat the procedure in steps 2, 3, and 4 for the new dividend to obtain

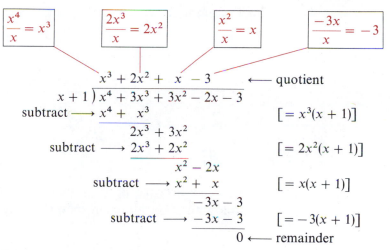

Therefore, the quotient is $x^3 + 2x^2 + x - 3$, and the remainder is 0.

Step 6. To check, we calculate

$$
\begin{aligned}
(\text{divisor})(\text{quotient}) + \text{remainder} &= (x + 1)(x^3 + 2x^2 + x - 3) + 0 \\
&= x^4 + 3x^3 + 3x^2 - 2x - 3 \\
&= \text{dividend}.
\end{aligned}
$$

EXAMPLE 5 Divide $2t^4 + 5 + 3t^3 - 3t$ by $2t - 1$.

SOLUTION First arrange both polynomials in descending powers of t, and write the missing term of the dividend with a zero coefficient to obtain:

$$\text{divisor} \longrightarrow 2t - 1 \,\overline{)\, 2t^4 + 3t^3 + 0t^2 - 3t + 5} \longleftarrow \text{dividend}$$

The remaining steps are shown as follows:

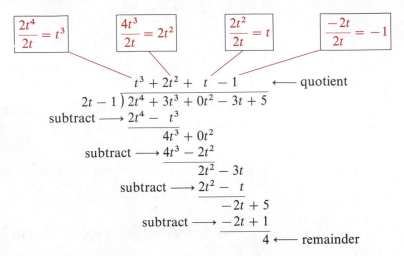

$$2t - 1\overline{\smash{\big)}\,2t^4 + 3t^3 + 0t^2 - 3t + 5}$$

quotient: $t^3 + 2t^2 + t - 1$

subtract ⟶ $2t^4 - t^3$

$4t^3 + 0t^2$

subtract ⟶ $4t^3 - 2t^2$

$2t^2 - 3t$

subtract ⟶ $2t^2 - t$

$-2t + 5$

subtract ⟶ $-2t + 1$

4 ⟵ remainder

Therefore, the quotient is $t^3 + 2t^2 + t - 1$, and the remainder is 4. To check, we calculate

$$\begin{aligned}
(\text{divisor})(\text{quotient}) + \text{remainder} &= (2t - 1)(t^3 + 2t^2 + t - 1) + 4 \\
&= (2t^4 + 3t^3 - 3t + 1) + 4 \\
&= 2t^4 + 3t^3 - 3t + 5 \\
&= \text{dividend}.
\end{aligned}$$

EXAMPLE 6   Divide $-3w^3 + 2w^4 + 5w^2 + 2w + 7$ by $-w + w^2 + 1$.

SOLUTION   First we arrange both polynomials in descending powers of w. Then we follow the procedures in the preceding examples to obtain

$$w^2 - w + 1\overline{\smash{\big)}\,2w^4 - 3w^3 + 5w^2 + 2w + 7}$$

quotient: $2w^2 - w + 2$

subtract ⟶ $2w^4 - 2w^3 + 2w^2$

$-w^3 + 3w^2 + 2w$

subtract ⟶ $-w^3 + w^2 - w$

$2w^2 + 3w + 7$

subtract ⟶ $2w^2 - 2w + 2$

$5w + 5$ ⟵ remainder

Therefore, the quotient is $2w^2 - w + 2$, and the remainder is $5w + 5$.

To check, we calculate

$$
\begin{aligned}
\text{(divisor)(quotient)} + \text{remainder} &= (w^2 - w + 1)(2w^2 - w + 2) + 5w + 5 \\
&= (2w^4 - 3w^3 + 5w^2 - 3w + 2) + 5w + 5 \\
&= 2w^4 - 3w^3 + 5w^2 + 2w + 7 \\
&= \text{dividend.}
\end{aligned}
$$

We can apply this method to the division of polynomials involving more than one variable. First we arrange the dividend and the divisor in descending powers of one of the variables, and then we divide as previously illustrated.

EXAMPLE 7 Divide $x^4 - y^4 + 3xy^3 - 3x^3y$ by $x + y$.

SOLUTION After arranging the dividend and the divisor in descending powers of x, we have

$$(x^4 - 3x^3y + 0x^2y^2 + 3xy^3 - y^4) \div (x + y).$$

Again, we follow the outlined procedure:

$$\dfrac{x^4}{x} = x^3 \qquad \dfrac{-4x^3y}{x} = -4x^2y \qquad \dfrac{4x^2y^2}{x} = 4xy^2 \qquad \dfrac{-xy^3}{x} = -y^3$$

$$
\begin{array}{r}
x^3 - 4x^2y + 4xy^2 - y^3 \\
x + y \overline{\smash{\big)}\ x^4 - 3x^3y + 0x^2y^2 + 3xy^3 - y^4} \\
\text{subtract} \longrightarrow x^4 + x^3y \\
\hline
-4x^3y + 0x^2y^2 \\
\text{subtract} \longrightarrow -4x^3y - 4x^2y^2 \\
\hline
4x^2y^2 + 3xy^3 \\
\text{subtract} \longrightarrow 4x^2y^2 + 4xy^3 \\
\hline
-xy^3 - y^4 \\
\text{subtract} \longrightarrow -xy^3 - y^4 \\
\hline
0
\end{array}
$$

Therefore, the quotient is

$$x^3 - 4x^2y + 4xy^2 - y^3,$$

and the remainder is 0.

To check, we calculate

$$
\begin{aligned}
\text{(divisor)(quotient)} + \text{remainder} &= (x + y)(x^3 - 4x^2y + 4xy^2 - y^3) + 0 \\
&= x^4 - 3x^3y + 3xy^3 - y^4 \\
&= \text{dividend.}
\end{aligned}
$$

PROBLEM SET 2.8

In problems 1–12, divide as indicated and simplify.

1. $6x^5$ by $2x^2$

2. $10y^7$ by $4y^3$

3. $12x^6y^7$ by $-4x^4y^9$

4. $-24t^5s^3$ by $-8t^2s$

5. $14u^3vw^4$ by $7u^5vw^2$

6. $30xy^3z^5$ by $-15x^3y^3z$

7. $9mn^2 - 6m^2$ by $3m$

8. $15u^2v^5 - 25u^4v^3$ by $-5uv^2$

9. $4x^2y^3 - 16xy^3 + 4xy$ by $2xy$

10. $4(a + b)^4 + 12(a + b)^3 - 8(a + b)$ by $2(a + b)$

11. $6a^3b^5 + 3a^2b^4 - 12ab + 9$ by $3a^2b^3$

12. $35(u - v)^3 - 15(u - v)^2 - 5(u - v)$ by $5(u - v)^2$

In problems 13–34, divide as indicated and check the result.

13. $x^2 - 7x + 10$ by $x - 5$

14. $y^2 - 5y + 8$ by $y - 2$

15. $2v^2 - 5v - 6$ by $2v - 1$

16. $1 - 4x + 4x^2$ by $2x - 1$

17. $w^3 + 3w^2 - 2w - 5$ by $w + 2$

18. $2x^4 + 3x^3 + 2x - 5x^2 - 1$ by $1 + x$

19. $3t^3 - 5t^2 + 2t + 2t^4 - 1$ by $t + 1$

20. $5y^5 - 2y^3 + 1$ by $y - 1$

21. $x^4 + 5x^3 + 9x^2 + 5x - 4$ by $x^2 + 2x - 1$

22. $5w^3 - 2w^2 + 3w - 4$ by $w^2 - 2w + 1$

23. $y^3 + 16y + 52 - 3y^2$ by $y^2 + 26 - 5y$

24. $m^3 + 2m - 4m^2 + 1$ by $m^2 - 1 - 3m$

25. $x^5 + 2x^2 - 24 + 3x^4$ by $x^3 - 2x^2 + x^4 + 6x - 12$

26. $3t^4 + 2t^2 - 28$ by $3t^3 + 6t^2 + 14t + 28$

27. $u^3 + 2v^3 - u^2v - uv^2$ by $u + v$

28. $3x^2y - 2xy^2 - 8y^3 + x^3$ by $x + 2y$

29. $2m^3 + mn^2 + m^2n + 4n^3$ by $m + n$

30. $w^2z - 6w^3 - 6z^3 - 12wz^2$ by $2w - 3z$

31. $2x^4 + 3x^3y - 6x^2y^2 + 4xy^3 - 7y^4$ by $2x - y$

32. $y^5 + z^5$ by $y + z$

33. $x^3 - y^3$ by $x - y$

34. $32x^5 - y^5$ by $2x - y$

REVIEW PROBLEM SET

In problems 1–6, identify the polynomial as a monomial, binomial, or trinomial. Give the degree and the coefficients.

1. $4y^2 - 3y + 2$

2. $8a - 5$

3. $-7x^2 + 3x$

4. $16m^3$

5. $10u^3v^2$

6. $\frac{1}{3}y^3 - \frac{2}{3}yz + 5z^2$

In problems 7–12, evaluate the polynomial for the given values of the variables.

7. $3x^2 + 5$, $x = -2$

8. $5t^2 - 3t + 2$, $t = 3$

9. $4w^2 - 3w - 6$, $w = 4$

10. $-2y^3 + 3y^2 - 4y + 2$, $y = 2$

11. $2u^2 - 3uv - v^2$, $u = 2$, $v = -1$

12. $5x^2 + 2xy + 3y^2$, $x = -2$, $y = 1$

In problems 13–20, perform the additions and subtractions.

13. $5w^2 + 7w^2 - 3w^2$

14. $-3xy^2 + 8xy^2 + xy^2$

15. $(2x^2 - 4) + (-5x^2 + 3)$

16. $(4x^3 + 3x^2) + (-5x^2 + 2x)$

17. $(3v^2 + 7v + 8) - (2v^2 + 3v + 2)$

18. $(-y^3 + 2y^2 + 3y - 7) - (-3y^3 + y^2 - 5y - 2)$

19. $(5x^2 - 3x + 2) + (2x^2 + 5x - 7) - (3x^2 - 4x - 1)$

20. $(7t^3 - 3t^2s + s^3) - (5t^3 - 4ts^2 + s^3) - (t^3 - s^3)$

In problems 21–36, use the properties of exponents to simplify each expression.

21. $x^3 x^8$

22. $(-t)^5(-t)^4$

23. $w^2 w^7 w^3$

24. $(-y)^2(-y)^3(-y)^7$

25. $(m^3)^9$

26. $(-x^4)^6$

27. $(-vu^2)^3$

28. $(3p^2q^3)^4$

29. $(2y^3z)^2$

30. $(-4x^2y^3z)^3$

31. $\left(\dfrac{3t}{s^3}\right)^2$

32. $\left(\dfrac{2x^3}{y^2}\right)^4$

33. $\dfrac{x^{12}}{x^9}$

34. $\dfrac{3y^2}{9y^5}$

35. $\dfrac{(3x^2z^4)^3}{(3x^3z)^2}$

36. $\dfrac{(-2ab^2c^4)^2}{(2a^2bc^3)^3}$

In problems 37–48, find each product.

37. $3t^2(2t^3 - 4t)$

38. $-4y^3(-5y^2 - 2y + 1)$

39. $-2xy^3(-3x^2y + 5xy^2 - y)$

40. $3mn^2p^3(7mn - 2m^2p + 5n)$

41. $(a - b)(a^2 - 2ab + b^2)$

42. $(u^2 - u + 1)(2u^2 - 3u + 2)$

43. $(w + 7)(w + 3)$

44. $(3t + 7)(2t + 5)$

45. $(2x - 3)(x + 5)$

46. $(4y + 9x)(3y - 9x)$

47. $(4u - 5v)(u - 3v)$

48. $(7 - 5pq)(4 - 3pq)$

In problems 49–62, use special products to find each product.

49. $(t + 8)(t - 8)$

50. $(2s - 3)^2$

51. $(3x + 7)^2$

52. $(2v - 1)^3$

53. $(t + 2)(t^2 - 2t + 4)$

54. $(m^2 + 5)(m^2 - 5)$

55. $(y + 3)^3$

56. $(3s - 4)(9s^2 + 12s + 16)$

57. $(4 - 3s)^2$

58. $(2p^2 + q^2)^2$

59. $(3w - 2)^3$

60. $(x^2 - y^2)^3$

61. $(2x - z)(4x^2 + 2xz + z^2)$

62. $(m^2 - 3)(m^4 + 3m^2 + 9)$

In problems 63–68, factor each expression by finding common factors.

63. $7x^2y - 21xy^3$

64. $8uv^3 + 16u^2v$

65. $26a^3b^2 + 39a^5b^4 - 52a^2b^3$

66. $25mnp^2 - 50mn^2 - 50mn^2p^2 + 75m^2n^2p^2$

67. $2y(y + z) - 4x(y + z)$

68. $7x(y - z) + 14x^2(y - z)$

In problems 69–74, use the method of grouping to factor each expression.

69. $3x - 3y + xz - yz$

70. $a^4 - b^4 - c^4 - 2b^2c^2$

71. $2ux + xv - 6uy - 3vy$

72. $2x^3 - 6x + x^2z - 3z$

73. $5am^2 - bn + 5an - bm^2$

74. $t^3 + t^2s + t^2 + ts + 2t + 2s$

In problems 75–82, use special products to factor each expression.

75. $25m^2 - 9n^2$

76. $16(u + v)^2 - 81$

77. $x^3 + 64$

78. $y^3 - 216$

79. $16t^4 - 81$

80. $x^8 - 256$

81. $(x + y)^2 - (z - 1)^2$

82. $121x^2y^4 - z^6$

In problems 83–92, use common factors, special products, and/or grouping to factor each expression.

83. $9u^3 - 81uv^2$

84. $100y^3w - 25yw^3$

85. $5st^3 + 320s$

86. $-32w^2 + 4w^5$

87. $64x^6 - y^6$

88. $m^6p^6 - q^6$

89. $x^2 + y^2 - w^2 - z^2 + 2xy - 2wz$

90. $9t^2 + s^2 - r^4 - 6ts$

91. $t^4 + t^2 + 1$

92. $x^4 - 3x^2 + 9$

In problems 93–106, factor each trinomial.

93. $x^2 + 2xy - 3y^2$

94. $t^2 + 7t - 18$

95. $m^2 - 5m - 36$

96. $z^2 - zy - 12y^2$

97. $3u^2 + 17u + 10$

98. $6x^2 - 29x + 35$

99. $2y^2 - y - 6$

100. $33w^2 + 14wz - 40z^2$

101. $20x^2 - 31xy + 12y^2$

102. $68 - 31y - 15y^2$

103. $w^3 + 9w^2 - 22w$

104. $2a^3b - 10a^2b^2 - 28ab^3$

105. $x^2yz - 6xy^2z - 16y^3z$

106. $2m^3np^2 + 7m^2n^2p^2 - 15mn^3p^2$

In problems 107–118, divide as indicated.

107. $18u^4$ by $3u^2$

108. $-12x^3y^4$ by $-6xy^3$

109. $32w^5z^3 - 16w^4z^2$ by $4w^3z$

110. $100t^3s^2 - 90t^4s^3 + 70t^5s^4$ by $10t^2s^3$

111. $5x^2 - 16x + 3$ by $x - 3$

112. $2y^2 + 17y + 21$ by $3 + 2y$

113. $x^3 - 6x^2 + 12x - 8$ by $x^2 - 4x + 4$

114. $x^2 + y^2 + 2xy - 6x - 6y + 9$ by $x + y - 3$

115. $t^4 + 3t^3s + 2t^2s^2 + ts^3 - s^4$ by $t^2 + ts + s^2$

116. $x^4 - 3xy^3 - 2y^4$ by $x^2 - xy - y^2$

117. $w^{15} + 1$ by $w^5 + 1$

118. $m^6 - n^6$ by $m - n$

CHAPTER 2 TEST

1. Evaluate the polynomial for the given values of the variables.
   (a) $7x^2 - 2x$, $x = -2$ (b) $2x^2 - 3xy + y^3$, $x = 2$, $y = -1$

2. Perform the additions and subtractions.
   (a) $(2m^2 - 3m + 7) + (3m^2 + 5m - 11)$
   (b) $(-4z^3 + 3z - 8) - (-z^3 + 4z^2 + 7z - 3)$
   (c) $(x^2 - 2x + 3) - (2x^2 + x - 5) - (-x^2 - 3x + 1)$

3. Use the properties of exponents to simplify each expression.

   (a) $(-x)^3(-x)^2$ (b) $\dfrac{6y^7}{2y^3}$ (c) $(x^2y^3)^4$ (d) $\dfrac{(-2x^2y)^3}{(2xy^3)^2}$

4. Find each product.
   (a) $3s^2(2s^3 - 4)$ (b) $(x^2 - xy + 3y^2)(x - y)$

5. Use special products to find each product.
   (a) $(2u - 3v)(2u + 3v)$ (b) $(3w - 7z)^2$ (c) $(x + 5y)^2$
   (d) $(2x - y)(4x^2 + 2xy + y^2)$ (e) $(m + 3n)^3$

6. Factor each expression by finding common factors.
   (a) $12x^2 - 30x^5$ (b) $15m^3n^5 - 25m^5n^2 + 35m^4n^3$

7. Use the method of grouping to factor each expression.
   (a) $ax + bx - ay - by$ (b) $2x^2z + 3y + 6x^2 + yz$

8. Use special products to factor each expression.

(a) $16x^2 - 9y^2$ (b) $27w^3 + 8$

9. Factor each trinomial.

(a) $x^2 - 5xy - 36y^2$ (b) $6y^2 - y - 15$

(c) $4x^2 - 12xz + 9z^2$ (d) $2x^3y + 10x^2y^2 - 28xy^3$

10. Divide as indicated.

(a) $24w^5$ by $-3w^2$ (b) $27x^3y - 18x^2y^2 + 9x^4y^3$ by $3x^2y$

(c) $z^4 + z^3 - 8z^2 + 9z - 10$ by $z - 2$ (d) $x^3 + 8$ by $x + 2$

3 The Algebra of Fractions

In this chapter, you will learn to use the rules that govern numerical fractions to reduce, multiply, divide, add, and subtract **algebraic fractions**—fractions that contain variables.

3.1 Rational Expressions

In Chapter 1, Section 1.2, we defined a rational number as a number that can be written in the form

$$\frac{a}{b}, \quad \text{where } a \text{ and } b \text{ are integers} \quad (b \neq 0).$$

We now define a **rational expression** as an algebraic expression that can be written in the form

$$\frac{P}{Q}, \quad \text{where } P \text{ and } Q \text{ are polynomials} \quad (Q \neq 0).$$

This expression is also called an **algebraic fraction** with **numerator** P and **denominator** Q. Throughout this book, we use the term "fraction" to mean either a common arithmetic fraction or an algebraic fraction.

Examples of rational expressions are

$$\frac{1}{x}, \qquad \frac{y+1}{5y-2}, \qquad \frac{5}{t^2-7}, \qquad \frac{3w^2+4w-1}{1}, \qquad \text{and} \qquad \frac{3c^2+c}{5c+1}.$$

Rational expressions have specific values when numbers are substituted for the variables (letters).

For example, the value of the rational expression

$$\frac{x}{x-2},$$

when 3 is substituted for x, is

$$\frac{3}{3-2} = \frac{3}{1} = 3.$$

The value of the expression, when x is replaced by -1, is

$$\frac{-1}{-1-2} = \frac{-1}{-3} = \frac{1}{3}.$$

However,

$$\frac{x}{x-2}$$

does not have a value when $x = 2$, because

$$\frac{2}{2-2} = \frac{2}{0}$$

is not defined. (The symbol 2/0 does not represent a number.)

We can make the following general statement:

A rational expression represents a real number when any values are assigned to the variable except those values that make the denominator zero. If the denominator of a rational expression is equal to zero, we say that the rational expression is *undefined* for that value of the variable.

For example, if we substitute $x = 1$ in the fraction

$$\frac{(x-1)^2}{(x-1)(x-2)}$$

we obtain

$$\frac{(1-1)^2}{(1-1)(1-2)} = \frac{0}{0(-1)} = \frac{0}{0}.$$

This expression is undefined. If we substitute $x = 2$ in the fraction, we have

$$\frac{(2-1)^2}{(2-1)(2-2)} = \frac{1}{1(0)} = \frac{1}{0}.$$

This expression is also undefined.

Thus,

$$\frac{(x-1)^2}{(x-1)(x-2)}$$

does not represent a real number when $x = 1$ or when $x = 2$.

EXAMPLE 1   For what value of x is

$$\frac{x-2}{x+3}$$

undefined?

SOLUTION   The expression

$$\frac{x-2}{x+3}$$

is undefined when the denominator is zero. That is, when

$$x+3=0 \quad \text{or} \quad x=-3.$$

In this text we assume that the variables in any fraction may not be assigned values that will result in a value of zero for the denominator (a division by zero). This assumption saves us the trouble of specifically stating this each time we work with a fraction that has variables in the denominator.

In Section 1.4 you learned how to determine the equality of rational numbers. For example, you know that

$$\frac{2}{3}=\frac{10}{15},$$

and that therefore $\frac{2}{3}$ can be used interchangeably with $\frac{10}{15}$ (or with $\frac{4}{6}$, or with $\frac{6}{9}$, etc.) in any expression in which $\frac{2}{3}$ appears. In order to extend this concept to rational expressions we must define the equality (or equivalence) of algebraic fractions.

We say that two algebraic fractions are **equivalent** if they give the same real numbers for every value assigned to their variables for which both fractions are defined. When two fractions are equivalent, we can represent this by writing an *equals* sign between them.

For example,

$$\frac{3y}{xy}=\frac{3}{x},$$

because both fractions have the same values for all nonzero values of x and y. However,

$$\frac{3y}{xy}\neq\frac{3y}{x},$$

because these fractions have different values for some nonzero values of x and y (for example, if $x=4$ and $y=2$).

Notice that in the two equalities,

$$\frac{2}{3} = \frac{10}{15} \quad \text{and} \quad \frac{3y}{xy} = \frac{3}{x},$$

the "cross products" are equal:

$$(2)(15) = (3)(10) \quad \text{and} \quad (3y)(x) = (3)(xy).$$

These examples illustrate the following property:

Property 1

Two fractions

$$\frac{P}{Q} \quad \text{and} \quad \frac{R}{S}$$

are **equivalent** when the cross products PS and QR are equal. That is,

$$\frac{P}{Q} = \frac{R}{S} \quad \text{if and only if} \quad PS = QR.$$

If two algebraic fractions are equal (or equivalent) they can be used interchangeably in any expression in which either fraction appears.

EXAMPLE 2   Indicate whether the given pair of fractions are equivalent.

(a) $\dfrac{4}{6}$ and $\dfrac{2}{3}$

(b) $\dfrac{14a}{56a^3}$ and $\dfrac{1}{4a^2}$

(c) $\dfrac{14y}{7y + 21}$ and $\dfrac{2y}{y + 3}$

(d) $\dfrac{15x}{10x + 4}$ and $\dfrac{3x}{2x + 1}$

SOLUTION   (a) Because $4 \cdot 3 = 6 \cdot 2 = 12$, it follows that $\dfrac{4}{6} = \dfrac{2}{3}$.

(b) Because $(14a)(4a^2) = (56a^3)(1) = 56a^3$, it follows that $\dfrac{14a}{56a^3} = \dfrac{1}{4a^2}$.

(c) Because $(14y)(y + 3) = (7y + 21)(2y) = 14y^2 + 42y$, it follows that

$$\frac{14y}{7y + 21} = \frac{2y}{y + 3}.$$

(d) $(15x)(2x + 1) = 30x^2 + 15x$ and $(10x + 4)(3x) = 30x^2 + 12x.$ Thus, $(15x)(2x + 1) \neq (10x + 4)(3x)$, and

$$\frac{15x}{10x + 4} \neq \frac{3x}{2x + 1}.$$

Property 1 verifies the following principle:

Fundamental Principle of Fractions

$$\text{If } Q \neq 0 \text{ and } K \neq 0, \text{ then } \frac{PK}{QK} = \frac{P}{Q}.$$

The fundamental principle of fractions states that when the numerator and the denominator of a given fraction are divided by the same *nonzero* expression, an equivalent fraction is obtained. In other words, we can "divide out" common factors from the numerator *and* the denominator of a fraction without changing the value of the fraction.

A fraction is said to be **reduced to lowest terms,** or **simplified,** if the numerator and the denominator have no common factors (other than 1 and -1). Thus, to **simplify** a fraction, first we factor both the numerator and the denominator. Then we divide both the numerator and the denominator by any factors they have in common.

For example, to reduce

$$\frac{9x + 15}{3x^2 + 5x}$$

to lowest terms, we begin by factoring both the numerator and the denominator:

$$\frac{9x + 15}{3x^2 + 5x} = \frac{3(3x + 5)}{x(3x + 5)}.$$

Next, we divide both the numerator and the denominator by $3x + 5$:

$$\frac{3(3x + 5)}{x(3x + 5)} = \frac{3(\cancel{3x + 5})}{x(\cancel{3x + 5})} = \frac{3}{x}.$$

The slanted lines drawn through $3x + 5$ in the numerator and the denominator indicate that we have divided out, or cancelled, the common factor, $3x + 5$.

In Examples 3–9, reduce each fraction to lowest terms.

EXAMPLE 3
$$\frac{36}{44}$$

SOLUTION
$$\frac{36}{44} = \frac{(9)(4)}{(11)(4)} = \frac{9}{11}$$

EXAMPLE 4
$$\frac{28x^3y}{21xy^2}$$

SOLUTION
$$\frac{28x^3y}{21xy^2} = \frac{4x^2(7xy)}{3y(7xy)} = \frac{4x^2}{3y}$$

EXAMPLE 5
$$\frac{y^2 - 3y}{y^2 - 9}$$

SOLUTION   Factoring the numerator and the denominator, we obtain

$$\frac{y^2 - 3y}{y^2 - 9} = \frac{y(y - 3)}{(y + 3)(y - 3)} = \frac{y}{y + 3}.$$

EXAMPLE 6
$$\frac{c^2 + 4c - 21}{c^2 - c - 6}$$

SOLUTION   Factoring, we obtain

$$\frac{c^2 + 4c - 21}{c^2 - c - 6} = \frac{(c + 7)(c - 3)}{(c + 2)(c - 3)} = \frac{c + 7}{c + 2}.$$

EXAMPLE 7
$$\frac{25 - w^2}{w^2 - 3w - 10}$$

SOLUTION   Factoring, we obtain

$$\frac{25 - w^2}{w^2 - 3w - 10} = \frac{(5 - w)(5 + w)}{(w - 5)(w + 2)}.$$

Notice that $5 - w = -(w - 5)$. Thus, we can write

$$\frac{(5 - w)(5 + w)}{(w - 5)(w + 2)} = \frac{-(w - 5)(5 + w)}{(w - 5)(w + 2)}.$$

Therefore,

$$\frac{25 - w^2}{w^2 - 3w - 10} = \frac{(5 - w)(5 + w)}{(w - 5)(w + 2)} = \frac{-(w - 5)(5 + w)}{(w - 5)(w + 2)} = \frac{-(5 + w)}{w + 2} = \frac{-5 - w}{w + 2}.$$

EXAMPLE 8 $\dfrac{6p^2 - 7p - 3}{4p^2 - 8p + 3}$

SOLUTION   Factoring, we obtain

$$\frac{6p^2 - 7p - 3}{4p^2 - 8p + 3} = \frac{(3p + 1)(2p - 3)}{(2p - 1)(2p - 3)} = \frac{3p + 1}{2p - 1}.$$

EXAMPLE 9 $\dfrac{x^2 y^4 - x^4 y^2}{x^2 y^4 + 2x^3 y^3 + x^4 y^2}$

SOLUTION   Factoring, we obtain

$$\frac{x^2 y^4 - x^4 y^2}{x^2 y^4 + 2x^3 y^3 + x^4 y^2} = \frac{x^2 y^2 (y^2 - x^2)}{x^2 y^2 (y^2 + 2xy + x^2)} = \frac{x^2 y^2 (y - x)(y + x)}{x^2 y^2 (y + x)(y + x)}$$
$$= \frac{y - x}{y + x}.$$

If we are given a fraction P/Q, and we multiply the numerator and the denominator of the fraction by the same nonzero expression K, we will always obtain a fraction that is equivalent to the original one, that is:

$$\boxed{\text{If } Q \neq 0 \text{ and } K \neq 0, \text{ then } \frac{P}{Q} = \frac{PK}{QK}.}$$

The following example illustrates this rule.

EXAMPLE 10   Find the missing expression for each of the equivalent fractions.

(a) $\dfrac{3}{4} = \dfrac{?}{20t}$ (b) $\dfrac{3a}{b} = \dfrac{?}{ba}$ (c) $\dfrac{x + 3}{x - 2} = \dfrac{?}{x^2 - 4}$

SOLUTION   (a) First we must determine by what factor the original denominator, 4, was multiplied to give the new denominator, $20t$. Then we multiply the original numerator, 3, by this same factor to obtain the missing numerator. Because $20t = 4 \cdot 5t$, we have

$$\frac{3}{4} = \frac{3 \cdot 5t}{4 \cdot 5t} = \frac{15t}{20t}$$

and the unknown expression is $15t$.

(b) The original denominator, b, must have been multiplied by a to get the new denominator, ba. Therefore, we multiply the original numerator, $3a$, by a:

$$\frac{3a}{b} = \frac{3a(a)}{b(a)} = \frac{3a^2}{ba},$$

and the unknown expression is $3a^2$.

(c) We can factor $x^2 - 4$ as $(x - 2)(x + 2)$. Therefore, we multiply the numerator and the denominator of the original fraction by $x + 2$:

$$\frac{x + 3}{x - 2} = \frac{(x + 3)(x + 2)}{(x - 2)(x + 2)} = \frac{x^2 + 5x + 6}{x^2 - 4},$$

and the unknown expression is $x^2 + 5x + 6$.

Signs of Fractions

Every fraction has three signs associated with it: the sign of the numerator, the sign of the denominator, and the sign of the fraction. Two of these signs may be changed without changing the value of the fraction. In general, we have the following rules:

$$\text{(i)} \quad \frac{-P}{Q} = \frac{P}{-Q} = -\frac{P}{Q}$$

$$\text{(ii)} \quad \frac{P}{Q} = -\frac{-P}{Q} = -\frac{P}{-Q}$$

$$\text{(iii)} \quad \frac{P}{Q} = \frac{-P}{-Q}$$

EXAMPLE 11   Write each of the following fractions as an equivalent fraction with denominator $x - y$.

(a) $\dfrac{1}{y - x}$ (b) $\dfrac{-xy}{y - x}$ (c) $\dfrac{a - b}{y - x}$ (d) $-\dfrac{1}{y - x}$ (e) $-\dfrac{-1}{y - x}$

SOLUTION   (a) $\dfrac{1}{y-x} = \dfrac{-1}{-(y-x)} = \dfrac{-1}{x-y}$ (by Rule iii)

(b) $\dfrac{-xy}{y-x} = \dfrac{xy}{-(y-x)} = \dfrac{xy}{x-y}$ (by Rule i)

(c) $\dfrac{a-b}{y-x} = \dfrac{-(a-b)}{-(y-x)} = \dfrac{b-a}{x-y}$ (by Rule iii)

(d) $-\dfrac{1}{y-x} = \dfrac{1}{-(y-x)} = \dfrac{1}{x-y}$ (by Rule i)

(e) $-\dfrac{-1}{y-x} = -\dfrac{1}{-(y-x)} = -\dfrac{1}{x-y}$ (by Rule ii)

PROBLEM SET 3.1

In problems 1–10, determine all values of the variable for which the fraction is not defined.

1. $\dfrac{7x}{x-3}$

2. $\dfrac{y+2}{y+6}$

3. $\dfrac{2}{y+10}$

4. $\dfrac{-11}{12-6x}$

5. $\dfrac{2x-4}{(x-6)(x+7)}$

6. $\dfrac{x^2-1}{(x+3)(x-2)}$

7. $\dfrac{t+2}{(t+2)(t-8)}$

8. $\dfrac{1-u}{(2-u)(6+u)}$

9. $\dfrac{v^2+2v-3}{(v+8)(v+9)(v-4)}$

10. $\dfrac{t^3+1}{(t+2)(t-5)(t+7)}$

In problems 11–20, use the property that $P/Q = R/S$ if and only if $PS = QR$, to determine whether the given pairs of fractions are equivalent.

11. $\dfrac{5}{4}$ and $\dfrac{10}{8}$

12. $\dfrac{-7}{6}$ and $\dfrac{21}{-18}$

13. $\dfrac{7}{9}$ and $\dfrac{8}{10}$

14. $\dfrac{14}{-9}$ and $\dfrac{-12}{7}$

15. $\dfrac{5}{x}$ and $\dfrac{15x}{3x^2}$

16. $\dfrac{a+b}{x}$ and $\dfrac{7a+b}{7x}$

17. $\dfrac{v+2}{v^2-4}$ and $\dfrac{3}{3v-12}$

18. $\dfrac{m+2n}{5}$ and $\dfrac{2m^2-8n^2}{10m-20n}$

19. $\dfrac{x^2-9}{x-3}$ and $\dfrac{2x+6}{2}$

20. $\dfrac{x+y}{1}$ and $\dfrac{x^3+y^3}{x^2-xy+y^2}$

In problems 21–40, reduce each fraction to lowest terms.

21. $\dfrac{15}{18}$

22. $\dfrac{65}{26}$

23. $\dfrac{25x^2y^5}{45x^3y^2}$

24. $\dfrac{14a^5b^2c^6}{7a^2b^6c^3}$

25. $\dfrac{m^2 + m}{m^2 - m}$

26. $\dfrac{x^2 - 9}{x^2 - 6x + 9}$

27. $\dfrac{t^2 + 2t - 3}{t^2 + 5t + 6}$

28. $\dfrac{y^2 + 4y}{y^2 + 6y + 8}$

29. $\dfrac{4x^2 - 9}{6x^2 - 9x}$

30. $\dfrac{4v^2 - 1}{8v^3 - 1}$

31. $\dfrac{v^2 + v - 12}{v^2 + 4v - 21}$

32. $\dfrac{x + x^2 - y - xy}{x^2 - 2xy + y^2}$

33. $\dfrac{3x^2 + 7x + 4}{3x^2 - 5x - 12}$

34. $\dfrac{10t^2 + 29t - 21}{4t^2 + 12t - 7}$

35. $\dfrac{4u^2 - 9}{8u^3 - 27}$

36. $\dfrac{7x^2 - 5xy}{49x^3 - 25xy^2}$

37. $\dfrac{3x^3 - 3xy^2}{3xy^2 + 3x^2y - 6x^3}$

38. $\dfrac{3 + 13m - 10m^2}{2m^2 + 5m - 12}$

39. $\dfrac{xz + xw - yz - yw}{xy + xz - y^2 - yz}$

40. $\dfrac{x^3 - 2x^2 + 5x - 10}{3x^5 + 15x^3 - x^2 - 5}$

In problems 41–56, find the missing numerator that will make the two fractions equivalent.

41. $\dfrac{9}{16} = \dfrac{?}{64}$

42. $\dfrac{13}{25} = \dfrac{?}{75}$

43. $\dfrac{5m^3y}{7my^3} = \dfrac{?}{28m^5y^6}$

44. $\dfrac{6}{11st} = \dfrac{?}{33s^4t}$

45. $\dfrac{3}{u + v} = \dfrac{?}{5(u + v)^2}$

46. $\dfrac{9(a - b)}{7a} = \dfrac{?}{7a^4(a - b)}$

47. $\dfrac{6}{x + 4} = \dfrac{?}{x^2 + 3x - 4}$

48. $\dfrac{mn}{m - n} = \dfrac{?}{m^3 - mn^2}$

49. $\dfrac{2t}{2t - 1} = \dfrac{?}{4t^2 - 1}$

50. $\dfrac{2x - 3}{x^2 + 3x + 9} = \dfrac{?}{x^3 - 27}$

51. $\dfrac{y - 3}{y^2 - y + 1} = \dfrac{?}{y^3 + 1}$

52. $\dfrac{2v}{u + 5v} = \dfrac{?}{2u^3 - 50uv^2}$

53. $\dfrac{x + 2}{x - 2} = \dfrac{?}{2x^2 - 7x + 6}$

54. $\dfrac{ts}{5t + s} = \dfrac{?}{10t^2 + 17ts + 3s^2}$

55. $\dfrac{2y + 3x}{3y + 5x} = \dfrac{?}{6y^2 + yx - 15x^2}$

56. $\dfrac{1}{x - y} = \dfrac{?}{x^4y - xy^4}$

In problems 57–62, write an equivalent fraction with the denominator $x - y$.

57. $\dfrac{-7}{y - x}$

58. $\dfrac{8}{-(x - y)}$

59. $-\dfrac{xy}{y - x}$

60. $\dfrac{c - d}{y - x}$

61. $\dfrac{3 - x}{y - x}$

62. $\dfrac{x + 2}{y - x}$

63. The height h of a closed right circular cylinder of radius r is given by the formula

$$h = \frac{S - 2\pi r^2}{2\pi r},$$

where S is the surface area. Find the height of a cylinder if the surface area is 56π square centimeters and the radius is 4 centimeters.

64. In business, the principal amount of investment can be determined by using the formula

$$P = \frac{S}{1 + rt},$$

where P is the principal amount of dollars invested at a rate of r percent and S is the amount of money accumulated after t years. Find P when $S = \$580$, $r = 8\%$, and $t = 2$ years.

3.2 Multiplication and Division of Fractions

The rules for multiplying and dividing rational expressions are the same as the rules for multiplying and dividing numerical fractions.

Multiplication of Fractions

The product of two fractions P/Q and R/S is a fraction whose numerator is the product of the two given numerators, $P \cdot R$, and whose denominator is the product of the two given denominators, $Q \cdot S$. That is,

$$\frac{P}{Q} \cdot \frac{R}{S} = \frac{P \cdot R}{Q \cdot S}$$

For example,

$$\frac{3}{14} \cdot \frac{35}{12} = \frac{(3)(35)}{(14)(12)} = \frac{105}{168} = \frac{(3)(7)(5)}{(2)(7)(3)(4)} = \frac{5}{8}$$

and

$$\frac{2x}{5y^3} \cdot \frac{25y^4}{4x^2} = \frac{(2x)(25y^4)}{(5y^3)(4x^2)} = \frac{50xy^4}{20x^2y^3} = \frac{(5y)(10xy^3)}{(2x)(10xy^3)} = \frac{5y}{2x}.$$

In practice, it is easier to divide out factors common to both the numerator and the denominator before performing the multiplication. In the last example, $2x$ and $5y^3$ are the common factors. Thus,

$$\frac{2x}{5y^3} \cdot \frac{25y^4}{4x^2} = \frac{2x}{5y^3} \cdot \frac{\overset{5y}{\cancel{25y^4}}}{\underset{2x}{\cancel{4x^2}}} = \frac{5y}{2x}.$$

In Examples 1–5, perform each multiplication and simplify the result.

EXAMPLE 1 $\dfrac{7t}{8} \cdot \dfrac{5t}{6}$

SOLUTION

$$\frac{7t}{8} \cdot \frac{5t}{6} = \frac{7t \cdot 5t}{8 \cdot 6} = \frac{35t^2}{48}.$$

EXAMPLE 2

$$\frac{a^3b^2}{a^2b} \cdot \frac{ab^3}{a^3b^2}$$

SOLUTION   We begin by finding the factors that are common to both the numerators and the denominators of the fractions, and then we divide out these factors:

$$\frac{a^3b^2}{a^2b} \cdot \frac{ab^3}{a^3b^2} = \frac{a^3b^2}{a^2b} \cdot \frac{ab^3}{a^3b^2} = \frac{b^2}{a}.$$

EXAMPLE 3

$$\frac{c}{c-1} \cdot \frac{c^2 - 1}{c^3}$$

SOLUTION   First we factor the numerators and the denominators of the fractions. Then we divide out the common factors:

$$\frac{c}{c-1} \cdot \frac{c^2 - 1}{c^3} = \frac{c}{c-1} \cdot \frac{(c-1)(c+1)}{c^3} = \frac{c+1}{c^2}.$$

EXAMPLE 4

$$\frac{6t - 6}{t^2 + 2t} \cdot \frac{t^2 + 4t + 4}{2t^2 + 2t - 4}$$

SOLUTION   First we factor the numerators and the denominators of the fractions. Then we divide out the common factors:

$$\frac{6t - 6}{t^2 + 2t} \cdot \frac{t^2 + 4t + 4}{2t^2 + 2t - 4} = \frac{6(t-1)}{t(t+2)} \cdot \frac{(t+2)(t+2)}{2(t+2)(t-1)} = \frac{3}{t}.$$

EXAMPLE 5

$$\frac{x^2 + xy - 2y^2}{xy} \cdot \frac{5xy^2}{x + 2y} \cdot \frac{1}{5x - 5y}$$

SOLUTION   We factor the numerators and the denominators of the fractions, and then we divide out the common factors:

$$\frac{x^2 + xy - 2y^2}{xy} \cdot \frac{5xy^2}{x + 2y} \cdot \frac{1}{5x - 5y} = \frac{(x+2y)(x-y)}{xy} \cdot \frac{5xy^2}{x+2y} \cdot \frac{1}{5(x-y)} = y.$$

Division of Fractions

To divide two numerical fractions, such as

$$\frac{4}{5} \div \frac{7}{13},$$

we invert the divisor (the fraction after the division sign) and multiply:

$$\frac{4}{5} \div \frac{7}{13} = \frac{4}{5} \cdot \frac{13}{7} = \frac{52}{35}.$$

This rule can be generalized to cover algebraic fractions:

> If P/Q and R/S are fractions, with $R/S \neq 0$, then
>
> $$\frac{P}{Q} \div \frac{R}{S} = \frac{P}{Q} \cdot \frac{S}{R} = \frac{PS}{QR}.$$

We must be careful *not* to divide out any factor common to the numerators and the denominators of the fractions we are going to divide until after the divisor has been inverted. Removing common factors can only be done after the operation has been converted to multiplication. Therefore,

> when you divide one fraction by another, first change the division to multiplication (invert the divisor), and then divide out the common factors.

In Examples 6–11, find the following quotients and simplify.

EXAMPLE 6 $\dfrac{34}{57} \div \dfrac{51}{95}$

SOLUTION   First invert the divisor, changing the division to multiplication:

$$\frac{34}{57} \div \frac{51}{95} = \frac{34}{57} \cdot \frac{95}{51}.$$

Next, divide out all the factors that are common to both the numerators and the denominators of the fractions, and then multiply:

$$\frac{34}{57} \cdot \frac{95}{51} = \frac{(2)(\cancel{17})}{(3)(\cancel{19})} \cdot \frac{(5)(\cancel{19})}{(3)(\cancel{17})} = \frac{10}{9}.$$

EXAMPLE 7 $\dfrac{3a^2}{5b} \div \dfrac{2a^3}{6b^3}$

SOLUTION   Use the division rule and divide out the common factors:

$$\frac{3a^2}{5b} \div \frac{2a^3}{6b^3} = \frac{3a^2}{5b} \cdot \frac{\overset{3b^2}{\cancel{6b^3}}}{\underset{a}{\cancel{2a^3}}} = \frac{9b^2}{5a}$$

EXAMPLE 8 $\dfrac{a + b}{2} \div \dfrac{(a + b)^2}{6}$

SOLUTION   We perform the division operation—invert the divisor, changing the division to multiplication—then we divide out the common factors and multiply:

$$\frac{a + b}{2} \div \frac{(a + b)^2}{6} = \frac{\cancel{a + b}}{\cancel{2}} \cdot \frac{\overset{3}{\cancel{6}}}{\underset{a + b}{\cancel{(a + b)^2}}} = \frac{3}{a + b}.$$

EXAMPLE 9 $\dfrac{x^2 + 5x + 6}{x^2 - 4} \div \dfrac{x^2 + 4x + 4}{x^2 - 4x + 4}$

SOLUTION   We begin by performing the division operation. Next, we factor each numerator and denominator and divide out the common factors:

$$\frac{x^2 + 5x + 6}{x^2 - 4} \div \frac{x^2 + 4x + 4}{x^2 - 4x + 4} = \frac{x^2 + 5x + 6}{x^2 - 4} \cdot \frac{x^2 - 4x + 4}{x^2 + 4x + 4}$$

$$= \frac{(x + 2)(x + 3)}{(x + 2)(x - 2)} \cdot \frac{(x - 2)(x - 2)}{(x + 2)(x + 2)}$$

$$= \frac{(x + 3)(x - 2)}{(x + 2)(x + 2)}.$$

EXAMPLE 10 $\dfrac{w^2 - w}{z^2 - z} \cdot \dfrac{z^2w - zw}{w - 1} \div \dfrac{w^2}{w - 1}$

SOLUTION   We perform the division operation, factor where possible, and divide out the common factors:

$$\frac{w^2 - w}{z^2 - z} \cdot \frac{z^2w - zw}{w - 1} \div \frac{w^2}{w - 1} = \frac{w^2 - w}{z^2 - z} \cdot \frac{z^2w - zw}{w - 1} \cdot \frac{w - 1}{w^2}$$

$$= \frac{w(w - 1)}{z(z - 1)} \cdot \frac{zw(z - 1)}{w - 1} \cdot \frac{w - 1}{w^2}$$

$$= \frac{w - 1}{1}$$

$$= w - 1.$$

EXAMPLE 11
$$\frac{2t - 1}{2t^2 + 2t} \div \left[\frac{6t^4 - 6}{4t^2 + t - 3} \cdot \frac{8t^2 - 10t + 3}{4t^3 - 4t} \right]$$

SOLUTION   First we perform the multiplication operation inside the brackets. Then we perform the division operation and divide out the common factors:

$$\frac{6t^4 - 6}{4t^2 + t - 3} \cdot \frac{8t^2 - 10t + 3}{4t^3 - 4t} = \frac{\overset{3}{\cancel{6}(t^2 - 1)}(t^2 + 1)}{(t + 1)\cancel{(4t - 3)}} \cdot \frac{\cancel{(4t - 3)}(2t - 1)}{\underset{2}{\cancel{4}t(t^2 - 1)}}$$

$$= \frac{3(t^2 + 1)(2t - 1)}{2t(t + 1)}.$$

Thus,

$$\frac{2t - 1}{2t^2 + 2t} \div \left[\frac{6t^4 - 6}{4t^2 + t - 3} \cdot \frac{8t^2 - 10t + 3}{4t^3 - 4t} \right] = \frac{2t - 1}{2t^2 + 2t} \div \frac{3(t^2 + 1)(2t - 1)}{2t(t + 1)}$$

$$= \frac{\cancel{2t - 1}}{\cancel{2t(t + 1)}} \cdot \frac{\cancel{2t(t + 1)}}{3(t^2 + 1)\cancel{(2t - 1)}}$$

$$= \frac{1}{3(t^2 + 1)}.$$

PROBLEM SET 3.2

In problems 1–22, perform each multiplication and reduce the answer to lowest terms.

1. $\dfrac{12}{13} \cdot \dfrac{39x}{60}$

2. $\dfrac{10}{17} \cdot \dfrac{34t}{25}$

3. $\dfrac{7x^3}{8y^4} \cdot \dfrac{16y}{21x^2}$

4. $\dfrac{3uv^3}{7u^2v} \cdot \dfrac{14u^5}{9uv^6}$

5. $\dfrac{5x^2y}{3t^2s} \cdot \dfrac{6ts}{10x^2}$

6. $\dfrac{15xyz}{16a^2} \cdot \dfrac{12a^3}{25x^2yz^2}$

7. $\dfrac{t + 4}{t^2} \cdot \dfrac{5t}{3t + 12}$

8. $\dfrac{m^2 + mn}{mn} \cdot \dfrac{5n}{m^2 - n^2}$

9. $\dfrac{3x + 6}{5x + 5} \cdot \dfrac{10x + 10}{x^2 - 6x - 16}$

10. $\dfrac{c + 2}{c^2 + 8c - 9} \cdot \dfrac{2c + 18}{2c^2 - 8}$

11. $\dfrac{1 - a^2}{a + 1} \cdot \dfrac{7a^2 - 5a - 2}{a^2 - 2a + 1}$

12. $\dfrac{3y^2 - 12}{3y^2 - 3} \cdot \dfrac{y - 1}{2y + 4}$

13. $\dfrac{v^2 - 1}{v - 3} \cdot \dfrac{3v^2 - 8v - 3}{v^2 - 10v + 9}$

14. $\dfrac{a^2 - 9b^2}{a^2 - b^2} \cdot \dfrac{5a - 5b}{a^2 + 6ab + 9b^2}$

15. $\dfrac{x^2 - 144}{x + 4} \cdot \dfrac{x^2 - 16}{x + 12}$

16. $\dfrac{t^2 + 7t + 10}{t^2 + 10t + 25} \cdot \dfrac{t + 5}{t + 2}$

17. $\dfrac{3y^2 - y - 2}{3y^2 + y - 2} \cdot \dfrac{3y^2 - 5y + 2}{3y^2 + 5y + 2}$

18. $\dfrac{u^2v^2 - 9}{w^2 - w - 2} \cdot \dfrac{w^3 - w^2 - 2w}{u^2v^2 - 6uv + 9}$

19. $\dfrac{3a^2 - 12b^2}{(a + b)^2} \cdot \dfrac{2a^2 - 2b^2}{6a^2 - 24ab + 24b^2}$

20. $\dfrac{2x^3 + 16}{5x^3 - 135} \cdot \dfrac{x^2 - 9}{x^2 - 2x + 4}$

21. $\dfrac{x^3 - y^3}{3x + 3y} \cdot \dfrac{6x^2 + 12xy + 6y^2}{2x^2 + 2xy + 2y^2}$

22. $\dfrac{8m^3 + n^3}{3m - 5n} \cdot \dfrac{9m^2 - 25n^2}{4m^2 - 2mn + n^2}$

In problems 23–42, perform each division and reduce the answer to lowest terms.

23. $\dfrac{9}{15} \div \dfrac{27}{25}$

24. $\dfrac{4}{11} \div \dfrac{24}{55}$

25. $\dfrac{4x^3}{y} \div \dfrac{2x}{3y^2}$

26. $\dfrac{5uv^3}{9a^2b} \div \dfrac{25u^3v}{18a^3b^2}$

27. $\dfrac{5t + 10}{t^3} \div \dfrac{t + 2}{t^4}$

28. $\dfrac{2x - y}{x^2 - y^2} \div \dfrac{4x^2 - y^2}{x + y}$

29. $\dfrac{12 - 6y}{7y - 21} \div \dfrac{2y - 4}{y^2 - 9}$

30. $\dfrac{9a^2 - 1}{a + 1} \div \dfrac{6a + 2}{2a + 2}$

31. $\dfrac{x^3 + 3x}{2x - 1} \div \dfrac{x^2 + 3}{x + 1}$

32. $\dfrac{2c}{c^3d^2 - c^2d^3} \div \dfrac{4c}{c^2 - d^2}$

33. $\dfrac{a^2 - 4a + 4}{a + 2} \div \dfrac{a^2 - 4}{3a + 6}$

34. $\dfrac{x^3 - 4x^2 + 3x}{x + 2} \div (x^2 - 3x)$

35. $\dfrac{15u + 15v}{u^2 + 6uv + 9v^2} \div \dfrac{5u^2 - 5v^2}{u^2 - 9v^2}$

36. $\dfrac{p^2 - 3}{p^3 - 4p} \div \dfrac{p^4 - 9}{p^2 - 4}$

37. $\dfrac{2x^2 - 5x - 3}{3x^2 - 5x - 2} \div \dfrac{2x^2 + 11x + 5}{x^2 + 3x - 10}$

38. $\dfrac{a^2 + 8a + 16}{a^2 - 8a + 16} \div \dfrac{a^3 + 4a^2}{a^2 - 16}$

39. $\dfrac{6y^2 + 7y - 3}{9y^2 - 25} \div \dfrac{12y^2 - y - 1}{12y^2 + 20y}$

40. $\dfrac{6x^2 + 11x - 10}{3x^2 - 5x - 12} \div \dfrac{2x^2 + 9x + 10}{3x^2 + 10x + 8}$

41. $\dfrac{8x^3 - 27}{9x^2 - 3x + 1} \div \dfrac{4x^2 + 6x + 9}{27x^3 + 1}$

42. $\dfrac{u^4 - 8u}{3u^2 + 6u + 12} \div \dfrac{u^4 - 4u^2}{u^2 + 4u + 4}$

In problems 43–50, perform the indicated operations and reduce the answer to lowest terms.

43. $\left[\dfrac{x - 1}{x^2 - 4} \cdot \dfrac{2x + 4}{x^2 - 1} \right] \div \dfrac{2x + 2}{x^2 - 4x + 4}$

44. $\left[\dfrac{t - 3}{t^2 + 2t - 3} \cdot \dfrac{t^2 - 2t + 1}{t^2 - 2t - 3} \right] \div \dfrac{t^2 - 9}{t^2 - 1}$

45. $\left[\dfrac{b - 1}{21 - 4a - a^2} \cdot \dfrac{b - 2}{b - b^3} \right] \div \dfrac{2 - b}{a^2 + 6a - 7}$

46. $\left[\left(\dfrac{x^2 - 1}{x^2} \right)^2 \cdot \dfrac{2}{x - 1} \right] \div \dfrac{x^2 + 2x + 1}{x^3}$

47. $\dfrac{7}{5v^2(v + 3)} \div \left[\dfrac{v^2 - 5v + 6}{8v^2} \cdot \dfrac{21}{5v^2 - 45} \right]$

48. $\dfrac{x^3 - 16x}{x^2 - 3x - 10} \div \left[\dfrac{x^3 + x^2 - 12x}{x^2 + 5x + 6} \cdot \dfrac{x^2 - 4}{x^2 - 9} \right]$

49. $\dfrac{2y^2 + 5y - 3}{6y^2 - 5y - 6} \div \left[\dfrac{2y^2 + 9y - 5}{12y^2 - y - 6} \cdot \dfrac{4y^2 + 9y - 9}{2y^2 + y - 6} \right]$

50. $\dfrac{3m^2 + m - 10}{8m^2 - 10m - 3} \div \left[\dfrac{2m^2 + 7m + 6}{8m^2 - 2m - 1} \div \dfrac{4m^2 - 9}{3m^2 + 10m - 25} \right]$

3.3 Addition and Subtraction of Fractions

In Section 1.5, we showed that when we add (or subtract) numerical fractions having the same denominators, we add (or subtract) the numerator and keep the denominator. If the denominators of the fractions are different, we replace the fractions with equivalent fractions that have the same denominators, and then we add or subtract the numerators. The same procedures are used for rational expressions.

To **add** (or **subtract**) fractions with the same denominators, add (or subtract) the numerators and keep the common denominator. Thus, if $Q \neq 0$, then:

$$\text{(i)} \quad \frac{P}{Q} + \frac{R}{Q} = \frac{P + R}{Q} \qquad\qquad \text{(ii)} \quad \frac{P}{Q} - \frac{P}{Q} = \frac{P - R}{Q}$$

In Examples 1–6, perform each operation and simplify the result.

EXAMPLE 1 $\dfrac{3}{7x} + \dfrac{2}{7x}$

SOLUTION

$$\frac{3}{7x} + \frac{2}{7x} = \frac{3 + 2}{7x} = \frac{5}{7x}.$$

equal denominators

EXAMPLE 2 $\dfrac{7}{9y} - \dfrac{4}{9y}$

SOLUTION

$$\frac{7}{9y} - \frac{4}{9y} = \frac{7 - 4}{9y} = \frac{3}{9y} = \frac{1}{3y}.$$

equal denominators

EXAMPLE 3 $\dfrac{7}{c^2} + \dfrac{13c}{c^2}$

SOLUTION

$$\frac{7}{c^2} + \frac{13c}{c^2} = \frac{7 + 13c}{c^2}.$$

Note that the fraction $(7 + 13c)/c^2$ cannot be reduced because we cannot factor the numerator.

EXAMPLE 4 $\dfrac{25}{5 - y} - \dfrac{y^2}{5 - y}$

SOLUTION

$$\frac{25}{5 - y} - \frac{y^2}{5 - y} = \frac{25 - y^2}{5 - y} = \frac{(5 - y)(5 + y)}{5 - y} = 5 + y.$$

EXAMPLE 5

$$\frac{3x-1}{x^2+x-12}+\frac{2x-14}{x^2+x-12}$$

SOLUTION

$$\frac{3x-1}{x^2+x-12}+\frac{2x-14}{x^2+x-12}=\frac{3x-1+2x-14}{x^2+x-12}=\frac{5x-15}{x^2+x-12}$$
$$=\frac{5(x-3)}{(x-3)(x+4)}=\frac{5}{x+4}.$$

EXAMPLE 6

$$\frac{8t+5}{6t^2-11t+3}-\frac{5t+6}{6t^2-11t+3}$$

SOLUTION

$$\frac{8t+5}{6t^2-11t+3}-\frac{5t+6}{6t^2-11t+3}=\frac{8t+5-(5t+6)}{6t^2-11t+3}=\frac{8t+5-5t-6}{6t^2-11t+3}$$
$$=\frac{3t-1}{(3t-1)(2t-3)}=\frac{1}{2t-3}.$$

To add (or subtract) fractions with different denominators, we change the fractions to equivalent fractions with the same denominators. We then follow the rule for adding (or subtracting) fractions with like denominators. For example:

$$\frac{3}{4}+\frac{5}{7}=\frac{3\cdot7}{4\cdot7}+\frac{5\cdot4}{7\cdot4}=\frac{21}{28}+\frac{20}{28}=\frac{21+20}{28}=\frac{41}{28}$$

and

$$\frac{3}{4}-\frac{2}{5}=\frac{3\cdot5}{4\cdot5}-\frac{2\cdot4}{5\cdot4}=\frac{15}{20}-\frac{8}{20}=\frac{15-8}{20}=\frac{7}{20}.$$

In general, if $Q \neq S$, we can add and subtract the fractions P/Q and R/S as follows:

Rule 1. $\dfrac{P}{Q}+\dfrac{R}{S}=\dfrac{PS+RQ}{QS}$	Rule 2. $\dfrac{P}{Q}-\dfrac{R}{S}=\dfrac{PS-RQ}{QS}$

In Examples 7–9 perform each operation.

EXAMPLE 7

$$\frac{2}{5}+\frac{5}{6}$$

SOLUTION

Following Rule 1 of addition, we have

$$\frac{2}{5}+\frac{5}{6}=\frac{2(6)+5(5)}{5(6)}=\frac{12+25}{30}=\frac{37}{30}.$$

EXAMPLE 8 $\dfrac{5}{x} - \dfrac{3}{y}$

SOLUTION Following Rule 2 of subtraction, we have

$$\frac{5}{x} - \frac{3}{y} = \frac{5y - 3x}{xy}.$$

EXAMPLE 9 $\dfrac{7}{10x} + \dfrac{8}{15x}$

SOLUTION Following Rule 1 of addition, we have

$$\frac{7}{10x} + \frac{8}{15x} = \frac{7(15x) + 8(10x)}{(10x)(15x)}$$

$$= \frac{105x + 80x}{150x^2} = \frac{185x}{150x^2} = \frac{\cancel{5}(37)\cancel{x}}{\cancel{5}(30)\cancel{x}^2} = \frac{37}{30x}.$$

We could have saved steps in Example 9 by changing each fraction to an equivalent fraction that had the **least common denominator** (LCD) of the original denominators. We can determine the LCD of two fractions by using the following procedure:

Finding the LCD

Step 1. Factor each denominator completely into a product of prime factors or into a product of prime factors and -1.

Step 2. List each prime factor the greatest number of times it appears in any one factored denominator. The product of the listed factors is the least common denominator, or LCD.

Returning to Example 9,

$$\frac{7}{10x} + \frac{8}{15x},$$

we determine the LCD by factoring the first denominator, $10x$, as

$$2 \cdot 5 \cdot x,$$

and the second denominator, $15x$, as

$$3 \cdot 5 \cdot x.$$

We list each factor the greatest number of times it appears in any one factored denominator to obtain the LCD:

$$2 \cdot 5 \cdot x \cdot 3 \quad \text{or} \quad 30x.$$

We must now multiply the numerator and the denominator of each of the original fractions by an expression that will make each denominator equal to the LCD:

$$\frac{7}{10x} + \frac{8}{15x} = \frac{7(3)}{10x(3)} + \frac{8(2)}{15x(2)} = \frac{21}{30x} + \frac{16}{30x}$$

$$= \frac{21 + 16}{30x} = \frac{37}{30x}.$$

In Examples 10–16, perform each operation.

EXAMPLE 10 $\dfrac{3}{4x} + \dfrac{5}{6xy}$

SOLUTION   First we find the LCD of the fractions. The following table is helpful in constructing the LCD:

Denominators	Powers of Prime Factors			
$4x$	2^2		x	
$6xy$	2	3	x	y
LCD	2^2	3	x	y

Thus the LCD is

$$2^2 \cdot 3 \cdot x \cdot y = 12xy.$$

Now we multiply the numerator and the denominator of each fraction by an expression that will make each denominator equal to the LCD:

$$\frac{3}{4x} + \frac{5}{6xy} = \frac{3(3y)}{4x(3y)} + \frac{5(2)}{6xy(2)}$$

$$= \frac{9y}{12xy} + \frac{10}{12xy}$$

$$= \frac{9y + 10}{12xy}.$$

EXAMPLE 11 $\dfrac{5}{a^2b} - \dfrac{3}{ab^3}$

SOLUTION   Note that the denominators are in factored form. Therefore, the LCD is

$$a^2b^3,$$

which is the product of the highest power of a and the highest power of b that occur in either denominator. Now we have

$$\frac{5}{a^2b} - \frac{3}{ab^3} = \frac{5(b^2)}{a^2b(b^2)} - \frac{3(a)}{ab^3(a)}$$

$$= \frac{5b^2 - 3a}{a^2b^3}.$$

EXAMPLE 12 $\dfrac{5}{y^2 - y} - \dfrac{4}{y^2 - 1}$

SOLUTION   The following table shows how the LCD of the fractions is obtained:

Denominators	Prime Factors		
$y^2 - y$	y	$y - 1$	
$y^2 - 1$		$y - 1$	$y + 1$
LCD	y	$y - 1$	$y + 1$

Thus, the LCD is

$$y(y - 1)(y + 1),$$

and we have

$$\frac{5}{y^2 - y} - \frac{4}{y^2 - 1} = \frac{5}{y(y - 1)} - \frac{4}{(y - 1)(y + 1)}$$

$$= \frac{5(y + 1)}{y(y - 1)(y + 1)} - \frac{4y}{(y - 1)(y + 1)y}$$

$$= \frac{5(y + 1) - 4y}{y(y - 1)(y + 1)}$$

$$= \frac{5y + 5 - 4y}{y(y - 1)(y + 1)}$$

$$= \frac{y + 5}{y(y - 1)(y + 1)}.$$

EXAMPLE 13
$$\frac{6}{w^2 - 2w - 8} + \frac{5}{w^2 + 2w}$$

SOLUTION

The following table shows how the LCD is obtained:

Denominators	Prime Factors		
$w^2 - 2w - 8$	$w + 2$	$w - 4$	
$w^2 + 2w$	$w + 2$		w
LCD	$w + 2$	$w - 4$	w

Thus, the LCD is

$$(w + 2)(w - 4)w,$$

and we have

$$\frac{6}{w^2 - 2w - 8} + \frac{5}{w^2 + 2w} = \frac{6}{(w + 2)(w - 4)} + \frac{5}{w(w + 2)}$$

$$= \frac{6w}{(w + 2)(w - 4)w} + \frac{5(w - 4)}{w(w + 2)(w - 4)}$$

$$= \frac{6w + 5(w - 4)}{(w + 2)(w - 4)w}$$

$$= \frac{6w + 5w - 20}{(w + 2)(w - 4)w}$$

$$= \frac{11w - 20}{(w + 2)(w - 4)w}.$$

EXAMPLE 14
$$\frac{5}{a - 3} - \frac{5}{3 - a}$$

SOLUTION

Here, we observe that the denominators are negatives of each other:

$$a - 3 = -(3 - a).$$

Therefore, we can change one denominator into the other by applying Rule (i), page 123, which states that

$$-\frac{P}{Q} = \frac{P}{-Q}.$$

$$\frac{5}{a-3} - \frac{5}{3-a} = \frac{5}{a-3} + \frac{5}{-(3-a)}$$

$$= \frac{5}{a-3} + \frac{5}{a-3}$$

$$= \frac{5+5}{a-3} = \frac{10}{a-3}.$$

EXAMPLE 15 $\dfrac{x}{x+1} - \dfrac{x}{x-1} + \dfrac{2}{x^2-1}$

SOLUTION   The following table shows how the LCD is obtained:

Denominators	Prime Factors	
$x+1$	$x+1$	
$x-1$		$x-1$
x^2-1	$x+1$	$x-1$
LCD	$x+1$	$x-1$

Therefore, the LCD is

$$(x-1)(x+1),$$

and we have

$$\frac{x}{x+1} - \frac{x}{x-1} + \frac{2}{x^2-1} = \frac{x}{x+1} - \frac{x}{x-1} + \frac{2}{(x-1)(x+1)}$$

$$= \frac{x(x-1)}{(x+1)(x-1)} - \frac{x(x+1)}{(x-1)(x+1)} + \frac{2}{(x-1)(x+1)}$$

$$= \frac{x(x-1) - x(x+1) + 2}{(x+1)(x-1)}$$

$$= \frac{x^2 - x - x^2 - x + 2}{(x+1)(x-1)}$$

$$= \frac{-2x+2}{(x+1)(x-1)}$$

$$= \frac{-2(x-1)}{(x+1)(x-1)}$$

$$= \frac{-2}{x+1}.$$

EXAMPLE 16

$$\frac{p}{p^2 + 6p + 5} - \frac{2}{p^2 + 4p - 5} + \frac{3}{p^2 - 1}$$

SOLUTION

The following table shows how the LCD of the fractions is obtained:

Denominators	Prime Factors		
$p^2 + 6p + 5$	$p + 1$	$p + 5$	
$p^2 + 4p - 5$		$p + 5$	$p - 1$
$p^2 - 1$	$p + 1$		$p - 1$
LCD	$p + 1$	$p + 5$	$p - 1$

Thus, the LCD is

$$(p + 1)(p + 5)(p - 1),$$

and we have

$$\frac{p}{p^2 + 6p + 5} - \frac{2}{p^2 + 4p - 5} + \frac{3}{p^2 - 1}$$

$$= \frac{p}{(p + 1)(p + 5)} - \frac{2}{(p + 5)(p - 1)} + \frac{3}{(p + 1)(p - 1)}$$

$$= \frac{p(p - 1)}{(p + 1)(p + 5)(p - 1)} - \frac{2(p + 1)}{(p + 5)(p - 1)(p + 1)} + \frac{3(p + 5)}{(p + 1)(p - 1)(p + 5)}$$

$$= \frac{p(p - 1) - 2(p + 1) + 3(p + 5)}{(p + 1)(p + 5)(p - 1)}$$

$$= \frac{p^2 - p - 2p - 2 + 3p + 15}{(p + 1)(p + 5)(p - 1)}$$

$$= \frac{p^2 + 13}{(p + 1)(p + 5)(p - 1)}.$$

PROBLEM SET 3.3

In problems 1–20, perform each operation and simplify the result.

1. $\dfrac{5}{8x} + \dfrac{1}{8x}$

2. $\dfrac{7}{12t} + \dfrac{5}{12t}$

3. $\dfrac{6}{5x} + \dfrac{14}{5x}$

4. $\dfrac{3}{2y^2} + \dfrac{7}{2y^2}$

5. $\dfrac{7}{12} - \dfrac{3}{12}$

6. $\dfrac{9}{14} - \dfrac{2}{14}$

7. $\dfrac{t - 1}{4} + \dfrac{t + 1}{4}$

8. $\dfrac{3x - 1}{5} + \dfrac{2x + 1}{5}$

9. $\dfrac{2x}{x^2-4} - \dfrac{4}{x^2-4}$

10. $\dfrac{2t}{4t^2-9} - \dfrac{-3}{4t^2-9}$

11. $\dfrac{36}{6-v} + \dfrac{-v^2}{6-v}$

12. $\dfrac{4}{16-x^2} + \dfrac{x}{16-x^2}$

13. $\dfrac{3m}{m^2+m-2} + \dfrac{6}{m^2+m-2}$

14. $\dfrac{12y}{2y^2+y-1} - \dfrac{6}{2y^2+y-1}$

15. $\dfrac{3x}{3x^2-10x-8} - \dfrac{12}{3x^2-10x-8}$

16. $\dfrac{4t}{4t^2+5t-6} + \dfrac{8}{4t^2+5t-6}$

17. $\dfrac{15}{8u^2-6u-5} - \dfrac{12u}{8u^2-6u-5}$

18. $\dfrac{4y}{21x^2+32xy-5y^2} - \dfrac{28x}{21x^2+32xy-5y^2}$

19. $\dfrac{8u^3}{4u^2+6uv+9v^2} - \dfrac{27v^3}{4u^2+6uv+9v^2}$

20. $\dfrac{16t^2}{64t^3+27s^3} - \dfrac{12ts-9s^2}{64t^3+27s^3}$

In problems 21–34, perform each operation by using Rule 1 or 2 from page 133. Simplify the result.

21. $\dfrac{5}{8x} + \dfrac{3}{7x}$

22. $\dfrac{2}{3t} + \dfrac{5}{8t}$

23. $\dfrac{7}{12t} - \dfrac{1}{5t}$

24. $\dfrac{15}{16y} - \dfrac{3}{7y}$

25. $\dfrac{2x}{3} + \dfrac{9x}{10}$

26. $\dfrac{3t}{4} + \dfrac{2t}{5}$

27. $\dfrac{5u}{7} - \dfrac{2u}{3}$

28. $\dfrac{3-x}{4} - \dfrac{x+1}{2}$

29. $\dfrac{9}{y-5} + \dfrac{6}{y-3}$

30. $\dfrac{2a}{a-3} + \dfrac{3}{a+7}$

31. $\dfrac{m}{m-3} - \dfrac{2}{m+3}$

32. $\dfrac{2y}{y-2} - \dfrac{y}{y+1}$

33. $\dfrac{x}{3x+2} + \dfrac{1}{x-4}$

34. $\dfrac{2u}{4u+3} + \dfrac{u}{2u+5}$

In problems 35–60, perform each operation and simplify the result.

35. $\dfrac{5}{t^2s} + \dfrac{3}{2ts^3}$

36. $\dfrac{7}{4x^3y^2} + \dfrac{5}{12xy^4}$

37. $\dfrac{12}{5m^3n} - \dfrac{4}{3m^2n^5}$

38. $\dfrac{9}{4u^4v^2} - \dfrac{3}{8v^5w^3}$

39. $\dfrac{3}{x^2+x} + \dfrac{2}{x^2-1}$

40. $\dfrac{3v}{4v^2-1} + \dfrac{1}{2v+1}$

41. $\dfrac{c}{c^2-9} - \dfrac{c-1}{c^2-5c+6}$

42. $\dfrac{t}{t^2+5t-6} + \dfrac{3}{t+6}$

43. $\dfrac{m-5}{m^2-5m-6} + \dfrac{m+4}{m^2-6m}$

44. $\dfrac{2}{a^2-4} + \dfrac{7}{a^2-4a-12}$

45. $\dfrac{x-2}{x^2+10x+16} + \dfrac{x+1}{x^2+9x+14}$

46. $\dfrac{u}{u^2-2uv+v^2} - \dfrac{v}{u^2-v^2}$

47. $\dfrac{3y}{y^2+7y+10} - \dfrac{y}{y^2+y-20}$

48. $\dfrac{1}{3x^2-x-2} - \dfrac{1}{2x^2-x-1}$

49. $\dfrac{4}{3t-2} + \dfrac{1}{2-3t}$

50. $\dfrac{12}{5y-7} - \dfrac{3}{7-5y}$

51. $\dfrac{3x+1}{2x^2-5x-12} - \dfrac{x+4}{6x^2+7x-3}$

52. $\dfrac{2t+3}{3t^2+2t-8} + \dfrac{3t+4}{2t^2+t-6}$

53. $\dfrac{v}{v+2} - \dfrac{v}{v-2} - \dfrac{v^2}{v^2-4}$

54. $\dfrac{x-3}{x+3} - \dfrac{x+3}{3-x} + \dfrac{x^2}{9-x^2}$

55. $\dfrac{2t^2 - t}{3t^2 - 27} - \dfrac{t - 3}{3t - 9} + \dfrac{6t^2}{9 - t^2}$

56. $\dfrac{4}{a^2 - 4} + \dfrac{2}{a^2 - 4a + 4} + \dfrac{1}{a^2 + 4a + 4}$

57. $\dfrac{1}{x^2 - x - 6} + \dfrac{2}{x^2 - 6x + 9} - \dfrac{1}{x^2 + 4x + 4}$

58. $\dfrac{2x^2}{x^4 - 1} + \dfrac{1}{x^2 + 1} + \dfrac{1}{x^2 - 1}$

59. $\dfrac{2y - 1}{3y^2 - y - 2} + \dfrac{y + 2}{y^2 + 2y - 3} - \dfrac{y - 3}{3y^2 + 11y + 6}$

60. $\dfrac{15c}{8c^2 - 26c + 15} - \dfrac{2c}{6c^2 - 13c - 5} - \dfrac{c}{12c^2 - 5c - 3}$

61. Write an expression for the sum of a number x and three times its reciprocal. Then simplify the resulting expression.

62. Write an expression for the sum of the reciprocals of two consecutive integers if the first integer is x. Then simplify the resulting expression.

63. John can do a job in x hours, while it takes Joe $x + 2$ hours to do the same job. What part of the job can John and Joe, working together, do in 1 hour? Simplify the resulting expression.

64. An inlet pipe can fill a pool in x hours, while an outlet pipe can empty it in $x + 3$ hours. If both pipes are open, what fraction of the pool is left unfilled after 1 hour? Simplify the resulting expression.

65. Optometrists use the formula

$$F = \frac{1}{p} + \frac{1}{q}$$

to help determine how strong to make the lenses for eyeglasses. If $p = x + 10$ and $q = x - 2$, find the corresponding value of F, and simplify the resulting expression.

66. The tens digit of a two-digit number is 5 more than the units digit. If the number is divided by the sum of its digits, the quotient is y and the remainder is 6. If x represents the units digit, express y in terms of x and simplify the resulting expression.

3.4 Complex Fractions

We have worked with fractions of the form P/Q, $Q \neq 0$, where P and Q are polynomial expressions (called simple fractions). In this section, we look at **complex fractions**—fractions in which the numerator or the denominator (or both) contain fractions. For example,

$$\frac{\frac{5}{7}}{3}, \quad \frac{\frac{2}{x}}{\frac{11}{x}}, \quad \frac{3 + \frac{3}{y}}{9}, \quad \frac{\frac{5}{9}}{\frac{w}{w - 5}}, \quad \frac{\frac{5}{t}}{7 + \frac{2}{t}}, \quad \text{and} \quad \frac{p + \frac{3}{p}}{\frac{2}{2 - p} - \frac{p}{2 - p}}$$

are complex fractions.

To simplify a complex fraction, we express it as a simple fraction in lowest terms. We do this by using one of the following methods:

Method 1. Express the complex fraction as a quotient of simple fractions: add and subtract as indicated in the numerator and denominator of the complex fraction; then divide and simplify the result.

Method 2. Multiply the numerator and the denominator of the complex fraction by the least common denominator (LCD) of all the fractions in the numerator and the denominator. Simplify the result.

In Examples 1–3, express each complex fraction as a simple fraction and simplify the result.

EXAMPLE 1

$$\dfrac{\dfrac{7}{11}}{\dfrac{5}{44}}$$

SOLUTION   Method 1:

$$\dfrac{\dfrac{7}{11}}{\dfrac{5}{44}} = \dfrac{7}{11} \div \dfrac{5}{44} = \dfrac{7}{11} \cdot \overset{4}{\dfrac{44}{5}} = \dfrac{28}{5}.$$

Method 2: The LCD is 44. Therefore,

$$\dfrac{\dfrac{7}{11}}{\dfrac{5}{44}} = \dfrac{\dfrac{7}{11} \cdot 44}{\dfrac{5}{44} \cdot 44} = \dfrac{7 \cdot 4}{5 \cdot 1} = \dfrac{28}{5}.$$

EXAMPLE 2

$$\dfrac{\dfrac{9}{xy^2}}{\dfrac{12}{x^2y}}$$

SOLUTION   Method 1:

$$\dfrac{\dfrac{9}{xy^2}}{\dfrac{12}{x^2y}} = \dfrac{9}{xy^2} \div \dfrac{12}{x^2y}$$

$$= \overset{3}{\dfrac{9}{xy^2}} \cdot \overset{x}{\dfrac{x^2y}{12}} = \dfrac{3x}{4y}.$$

Method 2: The LCD is x^2y^2, so

$$\dfrac{\dfrac{9}{xy^2}}{\dfrac{12}{x^2y}} = \dfrac{\dfrac{9}{xy^2} \cdot x^2y^2}{\dfrac{12}{x^2y} \cdot x^2y^2}$$

$$= \dfrac{\overset{3}{9x}}{\underset{4}{12y}} = \dfrac{3x}{4y}.$$

EXAMPLE 3
$$\frac{8 + \dfrac{4}{t}}{\dfrac{2t + 1}{12}}$$

SOLUTION

Method 1:

$$\frac{8 + \dfrac{4}{t}}{\dfrac{2t + 1}{12}} = \frac{\dfrac{8t + 4}{t}}{\dfrac{2t + 1}{12}}$$

$$= \frac{8t + 4}{t} \div \frac{2t + 1}{12}$$

$$= \frac{8t + 4}{t} \cdot \frac{12}{2t + 1}$$

$$= \frac{4(2t + 1)}{t} \cdot \frac{12}{2t + 1}$$

$$= \frac{48}{t}.$$

Method 2: The LCD is $12t$. So we have

$$\frac{8 + \dfrac{4}{t}}{\dfrac{2t + 1}{12}} = \frac{\left(8 + \dfrac{4}{t}\right) \cdot 12t}{\left(\dfrac{2t + 1}{12}\right) \cdot 12t}$$

$$= \frac{96t + 48}{2t^2 + t} = \frac{48(2t + 1)}{t(2t + 1)}$$

$$= \frac{48}{t}.$$

In the following examples, we use only Method 2. This method is usually easier to apply to more complicated problems.

In Examples 4–6, simplify each complex fraction.

EXAMPLE 4
$$\frac{c - \dfrac{1}{d}}{1 - \dfrac{c}{d}}$$

SOLUTION

We multiply both the numerator and the denominator of the fraction by d (their LCD):

$$\frac{c - \dfrac{1}{d}}{1 - \dfrac{c}{d}} = \frac{d\left(c - \dfrac{1}{d}\right)}{d\left(1 - \dfrac{c}{d}\right)} = \frac{d(c) - d\left(\dfrac{1}{d}\right)}{d(1) - d\left(\dfrac{c}{d}\right)}$$

$$= \frac{cd - 1}{d - c}.$$

EXAMPLE 5

$$\frac{\dfrac{a}{b} - \dfrac{b}{a}}{\dfrac{1}{a} + \dfrac{1}{b}}$$

SOLUTION The LCD is ab, we have

$$\frac{\dfrac{a}{b} - \dfrac{b}{a}}{\dfrac{1}{a} + \dfrac{1}{b}} = \frac{\left(\dfrac{a}{b} - \dfrac{b}{a}\right)ab}{\left(\dfrac{1}{a} + \dfrac{1}{b}\right)ab} = \frac{\left(\dfrac{a}{b}\right)ab - \left(\dfrac{b}{a}\right)ab}{\left(\dfrac{1}{a}\right)ab + \left(\dfrac{1}{b}\right)ab}$$

$$= \frac{a^2 - b^2}{b + a} = \frac{(a - b)(a + b)}{b + a} = a - b.$$

EXAMPLE 6

$$\frac{\dfrac{1}{t + 1} - 1}{1 - \dfrac{1}{t}}$$

SOLUTION The LCD is $t(t + 1)$, so we have

$$\frac{\dfrac{1}{t + 1} - 1}{1 - \dfrac{1}{t}} = \frac{\left(\dfrac{1}{t + 1} - 1\right)t(t + 1)}{\left(1 - \dfrac{1}{t}\right)t(t + 1)} = \frac{\left(\dfrac{1}{t + 1}\right)t(t + 1) - (1)t(t + 1)}{(1)t(t + 1) - \left(\dfrac{1}{t}\right)t(t + 1)}$$

$$= \frac{t - t(t + 1)}{t(t + 1) - (t + 1)} = \frac{t - t^2 - t}{t^2 + t - t - 1} = \frac{-t^2}{t^2 - 1} = \frac{t^2}{1 - t^2}.$$

PROBLEM SET 3.4

In problems 1–24, simplify each complex fraction.

1. $\dfrac{\dfrac{2}{3}}{\dfrac{4}{5}}$

2. $\dfrac{\dfrac{17}{6}}{\dfrac{34}{9}}$

3. $\dfrac{\dfrac{35x}{24y}}{\dfrac{7x}{9y^2}}$

4. $\dfrac{\dfrac{a^2b^2}{8}}{\dfrac{5ab^3}{16}}$

5. $\dfrac{3 + \dfrac{1}{c}}{2 - \dfrac{3}{c}}$

6. $\dfrac{5 + \dfrac{1}{x}}{1 - \dfrac{1}{x}}$

7. $\dfrac{\dfrac{1}{y}}{2 - \dfrac{1}{y}}$

8. $\dfrac{3 - \dfrac{x}{5}}{5 - \dfrac{x}{3}}$

9. $\dfrac{\dfrac{2}{1+v}}{3+\dfrac{1}{1+v}}$

10. $\dfrac{\dfrac{1}{t+1}-2}{3-\dfrac{1}{t+1}}$

11. $\dfrac{1+\dfrac{2}{m}}{1-\dfrac{4}{m^2}}$

12. $\dfrac{\dfrac{3}{y}-\dfrac{x}{y}}{\dfrac{3}{y}-1}$

13. $\dfrac{\dfrac{x}{y}-\dfrac{y}{x}}{\dfrac{x}{y}+\dfrac{y}{x}}$

14. $\dfrac{\dfrac{u}{v}+1+\dfrac{v}{u}}{\dfrac{u^3-v^3}{uv}}$

15. $\dfrac{\dfrac{1}{a+b}-\dfrac{1}{a-b}}{\dfrac{4}{a^2-b^2}}$

16. $\dfrac{\dfrac{a}{b}-2+\dfrac{b}{a}}{\dfrac{a}{b}+2+\dfrac{b}{a}}$

17. $\dfrac{\dfrac{3}{1-x}+\dfrac{x}{x-1}}{\dfrac{1}{1-x}}$

18. $\dfrac{\dfrac{c-3}{c-2}+\dfrac{c+1}{c+2}}{\dfrac{c^2-3c+6}{c^2-4}}$

19. $\dfrac{\dfrac{2}{t}+\dfrac{1}{t+1}}{\dfrac{3}{t+1}-\dfrac{1}{t}}$

20. $\dfrac{\dfrac{x^2-y^2}{x+y}}{\dfrac{x}{y}+1}$

21. $\dfrac{\dfrac{x-1}{x+1}-\dfrac{x+1}{x-1}}{\dfrac{x-1}{x+1}+\dfrac{x+1}{x-1}}$

22. $\dfrac{m+\dfrac{n}{n-m}}{n+\dfrac{m}{m-n}}$

23. $\dfrac{\dfrac{a}{b}-1+\dfrac{b}{a}}{\dfrac{a^3+b^3}{a^2b+ab^2}}$

24. $\dfrac{\dfrac{2u-v}{2u+v}+\dfrac{2u+v}{2u-v}}{\dfrac{2u-v}{2u+v}-\dfrac{2u+v}{2u-v}}$

25. In electronics, if two resistors of resistance R_1 and R_2 ohms are connected in parallel (Figure 1), then the resistance R of the combination is given by

$$R=\dfrac{1}{\dfrac{1}{R_1}+\dfrac{1}{R_2}}$$

Simplify the right side of the formula, and then find R when $R_1 = 10$ ohms and $R_2 = 15$ ohms.

Figure 1

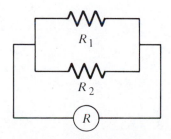

26. The finance charge F is used on the installment purchase of an automobile. If the loan is to be repaid in M dollars per installment in N monthly installments, then the approximate interest rate r is given by

$$r=\dfrac{1}{\dfrac{MN^2}{24F}+\dfrac{MN}{24F}}.$$

Simplify the right-hand side of the formula and then find M when $F = \$120$, $N = 12$ months, and $r = 12\%$.

REVIEW PROBLEM SET

In problems 1–4, determine all values of the variable for which the fraction is not defined.

1. $\dfrac{8 + t}{t - 5}$

2. $\dfrac{3x}{x + 11}$

3. $\dfrac{3x}{(x + 2)(x - 7)}$

4. $\dfrac{y^2 + 1}{(y - 9)(y + 4)}$

In problems 5–10, determine whether the given pair of fractions are equivalent.

5. $\dfrac{3}{uv}$ and $\dfrac{6u}{2u^2v}$

6. $\dfrac{9a}{11b^2}$ and $\dfrac{27a^3b^2}{33a^2b^4}$

7. $\dfrac{c - 4}{c^2 - 16}$ and $\dfrac{1}{c + 4}$

8. $\dfrac{2x - 6}{2}$ and $\dfrac{x^2 - 5x + 6}{x - 3}$

9. $\dfrac{y + 3}{2y + 2}$ and $\dfrac{y - 3}{2y - 2}$

10. $\dfrac{3a - b}{3}$ and $\dfrac{a^3 - b^3}{a^2 + ab + b^2}$

In problems 11–20, reduce each fraction to lowest terms.

11. $\dfrac{12u^3v^5}{18u^7v^2}$

12. $\dfrac{28ts^3}{35t^4s}$

13. $\dfrac{3m + 3n}{9m^2 - 9n^2}$

14. $\dfrac{4y^2 - 9}{2y^2 - 3y}$

15. $\dfrac{x^2 - 7x + 12}{x^2 + 3x - 18}$

16. $\dfrac{2a^2 - 2b^2}{5a^2 + 10ab + 5b^2}$

17. $\dfrac{2t^2 - 7t - 15}{4t^2 - 4t - 15}$

18. $\dfrac{3y^2 - 17y - 28}{3y^2 + 10y + 8}$

19. $\dfrac{x^4 - 8x}{x^4 + 2x^3 + 4x^2}$

20. $\dfrac{ax + ay - bx - by}{bx - by - ax + ay}$

In problems 21–28, find the missing numerator that will make the fractions equivalent.

21. $\dfrac{7x}{9y} = \dfrac{?}{27xy^3}$

22. $\dfrac{5u^2}{11vw} = \dfrac{?}{33v^2w^4}$

23. $\dfrac{3a}{2a - 2b} = \dfrac{?}{2a^2 - 2b^2}$

24. $\dfrac{2c}{3c + 1} = \dfrac{?}{3c^2 - 14c - 5}$

25. $\dfrac{2z + 1}{z - 2} = \dfrac{?}{3z^2 - 4z - 4}$

26. $\dfrac{xy}{x - 2y} = \dfrac{?}{2x^3 - 3x^2y - 2xy^2}$

27. $\dfrac{w}{w + 3} = \dfrac{?}{w^3 + 27}$

28. $\dfrac{3}{t - 2} = \dfrac{?}{3t^2 - 3t - 6}$

In problems 29–32, write an equivalent fraction with denominator $b - a$.

29. $\dfrac{4b}{a - b}$

30. $\dfrac{-5}{a - b}$

31. $-\dfrac{-3b}{a - b}$

32. $\dfrac{2b - a}{a - b}$

In problems 33–58, perform the indicated operations and simplify the result.

33. $\dfrac{5x^3y}{12uv^4} \cdot \dfrac{6u^3v}{25xy^2}$

34. $\dfrac{14m^3n^5}{9ab^4} \cdot \dfrac{3a^4b}{7mn^7}$

35. $\dfrac{b - 1}{b - 5} \cdot \dfrac{3}{2b - 2}$

36. $\dfrac{t^2 - 4}{5t + 15} \cdot \dfrac{5}{t + 2}$

37. $\dfrac{9 - y^2}{x^3 - x} \cdot \dfrac{x - 1}{y + 3}$

38. $\dfrac{a^2 + 8a + 16}{a^2 - 9} \cdot \dfrac{a - 3}{a + 4}$

39. $\dfrac{u^2 - v^2}{u^2 + 2uv + v^2} \cdot \dfrac{3u + 3v}{6u}$

40. $\dfrac{4x^2 - 64}{2x^2 - 8x} \cdot \dfrac{x - 4}{x + 4}$

41. $\dfrac{x^2 + 5x + 6}{x^2 + x - 2} \cdot \dfrac{x^2 + 3x - 4}{x^2 + 7x + 12}$

42. $\dfrac{4x^2 - 11x - 3}{6x^2 - 5x - 6} \cdot \dfrac{6x^2 - 13x + 6}{4x^2 + 13x + 3}$

43. $\dfrac{8t^3 + 1}{t^3 - 27} \cdot \dfrac{t^2 + 3t + 9}{4t^2 - 2t + 1}$

44. $\dfrac{a^3 - b^3}{a^2 - 4b^2} \cdot \dfrac{a^2 - ab - 2b^2}{a^2 + ab + b^2}$

45. $\dfrac{14x^2}{5b^2} \div \dfrac{21x^4}{15b}$

46. $\dfrac{5a}{12yz^2} \div \dfrac{25a^3}{18y^2z^3}$

47. $\dfrac{v + 7}{v + 2} \div \dfrac{v^2 - 49}{v^2 - 4}$

48. $\dfrac{4x^2 - 9}{9x^2 - 4} \div \dfrac{2x - 3}{3x + 2}$

49. $\dfrac{m^2 - m - 2}{m^2 - m - 6} \div \dfrac{m^2 - 2m}{2m + m^2}$

50. $\dfrac{y^2 - 5y + 6}{y^2 + y - 2} \div \dfrac{y^2 + y - 12}{y^2 + 3y - 4}$

51. $\dfrac{x^2 + 3xy + 2y^2}{x^2 - 2xy} \div \dfrac{x^2 + 4xy + 3y^2}{x^2 + xy - 6y^2}$

52. $\dfrac{t^3 + 1}{t^2 - 4u^2} \div \dfrac{t^2 - t + 1}{t - 2u}$

53. $\dfrac{2w^2 - w - 6}{3w^2 - 11w - 4} \div \dfrac{2w^2 + 5w + 3}{3w^2 + 7w + 2}$

54. $\dfrac{8x^2 + 18x + 9}{6x^2 - 7x + 2} \div \dfrac{4x^2 + 7x + 3}{2x^2 + 9x - 5}$

55. $\left[\dfrac{x^2 - 1}{x^2 + 4x + 4} \cdot \dfrac{x^2 - x - 6}{x^2 - 2x + 1} \right] \div \dfrac{x^2 + x}{x^2 - x}$

56. $\left[\dfrac{2y^2 - 3y - 2}{9y^2 - 1} \cdot \dfrac{3y^2 - 13y + 4}{2y^2 - y - 6} \right] \div \dfrac{2y^2 - 7y - 4}{3y^2 - 5y - 2}$

57. $\dfrac{25u^2 - 9}{2u^2 + 3u - 14} \div \left[\dfrac{5u^2 + 22u - 15}{u^2 + 7u + 10} \cdot \dfrac{5u^2 + 3u}{2u^2 + 7u} \right]$

58. $\dfrac{30x^2 + 6x}{40x^2 - 8x} \div \left[\dfrac{25x^2 + 10x + 1}{25x^2 - 10x + 1} \cdot \dfrac{25x^2 - 1}{20x + 4} \right]$

In problems 59–80, perform the indicated operations and simplify the result.

59. $\dfrac{5}{7x} + \dfrac{9}{7x}$

60. $\dfrac{3}{15y^2} + \dfrac{2}{15y^2}$

61. $\dfrac{3}{9 - t^2} - \dfrac{t}{9 - t^2}$

62. $\dfrac{3a}{a^2 - 4b^2} - \dfrac{6b}{a^2 - 4b^2}$

63. $\dfrac{3y}{y^2 - y - 2} + \dfrac{3}{y^2 - y - 2}$

64. $\dfrac{15t}{6t^2 - 5t - 6} + \dfrac{10}{6t^2 - 5t - 6}$

65. $\dfrac{y}{x - y} + \dfrac{x}{x + y}$

66. $\dfrac{t + 1}{2t - 1} - \dfrac{t - 1}{2t + 1}$

67. $\dfrac{2}{3 - t} - \dfrac{5}{t - 3}$

68. $\dfrac{x}{x + 2} + \dfrac{x}{x + 3}$

69. $\dfrac{y + 2}{y + 4} - \dfrac{y - 1}{y + 6}$

70. $\dfrac{z + 1}{3z + 2} - \dfrac{z - 3}{2z - 5}$

71. $\dfrac{4}{5m^2n} + \dfrac{3}{4mn^3}$

72. $\dfrac{3}{14x^3y^2} + \dfrac{5}{21x^2y^5}$

73. $\dfrac{3v}{4v^2 - 9} - \dfrac{v}{2v^2 + v - 6}$

74. $\dfrac{5}{x^2 + 8x + 15} - \dfrac{4}{x^2 + 2x - 3}$

75. $\dfrac{5}{x^2 - 7x + 12} + \dfrac{3}{2x^2 - 10x + 8}$

76. $\dfrac{7}{w^2 - z^2} + \dfrac{2}{w^2 - 2wz + z^2}$

77. $\dfrac{2}{2m^2 + m - 3} + \dfrac{4}{m^2 - 1} - \dfrac{1}{2m^2 + 5m + 3}$

78. $\dfrac{4c}{3c^2 + 5c - 2} - \dfrac{c}{2c^2 + 5c + 2} - \dfrac{2}{6c^2 + c - 1}$

79. $\dfrac{y}{x^2 - 9y^2} - \dfrac{x}{x^2 - 2xy - 3y^2} + \dfrac{x}{x^2 + 4xy + 3y^2}$

80. $\dfrac{3u}{2u^2 + 9u + 4} + \dfrac{2u}{4u^2 - 1} + \dfrac{u}{2u^2 + 7u - 4}$

In problems 81–90, simplify each complex fraction.

81. $\dfrac{\dfrac{34a}{4x^2 y}}{\dfrac{17a^3}{2xy}}$

82. $\dfrac{\dfrac{25x^2}{9y^2 z}}{\dfrac{100x^3}{27yz^2}}$

83. $\dfrac{\dfrac{3}{y} + \dfrac{x}{y}}{\dfrac{4}{y} + 2}$

84. $\dfrac{\dfrac{m}{n} + 1}{\dfrac{m^2}{n} - n}$

85. $\dfrac{2 + \dfrac{2}{y - 1}}{\dfrac{2}{y - 1}}$

86. $\dfrac{\dfrac{x^2 + y^2}{y} - 2x}{\dfrac{1}{y} - \dfrac{1}{x}}$

87. $\dfrac{1 - \dfrac{1}{t}}{t - 2 + \dfrac{1}{t}}$

88. $\dfrac{\dfrac{1}{a} - \dfrac{1}{b}}{\dfrac{a^2 - b^2}{ab}}$

89. $\dfrac{\dfrac{1}{2x - 3} - \dfrac{1}{2x + 3}}{\dfrac{x}{4x^2 - 9}}$

90. $\dfrac{\dfrac{y}{y^2 - 1} - \dfrac{1}{y + 1}}{\dfrac{y}{y - 1} + \dfrac{1}{y + 1}}$

CHAPTER 3 TEST

1. Determine all values of x for which the fraction $\dfrac{2x}{(x - 3)(3x + 5)}$ is not defined.

2. Determine whether the given pair of fractions are equivalent.

(a) $\dfrac{x + 3}{3y}$ and $\dfrac{2x^2 y + 6xy}{6xy^2}$ (b) $\dfrac{z - 3}{z + 1}$ and $\dfrac{2z + 3}{2z - 1}$

3. Reduce each fraction to lowest terms.

(a) $\dfrac{15x^2 y^3}{18xy^5}$ (b) $\dfrac{4x^2 - 4y^2}{2x^2 + 4xy + 2y^2}$

4. Find the missing numerator that will make the fractions equivalent.

(a) $\dfrac{7w^2}{13uv} = \dfrac{?}{39u^3 v^2 w}$ (b) $\dfrac{2x}{x - y} = \dfrac{?}{x^3 - y^3}$

5. Perform the indicated multiplications and divisions and simplify the result.

(a) $\dfrac{8x^3}{7y^5} \cdot \dfrac{14y^7}{16x}$ (b) $\dfrac{x^2 + 7x + 12}{x - 5} \div \dfrac{x^2 + 9x + 18}{x^2 - 7x + 10}$

(c) $\left(\dfrac{x + y}{x - y} \div \dfrac{x^2 + 2xy + y^2}{x^2 + 6xy + 9y^2} \right) \cdot \dfrac{x^2 - y^2}{x + 3y}$

6. Perform the indicated additions and subtractions and simplify the result.

(a) $\dfrac{5}{18xy} + \dfrac{1}{18xy}$ (b) $\dfrac{3x}{x - 4} + \dfrac{2}{x + 1}$

(c) $\dfrac{3}{2x^2 - x - 1} - \dfrac{2}{x^2 + x - 2}$

(d) $\dfrac{2z}{2z^2 + 5z + 2} + \dfrac{z}{3z^2 + 5z - 2} - \dfrac{1}{6z^2 + z - 1}$

7. Simplify each complex fraction.

(a) $\dfrac{\dfrac{3x^2}{5y}}{\dfrac{12x^3}{25y^2}}$ (b) $\dfrac{\dfrac{5}{x} + \dfrac{2y}{x}}{1 - \dfrac{3}{x}}$ (c) $\dfrac{y + \dfrac{x}{x - y}}{x + \dfrac{y}{y - x}}$

4 Linear Equations and Inequalities

Equations and inequalities are often used to describe situations and to solve problems. In this chapter, we apply the algebraic skills we have developed in the preceding chapters in order to solve algebraic equations and inequalities.

4.1 Equations

The following are examples of equations in one variable:

$$-2x + 37 = 39 \qquad 6y + 5 = 3(2y - 4) \qquad \sqrt{p - 2} = 5$$

$$u^2 - u - 6 = 0 \qquad \frac{2}{t} + \frac{1}{3} = \frac{2}{5} \qquad 2 + \frac{3}{3w + 1} = \frac{-4}{3w}.$$

An **equation** is a statement that two mathematical expressions (each representing a real number) are equal. The expressions on either side of the equals sign of an equation are called the **sides** or **members** of the equation. An equation produces a statement that is either true or false when a particular number is substituted for the variable. If a true statement results when we substitute a number for the variable, we say the number substituted **satisfies** the equation. Take, for instance, the equation $3x + 6 = 18$; if we substitute $x = 4$, we have a true statement: $12 + 6 = 18$. Thus, $x = 4$ satisfies the equation $3x + 6 = 18$, and we say that the number 4 is a **solution** or a **root** of the equation. To **solve** an equation we find all of its solutions.

An **identity** is an equation that is satisfied by *every* real number substituted for the variable for which both sides of the equation are defined. For example, the equation

$$(x + 2)^2 = x^2 + 4x + 4$$

is an identity. An equation that is not an identity is called a **conditional equation.** Both sides of a conditional equation are defined, but they are not equal when at least one real number is substituted for the variable. For instance, $7x = 14$ is a conditional equation because there is at least one substitution (say, $x = 5$) that produces a false statement.

150

Two equations are **equivalent** if they have exactly the same solutions. Thus, the equation $5x - 15 = 0$ is equivalent to $5x = 15$ because both equations have the same solution, $x = 3$. You can change an equation into an equivalent one by performing any of the following operations:

(i) **Addition and Subtraction Properties**

Add or subtract the same quantity on both sides of the equation. That is,

$$\text{if } P = Q, \text{ then } P + R = Q + R \text{ and } P - R = Q - R.$$

(ii) **Multiplication and Division Properties**

Multiply or divide both sides of the equation by the same nonzero quantity. That is,

$$\text{if } P = Q, \text{ then } PR = QR \text{ and } \frac{P}{R} = \frac{Q}{R}, \text{ where } R \neq 0.$$

(iii) **Symmetric Property**

Interchange the two sides of the equation. That is,

$$\text{if } P = Q, \text{ then } Q = P.$$

In this section, we use these properties to solve first-degree equations in one variable.

First-Degree or Linear Equations

Equations of the form

$$2x - 1 = 5, \qquad 3t - 5 = 7(1 + t), \qquad \text{and} \qquad 3u + 5 = 7u - 4,$$

are called **first-degree** or **linear equations.** These first-degree equations have a single variable that always has an exponent of 1. The most common technique for **solving** a first-degree (linear) equation is to write a sequence of equations (starting with the given equation) in which each equation is equivalent to the previous one. When we write this sequence of equations, we collect all the variable terms on one side of the equation, and the constant terms on the other side. Then we simplify each side of the equation, and we divide both sides by the coefficient of the variable to obtain the solution.

For example, to solve the equation

$$7x - 28 = 0,$$

we begin by adding 28 to both sides of the equation to obtain the equivalent equation:

$$7x - 28 + 28 = 0 + 28 \qquad \text{or} \qquad 7x = 28.$$

We then divide both sides of the equation by 7 to get the equivalent equation:

$$\frac{7x}{7} = \frac{28}{7} \qquad \text{or} \qquad x = 4.$$

Thus, the solution is 4.

In Examples 1–7, solve each equation.

EXAMPLE 1 $x + 7 = 13$

SOLUTION

$$x + 7 = 13$$
$$x + 7 - 7 = 13 - 7 \qquad \text{(We subtracted 7 from both sides.)}$$
$$x = 6.$$

Check We substitute 6 for x in the original equation:

$$6 + 7 = 13 \qquad \text{or} \qquad 13 = 13.$$

Thus, 6 is the solution.

EXAMPLE 2 $5t = 30$

SOLUTION

$$5t = 30$$
$$\frac{5t}{5} = \frac{30}{5} \qquad \text{(We divided both sides by 5.)}$$
$$t = 6.$$

Check We substitute $t = 6$ in the original equation:

$$5(6) = 30$$
$$30 = 30.$$

Therefore, the solution is 6.

EXAMPLE 3 $7y - 3 = 28$

SOLUTION

$$7y - 3 = 28$$
$$7y - 3 + 3 = 28 + 3 \qquad \text{(We added 3 to both sides.)}$$
$$7y = 31$$

$$\frac{7y}{7} = \frac{31}{7} \qquad \text{(We divided both sides by 7.)}$$

$$y = \frac{31}{7}.$$

Check Substitute $y = \frac{31}{7}$ in the original equation:

$$7\left(\frac{31}{7}\right) - 3 = 28$$

$$31 - 3 = 28$$

$$28 = 28.$$

Therefore, $\frac{31}{7}$ is the solution.

EXAMPLE 4 $5 - 4w = 21$

SOLUTION

$$5 - 4w = 21$$

$$5 - 4w - 5 = 21 - 5 \qquad \text{(We subtracted 5 from both sides.)}$$

$$-4w = 16$$

$$\frac{-4w}{-4} = \frac{16}{-4} \qquad \text{(We divided both sides by } -4.)$$

$$w = -4.$$

Check Substitute $w = -4$ in the original equation:

$$5 - 4(-4) = 21$$

$$5 + 16 = 21$$

$$21 = 21.$$

Therefore, -4 is the solution.

EXAMPLE 5 $12m + 1 = 25 - 12m$

SOLUTION

$$12m + 1 = 25 - 12m$$

$$12m + 1 + 12m = 25 - 12m + 12m \qquad \text{(We added } 12m \text{ to both sides.)}$$

$$24m + 1 = 25$$

$$24m + 1 - 1 = 25 - 1 \qquad \text{(We subtracted 1 from both sides.)}$$

$$24m = 24$$

$$\frac{24m}{24} = \frac{24}{24} \qquad \text{(We divided both sides by 24.)}$$

$$m = 1.$$

Check Substitute $m = 1$ in the original equation:

$$12(1) + 1 = 25 - 12(1)$$

$$12 + 1 = 25 - 12$$

$$13 = 13.$$

Therefore, 1 is the solution.

EXAMPLE 6 $4(u + 1) = 21 + 2u$

SOLUTION

$$4(u + 1) = 21 + 2u$$

$$4u + 4 = 21 + 2u \qquad \text{(We removed parentheses on the left side.)}$$

$$4u + 4 - 2u = 21 + 2u - 2u \qquad \text{(We subtracted } 2u \text{ from both sides.)}$$

$$2u + 4 = 21$$

$$2u + 4 - 4 = 21 - 4 \qquad \text{(We subtracted 4 from both sides.)}$$

$$2u = 17$$

$$\frac{2u}{2} = \frac{17}{2} \qquad \text{(We divided both sides by 2.)}$$

$$u = \frac{17}{2}.$$

Check Substitute $u = \frac{17}{2}$ in the original equation:

$$4\left(\frac{17}{2} + 1\right) = 21 + 2\left(\frac{17}{2}\right)$$

$$4\left(\frac{19}{2}\right) = 21 + 17$$

$$38 = 38.$$

Therefore, $\frac{17}{2}$ is the solution.

EXAMPLE 7 $5 + 8(x + 2) = 23 - 2(2x - 5)$

SOLUTION

$$5 + 8(x + 2) = 23 - 2(2x - 5)$$

$$5 + 8x + 16 = 23 - 4x + 10 \qquad \text{(We removed the parentheses on both sides.)}$$

$$4x + 5 + 8x + 16 = 4x + 23 - 4x + 10 \qquad \text{(We added } 4x \text{ to both sides.)}$$

$$12x + 21 = 33 \qquad \text{(We combined like terms on both sides.)}$$

$$12x + 21 - 21 = 33 - 21$$ (We subtracted 21 from both sides.)

$$12x = 12$$

$$\frac{12x}{12} = \frac{12}{12}$$ (We divided both sides by 12.)

$$x = 1.$$

Therefore, 1 is the solution. You can check this by substituting $x = 1$ in the original equation.

The procedure for solving first-degree or linear equations with one variable consists of one or all of the following steps:

Procedure for Solving Linear Equations

> Step 1. Remove all parentheses or grouping symbols on each side of the equation by using the distributive property.
>
> Step 2. Combine all like terms on each side of the equation.
>
> Step 3. Convert the equation to an equivalent equation in which all the variable terms are on one side of the equation and all the constant terms are on the other side. Combine like terms after each operation. (Addition and subtraction properties.)
>
> Step 4. Complete the solution by applying the multiplication or division property to bring the coefficient of the variable to 1.

The following example illustrates an interesting application of linear equations.

EXAMPLE 8   Express the repeating decimal $0.\overline{61}$ as a quotient of integers.

SOLUTION   Let $x = 0.\overline{61}$. Then $100x = 61.\overline{61}$. If we subtract $0.\overline{61}$ from $61.\overline{61}$, the repeating portion of the decimals cancels out:

$$100x - x = 61.\overline{61} - 0.\overline{61} = 61$$

$$99x = 61$$

$$x = \frac{61}{99}.$$

Therefore, $0.\overline{61} = 61/99$.

PROBLEM SET 4.1

In problems 1–50, solve each equation.

1. $x + 3 = 10$

2. $t + 2 = 8$

3. $u - 4 = 8$

4. $y - 5 = 2$

5. $w + 11 = 17$

6. $z + 12 = 18$

7. $8t = 24$

8. $7y = 35$

9. $-4u = 12$

10. $6x = 42$

11. $10c = -18$

12. $-8y = -32$

13. $15t = 75$

14. $14w = 42$

15. $16b = 30$

16. $9t = 54$

17. $-12b = 8$

18. $-3z = 1$

19. $3x - 6 = 15$

20. $7 - 5x = 11$

21. $6y + 7 = 31$

22. $7t - 3 = 18$

23. $5 + 4u = 17$

24. $3w - 2 = 10$

25. $3t - 5 = 20$

26. $-8t + 1 = 17$

27. $12u + 25 = 40$

28. $10w - 9 = 21$

29. $6x - 8 = 7 - x$

30. $18 - x = 3 + 4x$

31. $10 - 2m = 3m + 25$

32. $2u + 1 = 5u + 8$

33. $5 - 9z = -8z + 3$

34. $7c + 4 = c - 8$

35. $5y - 1 = 2y + 8$

36. $7y - 16 = y - 10$

37. $12t + 1 = 25 - 12t$

38. $17w - 2 = 12w + 8$

39. $3x - 2(x + 1) = 2(x - 1)$

40. $8(5x - 1) + 36 = -3(x + 5)$

41. $1 - 2(5 - 2y) = 26 - 3y$

42. $2(1 - 2y) = 3(2y - 4) + 94$

43. $7t - 3(9 - 5t) = 4t - 9t$

44. $7(w - 3) = 4(w + 5) - 47$

45. $6(c - 10) + 3(2c - 7) = -45$

46. $16 - 9(3 - u) + 4u = 15$

47. $13 - 5(2 - x) - 18 = 0$

48. $3(z - 2) + 5(z + 1) = 4(z - 1)$

49. $34 - 3y = 8(7 - y) + 23$

50. $11 - 7(1 - 2v) = 9(v + 1)$

In problems 51–58, express each repeating decimal as a quotient of integers.

51. $0.\overline{5}$

52. $0.\overline{46}$

53. $1.\overline{3}$

54. $0.0\overline{53}$

55. $0.4\overline{9}$

56. $-7.\overline{362}$

57. $-3.1\overline{28}$

58. $1.5\overline{821}$

C In problems 59–62, solve each equation with the aid of a calculator.

59. $41.03x + 49.37 = 0$

60. $271.73t + 839.41 = 972.82$

61. $0.1347y - 6.738 = 0.2814y - 1.813$

62. $2.719u - 3.482 = 6.432u - 1.713$

4.2 Equations Involving Fractions

We solved some first-degree (linear) equations in Section 4.1. Now we will solve linear equations involving fractions and fractional equations that can be changed to linear equations. We consider two types of equations. One type has fractions in which the denominators are constants. Examples of such equations are:

$$\frac{x}{5} = \frac{2}{3},$$

$$\frac{3t + 1}{10} + \frac{1}{2} = \frac{7}{8},$$

and $$\frac{u + 7}{10} - \frac{u}{25} = 1.$$

The other type of equation has fractions in which at least one denominator contains the variable. Examples of this type of equation are:

$$\frac{1}{3x} + \frac{1}{x} = \frac{1}{9},$$

$$\frac{1}{2y} - \frac{1}{6} = -\frac{1}{3y},$$

and $$\frac{4}{10 + w} + \frac{4}{10 - w} = \frac{5}{6}.$$

We must be careful when the variable appears in the denominator of a fraction. Because division by zero is undefined, we must never assign a value to the variable that would produce a zero in any denominator.

Procedure for Solving Equations Involving Fractions

Step 1. **Determine the LCD (least common denominator) of the fractions.**

Step 2. **Multiply both sides of the equation by the LCD in order to produce an equation that contains no fractions.**

Step 3. **Solve the resulting equation.**

Step 4. **If the original equation contains a variable in any denominator, you must check the proposed solution to see whether it should be accepted or rejected.**

In Examples 1–6, solve each equation.

EXAMPLE 1 $$\frac{2x}{3} + \frac{1}{2} = \frac{5}{6}$$

SOLUTION   The LCD of the fractions in the equation is 6. In order to produce an equation that contains no fractions, we multiply both sides of the equation by 6:

$$6\left(\frac{2x}{3}+\frac{1}{2}\right)=6\left(\frac{5}{6}\right)$$

$$6\left(\frac{2x}{3}\right)+6\left(\frac{1}{2}\right)=6\left(\frac{5}{6}\right) \qquad \text{(We used the distributive property.)}$$

$$4x+3=5$$

$$4x+3-3=5-3 \qquad \text{(We subtracted 3 from both sides.)}$$

$$4x=2$$

$$x=\frac{2}{4} \qquad \text{(We divided both sides by 4.)}$$

$$x=\frac{1}{2}$$

Therefore, $\frac{1}{2}$ is the solution.

EXAMPLE 2 $\dfrac{u-4}{3}=\dfrac{u}{5}+2$

SOLUTION   We multiply both sides of the equation by 15, the LCD of the fractions:

$$15\left(\frac{u-4}{3}\right)=15\left(\frac{u}{5}+2\right)$$

$$5(u-4)=15\left(\frac{u}{5}\right)+15(2) \qquad \text{(We applied the distributive property to the right side.)}$$

$$5u-20=3u+30 \qquad \text{(We applied the distributive property to the left side.)}$$

$$5u-3u=30+20 \qquad \text{(We subtracted } 3u \text{ from both sides and added 20 to both sides.)}$$

$$2u=50$$

$$u=\frac{50}{2} \qquad \text{(We divided both sides by 2.)}$$

$$u=25$$

Therefore, 25 is the solution.

EXAMPLE 3 $\dfrac{y-3}{4} - \dfrac{y-2}{3} = \dfrac{y-11}{12}$

SOLUTION   We multiply both sides of the equation by 12, the LCD of the fractions:

$$12\left(\dfrac{y-3}{4} - \dfrac{y-2}{3}\right) = 12\left(\dfrac{y-11}{12}\right)$$

$$12\left(\dfrac{y-3}{4}\right) - 12\left(\dfrac{y-2}{3}\right) = 12\left(\dfrac{y-11}{12}\right) \qquad \text{(We used the distributive property.)}$$

$$3(y-3) - 4(y-2) = y - 11$$

$$3y - 9 - 4y + 8 = y - 11 \qquad \text{(We used the distributive property.)}$$

$$-y - 1 = y - 11 \qquad \text{(We collected like terms.)}$$

$$-2y = -10 \qquad \text{(We subtracted } y \text{ from both sides, and added 1 to both sides.)}$$

$$y = \dfrac{-10}{-2} \qquad \text{(We divided both sides by } -2.)$$

$$y = 5.$$

Therefore, 5 is the solution.

EXAMPLE 4 $\dfrac{3}{5t} + 1 = \dfrac{4}{t} + \dfrac{18}{35}$

SOLUTION   We multiply both sides of the equation by $35t$, the LCD of the fractions:

$$35t\left(\dfrac{3}{5t} + 1\right) = 35t\left(\dfrac{4}{t} + \dfrac{18}{35}\right)$$

$$35t\left(\dfrac{3}{5t}\right) + 35t(1) = 35t\left(\dfrac{4}{t}\right) + 35t\left(\dfrac{18}{35}\right) \qquad \text{(We used the distributive property.)}$$

$$21 + 35t = 140 + 18t$$

$$35t - 18t = 140 - 21 \qquad \text{(We subtracted } 18t \text{ and 21 from both sides.)}$$

$$17t = 119 \qquad \text{(We collected like terms.)}$$

$$t = \dfrac{119}{17} \qquad \text{(We divided both sides by 17.)}$$

$$t = 7.$$

The original equation contains a variable in the denominator; therefore, we must check the solution.

Check If we substitute $t = 7$ in the original equation, we have

$$\frac{3}{5(7)} + 1 = \frac{4}{7} + \frac{18}{35}$$

$$\frac{3}{35} + 1 = \frac{4}{7} + \frac{18}{35}$$

$$\frac{38}{35} = \frac{38}{35}$$

Therefore, 7 is the solution.

EXAMPLE 5

$$\frac{2}{r - 1} + \frac{6}{r} = \frac{5}{r - 1}$$

SOLUTION   We multiply both sides of the equation by $r(r - 1)$, the LCD of the fractions:

$$r(r - 1)\left(\frac{2}{r - 1} + \frac{6}{r}\right) = r(r - 1)\left(\frac{5}{r - 1}\right)$$

$$r(r - 1)\left(\frac{2}{r - 1}\right) + r(r - 1)\left(\frac{6}{r}\right) = r(r - 1)\left(\frac{5}{r - 1}\right)$$ (We used the distributive property.)

$$2r + 6(r - 1) = 5r$$

$$2r + 6r - 6 = 5r$$ (We used the distributive property.)

$$8r - 6 = 5r$$ (We collected like terms.)

$$3r = 6$$ (We added 6 to both sides, and subtracted $5r$ from both sides.)

$$r = \frac{6}{3}$$

$$r = 2.$$

Check If we substitute $r = 2$ in the original equation, we have

$$\frac{2}{2 - 1} + \frac{6}{2} = \frac{5}{2 - 1}$$

$$\frac{2}{1} + 3 = \frac{5}{1}$$

$$5 = 5$$

Therefore, 2 is the solution.

EXAMPLE 6 $\dfrac{3}{w-3} + 4 = \dfrac{w}{w-3}$

SOLUTION   We multiply both sides of the equation by $w - 3$, the LCD of the fractions:

$$(w-3)\left(\frac{3}{w-3} + 4\right) = (w-3)\left(\frac{w}{w-3}\right)$$

$$(w-3)\left(\frac{3}{w-3}\right) + (w-3)4 = (w-3)\left(\frac{w}{w-3}\right) \quad \text{(We used the distributive property.)}$$

$$3 + (w-3)4 = w$$

$$3 + 4w - 12 = w \quad \text{(We used the distributive property.)}$$

$$4w - 9 = w$$

$$3w = 9 \quad \text{(We subtracted } w \text{ from both sides and added 9 to both sides.)}$$

$$w = 3.$$

Check If we substitute $w = 3$ in the original equation, we have

$$\frac{3}{3-3} + 4 = \frac{3}{3-3}$$

$$\frac{3}{0} + 4 = \frac{3}{0}$$

Because we cannot divide by zero, two of the terms are undefined. Therefore, we cannot substitute 3 for w—this does not produce a true statement—and there is *no* solution for this equation. We call 3 an **extraneous** root.

PROBLEM SET 4.2

In problems 1–44, solve each equation.

1. $\dfrac{2}{3} - \dfrac{5x}{3} = \dfrac{17}{3}$

2. $\dfrac{2x}{5} - \dfrac{4}{5} = \dfrac{9}{5}$

3. $\dfrac{y}{6} - \dfrac{1}{2} = \dfrac{2}{3}$

4. $\dfrac{5u}{4} + \dfrac{3}{16} = \dfrac{1}{2}$

5. $\dfrac{t}{6} - \dfrac{t}{7} = \dfrac{1}{42}$

6. $\dfrac{4z}{3} - \dfrac{5z}{6} = \dfrac{3}{4}$

7. $\dfrac{5w}{6} = \dfrac{52}{2} - \dfrac{w}{4}$

8. $\dfrac{3y}{7} - \dfrac{10}{7} = \dfrac{-2y}{3}$

9. $\dfrac{u-1}{2} + \dfrac{u}{7} = \dfrac{11}{14}$

10. $\dfrac{9z+1}{4} = z + \dfrac{1}{3}$

11. $\dfrac{5x-15}{7} - \dfrac{x}{3} = \dfrac{2}{5}$

12. $\dfrac{3w}{5} - \dfrac{13}{15} = \dfrac{w+2}{12}$

13. $\dfrac{5u}{4} - 1 = \dfrac{3u}{4} + \dfrac{1}{2}$

14. $z + \dfrac{16}{3} = \dfrac{3z}{2} + \dfrac{25}{6}$

15. $c - 1 = \dfrac{2c}{5} - \dfrac{7}{5}$

16. $\dfrac{2b + 1}{3} - \dfrac{b}{2} = \dfrac{b}{5}$

17. $\dfrac{1}{3}(3x - 2) + \dfrac{1}{2}(x - 3) = \dfrac{5}{6}$

18. $\dfrac{x - 14}{5} + 4 = \dfrac{x + 16}{10}$

19. $\dfrac{y + 9}{4} - \dfrac{6y - 9}{14} = 2$

20. $\dfrac{3u - 6}{4} - \dfrac{u + 6}{6} + \dfrac{2u}{3} = 5$

21. $\dfrac{z - 2}{3} - \dfrac{z - 3}{5} = \dfrac{13}{15}$

22. $\dfrac{8w + 10}{5} - \dfrac{6w + 1}{4} = \dfrac{3}{20}$

23. $\dfrac{1}{2x} + \dfrac{8}{5} = \dfrac{3}{x}$

24. $\dfrac{5}{x} + \dfrac{3}{8} = \dfrac{7}{16}$

25. $\dfrac{1}{y} + \dfrac{2}{y} = 3 - \dfrac{3}{y}$

26. $\dfrac{2}{3u} + \dfrac{1}{6u} = \dfrac{1}{4}$

27. $\dfrac{3}{8u} - \dfrac{1}{5u} = \dfrac{7}{10}$

28. $\dfrac{3}{4c} - \dfrac{1}{6} = \dfrac{4}{8c} + \dfrac{1}{2}$

29. $\dfrac{x}{x - 1} - \dfrac{3}{x + 1} = 1$

30. $\dfrac{8}{x - 3} = \dfrac{12}{x + 3}$

31. $\dfrac{1}{y} + \dfrac{1}{y - 1} = \dfrac{5}{y - 1}$

32. $\dfrac{t}{t + 1} + 2 = \dfrac{3t}{t + 2}$

33. $\dfrac{-4}{3u} = \dfrac{3}{3u + 1} + \dfrac{2}{u}$

34. $\dfrac{1}{y^2 - 2y} - \dfrac{1}{y} = \dfrac{1}{1 - y}$

35. $\dfrac{5}{3b + 1} - \dfrac{2}{2b - 1} = \dfrac{1}{2b - 1}$

36. $\dfrac{u + 1}{u - 4} - \dfrac{u}{u - 2} = \dfrac{3}{u - 6}$

37. $\dfrac{4}{y - 2} = \dfrac{5y}{y^2 - 4} - \dfrac{y + 3}{y^2 - 2y}$

38. $\dfrac{1}{2t + 5} - \dfrac{4}{2t - 1} = \dfrac{4t + 4}{(2t + 5)(2t - 1)}$

39. $\dfrac{1}{y - 3} - \dfrac{1}{3 - y} = \dfrac{1}{y^2 - 9}$

40. $\dfrac{2}{u - 2} + \dfrac{1}{u + 1} = \dfrac{1}{(u - 2)(u + 1)}$

41. $\dfrac{1}{t(t - 1)} - \dfrac{1}{t} = \dfrac{1}{t - 1}$

42. $\dfrac{5}{y - 5} = \dfrac{y}{y - 5} - 4$

43. $\dfrac{4}{5 - y} + \dfrac{6}{y + 5} = \dfrac{-40}{y^2 - 25}$

44. $\dfrac{y}{3y - 2} + \dfrac{2y^2 - 24y + 18}{(2 - 3y)(2y + 3)} = \dfrac{8}{2y + 3}$

4.3 Literal Equations and Formulas

The equations you have worked with so far have contained one variable or unknown (represented by a letter) and the constants have been known numbers. In this section we examine **literal equations**—equations containing more than one letter symbol in which one of the letter symbols represents the variable or unknown and the others represent constants. For instance, consider the literal equation

$$ax = b$$

in which x is the unknown and a and b represent unspecified constants. You may be asked to *solve* a literal equation for one of the letters. In this case, you express that unknown in terms of the other letters. Thus, to solve the equation $ax = b$ for x, you divide both sides of the equation by a ($a \neq 0$) to produce the equivalent equation

$$x = \frac{b}{a}.$$

In applied work, letters other than x are often used for a variable, because certain quantities (t for time, for example) are designated by conventional symbols. In each case, we specify which unknown we are solving for. For example, to solve the equation

$$3r + t = 6t + 5$$

for r, we can apply the procedure used in Section 4.1:

$$3r = 6t + 5 - t \qquad \text{(We subtracted } t \text{ from each side.)}$$

$$3r = 5t + 5 \qquad \text{(We collected like terms.)}$$

$$r = \frac{5t + 5}{3}. \qquad \text{(We divided both sides by 3.)}$$

Note that solving the literal equation $3r + t = 6t + 5$ for r means that the variable r will be left on one side of the equation and the other variable t and the constants will be left on the other side of the equation. Remember that r cannot be left on both sides of the equation.

In Examples 1–4, solve each equation for the indicated unknown.

EXAMPLE 1 $4x - a = x + 8a$, for x.

SOLUTION

$$4x - a = x + 8a$$

$$4x - x = 8a + a \qquad \text{(We subtracted } x \text{ from both sides and added } a \text{ to both sides.)}$$

$$3x = 9a \qquad \text{(We collected like terms.)}$$

$$x = \frac{9a}{3} \qquad \text{(We divided both sides by 3.)}$$

$$x = 3a$$

Check Substitute $x = 3a$ in the original equation:

$$4(3a) - a = 3a + 8a$$

$$12a - a = 3a + 8a$$

$$11a = 11a.$$

Therefore, the solution is $3a$.

EXAMPLE 2 $c(t - c) = d(t - d)$, for t.

SOLUTION

$$c(t - c) = d(t - d)$$

$$ct - c^2 = dt - d^2 \qquad \text{(We used the distributive property.)}$$

$$ct - dt = c^2 - d^2 \qquad \text{(We added } -dt \text{ and } c^2 \text{ to both sides.)}$$

$$t(c - d) = (c - d)(c + d) \qquad \text{(We factored both sides.)}$$

$$t = \frac{(c - d)(c + d)}{c - d} \qquad \text{(We divided both sides by } c - d.)$$

$$t = c + d.$$

EXAMPLE 3 $\dfrac{y - 3a}{b} = \dfrac{2a}{b} + y$, for y.

SOLUTION   We multiply both sides of the equation by b, the LCD:

$$b\left(\frac{y - 3a}{b}\right) = b\left(\frac{2a}{b} + y\right)$$

$$y - 3a = b\left(\frac{2a}{b}\right) + by$$

$$y - 3a = 2a + by$$

$$y - by = 2a + 3a$$

$$y(1 - b) = 5a$$

$$y = \frac{5a}{1 - b}.$$

EXAMPLE 4 $\dfrac{2x}{x - a} = 3 - \dfrac{x - a}{x}$, for x.

SOLUTION   We multiply both sides of the equation by $x(x - a)$, the LCD:

$$x(x - a)\left(\frac{2x}{x - a}\right) = x(x - a)\left(3 - \frac{x - a}{x}\right)$$

$$x(2x) = 3x(x - a) - (x - a)^2$$

$$2x^2 = 3x^2 - 3ax - (x^2 - 2ax + a^2)$$

$$2x^2 = 3x^2 - 3ax - x^2 + 2ax - a^2$$

$$2x^2 = 2x^2 - ax - a^2$$

$$ax = -a^2$$

$$x = -a.$$

In many fields, algebraic formulas are used to express relations among various quantities. These formulas often contain more than one unknown. For example, in physics, the formula

$$d = rt$$

gives the distance d in terms of the speed r and the time t. If we are asked to solve for t in terms of d and r, we divide both sides of the equation by r:

$$\frac{d}{r} = \frac{rt}{r}$$

or

$$\frac{d}{r} = t.$$

EXAMPLE 5   The formula

$$A = P + Prt$$

gives the total amount of money due at the end of t years if P (the principal) is the amount of money invested at a simple interest rate r. Solve for P.

SOLUTION

$$A = P + Prt$$

$$A = P(1 + rt) \qquad \text{(We factored out } P \text{ on the right side.)}$$

$$\frac{A}{1 + rt} = P \qquad \text{(We divided both sides by } 1 + rt.)$$

EXAMPLE 6   The formula

$$F = \frac{9}{5}C + 32$$

expresses the temperature F in degrees Fahrenheit in terms of the temperature C in degrees Celsius. Solve for C.

SOLUTION

$$F = \frac{9}{5}C + 32$$

$$5F = 5\left(\frac{9}{5}C + 32\right) \qquad \text{(We multiplied both sides by 5, the LCD.)}$$

$$5F = 9C + 160 \qquad \text{(We used the distributive property.)}$$

$$5F - 160 = 9C \qquad \text{(We subtracted 160 from both sides.)}$$

$$\frac{5F - 160}{9} = C \qquad \text{(We divided both sides by 9.)}$$

or

$$C = \frac{5}{9}(F - 32)$$

EXAMPLE 7   The formula

$$\frac{1}{R} = \frac{1}{R_1} + \frac{1}{R_2}$$

relates the values of the three resistances R, R_1, and R_2 in a certain type of electrical circuit. Solve for R_1.

SOLUTION

$$\frac{1}{R} = \frac{1}{R_1} + \frac{1}{R_2}$$

$$RR_1R_2\left(\frac{1}{R}\right) = RR_1R_2\left(\frac{1}{R_1} + \frac{1}{R_2}\right) \qquad \text{(We multiplied both sides by } RR_1R_2, \text{ the LCD.)}$$

$$R_1R_2 = RR_2 + RR_1 \qquad \text{(We used the distributive property.)}$$

$$R_1R_2 - RR_1 = RR_2 \qquad \text{(We subtracted } RR_1 \text{ from both sides.)}$$

$$R_1(R_2 - R) = RR_2 \qquad \text{(We factored out } R_1 \text{ on the left side.)}$$

$$R_1 = \frac{RR_2}{R_2 - R} \qquad \text{(We divided both sides by } R_2 - R.)$$

PROBLEM SET 4.3

In problems 1–34, solve each equation for the indicated unknown.

1. $6x + 7c = 37c$, for x.

2. $4u - 19a = 5u$, for u.

3. $at + b = c$, for t.

4. $ay - b = c$, for y.

5. $12z - 4b = 6z - 7b$, for z.

6. $13f + 6y = 8f - 9y$, for y.

7. $34c + 11n = 7c + 17n$, for n.

8. $18u + 11d = 14u - 19d$, for u.

9. $4x - 3a - (10x + 7a) = 0$, for x.

10. $27t - 4b - (15t - 6b) = 0$, for t.

11. $9z + 7h - (11h - 13z) = 6z$, for z.

12. $4(2k - 3m) - 3(5k - 7m) = 0$, for m.

13. $5(4r - 3c) - 2(7r - 9c) = 0$, for r.

14. $5(mu - 2d) - 3m(u - 4d) = 8d$, for u.

15. $8(w - 2b) - 3(5w + 11b) = 0$, for w.

16. $(ah + 7)(ah - 3) = ah(ah + 1)$, for h.

17. $3(a - 2b) + 4(b + a) = 5$, for a.

18. $a(y - a) = ab + 2b(y - b)$, for y.

19. $\dfrac{ay}{b} = c + \dfrac{d}{b}$, for y.

20. $\dfrac{ax}{7} - \dfrac{bc}{3} = \dfrac{2x}{3b^2}$, for x.

21. $\dfrac{b - x}{3} = \dfrac{2a - b}{4} - \dfrac{3x}{5}$, for x.

22. $\dfrac{y}{b^4} - \dfrac{3}{2b^4} = 2$, for y.

23. $\dfrac{u + 2c}{3} + \dfrac{u - 3c}{2} = \dfrac{5}{12}$, for u.

24. $\dfrac{m^2t}{4p} = \dfrac{2m^2t - 9}{5p}$, for t.

25. $\dfrac{2t + a}{4} - \dfrac{6t + 3a}{7} = \dfrac{15a}{28}$, for t.

26. $\dfrac{a^2r + 8}{b} = \dfrac{a^2r + 10}{3b}$, for r.

27. $\dfrac{a + b}{x} + \dfrac{a - b}{x} = 2a$, for x.

28. $\dfrac{8c^4 - 3y}{2y} + \dfrac{7}{2} = 0$, for y.

29. $\dfrac{3}{x} - \dfrac{4}{b} = \dfrac{5}{3b}$, for x.

30. $\dfrac{b}{x} - \dfrac{a}{3} = \dfrac{b - a}{3x}$, for x.

31. $\dfrac{3}{a - x} + \dfrac{a}{a + x} = \dfrac{1}{a^2 - x^2}$, for x.

32. $\dfrac{x + a}{2x - b} = \dfrac{x + b}{2x - a}$, for x.

33. $\dfrac{1}{y} + \dfrac{2}{y + a} = \dfrac{3}{y - a}$, for y.

34. $\dfrac{t - 2a}{2t + a} = \dfrac{2t - 7a}{4t - 3a}$, for t.

In problems 35–56, solve each formula for the indicated unknown.

35. $A = \frac{1}{2}bh$, for h. (area of a triangle)

36. $E = IR$, for I. (physics)

37. $V = \ell wh$, for ℓ. (volume of a box)

38. $V = \pi r^2 h$, for h. (volume of a cylinder)

39. $C = 2\pi r$, for r. (circumference of a circle)

40. $S = 2\pi rh$, for r. (surface area of a cylinder)

41. $V = gt$, for t. (physics)

42. $V = \frac{1}{3}\pi r^2 h$, for h. (volume of a cone)

43. $I = Prt$, for t. (simple interest)

44. $F = \dfrac{W}{g} - a$, for a. (physics)

45. $F = mx + b$, for m. (analytic geometry)

46. $E = I(R + r)$, for R. (physics)

47. $P = 2\ell + 2w$, for w. (perimeter of a rectangle)

48. $S = gt^2 + vt$, for v. (physics)

49. $S = \dfrac{n}{2}(a + \ell)$, for a.

50. $S = \dfrac{a}{1 - r}$, for r.

51. $S = \dfrac{n}{2}[2a + (n - 1)d]$, for d.

52. $S = \dfrac{a - r\ell}{\ell - r}$, for r.

53. $L = a + (m - 1)d$, for d.

54. $\dfrac{1}{u} + \dfrac{1}{v} = \dfrac{1}{f}$, for f.

55. $pv = k\left(1 + \dfrac{t}{m}\right)$, for t.

56. $I = \dfrac{E}{R + nr}$, for R.

4.4 Translating Verbal Expressions into Algebraic Expressions

When we describe problems of any kind, whether they have to do with geometry, business, the social sciences, or the physical sciences, we typically ask questions and supply facts in the form of words and sentences rather than letters and symbols. In order to solve these "word problems" or "story problems," we must put them into the form of algebraic expressions contained in equations. Often the most challenging step in solving a word problem is translating the situation described in the problem into algebraic form. Only after doing that can we put these algebraic expressions together in equations that can be solved. This translation process often requires some perseverance until you become familiar with it, but you will find that the necessary skills can readily be acquired with practice.

Table 1 summarizes many common algebraic expressions and their English equivalents. You should find the table helpful in translating word problems into algebraic forms.

Table 1

Verbal Expression	Algebraic Expression
The sum of two numbers x and y	$x + y$
The difference of two numbers x and y	$x - y$
The product of two numbers x and y	$x \cdot y$ or xy
The quotient of two numbers x and y	$\dfrac{x}{y}$
3 more than x	$3 + x$
Twice the sum of x and 7	$2(x + 7)$
The sum of twice x and 7	$2x + 7$
5 more than three times the number x	$5 + 3x$
5 less than three times the number x	$3x - 5$
The sum of two consecutive integers	$n + (n + 1)$
The product of two consecutive integers	$n(n + 1)$
x exceeds y by 4	$x - y = 4$ or $x = y + 4$ or $x - 4 = y$
The number of centimeters in x meters	$100x$
The number of inches in x feet	$12x$
6% of a number x	$6\% \, x$ or $0.06x$

In Examples 1–5, translate each statement into an algebraic form.

EXAMPLE 1   If x is a positive even integer, what are the next two larger consecutive positive even integers?

SOLUTION   Since x is the first positive even integer, then the next larger consecutive positive even integer is $x + 2$. The next larger consecutive positive even integer is $(x + 2) + 2$ or $x + 4$.

EXAMPLE 2   The larger of two numbers is 7 more than three times the smaller. What is the sum of the two numbers and 11?

SOLUTION   Let x represent the smaller number. Then the larger number is $3x + 7$. The sum of the two numbers and 11 is

$$x + (3x + 7) + 11 = 4x + 18.$$

EXAMPLE 3   Bill is twice as old as Tom. If Tom is x years old now, how old was Bill 5 years ago?

SOLUTION   Since Tom is x years old, and Bill is twice as old as Tom, then Bill's present age is $2x$. Five years ago, he was 5 years younger, or $(2x - 5)$ years old.

EXAMPLE 4   If a number is represented by y, what number is 15% of y more than y?

SOLUTION   Fifteen percent of a number y is the product $(15/100)y$ or $0.15y$. Therefore, the number we want is $y + 0.15y = 1.15y$.

EXAMPLE 5   How many cents are there in q quarters, d dimes, and n nickels?

SOLUTION   Since each quarter is 25 cents, then q quarters will contain q times 25 cents, or $25q$ cents. Each dime contains 10 cents, so d dimes will contain $10d$ cents. Also, since each nickel contains 5 cents, n nickels contain $5n$ cents. Taken together, there will be a total of

$$(25q + 10d + 5n) \text{ cents.}$$

PROBLEM SET 4.4

In problems 1–40, translate each statement into algebraic form.

1. If $x + 5$ is an integer, what is the next larger consecutive integer?

2. What number exceeds $x - 3$ by $x + 5$?

3. What number is $7x$ more than $3x - 7$?

4. What number is 3 more than 4 times another number represented by x?

5. If the first number is x, what number is 8 less than 5 times the first?

6. If the first number is $7x$, what number is one half the first, decreased by 3?

7. What is the sum of two consecutive positive odd integers and 7, if x represents the smallest positive odd integer?

8. What is the sum of three consecutive even integers and 9, if x represents the largest of these integers?

9. What is the sum of the squares of two consecutive odd integers, if the smallest integer is represented by x?

10. What number is 9 less than twice the sum of two consecutive integers, if the smallest integer is represented by x?

11. The larger of two numbers is 5 more than twice the smaller number. If the smaller number is x, what is the sum of the two numbers?

12. The larger of two numbers is 7 more than four times the smaller number. If the smaller number is x, what is the sum of the two numbers?

13. What number is $9x$ more than $11x + 7$?

14. Find the sum of three numbers if the second number is three times as large as the first number represented by x, and the third number is 4 greater than the second.

15. A father is six times as old as his son is now. If his son is x years old now, how old will the father be in 4 years?

16. Rose is twice as old as Lisa is now. If Lisa is x years old now, find the sum of their ages 7 years ago.

17. A woman has four daughters, each daughter being 2 years older than the next younger. If the youngest daughter is x years old now, find the sum of their ages 3 years from now.

18. Carlos is 2 years older than Linda and 3 years older than Karen. Find the sum of their ages 7 years ago, if x represents Carlos's age now.

19. How many cents are there in $(q + 2)$ quarters and $(2d + 5)$ dimes?

20. How many cents are there in d dollars, $(q - 3)$ quarters, and $(n + 4)$ nickels?

21. A toy savings bank contains $3n$ nickels, $4d$ dimes, and $5q$ quarters. How many cents are there in the bank?

22. George buys stamps at the post office. If he bought only f five-cent stamps, s six-cent stamps, and t ten-cent stamps, how many cents did George pay for the stamps?

23. A vending machine contains nickels, dimes, and quarters. There are d dimes, 12 fewer nickels than dimes, and as many quarters as there are nickels and dimes together. How many of each type of coin are there in the machine?

24. A newspaper stand coin box contains nickels, dimes, and quarters. There are x dimes, and 3 times as many nickels as dimes. If the number of quarters is twice the sum of nickels and dimes, how many of each type of coin are there in the coin box?

25. What number is 7 less than 30% of a number represented by x?

26. What number is $0.20x$ more than a number represented by x?

27. What number is 5 more than 30% of x?
28. What number exceeds y by 5% of y?
29. What number is 8 more than 80% of a number represented by t?
30. A clock radio originally marked at $\$X$ is discounted by 20%. What is the reduced price of the clock radio?
31. A baseball player batted 0.289 in a certain year and had x hits. How many times was he officially at bat?
32. If a cash register contains total receipts $\$x$ for the day, including 5% sales tax, what are the net receipts?
33. If the annual simple interest of 12% on a loan amounted to $\$(5x + 2)$ over a period of 4 years, how much money was borrowed initially?
34. A herbalist has x pounds of tea worth \$1.80 per pound and 3 pounds of tea worth \$2.40 per pound. How much is a blended mixture of both kinds of tea worth?
35. What is the width of a rectangle whose perimeter is p meters, if the length is 8 meters?
36. The length of a rectangle is 7 more than 3 times the width. If the width is x feet, what is the perimeter of the rectangle?
37. The width of a rectangle is 8 centimeters less than its length. If the length is x centimeters, what is the area of the rectangle?
38. Let x represent the length of a rectangle, whose width is 8 less than twice the length. If the length of the rectangle is increased by 5 and the width is tripled, what is the area of the new rectangle?
39. A women invested x dollars in her savings account, which pays 6% annual simple interest. How much money did she accumulate after 1 year?
40. Alice invested 70% of her money in one bank and 20% in another bank. If she invested x dollars in the second bank, how much did she invest in the first bank?

4.5 Applications of Linear Equations—Word Problems

In Section 4.4, we took our first steps in translating the situations described in simple word problems into algebraic expressions. In this section, we introduce a strategy for solving a wide variety of word problems—many of them practical problems and applications that you are likely to encounter outside of the classroom.

First, a step-by-step procedure that provides a useful guideline:

> Step 1. Read the problem carefully, and clearly identify the question or questions you must answer. Draw a diagram whenever possible to help interpret the given information.
>
> Step 2. List all the unknown quantities involved in the problem, and represent them in terms of a single algebraic symbol (say, x, y, etc.).
>
> Step 3. Use the information given in the problem to write algebraic relationships among the quantities identified in step 2.
>
> Step 4. Combine the algebraic relationships into a single equation.
>
> Step 5. Solve the equation for the unknown.
>
> Step 6. Check your answer to see if it agrees with the facts in the problem.

Number Problems

EXAMPLE 1   The sum of two numbers is 94, and the larger number is 5 less than twice the smaller number. Find the numbers.

SOLUTION   We follow the outlined procedure:

Step 1. Question: What are the two numbers?

Step 2. Unknown quantities:
Let
$$x = \text{the smaller number.}$$
Then
$$94 - x = \text{the larger number.}$$

Step 3. Information given:

$$\text{the larger number} = \text{twice the smaller number} - 5$$
$$= 2x - 5$$

Step 4. Equation: The relationship can be written as

$$94 - x = 2x - 5.$$

Step 5. We solve the equation as follows:

$$94 - x = 2x - 5$$
$$-x - 2x = -5 - 94$$
$$-3x = -99$$
$$x = 33.$$

Thus, the smaller number is 33 and the larger number is $94 - 33 = 61$.

Step 6. Check: Indeed,

$$33 + 61 = 94.$$

Is it also true that

$$94 - 33 = 2(33) - 5?$$

Yes,

$$61 = 66 - 5.$$

Age Problems

Many word problems involving age can be worked by solving first-degree equations. If x is the age of a person now, then the person's age after 5 years is $x + 5$, and the person's age 5 years ago is $x - 5$. The following problems give information about two people's ages at different times.

EXAMPLE 2   Janice is 7 years older than her brother, and 5 years from now the sum of their ages will be 63 years. How old is each now?

SOLUTION   Step 1. Question: What are the ages of Janice and her brother now?

Step 2. Unknown quantities:
   Let

$$x = \text{Janice's age now.}$$

   Then

$$x - 7 = \text{her brother's age now.}$$

Step 3. Information given:

	Janice	Brother
Ages now	x	$x - 7$
Ages 5 years from now	$x + 5$	$(x - 7) + 5$

Janice's age 5 years from now + her brother's age 5 years from now $= 63$.

Step 4. Equation:

$$x + 5 + [(x - 7) + 5] = 63.$$

Step 5. We solve the equation as follows:

$$x + 5 + [(x - 7) + 5] = 63$$
$$x + 5 + x - 7 + 5 = 63$$
$$2x + 3 = 63$$
$$2x = 60$$
$$x = 30.$$

Therefore, Janice's age now is 30, and her brother's age is $30 - 7 = 23$.

Step 6. Check:

$$(30 + 5) + (23 + 5) = 63.$$

Money-Value Problems

Linear equations are often used to solve problems involving a specific number of items, in which each item has a particular value. For example, in coin problems we consider the number and the value of each type of coin. For instance,

$$1 \text{ nickel} = 5¢ = \$0.05$$
$$1 \text{ dime} = 10¢ = \$0.10$$
$$1 \text{ quarter} = 25¢ = \$0.25$$

and

the total value of a quantity of each type of coin
= (the number of coins) × (the value of the coin).

EXAMPLE 3   Pedro has $6.80 in nickels, dimes, and quarters. He has the same number of coins of each kind. How many coins of each kind does he have?

SOLUTION   Let

d = the number of coins of each type Pedro has.

Information given:

Coins	Number of Coins	Individual Value (in dollars)	Total Value (in dollars)
Nickels	d	0.05	$0.05d$
Dimes	d	0.10	$0.10d$
Quarters	d	0.25	$0.25d$

Because the total value of the coins is $6.80, the equation and its solution are

$$0.05d + 0.10d + 0.25d = 6.80$$
$$5d + 10d + 25d = 680 \qquad \text{(We multiplied each side by 100.)}$$
$$40d = 680$$
$$d = 17.$$

Therefore, Pedro has 17 nickels, 17 dimes, and 17 quarters.

Check

$$0.05(17) + 0.10(17) + 0.25(17) = 0.85 + 1.70 + 4.25$$
$$= 6.80.$$

Finance and Investment Problems

Some problems deal with the sale of an item at a price that has been discounted from the original price. We may use the following formula to solve such problems:

sale price = original price − discount

$$S = P - D.$$

Other problems involve the investment of a sum of money (the principal) at a specified interest rate, over a given period of time. We may use the following formula to solve this type of problem:

simple interest = principal × rate × time

$$I = Prt.$$

EXAMPLE 4 A clothing store has discounted the price of a topcoat by 40%. If the sale price of the topcoat is $180, what was the original price?

SOLUTION Let

$$x = \text{the original price of the topcoat.}$$

Then

original price − discount = sale price = $180,

that is,

$$x - \text{discount} = 180$$

and

$$\text{discount} = 40\% \text{ of original price} = 0.40x.$$

The equation and its solution are

$$x - 0.40x = 180$$
$$0.60x = 180$$
$$x = 300.$$

Therefore, the original price of the topcoat was $300.

Check

$$40\% \text{ of } \$300 = 0.40(300)$$
$$= \$120 \text{ discount}$$

and

$$\$300 - \$120 = \$180 \text{ sale price.}$$

EXAMPLE 5   A businesswoman treated a client to a dinner and spent $76.16, which included the 4% tax and a 15% tip. What was the amount of the bill before tax and tip?

SOLUTION   Let

$$x = \text{the original amount of the bill before tax and tip.}$$

The following table summarizes the information given:

Amount of Bill Before Tax and Tip (in dollars)	Tax (in dollars)	Tip (in dollars)
x	$0.04x$	$0.15x$

Because the total bill is $76.16, the equation and its solution are

$$x + 0.04x + 0.15x = 76.16$$
$$1.19x = 76.16$$
$$x = 64.$$

Therefore, the amount of the bill before tax and tip was $64.00.

Check

$$\text{bill before tax and tip} = \$64$$

$$\text{tax: } 4\% \text{ of } \$64 = \$\ 2.56$$

$$\text{tip: } 15\% \text{ of } \$64 = \$\ 9.60$$

$$\text{total} = \overline{\$76.16}.$$

EXAMPLE 6   A small company had $24,000 to invest. It invested some of the money in a bank that paid 8% annual simple interest. The rest of the money was invested in stocks that paid dividends equivalent to 11% annual simple interest. At the end of 1 year, the combined income from these investments was $2,340. How much money was originally invested in stocks?

SOLUTION   In solving this problem, we use the formula

$$I = Prt.$$

Let

$$x = \text{amount of money (in dollars) invested in stocks.}$$

Since a total of $24,000 was invested,

$$24,000 - x = \text{the amount of money (in dollars) invested in the bank.}$$

The following table summarizes the information given about the quantities in the formula $I = Prt$:

Investment	Principal (in dollars)	Rate	Time (in years)	Simple Interest (in dollars)
Stocks	x	0.11	1	$0.11x$
Bank	$24,000 - x$	0.08	1	$0.08(24,000 - x)$

The combined simple interest is $2,340. The equation and its solution are as follows:

$$0.11x + 0.08(24,000 - x) = 2,340$$

$$11x + 8(24,000 - x) = 234,000$$

$$11x + 192,000 - 8x = 234,000$$

$$3x = 42,000$$

$$x = 14,000.$$

Therefore, $14,000 was invested in stocks.

Check

$$0.11(14{,}000) + 0.08(10{,}000) = 1{,}540 + 800$$
$$= 2{,}340.$$

Geometric Problems

Many problems involve perimeters and areas of squares and rectangles. To solve these problems, we use the following formulas for areas and perimeters:

	Length	Width	Area A	Perimeter P
Rectangle	ℓ	w	$A = \ell w$	$P = 2\ell + 2w$
Square	ℓ	ℓ	$A = \ell^2$	$P = 4\ell$

EXAMPLE 7   A 130-meter length of fence is used to enclose a rectangular garden. The length of the garden is 5 meters more than its width. Find the length and the width of the garden to be enclosed.

SOLUTION   Let

$$x = \text{the width of the garden (in meters).}$$

Then

$$x + 5 = \text{the length of the garden (in meters).}$$

Thus

$$2(\text{length}) + 2(\text{width}) = \text{length of fence}$$
$$= \text{perimeter of garden.}$$

The equation and its solution are

$$2(x + 5) + 2x = 130$$
$$2x + 10 + 2x = 130$$
$$4x + 10 = 130$$
$$4x = 120$$
$$x = 30.$$

Therefore, the width of the garden is 30 meters and the length is 35 meters.

Figure 1

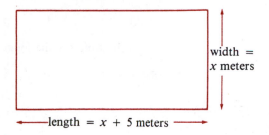

width =
x meters

length = $x + 5$ meters

Check

$$2(\text{length}) + 2(\text{width}) = 2(35) + 2(30)$$
$$= 70 + 60$$
$$= 130 = \text{length of fence.}$$

Motion Problems

Problems involving motion may also lead to first-degree equations. For example, if an object moves a distance d at a constant rate r (also called the **speed**) in t units of time, then

$$d = rt.$$

EXAMPLE 8   A bicycle rider travels 18 miles in the same amount of time that it takes a jogger to travel 10 miles. If the bicyclist goes 9.6 miles per hour faster than the jogger, how fast does each person travel?

SOLUTION   Let

$$r = \text{the jogger's rate (in miles per hour).}$$

Then

$$r + 9.6 = \text{the bicyclist's rate (in miles per hour).}$$

The following table summarizes the given information:

	Rate (in miles per hour)	Distance (in miles)	Time (in hours) = $\dfrac{\text{Distance}}{\text{Rate}}$
Jogger	r	10	$\dfrac{10}{r}$
Bicycle rider	$r + 9.6$	18	$\dfrac{18}{r + 9.6}$

We are told that

the time for the jogger = the time for the bicycle rider.

Therefore, the equation is

$$\frac{10}{r} = \frac{18}{r + 9.6}.$$

We solve the equation by multiplying each side by $r(r + 9.6)$, the LCD, and solve for r.

$$r(r + 9.6)\left(\frac{10}{r}\right) = r(r + 9.6)\left(\frac{18}{r + 9.6}\right)$$

$$10r + 96 = 18r$$

$$-8r = -96$$

$$r = 12.$$

Therefore, the speed of the jogger is 12 miles per hour and the speed of the bicycle rider is $12 + 9.6 = 21.6$ miles per hour.

Check

$$\frac{10}{12} = \frac{18}{21.6} = \frac{5}{6}.$$

Mixture Problems

Problems involving mixtures of substances can often be worked by solving first-degree equations.

EXAMPLE 9   A chemist has one solution containing a 16% concentration of acid and a second solution containing a 26% concentration of acid. How many milliliters of each should be mixed to obtain 30 milliliters of a solution containing an 18% concentration of acid?

SOLUTION   Let

x = the number of milliliters of the first solution.

Then

$30 - x$ = the number of milliliters of the second solution.

The following table summarizes the given information:

	Milliliters of Solution	Acid Concentration	Milliliters of Acid in Solution
First solution	x	0.16	$0.16x$
Second solution	$30 - x$	0.26	$0.26(30 - x)$
Mixture	30	0.18	$0.18(30) = 5.4$

Because the amount of acid in the mixture is the sum of the amounts of acid in the two solutions, the equation and its solution are

$$0.16x + 0.26(30 - x) = 5.4$$

$$0.16x + 7.8 - 0.26x = 5.4$$

$$-0.10x = -2.4$$

$$x = 24.$$

Therefore, 24 milliliters of the first solution and $30 - 24 = 6$ milliliters of the second solution should be mixed.

Check

$$16\% \text{ of } 24 = 0.16(24) = 3.84$$

$$26\% \text{ of } 6 = 0.26(6) = 1.56$$

so that

$$3.84 + 1.56 = 5.4 = 0.18(30).$$

Work Problems

Problems concerning a job that is done at a constant rate can be solved with the help of the following principle: If a job can be done in t hours, then $1/t$ of the job can be done in 1 hour.

EXAMPLE 10　Two cranes, operating together, can unload a cargo ship in 4 hours. If it takes one crane twice as long as the other to unload the ship, how long would it take each crane to unload the ship by itself?

SOLUTION   Let

> x = the number of hours for the faster crane to unload the ship.

Then

> $2x$ = the number of hours for the slower crane to unload the ship.

The given information is displayed in the following table:

Cranes	Part of Job Done in 1 Hour	Number of Hours	Part of Job Done in 4 Hours
Faster crane	$\dfrac{1}{x}$	4	$\dfrac{4}{x}$
Slower crane	$\dfrac{1}{2x}$	4	$\dfrac{4}{2x}$

When the two cranes operate together,

$$\left(\begin{matrix}\text{part of job done}\\\text{by faster crane}\end{matrix}\right) + \left(\begin{matrix}\text{part of job done}\\\text{by slower crane}\end{matrix}\right) = \left(\begin{matrix}\text{part of job done}\\\text{by both cranes in 4 hours}\end{matrix}\right)$$
$$= 1 \quad \text{(the entire job).}$$

The equation is

$$\frac{4}{x} + \frac{4}{2x} = 1.$$

To solve the equation, multiply both sides by $2x$, the LCD:

$$2x\left(\frac{4}{x} + \frac{4}{2x}\right) = 2x(1)$$

$$8 + 4 = 2x$$

$$2x = 12$$

$$x = 6.$$

Therefore, it would take the faster crane 6 hours to unload the ship by itself, and it would take $2 \cdot 6 = 12$ hours for the slower crane to unload the ship by itself.

Check

$$\frac{4}{6} + \frac{4}{12} = \frac{2}{3} + \frac{1}{3} = 1.$$

PROBLEM SET 4.5

c̄ In some of the following problems, a calculator may be useful to speed up the arithmetic.

1. Three more than twice a certain number is 57. Find the number.

2. Find two numbers whose sum is 18, if one number is 8 larger than the other.

3. Find two consecutive even integers such that seven times the first exceeds five times the second by 54.

4. Find three consecutive even integers such that the first plus twice the second plus four times the third equals 174.

5. One-fourth of a number is 3 greater than one-sixth of it. Find the number.

6. Two-thirds of a number plus five-sixths of the same number is equal to 42. What is the number?

7. Gus is 5 years younger than his brother, and 8 years from now he will be four-fifths as old as his brother is then. How old is each now?

8. Wendy's mother is three times as old as Wendy, and 14 years from now she will be twice as old as Wendy is then. How old is each now?

9. Raul is 3 years older than his brother, and 4 years from now the sum of their ages will be 33 years. How old is each now?

10. Jose is 5 years younger than his brother, and 3 years ago the sum of their ages was 23. How old is each now?

11. At a Christmas party, there are five times as many men as women. If 12 more women arrive, there will be only twice as many men as women. How many men are at the party?

12. Psychologists define an individual's intelligence quotient (IQ) to be 100 times the person's mental age divided by his or her chronological age. What is the chronological age of a person with an IQ of 160 and a mental age of 16?

13. A pay-phone slot receives quarters, dimes, and nickels. When the phone box was emptied, it yielded $6.50 in coins. If there were four more dimes than quarters and three times as many nickels as dimes, find the number of coins of each kind.

14. A cashier has four times as many nickels as quarters. She has $3.60. How many of each coin does she have?

15. A parking-meter slot receives dimes and nickels. When emptied, the box produced 70 coins worth $4.85. How many nickels and how many dimes were there?

16. A bank teller has $75 in $1 bills and $5 bills. He has three more $1 bills than $5 bills. How many bills of each kind does he have?

17. A vending machine receives dimes and quarters. When the machine is emptied, $12.50 worth of coins is found. If the number of dimes is 20 more than the number of quarters, find the number of each kind of coin.

18. Two electricians worked a total of 12 hours one day. One electrician earns $15 per hour and the other earns $18 per hour. If that day's payroll is $195, how many hours did each electrician work?

19. A pants suit on sale is discounted by 30%. If the selling price is $84, what was its original price?

20. A racquetball racket is on sale at 20% off. If the saving is $7.80, find the original price of the racket.

21. Joe has received an 11% increase in salary. His new weekly earnings are $222. What was his previous weekly salary?

22. Mary received an 18% increase in salary. If the increase was $54 per week, what was her weekly salary before the increase?

23. At the end of the year, a car dealer advertises that the list prices on all of last year's models have been discounted by 15%. What was the original price of a car that now carries a discounted price of $6,849.30?

24. Harry's stocks show a 38% increase in value for the year. If the stocks are now worth $27,000, what were they worth one year ago?

25. On his first dinner date with Lucy, Jack spent $19.04, which included the 4% tax and a 15% tip on the original bill. What was the amount of the bill before tax and tip?

26. John's new annual salary is $23,976, which includes a 7% pay raise and the addition of a 4% cost-of-living allowance. What was his original salary?

27. A tire was sold for $35.65, which included the 4% state sales tax and an 11% federal tax. What was the original price of the tire before the taxes?

28. A man has the first $1,200 exempted from his income tax, but he pays a 20% tax on the remainder of his income. If the total taxes he pays after the exemption are $5,440, find his total income.

29. A businesswoman had $18,000 to invest. She invested some of it in a bank certificate that paid 13% annual simple interest, and the rest in another certificate that paid 14% annual simple interest. At the end of 1 year, the combined income from these investments was $2,395. How much money did she originally invest at 13% interest?

30. A family invests a total of $85,000 in two tax-free municipal bonds to reduce its income tax. One bond pays 8% tax-free simple annual interest, and the other bond pays 8.5%. The total nontaxable income from both investments at the end of 1 year is $7,072. How much did the family invest in each bond?

31. A person invests part of $62,000 in a certificate that yields 13.2% simple annual interest, and puts the rest of the money in a certificate that yields 13.7% simple annual interest. At the end of 1 year, the combined interest on the two certificates is $8,354. How much money did the person invest in each certificate?

32. A retail store invested $25,000 in two kinds of toys. Over the course of 1 year, it made a profit of 15% from the first kind but lost 5% on the second kind. If the income from the two investments was a return of 8% on the entire amount invested, how much had the store invested in each kind of toy?

33. A certain amount of money is invested in a passbook savings account at 7% simple annual interest. In addition, $8,000 is invested in a certificate that pays 14% simple annual interest. The income from both investments amounts to 11.5% of their total. How much money is invested at 7%?

34. A small company takes advantage of a state income tax credit that provides 15% of the cost of installing solar-heating equipment, and 8% of the cost of upgrading insulation in a building. After spending a total of $6,510 on insulation and solar heating, the company receives a state income tax credit of $854. How much money was spent on solar heating?

35. The length of a rectangular house lot is twice the width, and the difference between the length and the width is 32 meters. What are the dimensions of the lot?

36. The length of a rectangular rug is 6 feet more than the width. The perimeter is 40 feet. Find the length and the width of the rug.

37. The length of a rectangle is four times its width. What are its dimensions if its perimeter is 150 meters?

38. The length of a rectangle is 17 centimeters less than three times the width, and the perimeter is 238 centimeters. Find the dimensions of the rectangle.

39. The length and width of a square are increased by 6 feet and 8 feet, respectively. The result is a rectangle whose area is 188 square feet more than the area of the square. Find the length of a side of the square.

40. The sides of two squares differ by 6 centimeters and their areas differ by 468 square centimeters. Find the lengths of the sides of the squares.

41. Carlos can walk 8 miles in the same time it takes him to jog 12 miles. His jogging rate is 5 miles per hour faster than his walking rate. At what rate does he walk? How long does it take him to walk 8 miles?

42. A bicycle rider travels 8 miles per hour faster than a jogger. It takes the bicycle rider half as much time as it takes the jogger to travel 16 miles. Find the jogger's speed.

43. If a freight train traveling at 30 miles per hour is 300 miles ahead of an express train traveling at 55 miles per hour, how long will it take the express train to catch up with the freight train?

44. An airplane travels 1,620 kilometers in the same time it takes a train to travel 180 kilometers. If the airplane goes 480 kilometers per hour faster than the train, find the rate of each.

45. Maria jogs 15 miles and bicycles back. The total trip requires 3 hours. If she bicycles twice as fast as she jogs, how fast does she bicycle?

46. Joshua rides his bicycle 5 miles from his home to the school bus stop at a rate of 8 miles per hour. He arrives in time to catch the bus, which travels at 25 miles per hour. If he spends $1\frac{1}{2}$ hours traveling from home to school, how far does he travel on the bus?

47. A chemist has one solution containing a 10% concentration of acid and a second solution containing a 15% concentration of acid. How many milliliters of each should be mixed in order to obtain 10 milliliters of a solution containing a 12% concentration of acid?

48. A chemist has 10 milliliters of a solution that contains a 30% concentration of acid. How many milliliters of pure acid must be added in order to increase the concentration to 50%?

49. A petroleum distributor has two gasohol storage tanks. The first contains 9% alcohol and the second contains 12% alcohol. The distributor receives an order for 300,000 gallons of gasohol containing 10% alcohol. How can this order be filled by mixing gasohol from the two storage tanks?

50. A car radiator contains 8 quarts of a mixture of water and antifreeze. If 40% of the mixture is antifreeze, how much of the mixture should be drained and replaced by pure antifreeze so that the resultant mixture will contain 60% antifreeze?

51. A grocer mixes two kinds of coffee to form a blend that sells for $5.85 per pound. He mixes coffee selling for $5.80 per pound with coffee selling for $6.00 per pound to get 100 pounds of the blend. How many pounds of each kind of coffee does he use?

52. A grocer has 100 pounds of candy worth $4.80 per pound. How many pounds of a different type of candy worth $5.20 per pound should he mix with the 100 pounds in order to obtain a mixture worth $5.00 per pound?

53. At a factory, smokestack A pollutes the air 1.25 times faster than smokestack B. How long would it take smokestack B, operating alone, to pollute the air by as much as both smokestacks do in 20 hours?

54. Two plumbers can do a job together in 6 days. If the first plumber can do the job alone in 10 days, how long will it take the second plumber to do the job?

55. A computer can do a payroll in 12 hours. A second computer can do the payroll in 6 hours. How long will it take to do the payroll if both computers operate at the same time?

56. One pipe can fill a tank in 18 minutes, and another pipe can fill it in 24 minutes. The drain pipe can empty the tank in 15 minutes. With all pipes open, how long will it take to fill the tank?

57. John can mow a lawn in 1 hour and 20 minutes. Tom can mow the same lawn in 2 hours. How long would it take John and Tom together to mow the lawn?

58. Jamal can fill the vending machines in 45 minutes. However, if his brother Gus helps, it takes them only 20 minutes. How long would it take Gus to fill the machines by himself?

4.6 Inequalities

Statements such as

$$2x + 5 < 3 \qquad 3y + 7 > 10$$
$$4x - 3 \leq 7x + 1 \qquad 7t - 13 \geq 4t + 2$$

are examples of inequalities. An **inequality** is a statement that an expression representing some real number is *greater than* (or *less than*) another expression representing a real number.

If the point with coordinate a lies to the left of the point with coordinate b on the number line (Figure 1), we say that a is **less than** b (or, equivalently, that b is **greater than** a), and by using the comparison symbols, we write $a < b$ (or $b > a$).

Figure 1

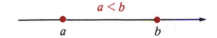

More formally, we have the following definition:

DEFINITION 1   **Inequality**

> $a < b$ (or $b > a$) means that $b - a$ is a positive number.
> A statement of the form $a < b$ (or $b > a$) is called an **inequality.**

For example, $3 < 7$ (or $7 > 3$) because $7 - 3 = 4$ (a positive number). Also, $-5 < -3$ (or $-3 > -5$) because $-3 - (-5) = 2$ (a positive number). Recall from Section 1.1 that other inequality signs are $\leq$, which means *less than or equal to*, and $\geq$, which means *greater than or equal to*. By definition,

$$a \leq b \qquad \text{if either} \qquad a < b \qquad \text{or} \qquad a = b$$

and

$$a \geq b \qquad \text{if either} \qquad a > b \qquad \text{or} \qquad a = b.$$

By writing one of the symbols $<$, $\leq$, $>$, or $\geq$ between two expressions, we obtain an inequality. The two expressions are called the **sides** or **members** of the inequality.

An inequality containing a variable will produce a statement that is either true or false when a particular number is substituted for the variable. If we substitute a number for the variable and obtain a true statement, we say that the number **satisfies** the inequality, and that the number is a **solution** of the inequality. The set of all solutions of an inequality is called the **solution set.**

To *graph* an inequality on a number line, we sketch the graph of its solution set.

In Examples 1–3, sketch the graph of each solution set on a number line.

EXAMPLE 1 $\{x \mid x < \frac{3}{2}\}$

SOLUTION   The graph of the solution set $\{x \mid x < \frac{3}{2}\}$, which is read as "the set of all x such that x is less than $\frac{3}{2}$," is shown by the colored arrow drawn to the left of $\frac{3}{2}$ to indicate all numbers less than $\frac{3}{2}$. The *parenthesis*,), at the point with coordinate $\frac{3}{2}$ indicates that $\frac{3}{2}$ is *excluded* from the solution set (Figure 2).

Figure 2

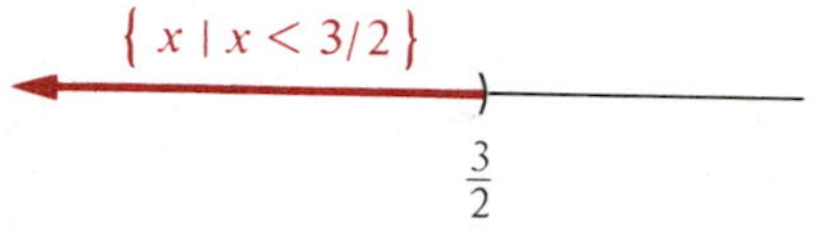

$\{ x \mid x < 3/2 \}$

$\frac{3}{2}$

EXAMPLE 2 $\{x \mid x \geq 4\}$

SOLUTION   The graph of the solution set of $\{x \mid x \geq 4\}$ is shown by the colored arrow drawn to the right of 4. The bracket, [, at the point with coordinate 4 indicates that 4 is *included* in the solution set (Figure 3).

Figure 3

$\{x \mid x \geqslant 4\}$

4

EXAMPLE 3 $\{x \mid x \leq -2\}$

SOLUTION   The graph of the solution set $\{x \mid x \leq -2\}$ consists of all real numbers less than and including -2 (Figure 4).

Figure 4

$\{x \mid x \leqslant -2\}$

-2

Now we graph inequalities connected with the word *and* or the word *or*. When there are two inequalities connected with the word "and" or the word "or," it is called a **compound inequality.** *A compound inequality connected with the word* and *is true if and only if both inequalities are true.* For example, to graph the solution set of

$$\{x \mid x > -1\} \quad \text{and} \quad \{x \mid x < 4\},$$

Figure 5

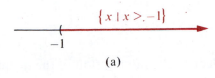

(a)

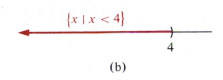

(b)

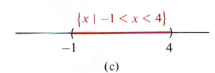

(c)

we first graph each inequality separately. Figure 5a shows the graph of $x > -1$. Figure 5b shows the graph of $x < 4$. Since the two inequalities are connected by the word *and*, we graph the part they have in common, that is, all numbers between -1 and 4 (Figure 5c).

Sometimes the compound inequality

$$x > -1 \quad \text{and} \quad x < 4$$

is written in the equivalent shorter form

$$-1 < x < 4.$$

We refer to the set of all elements common to the two sets $\{x \mid x > -1\}$ and $\{x \mid x < 4\}$ as the *intersection* of the two sets. Thus, in our example we see that the intersection of the sets $\{x \mid x > -1\}$ and $\{x \mid x < 4\}$ is $\{x \mid -1 < x < 4\}$. In general, the intersection of two sets is defined as follows:

DEFINITION 2   **The Intersection of Two Sets**

> The **intersection** of two sets A and B, written as $A \cap B$, is the set of all elements that are both in A and in B. Using set-builder notation, we write
>
> $$A \cap B = \{x \mid x \in A \quad \text{and} \quad x \in B\}.$$

EXAMPLE 4   Graph the solution set of

$$\{x \mid x \geq -1\} \cap \{x \mid x \leq 2\}$$

on a number line.

SOLUTION   The solution set of the compound inequality is the set of all real numbers x that are between and including the numbers -1 and 2 (Figure 6). The solution set $\{x \mid x \geq -1\} \cap \{x \mid x \leq 2\}$ is written as $\{x \mid -1 \leq x \leq 2\}$.

Figure 6

$$\{x \mid x \geqslant -1\} \cap \{x \mid x \leqslant 2\}$$

$$\underset{-1 \qquad\qquad 2}{\vdash\!\!\rule[0.5ex]{4em}{0.4pt}\!\!\dashv}$$

EXAMPLE 5   Graph the solution set

$$\{x \mid 1 \le x < 3\}$$

on a number line.

SOLUTION   The inequality $1 \le x < 3$ is equivalent to the intersection of the inequalities $1 \le x$ and $x < 3$. The solution set of the combined inequalities is the set of all real numbers in which x is between the numbers 1 and 3 and includes 1 (Figure 7).

Figure 7

Figure 8

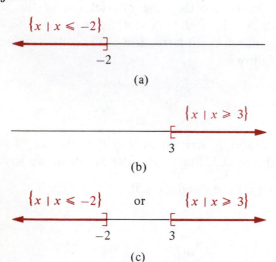

(a)

(b)

(c)

Now we graph the solution set of the compound inequality

$$\{x \mid x \le -2\} \quad \text{or} \quad \{x \mid x \ge 3\}.$$

We begin by graphing each inequality separately. Figure 8a shows the graph of $x \le -2$, and Figure 8b shows the graph of $x \ge 3$. Since the two inequalities are connected by the word *or* we graph all real numbers less than or equal to -2, along with all real numbers greater than or equal to 3 (Figure 8c). The solution set is written $\{x \mid x \le -2 \text{ or } x \ge 3\}$.

We refer to the set of all elements that are either in the set $\{x \mid x \le -2\}$ or in the set $\{x \mid x \ge 3\}$ as the *union* of the two sets. More formally we have:

DEFINITION 3   **The Union of Two Sets**

The *union* of two sets A and B, written $A \cup B$, is the set of all elements that are either in A or in B, or in both A and B. Using set-builder notation we write

$$A \cup B = \{x \mid x \in A \quad \text{or} \quad x \in B\}.$$

EXAMPLE 6   Graph the solution set of

$$\{x \mid x \le -5\} \cup \{x \mid x > 1\}.$$

SOLUTION   The graph of the solution set consists of all real numbers less than or equal to -5, along with all real numbers that are greater than 1 (Figure 9). Sometimes we omit the set-builder notation and write the solution set as

$$x \leq -5 \quad \text{or} \quad x > 1.$$

Figure 9

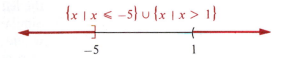

$$\{x \mid x \leq -5\} \cup \{x \mid x > 1\}$$

EXAMPLE 7   Graph the solution set of

$$\{x \mid x \leq -1\} \cup \{x \mid 2 < x < 5\}$$

on a number line.

SOLUTION   The graph of the solution set of the combined inequalities is the set of all real numbers that are either less than or equal to -1, or between 2 and 5 (Figure 10).

Figure 10

$$\{x \mid x \leq -1\} \cup \{x \mid 2 < x < 5\}$$

Properties of Inequalities

The following properties regulate how we work with inequalities.

Let us consider the effect of adding the same number to both sides of an inequality. For instance, suppose 3 is added to both sides of the inequality $2 < 4$. We have $2 + 3 < 4 + 3$, or, equivalently, $5 < 7$. The result of adding is illustrated in Figure 11. By adding 3 to both sides, we have moved (also called translated) both points in the same direction and the same distance along the number line. The relation between the sums will be the same as between the original two numbers. This example can be generalized as follows:

Figure 11

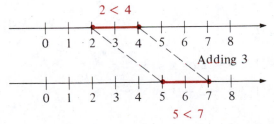

1 Addition Property for Inequalities

For any real numbers a, b and c, if $a < b$, then $a + c < b + c$.

Figure 12

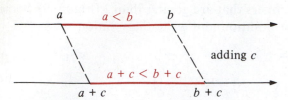

Property 1 is illustrated in Figure 12 where $c > 0$. We see in this figure that, if a lies to the left of b, then by moving from a and b the same number of units in the same direction along the number line, the resulting point $a + c$ must lie to the left of the point $b + c$; that is, $a + c < b + c$. Similarly, if $c < 0$, the points $a + c$ and $b + c$ would move to the left of the points a and b, respectively; and $a + c < b + c$ holds.

Since subtraction is defined in terms of addition, Property 1 holds for subtraction as well as addition. That is, if the same number is subtracted from both sides of an inequality the inequality is preserved. Thus,

> If a and b are real numbers, and $a < b$, then for any real number c,
>
> $$a - c < b - c.$$

The Addition Property is also true for an inequality that contains the symbols $\leq$ or $\geq$.

In Examples 8 and 9, insert the symbol $<$ or $>$ in the blank so that the resulting statement is true.

EXAMPLE 8

(a) Since $-3 < 4$, then $-3 + 5$ ____ $4 + 5$.

(b) Since $3 > 2$, then $3 - 7$ ____ $2 - 7$.

SOLUTION

(a) Since $-3 < 4$, then $-3 + 5 < 4 + 5$, or $2 < 9$.

(b) Since $3 > 2$, then $3 - 7 > 2 - 7$, or $-4 > -5$.

EXAMPLE 9

(a) If $x + 2 < 5$, then x ____ 3.

(b) If $x - 3 > -2$, then x ____ 1.

SOLUTION

(a) If $x + 2 < 5$, then $x + 2 - 2 < 5 - 2$ or $x < 3$.

(b) If $x - 3 > -2$, then $x - 3 + 3 > -2 + 3$ or $x > 1$.

Now consider the effect of multiplying both sides of an inequality by any real number. We shall see that the result of multiplication of both sides of an inequality by a number can either preserve or reverse the sign of the original inequality, depending on whether we multiply by a positive number or a negative number.

Figure 13

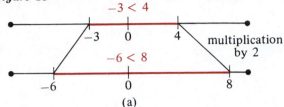

multiplication by 2

(a)

multiplication by −2

(b)

We use a specific example to illustrate.

Consider $-3 < 4$ and multiply both sides by 2 (Figure 13a). We see from the figure that multiplying both sides of the inequality $-3 < 4$ by 2 produces a new inequality $(-3) \cdot 2 < (4) \cdot 2$, or $-6 < 8$. The original inequality was preserved under the operation of multiplication by a positive number.

On the other hand, if we multiply both sides of the inequality $-3 < 4$ by -2, we have $(-3)(-2) > 4(-2)$, or $6 > -8$, as shown in Figure 13b. Here multiplication by -2 reverses the inequality. This can be generalized as follows:

2 Multiplication Properties for Inequalities

(i) If $a < b$ and $c > 0$, then $ac < bc$.

(ii) If $a < b$ and $c < 0$, then $ac > bc$.

In Figure 14, we illustrate Properties 2(i) and 2(ii), by letting $a = 3$, $b = 6$, $c = 2$ (Figure 14a) and $c = -2$ (Figure 14b).

Figure 14

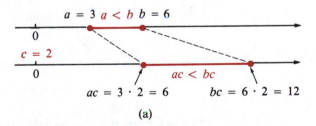

(a)

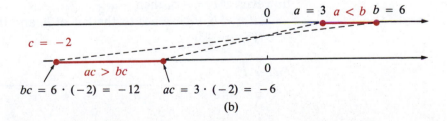

(b)

In Examples 10 and 11, insert the symbol < or > in the blank so that the resulting statement is true.

EXAMPLE 10   (a) Since $-4 < -3$, then $(-4)5$ ___ $(-3)5$.

(b) Since $4 > 3$, then $4(-7)$ ___ $3(-7)$.

SOLUTION   (a) Since $-4 < -3$, then $(-4)5 < (-3)5$ or $-20 < -15$.

(b) Since $4 > 3$, then $4(-7) < 3(-7)$ or $-28 < -21$.

EXAMPLE 11   (a) If $\frac{1}{2}x < 3$, then x ___ 6.

(b) If $(-1/5)x > 2$, then x ___ -10.

SOLUTION   (a) If $\frac{1}{2}x < 3$, then $2(\frac{1}{2}x) < 2(3)$ or $x < 6$.

(b) If $(-1/5)x > 2$, then $-5[(-1/5)x] < -5(2)$ or $x < -10$.

Since division by a nonzero real number can be expressed as a multiplication, we can conclude from Property 2 that dividing both sides of an inequality by a positive number does not change the order of the inequality, whereas dividing both sides of an inequality by a negative number reverses the inequality; that is:

3 Division Properties

For a, b and c real numbers

(i) If $a < b$ and $c > 0$, then $\dfrac{a}{c} < \dfrac{b}{c}$.

(ii) If $a < b$ and $c < 0$, then $\dfrac{a}{c} > \dfrac{b}{c}$.

For example, if $a = 3$ and $b = 4$, we know that $3 < 4$. If $c = 5$, then $\frac{3}{5} < \frac{4}{5}$; however, if $c = -5$, then $-\frac{3}{5} > -\frac{4}{5}$.

If $a < b$, then a lies to the left of b, and if $b < c$, then b lies to the left of c (Figure 15).

Figure 15

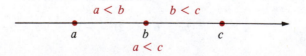

We see from this figure that if a lies to the left of b and b lies to the left of c, then a must lie to the left of c; that is, $a < c$. For example, since $3 < 5$ and $5 < 7$, then $3 < 7$. This notion is called the transitive property and can be stated as follows:

4 Transitive Property

If a, b, and c are real numbers such that $a < b$ and $b < c$, then $a < c$.

For example, since $4 < 8$ and $8 < 12$, then $4 < 12$. Also, since $2 > -4$ and $-4 > -6$, then $2 > -6$.

It is important to note that the properties of inequalities also hold when the symbol $<$ is replaced by $>$, $\leq$, or $\geq$. For example, the addition property could be stated as: if $a \geq b$, then $a + c \geq b + c$ for any real number c.

EXAMPLE 12   State the property of the inequality that justifies each statement.

(a) $2 < 3$, so it follows that $2 + 5 < 3 + 5$.

(b) $-2 < -1$, so it follows that $-2 + (-3) < -1 + (-3)$.

(c) $3 < 7$, so it follows that $3(4) < 7(4)$

(d) $x < y$, so it follows that $\dfrac{x}{-3} > \dfrac{y}{-3}$.

(e) $-2 < -1$ and $-1 < 5$, so it follows that $-2 < 5$.

SOLUTION   (a) The addition property   (b) The addition property

(c) The multiplication property   (d) The division property

(e) The transitive property

PROBLEM SET 4.6

In problems 1–12, graph each solution set on a number line.

1. $\{x \mid x \leq 1\}$ **2.** $\{x \mid x \geq -3\}$ **3.** $\{x \mid x \geq -\frac{3}{4}\}$ **4.** $\{x \mid x < \frac{5}{2}\}$
5. $\{x \mid x \geq -4\}$ **6.** $\{x \mid -3 < x\}$ **7.** $\{x \mid 3 > x\}$ **8.** $\{x \mid 4 \geq x\}$
9. $\{x \mid x \geq 0\}$ **10.** $\{x \mid x < 0\}$ **11.** $\{x \mid -5 > x\}$ **12.** $\{x \mid -5 \leq x\}$

In problems 13–24, graph the solution set of each compound inequality on a number line.

13. $\{x \mid x \geq -3\} \cap \{x \mid x \leq 3\}$ **14.** $\{x \mid x \geq -1\} \cap \{x \mid x \leq 1\}$ **15.** $\{x \mid x < 1\} \cap \{x \mid x \geq -2\}$

16. $\{x \mid x \geq -2\} \cap \{x \mid x \leq 4\}$

17. $\{x \mid -4 < x\} \cap \{x \mid x \leq 1\}$

18. $\{x \mid x \geq -3\} \cap \{x \mid x \leq 0\}$

19. $\{x \mid x < -2\} \cup \{x \mid x \geq 2\}$

20. $\{x \mid x \geq 0\} \cup \{x \mid x \leq -2\}$

21. $\{x \mid x < -\frac{1}{2}\} \cup \{x \mid x \geq \frac{5}{3}\}$

22. $\{x \mid x < 0\} \cup \{x \mid x > 5\}$

23. $\{x \mid x > \frac{7}{3}\} \cup \{x \mid x < \frac{2}{3}\}$

24. $\{x \mid x > -\frac{3}{2}\} \cup \{x \mid x < -3\}$

In problems 25–34, graph the solution set on a number line.

25. $\{x \mid -1 \leq x \leq 3\}$

26. $\{x \mid -3 < x < 2\}$

27. $\{x \mid -2 \leq x < 5\}$

28. $\{x \mid 2 \leq x \leq 6\}$

29. $\{x \mid -4 < x < 2\}$

30. $\{x \mid -4 \leq x \leq 0\}$

31. $\{x \mid x < -2\} \cup \{x \mid 1 < x < 3\}$

32. $\{x \mid -2 < x < 0\} \cup \{x \mid x > 4\}$

33. $\{x \mid -2 < x < 7\} \cup \{x \mid -1 \leq x \leq 3\}$

34. $\{x \mid -3 < x < -1\} \cup \{x \mid 1 \leq x \leq 2\}$

In problems 35–48, insert the symbol > or < in the blank so that each resulting statement is true.

35. Since $2 < 3$, then $2 + 3$ ___ $3 + 3$.

36. Since $2 > -5$, then $2 - 6$ ___ $-5 - 6$.

37. Since $-4 < -3$, then $(-4)5$ ___ $(-3)5$.

38. Since $-1 > -2$, then $(-1)(-4)$ ___ $(-2)(-4)$.

39. Since $5 > 2$, then $5 - 1$ ___ $2 - 1$.

40. Since $-3 < 2$, then $(-3)/5$ ___ $2/5$.

41. Since $1 < 2$ and $2 < 5$, then 1 ___ 5.

42. Since $-2 < -1$ and $-1 < 3$, then -2 ___ 3.

43. If $x > 4$, then $x + 3$ ___ $4 + 3$.

44. If $x < -3$, then $x - 2$ ___ $-3 - 2$.

45. If $x < 5$, then $2x$ ___ 10.

46. If $x > -2$, then $-3x$ ___ 6.

47. If $-4x < 16$, then x ___ -4.

48. If $(-1/3)x < 2$, then x ___ -6.

In problems 49–58, state the property of inequalities that justifies each statement.

49. $2 < 4$, so it follows that $2 + 3 < 4 + 3$.

50. $-3 < 0$, so it follows that $-3 + (-1) < 0 + (-1)$.

51. $-5 < -3$ and $-3 < 4$, so it follows that $-5 < 4$.

52. $7 > -2$ and $-2 > -4$, so it follows that $7 > -4$.

53. If $x > -4$, then $-5x < 20$.

54. If $t < 10$, than $\dfrac{t}{-2} > -5$.

55. If $a - 3 > 2$, then $a > 5$.

56. If $y - 2 < 7$, then $y < 9$.

57. If $\dfrac{-x}{3} \leq -5$, then $x \geq 15$.

58. If $a > 1$, then $a^3 > a$.

4.7 Linear Inequalities

A **first-degree** or **linear inequality** is an inequality of the form

$$ax + b < c,$$

where a, b, and c are constant real numbers with $a \neq 0$ and x is a variable.

The symbol $<$ can be replaced with any of the other three inequality symbols, $\leq$, $>$, or $\geq$. Two examples are

$$5(x - 3) \geq 4 \qquad \text{and} \qquad 3 - 5x < 2.$$

To **solve** a linear inequality—that is, to **find its solution set**—we use a technique similar to the step-by-step procedure used to solve linear equations (page 155). However, we must be careful when using the multiplication (or division) property (discussed in Section 4.6), since multiplying or dividing both sides of an inequality by a negative number *always reverses* the sense of the inequality.

To solve the inequality

$$x + 3 \geq 7$$

we apply the subtraction property, so that

$$x + 3 - 3 \geq 7 - 3$$

or

$$x \geq 4.$$

The solution set is $\{x \mid x \geq 4\}$. We graph the solution set on a number line as shown in Figure 1.

Figure 1

In Examples 1–4, solve each inequality and graph the solution set on a number line.

EXAMPLE 1 $x + 2 < 4$

SOLUTION

$$x + 2 < 4$$
$$x + 2 - 2 < 4 - 2 \qquad \text{(We subtracted 2 from both sides.)}$$
$$x < 2.$$

Thus, the solution set is $\{x \mid x < 2\}$ (Figure 2).

Figure 2

EXAMPLE 2 $3x \geq 12$

SOLUTION

$$3x \geq 12$$

$$\frac{3x}{3} \geq \frac{12}{3} \qquad \text{(We divided both sides by 3, a positive number.)}$$

$$x \geq 4.$$

Thus, the solution set is $\{x \mid x \geq 4\}$ (Figure 3).

Figure 3

EXAMPLE 3 $-5x \geq 15$

SOLUTION

$$-5x \geq 15$$

$$\frac{-5x}{-5} \leq \frac{15}{-5} \qquad \text{(We divided both sides by } -5 \text{ and reversed the inequality because } -5 \text{ is a negative number.)}$$

$$x \leq -3.$$

Thus, the solution set is $\{x \mid x \leq -3\}$ (Figure 4).

Figure 4

EXAMPLE 4 $2x + 3 < 11 - 2x.$

SOLUTION

$$2x + 3 < 11 - 2x$$

$$2x + 3 + 2x < 11 - 2x + 2x \qquad \text{(We added } 2x \text{ to both sides.)}$$

$$4x + 3 < 11$$

$$4x + 3 - 3 < 11 - 3 \qquad \text{(We subtracted 3 from both sides.)}$$

$$4x < 8$$

$$\frac{4x}{4} < \frac{8}{4} \qquad \text{(We divided both sides by 4.)}$$

$$x < 2.$$

Thus, the solution set is $\{x \mid x < 2\}$ (Figure 5).

Figure 5

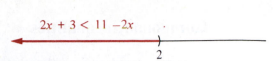

$$2x + 3 < 11 - 2x$$

2

When the inequalities become more complicated, we simplify each side of the inequality before applying the addition property or multiplication properties.

In Examples 5 and 6, solve each inequality and graph the solution on a number line.

EXAMPLE 5 $4(x - 3) \geq 3(x - 2)$

SOLUTION

$$4(x - 3) \geq 3(x - 2)$$

$$4x - 12 \geq 3x - 6$$ (We used the distributive property to remove parentheses.)

$$4x - 3x - 12 \geq 3x - 3x - 6$$ (We subtracted $3x$ from both sides.)

$$x - 12 + 12 \geq -6 + 12$$ (We added 12 to both sides.)

$$x \geq 6.$$

Thus, the solution set is $\{x \mid x \geq 6\}$ (Figure 6).

Figure 6

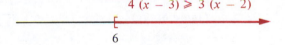

$$4(x - 3) \geq 3(x - 2)$$

6

EXAMPLE 6 $\frac{1}{2}(x + 1) \leq \frac{1}{3}(x - 5)$

SOLUTION

$$\frac{1}{2}(x + 1) \leq \frac{1}{3}(x - 5)$$

$$6(\tfrac{1}{2})(x + 1) \leq 6(\tfrac{1}{3})(x - 5)$$ (We multiplied both sides by 6, the LCD of the fractional coefficients.)

$$3(x + 1) \leq 2(x - 5)$$

$$3x + 3 \leq 2x - 10$$ (We removed parentheses.)

$$3x - 2x + 3 \leq 2x - 2x - 10$$ (We subtracted $2x$ from both sides.)

$$x + 3 - 3 \leq -10 - 3$$ (We subtracted 3 from both sides.)

$$x \leq -13.$$

The solution set is $\{x \mid x \leq -13\}$ (Figure 7).

Figure 7

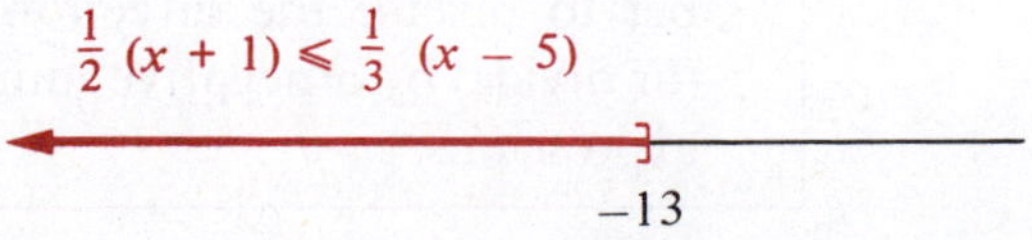

$$\tfrac{1}{2}(x + 1) \leq \tfrac{1}{3}(x - 5)$$

-13

Compound Inequalities

Now we extend our work with the properties of inequalities to include the solution of compound inequalities. Consider the compound inequality

$$12 < 5x - 3 < 17,$$

which consists of

$$12 < 5x - 3 \quad \text{and} \quad 5x - 3 < 17.$$

The word *and* indicates that we want to find the *intersection* of the two solution sets.

Solving the inequalities, we have

$$12 < 5x - 3 \quad \text{and} \quad 5x - 3 < 17$$
$$15 < 5x \quad \text{and} \quad 5x < 20$$
$$3 < x \quad \text{and} \quad x < 4.$$

The solution set of the inequality $12 < 5x - 3$ is $\{x \mid x > 3\}$ (Figure 8a). The solution set of the inequality $5x - 3 < 17$ is $\{x \mid x < 4\}$ (Figure 8b). The solution set of the original compound inequality is $\{x \mid x > 3\} \cap \{x \mid x < 4\} = \{x \mid 3 < x < 4\}$ (Figure 8c).

Figure 8   (a) $12 < 5x - 3; 3 < x$

(b) $5x - 3 < 17; x < 4$

(c) $3 < x < 4$

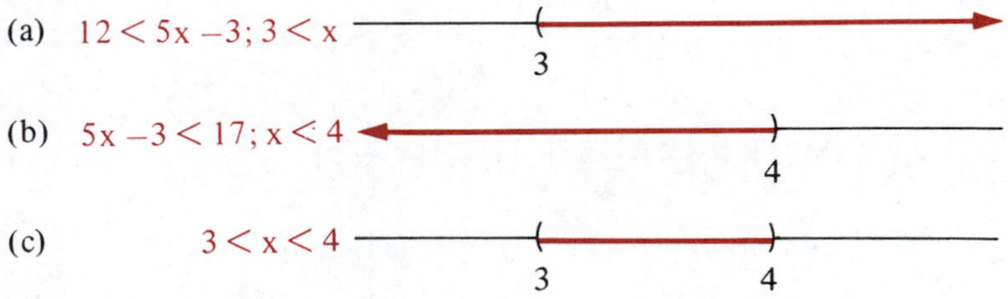

This procedure can be shortened as follows:

Guidelines for Solving Compound Inequalities

1. If we add (or subtract) a number to (from) the middle expression, we must add (or subtract) the same number to (from) the outside expressions.

2. If we multiply (or divide) the middle expression by a number [remember to reverse the direction of the inequality symbol if we multiply (or divide) by a negative number], we must do the same to the outside expressions.

Thus, we can solve the inequality

$$12 < 5x - 3 < 17$$

as follows:

$$12 < 5x - 3 < 17$$

$$15 < 5x < 20 \qquad \text{(Add 3 to all three members.)}$$

$$3 < x < 4 \qquad \text{(Divide all three members by 5.)}$$

In Examples 7 and 8, solve the inequality and graph the solution set on a number line.

EXAMPLE 7 $-11 \leq 2x - 3 \leq 7$

SOLUTION   Using the guidelines just listed, we have

$$-11 \leq 2x - 3 \leq 7$$

$$-11 + 3 \leq 2x - 3 + 3 \leq 7 + 3 \qquad \text{(We added 3 to each member of the inequality.)}$$

$$-8 \leq 2x \leq 10 \qquad \text{(We simplified.)}$$

$$-\tfrac{8}{2} \leq \tfrac{2}{2}x \leq \tfrac{10}{2} \qquad \text{(We divided all members by 2.)}$$

$$-4 \leq x \leq 5. \qquad \text{(We simplified.)}$$

The solution set is $\{x \mid -4 \leq x \leq 5\}$ (Figure 9).

Figure 9

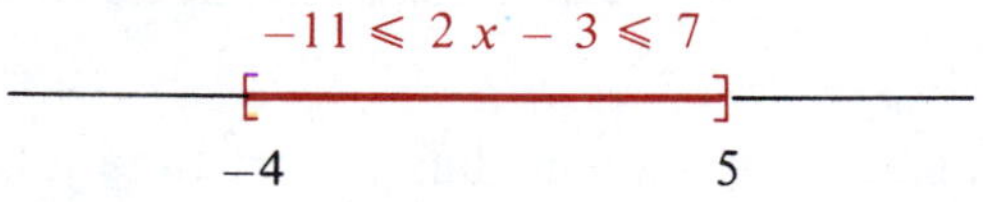

EXAMPLE 8 $2x + 3 \leq -7$ or $2x + 3 \geq 7$

SOLUTION   Here we want to find the *union* of the solution sets because the statements are connected by the word *or*. To obtain the solution sets, we have

$$2x + 3 \leq -7 \qquad \text{or} \qquad 2x + 3 \geq 7$$

$$2x + 3 - 3 \leq -7 - 3 \qquad \text{or} \qquad 2x + 3 - 3 \geq 7 - 3 \qquad \text{(We subtracted 3.)}$$

$$2x \leq -10 \qquad \text{or} \qquad 2x \geq 4 \qquad \text{(We simplified.)}$$

$$x \leq -5 \qquad \text{or} \qquad x \geq 2. \qquad \text{(We divided by 2.)}$$

The solution set is $\{x \mid x \leq -5\} \cup \{x \mid x \geq 2\}$ (Figure 10).

Figure 10

$$2x + 3 \leq -7 \text{ or } 2x + 3 \geq 7$$

$$-5 \qquad 2$$

It should be noted that it is *incorrect* to write the compound inequality $\{x \mid x \leq -5\} \cup \{x \mid x \geq 2\}$ as $\{x \mid -5 \geq x \geq 2\}$.

Applications of Linear Inequalities

The step-by-step procedure on page 172 for solving word problems applies equally to word problems that give rise to linear inequalities. Table 2 should be helpful in translating inequality statements into symbols.

Table 2

Word Statement	Algebraic Statement
x is at least 20	$x \geq 20$
x is at most 15	$x \leq 15$
x is no more than 10	$x \leq 10$
x is not less than 7	$x \geq 7$
x is less than 5	$x < 5$
x ranges from a to b	$a \leq x \leq b$
x is between 6 and 7	$6 < x < 7$

EXAMPLE 9 A student has a part-time job that requires her to work at least 15 hours but no more than 20 hours each week. Write an inequality, using the letter t, that gives the number of hours she works each week.

SOLUTION If t is at least 15 but no more than 20, then $15 \leq t$ and $t \leq 20$, or equivalently,

$$15 \leq t \leq 20.$$

EXAMPLE 10 Three times a number, diminished by 5, is less than 8. Write an inequality, using the letter x to represent the number.

SOLUTION

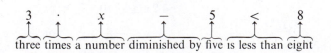

$$\underbrace{3 \quad \cdot \quad x}_{\text{three times a number}} \quad \underbrace{- \quad 5}_{\text{diminished by five}} \quad \underbrace{<}_{\text{is less than}} \quad \underbrace{8}_{\text{eight}}$$

or

$$3x - 5 < 8.$$

EXAMPLE 11   A telephone company offered its customers a choice between two schedules of billing for its services: a fixed $25.00 monthly charge for unlimited local calls, or a base rate of $7.00 per month plus 6¢ per message unit. Above what level of usage (number of message units) does it cost less to choose the unlimited service?

SOLUTION   Let

$x =$ the number of message units above
which the fixed monthly charge is preferable.

The first choice is to pay $25.00 a month. The second choice is to pay $7.00 plus 6¢ ($= 0.06) for each message unit, that is,

the second choice $= 7 + 0.06x.$

To determine when the first choice produces a lower cost than the second choice, we must solve the inequality

$$7 + 0.06x > 25.$$

We have

$$0.06x > 18$$

$$x > \frac{18}{0.06}$$

$$x > 300.$$

Any customer whose anticipated use exceeds 300 message units should select the first choice.

PROBLEM SET 4.7

In problems 1–54, solve each inequality and graph the solution set on a number line.

1. $x - 2 < 4$
2. $3 - x < 2$
3. $x + 2 \geq 3$
4. $x - 4 \geq -2$
5. $4 + x \leq -1$
6. $5 - x \leq 7$
7. $2x \geq 6$
8. $3x \leq 9$
9. $-2x < -3$
10. $-4x > 6$
11. $5x > -15$
12. $-7x \leq 21$
13. $7x \leq -2$
14. $9x \geq -1$
15. $-6x > 12$
16. $-4x \leq 8$
17. $-5x < -1$
18. $-9x > -2$
19. $-11x \geq -2$
20. $-8x \leq -4$

21. $4x - 1 \geq 11$ **22.** $5 + 3x \leq 8$ **23.** $3x + 4 < 7$ **24.** $2x - 1 \leq 5$

25. $-5x + 2 > 12$ **26.** $-7x - 1 > 13$ **27.** $3x - 4 \geq 6x$ **28.** $3x - 2 + x \leq 5x$

29. $5 - 3x \geq 7$ **30.** $x + 6 \leq 4 - 3x$ **31.** $5 + x < -x + 3$ **32.** $3x - 4 > 2x - 9$

33. $-4x > -21 + 3x$ **34.** $2x + 1.3 > -4.1$ **35.** $0 \leq 2x + 7 - 9x$ **36.** $2 - 8x \leq 5 + x$

37. $4(3 - x) \geq 2(x - 1)$ **38.** $5x \geq -3(x - 2)$ **39.** $-3(x + 1) < -4(2x - 1)$

40. $4(-x + 2) - (1 - 5x) \geq -8$ **41.** $7(x - 3) \leq 4(x + 5) - 47$ **42.** $6(x - 10) + 3(2x - 7) < -45$

43. $\frac{1}{4}x \leq 2$ **44.** $\frac{3}{5}x \leq \frac{8}{10}$ **45.** $\frac{3x}{2} \geq -6 - \frac{x}{2}$

46. $\frac{x}{3} + 2 < \frac{x}{4} - 2x$ **47.** $\frac{1}{3}(4x - 3) \geq 5$ **48.** $\frac{1}{2}(x - 3) \leq -15$

49. $\frac{1}{3}(2x + 3) \leq \frac{3}{4}x$ **50.** $\frac{2}{3}(2x - 1) - \frac{2}{5}x \leq 4$ **51.** $\frac{3}{5}(3x - 2) \geq \frac{1}{10}(6x + 7)$

52. $\frac{1}{6}(2x - 7) > \frac{1}{2}(x + 1)$ **53.** $\frac{3x + 7}{7} - \frac{2x - 1}{3} \leq 1$ **54.** $\frac{2x - 6}{4} - \frac{5x + 1}{7} \leq 4$

In problems 55–76, solve each inequality and graph the solution set on a number line.

55. $-5 \leq 4x + 1 \leq 1$ **56.** $-4 < x + 1 < 4$ **57.** $-5 \leq 3x - 1 \leq 5$ **58.** $-3 \leq 4x - 1 \leq 5$

59. $1 < 8 - 3x < 12$ **60.** $0 < 9 - 5x < 29$ **61.** $-1 < 5 - x < 3$ **62.** $-3 \leq 6 - x \leq -2$

63. $1 \leq 4(x - 3) + 1 \leq 5$ **64.** $-5 \leq -3(x - 2) - 2 \leq 11$ **65.** $2x - 1 < 3x + 7 \leq x + 9$

66. $2x - 3 \leq 3x + 1 \leq 4x - 5$ **67.** $5x - 3 \geq 7$ or $5x - 3 \leq -7$ **68.** $2x + 9 \geq 5$ or $2x + 9 \leq -5$

69. $6 - x > 5$ or $6 - x < -5$ **70.** $3x - 4 > 4$ or $3x - 4 < -4$ **71.** $2x + 7 > 1$ or $2x + 7 < -1$

72. $4x - 5 < -3$ or $4x - 5 > 3$ **73.** $5x - 4 > 1$ or $5x - 4 < -1$ **74.** $4 - 3x > 2$ or $4 - 3x < -2$

75. $5 - 4x > 2$ or $5 - 4x < -2$ **76.** $4 - 7x \geq 3$ or $4 - 7x \leq -3$

77. Five less than 4 times a number is less than 13. Write an inequality to satisfy this condition, using the letter x to represent the number.

78. A cashier is required to work at least 30 hours but less than 40 hours per week. Write an inequality using t that gives the number of hours the cashier works each week.

79. A health club employs two workers per week. The total wages paid to these employees ranges from \$660 to \$895 per week. If one employee will earn \$65 more than the other, what are the possible amounts of money earned by each worker per week?

80. Nine less than 7 times a number is less than 3 times the number. Write an inequality that expresses these conditions, using the letter x to represent the number.

81. The formula $F = \frac{9}{5}C + 32$ expresses the temperature F (in degrees Fahrenheit) in terms of the temperature C (in degrees Celsius). If the temperature range on a certain day is 66° to 84° Fahrenheit (that is, $66 \leq F \leq 84$), what is the temperature range in degrees Celsius?

82. A bank teller is entitled to a 2-week vacation for the first year of employment. Thereafter, she is entitled to a 3-week vacation for each year she works. How many years must she work without taking a vacation to entitle her to a vacation of 30 weeks or more?

83. A taxpayer has the following choices: either to pay a 30% tax on his gross income, or to pay a 35% tax on the difference between his gross income and $5,000. The taxpayer should elect to pay at the 30% rate when his income is above what level?

84. An investor had $10,000 to invest. She invested some of the money in a savings account at a simple annual rate of 5% and the rest in a commercial paper that paid interest at a simple annual rate of 15%. If her income from both investments for 1 year was at least $700, how much did she invest in the savings account?

4.8 Equations Involving Absolute Values

We learned in Section 1.4 that the absolute value notation $|x|$ is used to represent the distance on the number line between x and 0. The number of units of distance between the point with coordinate x and the origin is $|x|$, regardless of whether the point is to the right or to the left of the origin (Figure 1).

Figure 1

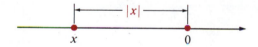

Now consider the two points 3 and 7 on a number line (Figure 2). We see from this figure that the distance d between 3 and 7 is four units:

$$d = 7 - 3 = 4 \qquad \text{or} \qquad d = |7 - 3| = |4| = 4.$$

Figure 2

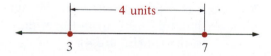

We may also state that the distance d between 3 and 7 is given by

$$d = |3 - 7| = |-4| = 4.$$

In this example you can see that the distance between these two points on the number line can be found by determining the absolute value of the difference of the coordinates of the two points.

In general, the distance d between the points whose coordinates are a and b is given by:

$$d = |a - b|.$$

This formula holds no matter which point is to the left of the other (Figure 3a and b).

Figure 3

(a)   (b)

EXAMPLE 1 Find the distance d between the points whose coordinates are the given pairs of numbers.

(a) 7 and -1 (b) 5 and 0   (c) -4 and 0

SOLUTION We substitute the given numbers for a and b in the formula $d = |a - b|$:

(a) $d = |7 - (-1)| = |7 + 1| = |8| = 8$ (b) $d = |5 - 0| = |5| = 5$

(c) $d = |-4 - 0| = |-4| = 4$

In this section we use the idea that the absolute value of an expression is its distance from the origin, 0, to solve equations involving absolute values. Consider the equation

$$|x| = 3.$$

Geometrically, the equation $|x| = 3$ means that the point with coordinate x is 3 units from 0 on the number line. Obviously, there are two points that are 3 units from the origin—one to the right of the origin and the other to the left (Figure 4).

Figure 4

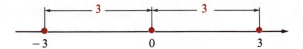

Thus, the equation $|x| = 3$ has two solutions, $x = 3$ or $x = -3$. More generally, we have the following property:

Property 1

> Let a be a real number.
>
> (i) If a is positive, then the equation
>
> $$|u| = a$$
>
> is equivalent to $u = -a$ or $u = a$.
>
> (ii) If a is negative, then the equation
>
> $$|u| = a$$
>
> has no solution.
>
> (iii) If $a = 0$, then the equation
>
> $$|u| = 0$$
>
> is equivalent to $u = 0$.

Some examples of equations involving absolute values are:

$$|2x| = 8, \qquad |3t| - 1 = 15, \qquad \text{and} \qquad |5y + 2| = 7.$$

These equations can be solved by using Property 1.

In Examples 2–6, solve each equation.

EXAMPLE 2 $|3x| = 12$

SOLUTION   Using part (i) of Property 1 with $u = 3x$ and $a = 12$, we have

$$u = -a \qquad \text{or} \qquad u = a$$
$$3x = -12 \qquad \text{or} \qquad 3x = 12$$
$$x = -4 \qquad \text{or} \qquad x = 4.$$

EXAMPLE 3 $|4t| - 1 = 7$

SOLUTION   The equation is equivalent to $|4t| = 8$. Using part (i) of Property 1 with $u = 4t$ and $a = 8$, we have

$$u = -a \qquad \text{or} \qquad u = a$$
$$4t = -8 \qquad \text{or} \qquad 4t = 8$$
$$t = -2 \qquad \text{or} \qquad t = 2.$$

EXAMPLE 4 $3|y - 2| = 15$

SOLUTION   We divide both sides of the equation by 3:

$$|y - 2| = 5,$$

which is equivalent to $|u| = a$ in which $u = y - 2$ and $a = 5$.

$$
\begin{array}{lcl}
u = -a & \text{or} & u = a \\
y - 2 = -5 & \text{or} & y - 2 = 5 \\
y = -3 & \text{or} & y = 7.
\end{array}
$$

EXAMPLE 5 $|3 - 2t| = 15$

SOLUTION   Let $u = 3 - 2t$ and $a = 15$.

$$
\begin{array}{lcl}
u = -a & \text{or} & u = a \\
3 - 2t = -15 & \text{or} & 3 - 2t = 15 \\
-2t = -18 & \text{or} & -2t = 12 \\
t = 9 & \text{or} & t = -6.
\end{array}
$$

EXAMPLE 6 $\left|1 - \dfrac{2}{3}w\right| = -7$

SOLUTION   Using part (ii) of Property 1, we see that $|1 - (2/3)w| = -7$ has no solution, since $|1 - (2/3)w|$ must be nonnegative.

An equation of the form

$$|u| = |v|$$

is equivalent to $u = -v$ or $u = v$.

EXAMPLE 7   Solve the equation $|3x - 2| = |x + 6|$.

SOLUTION   Using $u = 3x - 2$ and $v = x + 6$, the given equation is equivalent to

$$
\begin{array}{lcl}
3x - 2 = -(x + 6) & \text{or} & 3x - 2 = x + 6 \\
3x - 2 = -x - 6 & \text{or} & 3x - x = 6 + 2 \\
4x = -4 & \text{or} & 2x = 8 \\
x = -1 & \text{or} & x = 4.
\end{array}
$$

PROBLEM SET 4.8

In problems 1–12, find the distance d between the points whose coordinates are the given pair of numbers.

1. 3 and 8
2. 15 and 3
3. −4 and 6
4. 19 and −8
5. −8 and −1
6. −11 and −39
7. 14 and 6
8. 3.6 and 2.7
9. 3.5 and −1.7
10. −7.3 and 5.7
11. 0 and −14
12. −6.8 and −9.2

In problems 13–68, solve each equation.

13. $|x| = 2$
14. $|y| = 7$
15. $|t| = \frac{5}{2}$
16. $|t| - 1 = 3$

17. $|w| + 2 = 5$
18. $|y| + 3 = 1$
19. $|5x| = 15$
20. $|-4t| = 16$

21. $|-7y| = 14$
22. $3|2x| = 12$
23. $3|6w| = 24$
24. $-4|5c| = 2$

25. $|3z| - 1 = 14$
26. $|5w| - 2 = 13$
27. $5 - |2u| = 3$
28. $8 + |3w| = 17$

29. $|2y| = -4$
30. $|-5t| = |-20|$
31. $|x - 2| = 3$
32. $|p - 3| = 3$

33. $|5 - q| = 6$
34. $|3 - y| = 4$
35. $3|2w - 3| = 7$
36. $2|5 - 2t| = 1$

37. $4|2 - 7x| = 16$
38. $3|3 - 2y| = 7$
39. $|3w + 2| = -1$
40. $|\frac{1}{3} - 2x| = -\frac{2}{3}$

41. $|3t + 2| = 0$
42. $|5 - 4x| = 0$
43. $\left|1 - \frac{x}{2}\right| = 4$
44. $\left|2 - \frac{x}{3}\right| = 8$

45. $\left|\frac{4}{5}x - 5\right| = 10$
46. $\left|1 - \frac{3}{7}y\right| = 2$
47. $|x + 1| + 2 = 4$
48. $|p - 1| - 3 = 4$

49. $|5x + 3| + 2 = 8$
50. $|2(y + 4) - 3| = 5$
51. $|3(x + 7)| - 8 = 4$

52. $|1 + 5(x - 2)| = 3$
53. $|2x + 3(x - 1) + 1| = 4$
54. $|2(x + 1) - 7x + 3| = 5$

55. $|4 + \frac{1}{2}p| - 7 = 2$
56. $1 = -4 + \left|3 - \frac{t}{5}\right|$
57. $|2x - 1| = |4x - 1|$

58. $|1 + 5x| = |3x - 2|$
59. $|y - 2| = |y + 3|$
60. $|7 - x| = |9 - 2x|$

61. $|4x + 1| = |4x - 1|$
62. $|3x - 8| = |3x + 8|$
63. $|3 - y| = |y + 5|$

64. $|5 - t| = |8 - 3t|$
65. $|x - 4| = |4 - x|$
66. $|y - 7| = |7 - y|$

67. $|3 - x| = |x - 3|$
68. $|11 - y| = |y - 11|$

69. List all the numbers in the set $\{-4, -3, -2, -1, 0, 1, 2, 3, 4\}$ that are solutions of the equation $|2x - 7| = 7 - 2x$.

70. List all the numbers in the set $\{-2, -1, 0, 1, 5, 7, 10, 13, 18\}$ that are solutions of the equation $|2x - 5| = x + 8$.

4.9 **Inequalities Involving Absolute Values**

In Section 4.8, we solved equations involving absolute values. The same procedure allows us to solve inequalities involving absolute values, such as the following:

$$|x| < 3, \qquad |2x - 1| \leq 5 \qquad \text{and} \qquad |3x - 2| > 4.$$

To solve the inequality $|x| < 3$, we begin by examining the number line in Figure 1. We see that the inequality

$$|x| < 3$$

Figure 1

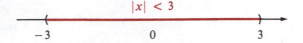

describes all real numbers whose distance from the origin is less than 3 (Figure 1). These are the numbers between -3 and 3. Thus,

$$|x| < 3 \quad \text{is equivalent to} \quad -3 < x < 3.$$

(Recall that $-3 < x < 3$ is a shorthand notation for $-3 < x$ and $x < 3$.) The solution set is $\{x \mid -3 < x < 3\}$.

More generally, we have the following property:

Property 1

> If $a > 0$, then
>
> (i) $|u| < a$ is equivalent to $-a < u < a$.
>
> (ii) $|u| \leq a$ is equivalent to $-a \leq u \leq a$.

In Examples 1–3, solve each inequality and graph its solution set on a number line.

EXAMPLE 1 $|2x| < 10$

SOLUTION   Using Property 1(i) with $u = 2x$ and $a = 10$, we have

$$-a < u < a$$
$$-10 < 2x < 10$$
$$-5 < x < 5. \qquad \text{(We divided all three members by 2.)}$$

Thus, the solution set is $\{x \mid -5 < x < 5\}$ (Figure 2).

Figure 2

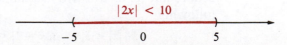

EXAMPLE 2 $|7x - 2| \leq 9$

SOLUTION   Using Property 1(ii) with $u = 7x - 2$ and $a = 9$, we have

$$-a \leq u \leq a$$

$$-9 \leq 7x - 2 \leq 9$$

$$-7 \leq 7x \leq 11 \qquad \text{(We added 2 to all members.)}$$

$$-1 \leq x \leq \tfrac{11}{7}. \qquad \text{(We divided all members by 7.)}$$

Thus, the solution set is $\{x \mid -1 \leq x \leq \tfrac{11}{7}\}$ (Figure 3).

Figure 3

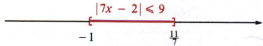

EXAMPLE 3 $|2x - 1| - 2 < 3$

SOLUTION   Before we apply Property 1, we must isolate the absolute value on one side of the inequality. To do so, we add 2 to each side:

$$|2x - 1| - 2 + 2 < 3 + 2$$

$$|2x - 1| < 5.$$

Now we use Property 1(i) with $u = 2x - 1$ and $a = 5$. We have

$$-a < u < a$$

$$-5 < 2x - 1 < 5$$

$$-4 < 2x < 6 \qquad \text{(We added 1 to all members.)}$$

$$-2 < x < 3. \qquad \text{(We divided by 2.)}$$

Thus, the solution set is $\{x \mid -2 < x < 3\}$ (Figure 4).

Figure 4

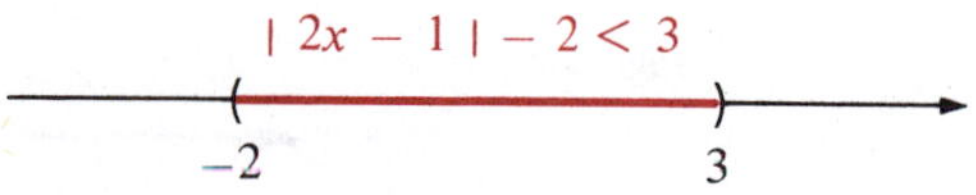

$$| 2x - 1 | - 2 < 3$$

$$\overset{(}{-2} \qquad\qquad \overset{)}{3}$$

Next, let us solve the inequality

$$|x| > 3.$$

In this case, we need to find the real numbers whose absolute value is greater than 3. On a number line this calls for all real numbers whose distance from the origin is more than 3 units (Figure 5). These numbers are either less than -3 or greater than 3. That is,

$$|x| > 3 \qquad \text{is equivalent to} \qquad x < -3 \qquad \text{or} \qquad x > 3.$$

Figure 5

Therefore, the solution set is the union of two separate inequalities. That is, the solution set is

$$\{x \mid x < -3\} \cup \{x \mid x > 3\} \qquad \text{or} \qquad \{x \mid x < -3 \ \text{or} \ x > 3\}.$$

It should be noted that the solution set of $|x| > 3$ must be written as two separate inequalities.

More generally, we have the following property:

Property 2

> If $a > 0$, then
>
> (i) $|u| > a$ is equivalent to $u < -a$ or $u > a$.
>
> (ii) $|u| \geq a$ is equivalent to $u \leq -a$ or $u \geq a$.

In Examples 4–7, solve each inequality and sketch the solution set on a number line.

EXAMPLE 4 $|x + 1| > 2$

SOLUTION   Using Property 2(i) with $u = x + 1$ and $a = 2$, we have

$$
\begin{array}{ccc}
u < -a & \text{or} & u > a \\
x + 1 < -2 & \text{or} & x + 1 > 2 \\
x < -3 & \text{or} & x > 1. \quad \text{(We subtracted 1.)}
\end{array}
$$

The solution set is $\{x \mid x < -3 \ \text{or} \ x > 1\}$ (Figure 6).

Figure 6

EXAMPLE 5 $|7x| + 1 > 15$

SOLUTION   Before we apply Property 2, we must isolate the absolute value on one side of the inequality. To do so, we subtract 1 from each side:

$$|7x| + 1 - 1 > 15 - 1$$
$$|7x| > 14.$$

Now using Property 2(i) with $u = 7x$ and $a = 14$, we have

$$u < -a \qquad \text{or} \qquad u > a$$
$$7x < -14 \qquad \text{or} \qquad 7x > 14$$
$$x < -2 \qquad \text{or} \qquad x > 2. \qquad \text{(We divided by 7.)}$$

Therefore, the solution set is $\{x \mid x < -2 \text{ or } x > 2\}$ (Figure 7).

Figure 7

$|7x| + 1 > 15$

EXAMPLE 6 $|2x + 7| \geq 11$

SOLUTION   Using Property 2(ii) with $u = 2x + 7$ and $a = 11$, we have

$$u \leq -a \qquad \text{or} \qquad u \geq a$$
$$2x + 7 \leq -11 \qquad \text{or} \qquad 2x + 7 \geq 11$$
$$2x \leq -18 \qquad \text{or} \qquad 2x \geq 4 \qquad \text{(We subtracted 7.)}$$
$$x \leq -9 \qquad \text{or} \qquad x \geq 2. \qquad \text{(We divided by 2.)}$$

Thus, the solution set is $\{x \mid x \leq -9 \text{ or } x \geq 2\}$ (Figure 8).

Figure 8

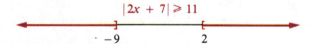

$|2x + 7| \geq 11$

EXAMPLE 7 $|2 - 3x| > 5$

SOLUTION   Using Property 2(i) with $u = 2 - 3x$ and $a = 5$, we have

$$u < -a \qquad \text{or} \qquad u > a$$
$$2 - 3x < -5 \qquad \text{or} \qquad 2 - 3x > 5$$
$$-3x < -7 \qquad \text{or} \qquad -3x > 3 \qquad \text{(We subtracted 2.)}$$
$$x > \tfrac{7}{3} \qquad \text{or} \qquad x < -1 \qquad \text{(We divided by } -3 \text{ and reversed}$$
$$\text{the sense of the inequalities).}$$

Therefore, the solution set is $\{x \mid x < -1 \text{ or } x > \tfrac{7}{3}\}$ (Figure 9).

Figure 9

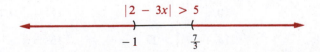

$|2 - 3x| > 5$

Sometimes we encounter special inequalities like the following:

$$|x + 2| < -1, \quad |x - 4| > -2, \quad |x - 3| \geq 0, \quad \text{or} \quad |x - 2| < 0.$$

We examine their solutions in the following examples.

In Examples 8 and 9, solve each inequality (if possible).

EXAMPLE 8   (a) $|x + 2| < -1$ (b) $|x - 4| > -2$

SOLUTION   Before starting to solve these inequalities, we must examine each statement carefully.

(a) The left side of this inequality cannot be negative, because absolute value means the distance from the origin, 0. Since $|x + 2| = |x - (-2)|$, we ask: Can the distance between x and -2 be less than -1? The answer is no, because a positive quantity or zero cannot be less than a negative quantity. Therefore, the solution set is $\emptyset$.

(b) The left side of the inequality represents the distance between x and 4, and as such it is always positive or zero, and the right side is negative. In this case, every real number we choose for x gives us a true statement. Therefore, the solution set is the set of all real numbers $\mathbb{R}$.

EXAMPLE 9   (a) $|x - 2| < 0$ (b) $|x - 3| \geq 0$

SOLUTION   We examine each inequality as we did in Example 8.

(a) In this case, the solution set is $\emptyset$, because the absolute value of a number (or expression) can never be less than 0.

(b) Since the absolute value of every number is always greater than or equal to 0, the solution set is the set of all real numbers $\mathbb{R}$.

PROBLEM SET 4.9

In problems 1–58, solve each inequality and graph the solution set on a number line.

1. $|x| < 1$ **2.** $|x| \leq 7$ **3.** $|x| > 5$ **4.** $|x| \geq 4$

5. $|3x| < 15$ **6.** $|4x| < |-8|$ **7.** $|5x| \leq 20$ **8.** $|-7x| \leq 14$

9. $\left|\dfrac{1}{3}x\right| \leq 2$ **10.** $\left|-\dfrac{1}{5}x\right| < 1$ **11.** $|4x| > 12$ **12.** $|5x| > 25$

13. $|-6x| \geq 18$ **14.** $|-9x| \geq 27$ **15.** $|3x| \geq |-12|$ **16.** $|7x| > |-28|$

17. $|x - 1| < 3$

18. $|x + 2| < 5$

19. $|x - 1| \leq 1$

20. $|x - 3| \leq 2$

21. $|x - 4| > 5$

22. $|x - 2| \geq 2$

23. $|x - 3| \geq 4$

24. $|x - 5| > 2$

25. $|3x - 4| < 7$

26. $|2x - 1| \geq 3$

27. $|2x - 5| \geq 1$

28. $|3x - 1| > 8$

29. $|8x + 3| \geq 7$

30. $|10 - 3x| > 13$

31. $|2x - 1| \leq 3$

32. $|5x - 3| > 12$

33. $|4x - 1| \geq 11$

34. $|3 - 2x| < 5$

35. $\left|\frac{2}{3} - x\right| \leq \frac{1}{3}$

36. $|5x + 1| \geq -2$

37. $\left|\frac{1}{3} - \frac{x}{6}\right| \geq \frac{1}{2}$

38. $\left|\frac{3x}{7} + \frac{2}{21}\right| < 0$

39. $|5x + 3| + 2 > 8$

40. $|1 - 2x| - 5 \leq 7$

41. $\frac{1}{4}|2x - 1| \leq 3$

42. $\frac{1}{2}|3x + 2| \geq 4$

43. $|7x - 1| \leq 2$

44. $|1 - 6x| \leq 6$

45. $|2x + 1| + 4 \leq 7$

46. $|6x - 1| - 41 \geq 2$

47. $|5x - 3| - 4 \geq 3$

48. $|3x + 1| + 2 \leq 7$

49. $|8 - 7x| + 9 > 1$

50. $\left|4 - \frac{2}{7}x\right| + 2 \geq 14$

51. $|4x - 1| \leq -2$

52. $|5x + 2| > -5$

53. $|2x + 3| > -3$

54. $|8x + 17| < -6$

55. $|3x + 8| \geq 0$

56. $|7x - 1| < 0$

57. $|3 - 5x| < 0$

58. $|9 + 4x| > 0$

59. Use Property 1(ii) to write the inequality $-5 \leq x \leq 5$ as a single inequality involving absolute value.

60. Use Property 2(ii) to write the inequality $x \leq -8$ or $x \geq 8$ as a single inequality involving absolute value.

REVIEW PROBLEM SET

In problems 1–34, solve each equation.

1. $y + 1 = 3$

2. $z - 11 = 4$

3. $u + 8 = 21$

4. $m - 7 = 53$

5. $-17u = 51$

6. $-8t = 24$

7. $2y + 13 = -1$

8. $10 = 2 - 3m$

9. $-4 + 5z = 16$

10. $5x + 2 = 32$

11. $-4x - 7 = 21$

12. $4y - 16 = 6y$

13. $3w - 4 = w - 12$

14. $17 - 3c = c + 5$

15. $10 - 8(3 - y) = 0$

16. $3(3t - 1) + 2 = 4(2 - t)$

17. $2(x - 1) = 3 - 3(x - 5)$

18. $12(y - 2) + 8 = 5(y - 1) + 2y$

19. $z - 7(4 + z) = 5z - 6(3 - 4z)$

20. $5(w + 2) - 3(w - 1) = 4(w + 1) + 15$

21. $2(u - 1) + 3(u - 2) = -4(u - 3)$

22. $m - 2(3 - m) = 2(m + 3) - (m - 2)$

23. $\frac{2z}{3} + z = 5$

24. $\frac{3t}{6} - \frac{2t}{8} = 8$

25. $\frac{x}{3} - \frac{x}{6} = 1$

26. $\frac{2x - 1}{3} = \frac{x}{5} + \frac{2}{15}$

27. $\frac{5}{y} + \frac{3}{8} = \frac{7}{16}$

28. $\frac{t + 4}{3} - \frac{t + 2}{2} = \frac{t - 2}{5}$

29. $\frac{13}{14z} - \frac{1}{2z} = \frac{1}{7}$

30. $\frac{2}{w} + \frac{w - 1}{3w} = \frac{2}{5}$

31. $\frac{3}{y - 2} = \frac{5}{y + 2}$

32. $\frac{10 - w}{w} + \frac{3w + 3}{3w} = 3$

33. $\frac{t + 1}{t - 1} - \frac{t}{t + 1} = \frac{5 - t}{t^2 - 1}$

34. $\frac{12}{c^2 - 25} = \frac{1}{c + 5} + \frac{2}{c - 5}$

In problems 35–38, express each repeating decimal as a quotient of integers.

35. $0.\overline{37}$ **36.** $-0.\overline{03}$ **37.** $0.04\overline{8}$ **38.** $1.\overline{431}$

In problems 39–52, solve each equation for the indicated unknown.

39. $a - 2x = 5a$, for x.   **40.** $5u - 17b = 39u$, for u.   **41.** $a - bt = 3a$, for t.

42. $a - y = a(y - 1)$, for y.   **43.** $az = b + cz$, for z.   **44.** $y - a = m(x - b)$, for x.

45. $6(a - bc) = a^3 - 6a$, for c.   **46.** $a^2t - 2at = 1 - t$, for t.

47. $3(ax - 2b) = 4(2b - ax)$, for a.   **48.** $7(a^2b + c) - 4a^2(b + 3a^2) = 10c$, for b.

49. $\dfrac{x - 1}{a} + \dfrac{2}{a} = 3$, for x.   **50.** $\dfrac{u}{5b} - \dfrac{u}{7b} = 3$, for u.

51. $\dfrac{c}{a} - \dfrac{3}{a - b} = 1$, for b.   **52.** $\dfrac{2w - a}{5} + \dfrac{w}{3} - 3a = 1$, for w.

In problems 53–60, solve each formula for the indicated unknown.

53. $E = mc^2$, for m.   **54.** $S = 2A + ph$, for A.   **55.** $S = 2\pi rh + 2\pi r^2$, for h.

56. $A = \pi r^2 + 2\pi rs$, for s.   **57.** $G = \dfrac{m}{r^3}$, for m.   **58.** $S = \dfrac{ax^n - a}{x - 1}$, for a.

59. $g = \dfrac{v^2}{R + h}$, for h.   **60.** $En = RI + \dfrac{nr\ell}{m}$, for ℓ.

In problems 61–74, translate each statement into algebraic form.

61. What number is 5 more than twice another number represented by x?

62. If the first number is represented by $3y$, what number is 7 less than one-third the first?

63. Sam is 3 years older than Susan. Represent the sum of their ages in 5 years if x represents Sam's age now.

64. A vending machine contains nickels, dimes, and quarters. If the number of nickels is x, and there are 10 more dimes than nickels and 3 times as many quarters as dimes, what is the expression that represents the total number of coins in the machine?

65. In a Red Cross drive, Dawn contributed twice as much as Joe, and Mike contributed $4 more than Dawn. How much did each person give if their donations totaled $79?

66. A billfold contains $223 in $10, $5, and $1 bills. There are 47 bills in all, and there are five more $5 bills than $10 bills. How many bills of each kind are there?

67. A woman used 420 meters of wire to fence a rectangular garden. If the length of the garden is four times the width, what are the length and the width of the garden?

68. How many pounds of water must be evaporated from 50 pounds of a 3% salt solution so that the remaining portion will be a 5% solution?

69. A retail store invested $25,000 in two kinds of radios. All the radios had been sold and the store made a profit of 15% from the first type, but lost 5% on the second. If the two investments provided a return of 10% on the entire amount invested, how much had the store invested in each kind of radio?

70. A chemist has a gallon containing a 4% solution of a certain chemical. How much water should be added so that the resulting mixture will be a 1% solution?

71. A jogger traveling at 10 miles per hour leaves a starting point. Twenty minutes later a bicycle rider traveling at 12 miles per hour leaves the same starting point and follows the same route. How long does it take the bicyclist to overtake the jogger?

72. In a predator–prey model, the total population is 34,100, and the number of the prey is 10 times the number of predators. What is the predator population?

73. If a taxicab fare is $0.80 for the first mile and $0.50 for each additional mile, the cost C for n miles is represented by the formula

$$C = 80 + 50(n - 1).$$

How many miles are traveled in a trip that cost $3.75?

74. A rocket is fired straight up at an initial velocity of v feet per second. The rocket's distance of s feet above the ground t seconds after being fired is given by the formula $s = vt - 16t^2$. Solve this formula for v.

In problems 75–84, graph each solution set on a number line.

75. $\{x \mid x < 7\}$

76. $\{x \mid x > -5\}$

77. $\{x \mid -5 \leq x\}$

78. $\{x \mid x \leq 11\}$

79. $\{x \mid x > -4\} \cap \{x \mid x < 4\}$

80. $\{x \mid x \leq 7\} \cap \{x \mid x \geq -4\}$

81. $\{x \mid x \leq 2\} \cup \{x \mid x \geq 5\}$

82. $\{x \mid x < -2\} \cup \{x \mid x \geq 2\}$

83. $\{x \mid -4 \leq x \leq 1\}$

84. $\{x \mid -8 < x < 6\}$

In problems 85–96, solve each inequality and graph the solution set on a number line.

85. $3x > 6$

86. $-3x \leq 12$

87. $2x + 1 > 13$

88. $3 - 2x \leq 9$

89. $3x + 5 < 2x + 1$

90. $3(x - 4) < 2(4 - x)$

91. $2(1 - 2x) \geq 3(4 - x)$

92. $5x + 3(3x - 1) \geq -1 - 5(x - 3)$

93. $\frac{1}{19}(3x - 5) \geq 1$

94. $\frac{1}{4}(2x + 3) \geq 1 - x$

95. $\frac{1}{3}(1 - 3x) \leq \frac{1}{4}(1 + 2x)$

96. $\frac{1}{5}(2x + 1) - 2 \leq \frac{1}{2}(3x + 1)$

97. A postal service charges a fixed fee of $0.70 plus an additional $0.10 for every ounce over the first pound. What restriction in weight is there on a package that is to be delivered for less than $4.35?

98. Judy asks for a $7 weekly payroll deduction from her salary in order to save money for a new camera. How many weeks will it take for her payroll deductions to add up to at least $437 (the amount she needs for the camera)?

In problems 99–110, solve each equation.

99. $|7x| = 35$

100. $|-11x| = 55$

101. $|y + 2| = 31$

102. $|4 - w| = 13$

103. $|4 - 2z| = -2$

104. $|3t + 17| = -1$

105. $|4u + 3| = 13$

106. $|2(y - 3)| = 8$

107. $|7 - 11x| = 29$

108. $|3t - 1| + 4 = 8$

109. $|2x + 1| = |3 - x|$

110. $|4x - 5| = |4x + 5|$

In problems 111–122, solve each inequality and graph the solution set on a number line.

111. $|x| < 11$

112. $|x| \geq 12$

113. $|13x| \geq 52$

114. $|17x| < 51$

115. $|x + 40| \leq 43$

116. $|8 - 3x| \leq 6$

117. $|4x + 7| \geq 9$

118. $\left|\dfrac{3x}{4} - 1\right| > 2$

119. $\left|\frac{1}{5}(1 - x)\right| < 3$

120. $\left|\frac{1}{3}(2 - x)\right| \leq 5$

121. $|3x - 2| + 5 \leq 11$

122. $|3 - 9x| < 0$

CHAPTER 4 TEST

1. Solve each of the following equations
   (a) $15 - 4x = 2x - 1$
   (b) $4(2y - 3) + 55 = 5(4 - 3y)$
   (c) $\dfrac{5w}{7} - \dfrac{3w}{2} = 11$
   (d) $\dfrac{2}{t + 3} + \dfrac{3}{t - 3} = \dfrac{18}{t^2 - 9}$

2. Solve each equation for the indicated unknown.
   (a) $3b - 2cx = 5b$, for x
   (b) $\dfrac{3w - a}{4} + \dfrac{w}{2} - 1 = 5a$, for w

3. Solve each formula for the indicated unknown.
   (a) $S = 2A + ph$, for h
   (b) $g = \dfrac{v^2}{R + h}$, for R

4. Translate the following statement into algebraic form: Mary is 4 times as old as her son John. Represent their ages in 6 years if John's present age is represented by x.

5. In a school candy sale, Joe sold twice as many candy bars as Jim, and 10 more candy bars than Sue. If they sold 50 candy bars together, how many did each sell?

6. Graph each solution set on a number line.
   (a) $\{x \mid x \leq -3\}$
   (b) $\{x \mid x > -2\} \cap \{x \mid x < 3\}$
   (c) $\{x \mid x \leq 5\} \cup \{x \mid x > 8\}$
   (d) $\{x \mid -6 \leq x < 0\}$

7. Solve each inequality.
   (a) $3x - 7 \geq 5$
   (b) $3 - 2y < 8$
   (c) $\frac{1}{2}(2 - 3x) \leq \frac{1}{3}(2x + 1)$

8. Solve each of the following equations.
   (a) $|3y| = 24$
   (b) $|2 - 3x| = 7$

9. Solve each inequality and graph the solution on a number line.
   (a) $|x - 3| < 5$
   (b) $|2(x - 1)| \geq 3$

Exponents, Radicals, and Complex Numbers

Exponents and radicals are often used in applied mathematics and the sciences. In dealing with very large or very small numbers on computers or calculators, for example, these numbers are typically expressed in exponential notation. Scientific problems are commonly formulated as mathematical equations that involve exponents and/or radicals.

We begin this chapter by extending the properties of exponents, discussed in Chapter 2, beyond the positive integers. We then translate the properties of exponents to radicals. The chapter concludes with a discussion of complex numbers.

5.1 Zero and Negative Integer Exponents

In Section 2.3, we established the properties of positive integer exponents that follow from the definition

$$a^n = \overbrace{a \cdot a \cdot a \cdots a}^{n \text{ factors}},$$

where n is a positive integer. We review these properties, together with illustrative examples alongside.

Properties	Examples
(i) $a^m \cdot a^n = a^{m+n}$	1. $x^7 \cdot x^4 = x^{11}$
(ii) $(a^m)^n = a^{mn}$	2. $(x^4)^5 = x^{20}$
(iii) $(ab)^n = a^n b^n$	3. $(xy)^4 = x^4 y^4$
(iv) $\left(\dfrac{a}{b}\right)^n = \dfrac{a^n}{b^n}, \qquad b \neq 0$	4. $\left(\dfrac{x}{y}\right)^7 = \dfrac{x^7}{y^7}$
(v) $\dfrac{a^m}{a^n} = \begin{cases} a^{m-n}, & \text{if } m > n \\ 1, & \text{if } m = n \\ 1/a^{n-m}, & \text{if } m < n \end{cases} a \neq 0$	5. $\begin{cases} \dfrac{x^5}{x^2} = x^3 \\ \dfrac{x^4}{x^4} = 1 \\ \dfrac{x^3}{x^8} = \dfrac{1}{x^5} \end{cases}$

We now extend this definition so that zero and the negative integers can also be used as exponents, and do so in such a way that the properties of positive exponents continue to hold.

We begin with a question: How can we define a^0? If the multiplication property, $a^m \cdot a^n = a^{m+n}$, continues to hold for $m = 0$, we have

$$a^0 \cdot a^n = a^{0+n}$$
$$= a^n.$$

This equation is true only if $a \neq 0$ and if $a^0 = 1$, which leads us to the following definition.

DEFINITION 1   **Zero as an Exponent**

> If a is any nonzero real number, we define $a^0 = 1$.

For example,

$$3^0 = 1, \qquad \left(-\frac{4}{7}\right)^0 = 1, \qquad \text{and} \qquad (1 + y^2)^0 = 1.$$

Notice that in Definition 1, $a \neq 0$. For if $a = 0$, then what does 0^0 mean? Consider the following statement:

$$0^0 = 0^{8-8} = \frac{0^8}{0^8} = \frac{0}{0}.$$

Therefore, 0^0 is equivalent to $\frac{0}{0}$, where the denominator is zero. Since division by zero is not defined, then 0^0 is an undefined expression. For example,

$$(-5x + 3)^0 \text{ is not defined, when } x = \frac{3}{5}.$$

If the multiplication property $a^m \cdot a^n = a^{m+n}$ continues to hold for $m = -n$, then we have

$$a^{-n} \cdot a^n = a^{-n+n} = a^0 = 1, \qquad \text{for } a \neq 0.$$

Dividing both sides of the equation $a^{-n} \cdot a^n = 1$, first by a^n and then by a^{-n} produces the following statements:

$$a^{-n} = \frac{1}{a^n} \qquad \text{and} \qquad a^n = \frac{1}{a^{-n}}.$$

This procedure leads to the following definition:

DEFINITION 2 **Negative Integer Exponents**

If a is a nonzero real number and n is any positive integer, then

$$a^{-n} = \frac{1}{a^n} \quad \text{and} \quad a^n = \frac{1}{a^{-n}}.$$

For instance,

$$10^{-4} = \frac{1}{10^4}, \qquad 3^{-4} = \frac{1}{3^4}, \qquad \frac{1}{5^{-2}} = 5^2, \qquad \text{and} \qquad \frac{1}{b^{-4}} = b^4, \qquad \text{for} \quad b \neq 0.$$

It should be noticed that the negative sign in the exponent has nothing to do with the algebraic sign of the expression. For example,

$$(-7)^{-1} = \frac{1}{-7} = -\frac{1}{7}, \text{ a negative result,}$$

whereas

$$(7)^{-1} = \frac{1}{7}, \text{ a positive result.}$$

In Examples 1 and 2, rewrite each expression so that it contains only positive exponents, and simplify the results. (Assume all variables that appear in a base with a zero or a negative exponent represent real numbers that will make the base nonzero.)

EXAMPLE 1

(a) 6^{-2} (b) $(2x)^{-3}$ (c) $(3y^0)^{-4}$ (d) $1 + (3x)^0$

(e) $\dfrac{1}{7(w+4)^{-3}}$ (f) $\dfrac{1}{(-5)^{-3}}$ (g) $5t^{-4}$ (h) $\left(\dfrac{p^4}{r^{-2}}\right)^0$

(i) $\dfrac{10^0}{10^{-2}}$ (j) $\dfrac{1}{x^{-3}}$

SOLUTION

(a) $6^{-2} = \dfrac{1}{6^2} = \dfrac{1}{36}$ (by Definition 2)

(b) $(2x)^{-3} = \dfrac{1}{(2x)^3}$ (by Definition 2)

$\qquad = \dfrac{1}{2^3 \cdot x^3} = \dfrac{1}{8x^3}$ (why?)

(c) $(3y^0)^{-4} = (3 \cdot 1)^{-4}$ (by Definition 1)

$\qquad = 3^{-4} = \dfrac{1}{3^4} = \dfrac{1}{81}$ (by Definition 2)

(d) $1 + (3x)^0 = 1 + 1 = 2$ $\qquad$ (by Definition 1)

(e) $\dfrac{1}{7(w+4)^{-3}} = \dfrac{(w+4)^3}{7}$ $\qquad$ (by Definition 2)

(f) $\dfrac{1}{(-5)^{-3}} = (-5)^3 = -125$ $\qquad$ (by Definition 2)

(g) $5t^{-4} = 5(t^{-4}) = 5\left(\dfrac{1}{t^4}\right) = \dfrac{5}{t^4}$ $\qquad$ (by Definition 2)

(h) $\left(\dfrac{p^4}{r^{-2}}\right)^0 = 1$ $\qquad$ (by Definition 1)

(i) $\dfrac{10^0}{10^{-2}} = \dfrac{1}{10^{-2}} = 10^2 = 100$ $\qquad$ (why?)

(j) $\dfrac{1}{x^{-3}} = x^3$ $\qquad$ (by Definition 2)

EXAMPLE 2 (a) $3^{-2} - 5^{-2}$ $\qquad$ (b) $\dfrac{1}{2^{-2} + 3^{-2}}$ $\qquad$ (c) $\dfrac{1}{(2+3)^{-2}}$

(d) $a^{-4} - a^{-6}$ $\qquad$ (e) $(4 + 4^{-1})^{-1}$ $\qquad$ (f) $\dfrac{u^3}{v^{-3}} - \dfrac{v^3}{u^{-3}}$

(g) $\dfrac{(p^2 + q^2)^0}{p^{-1} - q^{-1}}$ $\qquad$ (h) $\dfrac{x^{-1} + y^{-1}}{x + y}$

SOLUTION (a) $3^{-2} - 5^{-2} = \dfrac{1}{3^2} - \dfrac{1}{5^2} = \dfrac{1}{9} - \dfrac{1}{25} = \dfrac{25 - 9}{225} = \dfrac{16}{225}$

(b) $\dfrac{1}{2^{-2} + 3^{-2}} = \dfrac{1}{\dfrac{1}{2^2} + \dfrac{1}{3^2}} = \dfrac{1}{\dfrac{1}{4} + \dfrac{1}{9}}$

$\qquad = \dfrac{1}{\dfrac{9 + 4}{36}} = \dfrac{1}{\dfrac{13}{36}} = \dfrac{36}{13}$

(c) $\dfrac{1}{(2+3)^{-2}} = \dfrac{1}{5^{-2}} = 5^2 = 25$

(d) $a^{-4} - a^{-6} = \dfrac{1}{a^4} - \dfrac{1}{a^6} = \dfrac{a^2}{a^6} - \dfrac{1}{a^6} = \dfrac{a^2 - 1}{a^6}$

(e) $(4 + 4^{-1})^{-1} = \left(4 + \dfrac{1}{4}\right)^{-1} = \left(\dfrac{16 + 1}{4}\right)^{-1} = \left(\dfrac{17}{4}\right)^{-1} = \dfrac{4}{17}$

(f) $\dfrac{u^3}{v^{-3}} - \dfrac{v^3}{u^{-3}} = u^3 v^3 - v^3 u^3 = 0$

(g) $\dfrac{(p^2 + q^2)^0}{p^{-1} - q^{-1}} = \dfrac{1}{\dfrac{1}{p} - \dfrac{1}{q}} = \dfrac{1}{\dfrac{q - p}{pq}} = \dfrac{pq}{q - p}$

(h) $\dfrac{x^{-1} + y^{-1}}{x + y} = \dfrac{\dfrac{1}{x} + \dfrac{1}{y}}{x + y} = \dfrac{\dfrac{y + x}{xy}}{x + y} = \dfrac{y + x}{xy} \cdot \dfrac{1}{x + y} = \dfrac{1}{xy}$

Definitions 1 and 2 enable you to use any *integer*—positive, negative, or zero—as an exponent, with the exception that 0^n is defined only when n is positive, so that Properties (i) through (v) listed below hold for *all* integer exponents.

Properties of Integer Exponents

Let a and b be real numbers, and let m and n be integers. Then each of the following is true for all values of a and b for which both sides of the equation are defined:

(i) $a^m \cdot a^n = a^{m+n}$ (ii) $(a^m)^n = a^{mn}$ (iii) $(ab)^m = a^m b^m$

(iv) $\left(\dfrac{a}{b}\right)^m = \dfrac{a^m}{b^m}$ (v) $\dfrac{a^m}{a^n} = a^{m-n}$

The following properties are direct applications of the preceding properties of integer exponents.

If $a \neq 0$ and $b \neq 0$, then

(1) $\left(\dfrac{a}{b}\right)^{-n} = \left(\dfrac{b}{a}\right)^n.$ (2) $\dfrac{a^{-n}}{b^{-m}} = \dfrac{b^m}{a^n}.$

For example,

$$\left(\dfrac{p}{q}\right)^{-3} = \left(\dfrac{q}{p}\right)^3 = \dfrac{q^3}{p^3} \quad \text{and} \quad \dfrac{a^{-1}}{c^{-3}} = \dfrac{c^3}{a}.$$

In Examples 3 and 4, use the properties of exponents to rewrite each expression so that it contains only nonnegative exponents, and simplify the results. (Assume all variables that appear with a negative exponent represent real numbers that will make the base nonzero.)

EXAMPLE 3   (a) $x^6 \cdot x^{-2}$ (b) $(p^4)^{-3}$ (c) $(6y^3)^{-2}$

(d) $\left(\dfrac{3}{5}\right)^{-4}$ (e) $\dfrac{w^3}{w^{-6}}$ (f) $(x^2 y^{-3})^{-1}$

SOLUTION

(a) $x^6 \cdot x^{-2} = x^{6-2}$ [by Property (i)]

$= x^4$

(b) $(p^4)^{-3} = p^{4(-3)}$ [by Property (ii)]

$= p^{-12} = \dfrac{1}{p^{12}}$

(c) $(6y^3)^{-2} = 6^{-2} \cdot (y^3)^{-2}$ [by Property (iii)]

$= \dfrac{1}{6^2} \cdot \dfrac{1}{(y^3)^2}$

$= \dfrac{1}{36} \cdot \dfrac{1}{y^6} = \dfrac{1}{36y^6}$

(d) $\left(\dfrac{3}{5}\right)^{-4} = \dfrac{3^{-4}}{5^{-4}}$ [by Property (iv)]

$= \dfrac{5^4}{3^4}$ $\left(\text{by the Property } \dfrac{a^{-n}}{b^{-m}} = \dfrac{b^m}{a^n}\right)$

$= \dfrac{625}{81}$

(e) $\dfrac{w^3}{w^{-6}} = w^{3-(-6)}$ [by Property (v)]

$= w^{3+6} = w^9$

(f) $(x^2 y^{-3})^{-1} = (x^2)^{-1} \cdot (y^{-3})^{-1}$ [by Property (iii)]
$= x^{-2} \cdot y^3$ [by Property (ii)]

$= \dfrac{y^3}{x^2}$

EXAMPLE 4   (a) $\dfrac{3ab^{-2}}{c^3 d^{-4}}$ (b) $\left[\dfrac{x^{-4}(-3x)}{(-5x)^{-2}}\right]^{-1}$ (c) $\dfrac{5c^{-3}(r+s)^2}{15c^4(r+s)^{-5}}$

SOLUTION   (a) $\dfrac{3ab^{-2}}{c^3d^{-4}} = \dfrac{3a}{c^3}\left(\dfrac{b^{-2}}{d^{-4}}\right)$ (why?)

$\qquad\qquad = \dfrac{3a}{c^3}\left(\dfrac{d^4}{b^2}\right)$ [by Property (2)]

$\qquad\qquad = \dfrac{3ad^4}{c^3b^2}$

(b) $\left[\dfrac{x^{-4}(-3x)}{(-5x)^{-2}}\right]^{-1} = \left[\dfrac{x^{-4}}{(-5x)^{-2}}(-3x)\right]^{-1}$

$\qquad\qquad = \left[\dfrac{(-5x)^2(-3x)}{x^4}\right]^{-1}$ [by Property (2)]

$\qquad\qquad = \left[\dfrac{25x^2(-3x)}{x^4}\right]^{-1} = \left(\dfrac{-75x^3}{x^4}\right)^{-1}$

$\qquad\qquad = \left(\dfrac{-75}{x^{4-3}}\right)^{-1} = \left(-\dfrac{75}{x}\right)^{-1} = -\dfrac{x}{75}$

(c) $\dfrac{5c^{-3}(r+s)^2}{15c^4(r+s)^{-5}} = \dfrac{(r+s)^2(r+s)^5}{3c^4c^3}$ (why?)

$\qquad\qquad = \dfrac{(r+s)^{2+5}}{3c^{4+3}} = \dfrac{(r+s)^7}{3c^7}$ [by Property (i)]

Car and Home Mortgage Payments

Exponents can be used to calculate a regular monthly car payment or home mortgage payment. The formula

$$m = \dfrac{Pr}{12[1 - (1 + (r/12))^{-12t}]}$$

is used to calculate the size m of the successive equal monthly **car** or **home mortgage payments** required to pay off a borrowed principal amount of P dollars, at an annual interest rate of r, paid over a term of t years.

The total amount of **interest, I,** charged (in dollars) is given by the formula

$$I = 12m\left[t - \dfrac{1 - (1 + (r/12))^{-12t}}{r}\right].$$

EXAMPLE 5   ☐ Suppose you want to buy a car and need to borrow $8,000 from your local bank. The bank charges an annual interest of 11.5% on automobile loans with monthly payments over a term of 36 months (3 years). Calculate (a) your monthly payment, and (b) the total interest charge.

SOLUTION

(a) Substituting $P = 8,000$, $r = 0.115$, and $t = 3$ into the monthly car payment formula, we have

$$m = \frac{Pr}{12[1 - (1 + (r/12))^{-12t}]} = \frac{(8,000)(0.115)}{12[1 - (1 + (0.115/12))^{-12(3)}]}$$
$$= 263.81.$$

Therefore, the monthly car payment is $263.81.

(b) Substitute $m = 263.81$, $r = 0.115$, and $t = 3$ in the formula

$$I = 12m\left[t - \frac{1 - (1 + (r/12))^{-12t}}{r}\right]$$
$$= (12)(263.81)\left[3 - \frac{1 - (1 + (0.115/12))^{-12(3)}}{0.115}\right]$$
$$= 1,497.10.$$

Therefore, the total interest charge is $1,497.10.

Suppose that you borrow P dollars from a bank for a home mortgage at an annual interest rate r payable in successive equal monthly payments of m dollars each over a term of t years. The **balance due** to the bank at the beginning of **the kth month**, P_k dollars, is given by the formula

$$P_k = P\left[\frac{(1 + (r/12))^{12t} - (1 + (r/12))^{k-1}}{(1 + (r/12))^{12t} - 1}\right].$$

EXAMPLE 6   © A family is buying a new house costing $120,000, paying 20% down, and taking out a 30-year mortgage on the remaining amount at an annual interest rate of 10.5%. (a) What is the monthly payment on the mortgage? (b) After 15 years, how much of the original mortgage will be paid off?

SOLUTION

(a) The down payment is 20% of $120,000 = (0.20)($120,000) = $24,000. So the remaining amount is $120,000 − $24,000 = $96,000. Substitute $P = 96,000$, $r = 0.105$, and $t = 30$ in the formula

$$m = \frac{Pr}{12[1 - (1 + (r/12))^{-12t}]} = \frac{(96,000)(0.105)}{12[1 - (1 + (0.105/12))^{-12(30)}]}$$
$$= 878.15.$$

Therefore the monthly house payment is $878.15.

(b) $15(12) = 180$. Substitute $k = 181$, $r = 0.105$, $t = 30$, and $P = 96,000$ in the formula

$$P_k = P\left[\frac{(1 + (r/12))^{12t} - (1 + (r/12))^{k-1}}{(1 + (r/12))^{12t} - 1}\right]$$

$$= 96,000\left[\frac{(1 + (0.105/12))^{12(30)} - (1 + (0.105/12))^{181-1}}{(1 + (0.105/12))^{12(30)} - 1}\right]$$

$$= 79,441.88$$

Therefore, the balance due after 15 years is $79,441.88, and the amount of the original mortgage paid off is $96,000 - \$79,441.88 = \$16,558.12$.

Scientific Notation

Recall from Section 1.5 that a large number x greater than 10 is written in scientific notation as

$$x = s \times 10^n,$$

where $1 \le s < 10$, and n is a positive integer. Now that we have been given the meaning of zero and negative integer exponents, we can write positive numbers less than 1 in scientific notation.

A positive number x is written in scientific notation if x has the form

$$x = s \times 10^n,$$

where $1 \le s < 10$, and n is *any* integer. To change a small number from ordinary decimal form to scientific notation, we move the decimal point n places to the *right* to obtain a number between 1 and 10. Then we multiply this number by 10^{-n}.

For example, to change the number 0.00573 to scientific notation, we move the decimal point 3 places to the right. Then we multiply this number by 10^{-3}. That is,

$$0.00573 = 5.73 \times 10^{-3}.$$

The following table lists additional examples of numbers written in ordinary decimal form and in scientific notation.

Numbers in Ordinary Decimal Form	Numbers in Scientific Notation
6	6×10^0
0.35	3.5×10^{-1}
0.078	7.8×10^{-2}
0.00567	5.67×10^{-3}
0.000384	3.84×10^{-4}
0.000083	8.3×10^{-5}

The properties of exponents can be used to simplify expressions involving numbers written in scientific notation.

EXAMPLE 7 $\boxed{\text{C}}$ Simplify the expression and write the answer in scientific notation:

$$\frac{(6.75 \times 10^{-5}) \cdot (8.25 \times 10^{-4})}{13.50 \times 10^{-11}}.$$

SOLUTION

$$\frac{(6.75 \times 10^{-5}) \cdot (8.25 \times 10^{-4})}{13.50 \times 10^{-11}} = \frac{(6.75) \cdot (8.25)}{13.50} \times \frac{10^{-5} \cdot 10^{-4}}{10^{-11}}$$

$$= 4.125 \times 10^2$$

EXAMPLE 8 $\boxed{\text{C}}$ Write each number in scientific notation, and then simplify. Express the answer in scientific notation

$$\frac{681{,}000 \times 0.000000397}{0.000781 \times 0.00173}$$

SOLUTION

First we express each number in scientific notation as follows:

$$681{,}000 = 6.81 \times 10^5, \qquad 0.000000397 = 3.97 \times 10^{-7}$$

$$0.000781 = 7.81 \times 10^{-4}, \qquad 0.00173 = 1.73 \times 10^{-3}$$

so that

$$\frac{681{,}000 \times 0.000000397}{0.000781 \times 0.00173} = \frac{(6.81 \times 10^5) \cdot (3.97 \times 10^{-7})}{(7.81 \times 10^{-4}) \cdot (1.73 \times 10^{-3})}$$

$$= \frac{(6.81) \cdot (3.97)}{(7.81) \cdot (1.73)} \times \frac{10^5 \cdot 10^{-7}}{10^{-4} \cdot 10^{-3}}$$

$$= 2.00097 \times 10^5.$$

PROBLEM SET 5.1

In problems 1–48, rewrite each expression so that it contains only positive exponents and simplify the results (if possible). (Assume all variables that appear in a base with a zero or a negative exponent represent real numbers that make the base nonzero.)

1. 4^0

2. $(-3)^0$

3. $(1 + 7y^{-3})^0$

4. $(3 + 5x^2)^0$

5. $(-3y^2)^0$

6. $(7^{-1} + 3^{-2})^0$

7. $(3 - 3)^0$

8. $2 + (x - x)^0$

9. 5^{-2}

10. 2^{-4}

11. $(-7)^{-2}$

12. $(-3)^{-4}$

13. $(-10)^{-1}$

14. $(-25)^{-1}$

15. $\left(\dfrac{4}{3}\right)^{-3}$

16. $\left(\dfrac{1}{3}\right)^{-4}$

17. p^{-5}

18. y^{-3}

19. $\left(\dfrac{1}{x}\right)^{-1}$

20. $\left(\dfrac{1}{y}\right)^{-2}$

21. $u^{-3}v^{-6}$

22. $x^4 y^{-7}$

23. $4a^{-7}b^2$

24. $5k^{-2}r^{-3}$

25. $\dfrac{10^0}{10^{-3}}$

26. $\dfrac{7^0}{7^{-2}}$

27. $\dfrac{(-6)^{-3}}{(-3)^{-2}}$

28. $\dfrac{a^{-2}}{8p^{-4}}$

29. $2^2 \cdot 3^2 \cdot 6^{-3}$

30. $3^2 \cdot 4^2 \cdot 12^{-2}$

31. $\dfrac{y^{-5}}{7x^{-2}}$

32. $\dfrac{2}{-3x^4 y^{-2}}$

33. $\dfrac{2^{-2}+5^{-2}}{10^{-3}}$

34. $\dfrac{a^{-1}+b^{-1}}{(a+b)^{-1}}$

35. $\dfrac{x^{-1}-y^{-1}}{(xy)^{-1}}$

36. $x^{-1}y - xy^{-1}$

37. $\dfrac{5^{-1}-3^{-2}}{(45)^{-1}}$

38. $\dfrac{1}{4^{-2}-3^{-2}}$

39. $\dfrac{1}{(4+3)^{-2}}$

40. $\dfrac{1}{u^{-2}+v^{-2}}$

41. $\dfrac{1}{x^{-2}}-x^{-2}$

42. $\dfrac{3}{(a-b)^{-2}}$

43. $\dfrac{a^2}{b^{-2}}+\dfrac{b^2}{a^{-2}}$

44. $\dfrac{t^3}{u^{-3}}-\dfrac{u^3}{t^{-3}}$

45. $\dfrac{4-x^{-2}}{2+x^{-1}}$

46. $\dfrac{(a^{-2}+b^{-1})^{-1}}{(a^2+b)^{-1}}$

47. $\dfrac{3x^{-2}+7x^{-1}+4}{3x^{-2}-5x^{-1}-12}$

48. $\dfrac{4a^{-2}-9b^{-2}}{8a^{-3}-27b^{-3}}$

In problems 49–76, use the properties of exponents to rewrite each expression so that it contains only positive exponents, and simplify the results. (Assume all variables that appear in a base with a negative exponent represent real numbers that make the base nonzero.)

49. $7^{-1} \cdot 7^3$

50. $5^{-3} \cdot 5^{-2}$

51. $9^{-3} \cdot 9$

52. $3^{-4} \cdot 3^7$

53. $x^{-3} \cdot x^{-2}$

54. $y^{-7} \cdot y^{-2}$

55. $(2^3)^{-2}$

56. $(5^2)^{-1}$

57. $[(-4)^2]^{-2}$

58. $(p^{-2})^{-4}$

59. $(3x)^{-4}$

60. $(3y^{-4})^2$

61. $(5^{-1}p^4)^{-2}$

62. $(p^{-1}y^{-2})^5$

63. $\left(\dfrac{5}{x}\right)^{-2}$

64. $\left(\dfrac{3}{p^{-1}}\right)^{-1}$

65. $\left(\dfrac{p^2}{q^2}\right)^{-2}$

66. $\left(\dfrac{x^{-3}}{y^{-2}}\right)^{-1}$

67. $\dfrac{8^{-3}}{8^{-5}}$

68. $\dfrac{7^{-3}}{7^{-8}}$

69. $\dfrac{x^{-3}}{x^5}$

70. $\dfrac{y}{y^{-6}}$

71. $\dfrac{xy^{-7}}{xy^{-4}}$

72. $\dfrac{(ab)^{-4}}{(ab)^{-7}}$

73. $\dfrac{(a+b)^{-2}}{(a+b)^{-3}}$

74. $x^{-2n} \cdot x^{2n+5}$

75. $x^{3+n} \cdot x^{2-n}$

76. $\dfrac{(x+3y)^{-12}}{(x+3y)^{10}}$

In problems 77–86, rewrite each expression without negative exponents, and simplify. (Assume all variables that appear in a base with a negative exponent represent real numbers that make the base nonzero.)

77. $\left(\dfrac{b^{-1}}{3a^{-1}}\right)\left(\dfrac{3a}{b}\right)^{-1}$

78. $\left(\dfrac{a^{-1}c^2}{a^{-2}c^{-1}}\right)^{-2}$

79. $\left(\dfrac{x^{-1}y^{-2}}{x^{-2}y^3}\right)^2$

80. $\left(\dfrac{p^{-2}q^{-1}r^{-3}}{q^3 r^{-1}}\right)^{-4}$

81. $\left(\dfrac{a^2c^{-2}b^{-1}}{(abc)^{-1}}\right)^{-3}$

82. $\left(\dfrac{a^{-4}b^2c^{-6}}{(ab)^{-2}(bc)^{-3}}\right)^{-1}$

83. $\left(\dfrac{u^{-4}v^3z^{-3}}{u^3(vz)^{-2}}\right)^{-4}$

84. $\left(\dfrac{(x^{-2})^{-1}(y^{-2})^3}{(x^{-1})^2(y^{-1})^2}\right)^{-2}$

85. $\left[\left(\dfrac{a^{-1}}{b}\right)^{-1}\cdot\left(\dfrac{b^{-1}}{a}\right)^{-1}\right]^{-2}$

86. $\left(\dfrac{3^{-1}x^{-2}y^{-5}}{9^{-1}x^{-1}y^{-3}}\right)^{-3}$

© In problems 87–90, simplify each expression and write each answer in scientific notation.

87. $\dfrac{(9.75\times10^8)\cdot(1.5\times10^{-3})}{(7.5\times10^{-2})\cdot(1.1\times10^{-4})}$

88. $\dfrac{(5.7\times10^4)^2\cdot(6.6\times10^{-5})^3}{(3.3\times10^{-7})\cdot(3.8\times10^{-6})^4}$

89. $\dfrac{(1.12\times10^{-3})\cdot(8.25\times10^{-5})}{(2.35\times10^{-6})^2}$

90. $\dfrac{(1.86\times10^5)\cdot(2.4\times10^{-9})}{(3.6\times10^{-7})\cdot(4.8\times10^{-11})^2}$

In problems 91–94, suppose you want to buy a car and need to borrow P dollars from a local bank. The bank charges an annual interest rate r on automobile loans with monthly payments over t years. Find (a) the monthly payment m and (b) the total interest charge I.

91. $P = \$6{,}000$, $r = 11.5\%$, and $t = 3$ years.

92. $P = \$10{,}000$, $r = 10.8\%$, and $t = 4$ years.

93. $P = \$12{,}000$, $r = 12\%$, and $t = 4$ years.

94. $P = \$9{,}000$, $r = 11.6\%$, and $t = 4$ years.

In problems 95–98, suppose that you borrow P dollars from a bank for a home mortgage at an annual interest rate r payable in successive equal monthly payments each over a term of t years. Assume that the down payment is 20% of the original cost C. Let P_k be the balance due to the bank at the beginning of the kth month. Find (a) the monthly payment m, and (b) the amount of the original that will be paid off after x years.

95. $C = \$65{,}000$, $r = 9\%$, $t = 25$ years, and $x = 12$ years.

96. $C = \$80{,}000$, $r = 9.5\%$, $t = 30$ years, and $x = 15$ years.

97. $C = \$50{,}000$, $r = 10\%$, $t = 30$ years, and $x = 10$ years.

98. $C = \$40{,}000$, $r = 10\%$, $t = 30$ years, and $x = 18$ years.

© **99.** An electronic computer can do an arithmetic operation in 0.00000036 second. If the electricity travels in a computer circuit at the speed of light (186,000 miles per second), how far will the electricity travel in the time it takes the computer to complete 7 arithmetic operations? Express the answer in scientific notation.

© **100.** A certain virus is shaped like a sphere. If its radius is 0.000013 centimeter, find the volume of the virus. Express the answer in scientific notation.

© **101.** A satellite leaves the earth traveling at a uniform speed of 310,000 miles per hour. How long does it take for the satellite to reach the sun, if the average distance from the earth to the sun is 93,000,000 miles? Express the answer in scientific notation.

© **102.** One gram of hydrogen contains 6.023×10^{23} atoms. What is the weight of one atom of hydrogen? Express the answer in scientific notation.

5.2 Roots and Rational Exponents

Suppose that the area A of a square is 25 square units. We ask: What is the length x of one side of the square? If we substitute 25 for A in the equation $A = x^2$, we have

$$25 = x^2.$$

To find x, we must find a number whose square is 25. Two numbers will satisfy this criterion: 5 and -5, because

$$5^2 = 25 \quad \text{and} \quad (-5)^2 = 25.$$

Since x represents the length of one side of a square, our answer must be positive. Therefore, we must reject -5.

The process of finding a number whose square is 25 is called finding a **square root** of 25. That is, because $5^2 = 25$, we say that 5 is a **square root** of 25.

Similarly, suppose that the volume V of a cube is 8 cubic units. Substituting 8 for V in the equation $V = x^3$, where x units is the length of a side of a cube, we have

$$8 = x^3.$$

To find x, we must find a number whose cube is 8. The number 2 is the only real number whose cube is 8. That is, because

$$2^3 = 8,$$

we say that 2 is a **cube root** of 8.

As you can see, since

$$3^4 = 81,$$

we reverse the procedure to find a fourth root of 81, and we say that 3 is a **fourth root** of 81. In general, we have the following definition:

DEFINITION 1

The nth Root of a Number

> If a and b are real numbers, and n is a positive integer greater than 1, such that
> $$b^n = a,$$
> then b is called an **nth root** of a.

Radical notation is commonly used to indicate a root. We write $\sqrt[n]{a}$ to represent the nth root of a number a.

In the expression $\sqrt[n]{a}$, the symbol $\sqrt{\ }$ is called the **radical.** The positive integer n is called the **index** (if the index is not written, it is understood to be 2), and the real number under the radical is called the **radicand:**

index ↘
radical ↗ $\sqrt[n]{a}$ ← radicand

A number may have both a positive and a negative nth root. When this occurs, the *positive* nth root of the number is called the **principal nth root,** and is denoted by $\sqrt[n]{a}$. For example, although it is correct to say that both 3 and -3 are square roots of 9, it is *incorrect* to write $\sqrt{9} = -3$ (but *correct* to write $\sqrt{9} = 3$; we usually write $\sqrt{\ }$ rather than $\sqrt[2]{\ }$). If we wish to designate the negative square root of 9 using radical notation, we write $-\sqrt{9} = -3$.
  Similarly,

$$\sqrt[4]{81} = 3, \quad \text{since} \quad 3^4 = 81 \quad \text{and} \quad 3 > 0,$$

and

$$\sqrt[6]{64} = 2, \quad \text{since} \quad 2^6 = 64 \quad \text{and} \quad 2 > 0.$$

If a number a has *only* a negative nth root, the negative root is considered the principal nth root of a. Thus,

$$\sqrt[3]{-8} = -2, \quad \text{since} \quad (-2)^3 = -8, \text{ and } -2 \text{ is the only cube root of } -8,$$

and

$$\sqrt[5]{-32} = -2, \quad \text{since} \quad (-2)^5 = -32, \text{ and } -2 \text{ is the only fifth root of } -32.$$

Of course, if a number a has *only* a positive nth root, then the positive root is considered to be the principal nth root of a. For example, $\sqrt[3]{27} = 3$, the only cube root of 27.
  Following is a list of the most common roots. These are used in the remainder of the chapter and should be memorized.

Principal Square Roots		Principal Cube Roots	Principal Fourth Roots	Principal Fifth Roots
$\sqrt{1} = 1$	$\sqrt{49} = 7$	$\sqrt[3]{1} = 1$	$\sqrt[4]{1} = 1$	$\sqrt[5]{1} = 1$
$\sqrt{4} = 2$	$\sqrt{64} = 8$	$\sqrt[3]{8} = 2$	$\sqrt[4]{16} = 2$	$\sqrt[5]{32} = 2$
$\sqrt{9} = 3$	$\sqrt{81} = 9$	$\sqrt[3]{27} = 3$	$\sqrt[4]{81} = 3$	$\sqrt[5]{243} = 3$
$\sqrt{16} = 4$	$\sqrt{100} = 10$	$\sqrt[3]{64} = 4$	$\sqrt[4]{256} = 4$	
$\sqrt{25} = 5$	$\sqrt{121} = 11$	$\sqrt[3]{125} = 5$	$\sqrt[4]{625} = 5$	
$\sqrt{36} = 6$	$\sqrt{144} = 12$	$\sqrt[3]{216} = 6$		

We can now formulate the following definition:

DEFINITION 2   **The Principal nth Root of a Number**

> Let n be a positive integer greater than 1, and let a be a real number. Then $\sqrt[n]{a}$—the **principal nth root** of a—is defined as follows:
>
> (i) If $a > 0$, $\sqrt[n]{a}$ is the positive nth root of a.
>
> (ii) If $a < 0$ and n is odd, $\sqrt[n]{a}$ is the negative nth root of a.
>
> (iii) If $a = 0$, then $\sqrt[n]{0} = 0$. Notice that $\sqrt[n]{a}$ is not defined as a real number when $a < 0$ and n is even.

EXAMPLE 1   Evaluate those radicals that are real numbers and check your conclusion. If an expression is not a real number, give a reason.

(a) $\sqrt{169}$ (b) $\sqrt[3]{-125}$

(c) $\sqrt[4]{\dfrac{16}{81}}$ (d) $\sqrt[4]{-16}$

(e) $\sqrt{y^4}$

SOLUTION   (a) $\sqrt{169} = 13$, because $13^2 = 169$.

(b) $\sqrt[3]{-125} = -5$, because $(-5)^3 = -125$.

(c) $\sqrt[4]{\dfrac{16}{81}} = \dfrac{2}{3}$, because $\left(\dfrac{2}{3}\right)^4 = \dfrac{2^4}{3^4} = \dfrac{16}{81}$.

(d) $\sqrt[4]{-16}$ is undefined (not a real number), since $-16 < 0$, and $n = 4$ is even.

(e) $\sqrt{y^4} = y^2$, because $(y^2)^2 = y^4$ and $y^2 \geq 0$.

Rational Exponents

In Section 5.1, we extended the definition of exponents to include negative integers and zero. Now we will expand the definition of exponents to include rational numbers. Radicals can be used to define rational exponents. Let m

and n be integers with $n \neq 0$. We wish to define $a^{m/n}$. If Property (ii) of Section 5.1 (page 219) is true, we must have

$$a^{m/n} = a^{(1/n)m} = (a^{1/n})^m.$$

The basic question, therefore, is: How do we define $a^{1/n}$? Again, if Property (ii) is true, then

$$(a^{1/n})^n = a^{(1/n)n} = a^1 = a.$$

In other words, $a^{1/n}$ must be defined as the **principal nth root** of a. Therefore, we have the following definition:

DEFINITION 3   **Rational Exponents of the Form $a^{1/n}$**

> Let n be a positive integer, $n \geq 2$. If a is a real number such that $\sqrt[n]{a}$ is defined, then
>
> $$a^{1/n} = \sqrt[n]{a}$$
>
> is the *principal nth root* of a.

For example,

$$4^{1/2} = \sqrt{4} = 2, \qquad 27^{1/3} = \sqrt[3]{27} = 3, \qquad \text{and} \qquad 16^{1/4} = \sqrt[4]{16} = 2.$$

Similarly,

$$(-8)^{1/3} = \sqrt[3]{-8} = -2, \qquad (-32)^{1/5} = \sqrt[5]{-32} = -2, \qquad \text{and} \qquad 0^{1/7} = \sqrt[7]{0} = 0.$$

Since $\sqrt{-9}$ is not a real number, then $(-9)^{1/2}$ is not defined. In fact, $a^{1/n}$ is not a real number when $a < 0$ and n is even.

Having, defined expressions of the form $a^{1/n}$, we are now able to extend our definition to include expressions of the form $a^{m/n}$, where m/n is any rational number. Observe, for example, the two ways that $27^{2/3}$ can be evaluated on the assumption that the properties of exponents are to hold for rational numbers:

$$27^{2/3} = (27^{1/3})^2 = (\sqrt[3]{27})^2 = 3^2 = 9$$

or

$$27^{2/3} = (27^2)^{1/3} = \sqrt[3]{27^2} = \sqrt[3]{729} = 9.$$

This leads to the following definition:

DEFINITION 4   **Rational Exponents of the Form $a^{m/n}$**

> Suppose that m and n are integers with $n \geq 2$, and that the fraction m/n is reduced to lowest terms. If a is a nonzero real number such that $\sqrt[n]{a}$ exists, then $a^{m/n} = (a^{1/n})^m = (\sqrt[n]{a})^m$ or $a^{m/n} = (a^m)^{1/n} = \sqrt[n]{a^m}$.

EXAMPLE 2   Rewrite each expression in terms of rational exponents.

(a) $\sqrt{7}$ (b) $\sqrt[3]{-x^2}$

(c) $\sqrt[5]{11p^4}$ (d) $\sqrt[7]{(a+b)^5}$

SOLUTION   Using Definitions 3 and 4, we have

(a) $\sqrt{7} = 7^{1/2}$

(b) $\sqrt[3]{-x^2} = (-x^2)^{1/3} = -(x^2)^{1/3} = -x^{2/3}$

(c) $\sqrt[5]{11p^4} = (11p^4)^{1/5} = 11^{1/5}p^{4/5}$

(d) $\sqrt[7]{(a+b)^5} = (a+b)^{5/7}$

EXAMPLE 3   Rewrite each expression in terms of radicals. (Assume that variables are restricted to values for which all expressions are defined.)

(a) $5^{3/4}$ (b) $(6x)^{3/2}$

(c) $(-11y)^{2/3}$ (d) $(8-r)^{2/5}$

SOLUTION   We use $a^{m/n} = \sqrt[n]{a^m}$:

(a) $5^{3/4} = \sqrt[4]{5^3} = \sqrt[4]{125}$ (b) $(6x)^{3/2} = \sqrt{(6x)^3} = \sqrt{216x^3}$

(c) $(-11y)^{2/3} = \sqrt[3]{(-11y)^2} = \sqrt[3]{121y^2}$ (d) $(8-r)^{2/5} = \sqrt[5]{(8-r)^2}$

Remember, Definition 4 is to be used *only* when $n \geq 2$ and when m/n *is reduced to lowest terms*. For example, it is incorrect to write $(-8)^{2/6} = (\sqrt[6]{-8})^2$, since we cannot take the sixth root of -8 and $\frac{2}{6}$ is not reduced to lowest terms. First we must reduce $\frac{2}{6}$ to lowest terms:

$$(-8)^{2/6} = (-8)^{1/3}.$$

Then we can apply Definition 3:

$$(-8)^{1/3} = \sqrt[3]{-8} = -2.$$

EXAMPLE 4   Find the value of each expression.

(a) $27^{4/3}$ (b) $32^{3/5}$ (c) $(-4)^{3/2}$ (d) $(-64)^{8/12}$

SOLUTION   We use $a^{m/n} = (\sqrt[n]{a})^m$ of Definition 4:

(a) $27^{4/3} = (\sqrt[3]{27})^4 = 3^4 = 81$. Also, Definition 4 tells us that
$27^{4/3} = (3^3)^{4/3} = 3^{3 \cdot 4/3} = 3^4 = 81$.

(b) $32^{3/5} = (\sqrt[5]{32})^3 = 2^3 = 8$

(c) $(-4)^{3/2}$ is undefined, because $\sqrt{-4}$ does not exist (as a real number).

(d) $(-64)^{8/12} = (-64)^{2/3} = (\sqrt[3]{-64})^2 = (-4)^2 = 16$

Suppose that m and n are positive integers and that m/n is reduced to lowest terms. Then $-(m/n)$ represents a negative rational number. If Property (ii) (page 223) continues to hold true for negative rational exponents, we have

$$a^{-(m/n)} = (a^{1/n})^{-m}.$$

We use Definition 2 of a negative integer exponent (page 221). If $a \neq 0$, then

$$(a^{1/n})^{-m} = \frac{1}{(a^{1/n})^m}.$$

We have the following definition:

DEFINITION 5   **Negative Rational Exponents**

Let a be a nonzero real number. Suppose that m and n are positive integers, and that m/n is in lowest terms. Then, if $\sqrt[n]{a}$ exists,

$$a^{-m/n} = \frac{1}{a^{m/n}}.$$

EXAMPLE 5   Find the value of each expression.

(a) $16^{-3/4}$ (b) $(-8)^{-2/3}$ (c) $-8^{-2/3}$ (d) $64^{-5/6}$ (e) $(-125)^{-1/3}$

SOLUTION   We use Definition 5:

(a) $16^{-3/4} = \dfrac{1}{16^{3/4}} = \dfrac{1}{(\sqrt[4]{16})^3} = \dfrac{1}{2^3} = \dfrac{1}{8}$

(b) $(-8)^{-2/3} = \dfrac{1}{(-8)^{2/3}} = \dfrac{1}{(\sqrt[3]{-8})^2} = \dfrac{1}{(-2)^2} = \dfrac{1}{4}$

(c) $-8^{-2/3} = -\dfrac{1}{8^{2/3}} = -\dfrac{1}{(\sqrt[3]{8})^2} = -\dfrac{1}{2^2} = -\dfrac{1}{4}$

(d) $64^{-5/6} = \dfrac{1}{64^{5/6}} = \dfrac{1}{(\sqrt[6]{64})^5} = \dfrac{1}{2^5} = \dfrac{1}{32}$

(e) $(-125)^{-1/3} = \dfrac{1}{(-125)^{1/3}} = \dfrac{1}{\sqrt[3]{-125}} = \dfrac{1}{-5} = -\dfrac{1}{5}$

It is true that rational exponents (positive, negative, or zero) satisfy the properties of exponents listed in Section 5.1. We can therefore restate these properties to indicate that they hold for all rational exponents:

Properties of Rational Exponents

> Let a and b be real numbers, and let p and q be rational numbers. Then, if all expressions are defined (as real numbers):
>
> (i) $a^p a^q = a^{p+q}$ (ii) $(a^p)^q = a^{pq}$ (iii) $(ab)^p = a^p b^p$
>
> (iv) $\left(\dfrac{a}{b}\right)^p = \dfrac{a^p}{b^p}$ (v) $\dfrac{a^p}{a^q} = a^{p-q}$

The following examples illustrate how these properties are used to simplify algebraic expressions involving rational exponents.

In Examples 6 and 7, rewrite each expression so that it contains only positive exponents, and simplify. You may assume that variables are restricted to values for which all expressions represent real numbers.

EXAMPLE 6   (a) $7^{-1/2} \cdot 7^{5/2}$ (b) $(x^{-3/4})^{-8/3}$ (c) $(125x^{-18})^{-4/3}$

(d) $\left(\dfrac{32}{x^{-5}}\right)^{-2/5}$ (e) $\dfrac{x^{2/3}}{x^{-4/5}}$

SOLUTION   (a) $7^{-1/2} \cdot 7^{5/2} = 7^{-1/2 + 5/2}$ [by Property (i)]

$= 7^{4/2} = 7^2 = 49$

(b) $(x^{-3/4})^{-8/3} = x^{(-3/4)(-8/3)} = x^2$ [by Property (ii)]

(c) $(125x^{-18})^{-4/3} = 125^{-4/3}(x^{-18})^{-4/3}$ [by Property (iii)]

$= (5^3)^{-4/3}(x^{-18})^{-4/3}$

$= 5^{3(-4/3)}x^{(-18)(-4/3)}$ [by Property (ii)]

$= 5^{-4}x^{24} = \dfrac{x^{24}}{5^4} = \dfrac{x^{24}}{625}$

(d) $\left(\dfrac{32}{x^{-5}}\right)^{-2/5} = \dfrac{32^{-2/5}}{(x^{-5})^{-2/5}}$ [by Property (iv)]

$\qquad = \dfrac{(2^5)^{-2/5}}{(x^{-5})^{-2/5}} = \dfrac{2^{5(-2/5)}}{x^{(-5)(-2/5)}}$

$\qquad = \dfrac{2^{-2}}{x^2} = \dfrac{1}{2^2 x^2} = \dfrac{1}{4x^2}$

(e) $\dfrac{x^{2/3}}{x^{-4/5}} = x^{2/3 - (-4/5)}$ [by Property (v)]

$\qquad = x^{10/15 - (-12/15)}$

$\qquad = x^{22/15}$

EXAMPLE 7   (a) $(2r^{1/3}t^{3/2})^6$ (b) $\left(\dfrac{2p^{2/3}}{q^{1/2}}\right)^2 \left(\dfrac{4p^{-5/3}}{q^{2/3}}\right)$ (c) $\left(\dfrac{125a^{-9}b^{-12}}{8c^{-15}}\right)^{2/3}$

SOLUTION   (a) $(2r^{1/3}t^{3/2})^6 = 2^6(r^{1/3})^6(t^{3/2})^6 = 64r^{(1/3)(6)}t^{(3/2)(6)} = 64r^2t^9$

(b) $\left(\dfrac{2p^{2/3}}{q^{1/2}}\right)^2 \left(\dfrac{4p^{-5/3}}{q^{2/3}}\right) = \left(\dfrac{2^2(p^{2/3})^2}{(q^{1/2})^2}\right)\left(\dfrac{4p^{-5/3}}{q^{2/3}}\right)$

$\qquad = \left(\dfrac{4p^{4/3}}{q}\right)\left(\dfrac{4p^{-5/3}}{q^{2/3}}\right) = \dfrac{16p^{4/3 - 5/3}}{q^{1 + 2/3}}$

$\qquad = \dfrac{16p^{-1/3}}{q^{5/3}} = \dfrac{16}{p^{1/3}q^{5/3}}$

(c) $\left(\dfrac{125a^{-9}b^{-12}}{8c^{-15}}\right)^{2/3} = \dfrac{125^{2/3}(a^{-9})^{2/3}(b^{-12})^{2/3}}{8^{2/3}(c^{-15})^{2/3}}$

$\qquad = \dfrac{(5^3)^{2/3}(a^{-9})^{2/3}(b^{-12})^{2/3}}{(2^3)^{2/3}(c^{-15})^{2/3}}$

$\qquad = \dfrac{5^2 a^{-6}b^{-8}}{2^2 c^{-10}} = \dfrac{25c^{10}}{4a^6 b^8}$

The next two examples illustrate how algebraic procedures used with polynomial expressions apply to expressions with rational exponents.

EXAMPLE 8   Perform each multiplication. Write each answer with positive exponents. (Assume that variables are restricted to values for which all expressions represent real numbers.)

(a) $x^{-2/3}(x^3 - 5x^{2/3})$ (b) $(a^{1/2} - 2b^{-1/2})(a^{1/2} + 2b^{-1/2})$

(c) $(x^{3/2} + 5)^2$

SOLUTION   (a) Using the distributive property, we have

$$x^{-2/3}(x^3 - 5x^{2/3}) = x^{-2/3} \cdot x^3 - x^{-2/3} \cdot 5x^{2/3}$$
$$= x^{-2/3+3} - 5x^{-2/3+2/3}$$
$$= x^{7/3} - 5.$$

(b) Using the special product $(a + b)(a - b) = a^2 - b^2$, we have

$$(a^{1/2} - 2b^{-1/2})(a^{1/2} + 2b^{-1/2}) = (a^{1/2})^2 - (2b^{-1/2})^2$$
$$= a^1 - 4b^{-1} = a - \frac{4}{b} = \frac{ab - 4}{b}.$$

(c) Using the special product $(a + b)^2 = a^2 + 2ab + b^2$, we have

$$(x^{3/2} + 5)^2 = (x^{3/2})^2 + 2(x^{3/2})(5) + 5^2$$
$$= x^3 + 10x^{3/2} + 25.$$

EXAMPLE 9   Perform the subtraction and simplify:

$$(x^2 + 9)^{1/2} - x^4(x^2 + 9)^{-3/2}.$$

SOLUTION   First, we write each expression without negative exponents, and then we perform the subtraction, so that

$$(x^2 + 9)^{1/2} - x^4(x^2 + 9)^{-3/2}$$

$$= (x^2 + 9)^{1/2} - \frac{x^4}{(x^2 + 9)^{3/2}}$$

$$= \frac{(x^2 + 9)^{1/2} \cdot (x^2 + 9)^{3/2}}{(x^2 + 9)^{3/2}} - \frac{x^4}{(x^2 + 9)^{3/2}} \qquad \text{[LCD is } (x^2 + 9)^{3/2}]$$

$$= \frac{(x^2 + 9)^{4/2} - x^4}{(x^2 + 9)^{3/2}} = \frac{(x^2 + 9)^2 - x^4}{(x^2 + 9)^{3/2}}$$

$$= \frac{x^4 + 18x^2 + 81 - x^4}{(x^2 + 9)^{3/2}} = \frac{18x^2 + 81}{(x^2 + 9)^{3/2}}.$$

PROBLEM SET 5.2

In problems 1–8, find the principal root (if it is defined).

1. $\sqrt{16}$ 2. $\sqrt{144}$ 3. $\sqrt{441}$ 4. $\sqrt[3]{-8}$

5. $\sqrt[3]{-27}$ 6. $\sqrt[4]{-81}$ 7. $\sqrt[6]{-64}$ 8. $\sqrt[12]{-1}$

In problems 9–16, rewrite each expression in terms of rational exponents.

9. $\sqrt{13}$ 10. $\sqrt[3]{-17}$ 11. $\sqrt[5]{3x^2}$ 12. $\sqrt[7]{(x + y)^3}$

13. $\sqrt[4]{a^3b^2}$ 14. $\sqrt[9]{(3a + b)^2}$ 15. $\sqrt[5]{-3a^2b^4}$ 16. $\sqrt[11]{-2x^9}$

In problems 17–22, rewrite each expression in terms of radicals.

17. $3^{5/6}$ **18.** $11^{2/3}$ **19.** $(7y)^{3/2}$ **20.** $(3a + b)^{4/5}$

21. $(4xy)^{5/7}$ **22.** $(-7x^2y^3)^{5/9}$

In problems 23–48, find the value of each expression.

23. $36^{1/2}$ **24.** $49^{-1/2}$ **25.** $-16^{-1/2}$ **26.** $-8^{-1/3}$

27. $\left(\dfrac{-8}{27}\right)^{-1/3}$ **28.** $\left(\dfrac{81}{25}\right)^{-1/2}$ **29.** $\left(\dfrac{9}{16}\right)^{1/2}$ **30.** $\left(\dfrac{-64}{125}\right)^{-1/3}$

31. $8^{2/3}$ **32.** $-8^{4/3}$ **33.** $(-8)^{4/3}$ **34.** $(-27)^{2/3}$

35. $(-32)^{3/5}$ **36.** $128^{2/7}$ **37.** $81^{3/4}$ **38.** $(-64)^{4/3}$

39. $(-16)^{3/2}$ **40.** $(-36)^{5/2}$ **41.** $27^{8/12}$ **42.** $25^{7/14}$

43. $(-8)^{-4/3}$ **44.** $(-27)^{-2/3}$ **45.** $9^{-3/2}$ **46.** $(-32)^{-4/5}$

47. $(-128)^{-5/7}$ **48.** $(0.216)^{-4/6}$

In problems 49–78, rewrite each expression so that it contains only positive exponents, and simplify. (Assume that variables are restricted to values for which all expressions represent real numbers.)

49. $2^{1/3}2^{2/3}$ **50.** $5^{1/2}5^{-3/2}$ **51.** $x^{-2/3}x^{5/3}$ **52.** $y^{2/15}y^{-7/60}$

53. $(5^{1/7})^{14}$ **54.** $(8^{-2/3})^{-6}$ **55.** $(x^{-7/9})^{18/7}$ **56.** $(y^{-2})^{-15/2}$

57. $(8p^9)^{4/3}$ **58.** $(32u^{-5})^{-3/5}$ **59.** $(2^{-1/3}y^{-1/7})^{-21}$ **60.** $(81m^{12})^{-3/4}$

61. $\left(\dfrac{125}{y^3}\right)^{-1/3}$ **62.** $\left(\dfrac{x^{-5}}{32}\right)^{4/5}$ **63.** $\dfrac{5^{2/3}}{5^{-1/7}}$

64. $\dfrac{(x^{-3/4})^{-2}}{(x^2)^{-5/8}}$ **65.** $\dfrac{x^{1/3}}{x^{-1/6}}$ **66.** $\dfrac{y^{3/2}}{y^{-7/2}}$

67. $(4c^{2/3}d^{3/4})(2c^{-5/3}d^{1/4})$ **68.** $(5x^{1/3}y^{2/3})(3x^{-4/3}y^{4/3})$ **69.** $(6a^{7/2}b^{-3/2})^2(4a^{-1/3}b^{-2/3})^3$

70. $\left(\dfrac{x^{-10}y^8}{x^{-12}y^{-4}}\right)^{-1/2}$ **71.** $\left(\dfrac{5^2 4^6}{5^{-4}4^7}\right)^{-1/2}$ **72.** $\left(\dfrac{2x^{3/2}y^{7/2}}{4x^2y^{-1}}\right)^{-4}$

73. $\dfrac{(125x^7y^{-5})^{-1/3}}{(64x^2y^8)^{-1/6}}$ **74.** $\left(\dfrac{a^3b^{3/2}}{a^{-3}b^{1/2}}\right)^{1/6}$ **75.** $\dfrac{(p^{2/5}q^2)^5}{(p^{1/6}q^{1/3})^{12}}$

76. $\left[\dfrac{(x^{5/7}y^{3/2})(x^{2/7}y^{-5/2})}{x^2y^{-1/2}}\right]^{-2}$ **77.** $\left(\dfrac{81p^{-12}}{q^{16}}\right)^{-1/4}\cdot\left(-\dfrac{p^{-2/3}}{q^{1/3}}\right)^3$ **78.** $\left[\dfrac{(5x^2y^3)^{3/4}(5x^2y^3)^{1/4}}{(x^2y^4)^{-1/2}}\right]^{-4}$

In problems 79–88, perform each multiplication. Write each answer with positive exponents. (Assume that variables are restricted to values for which all expressions represent real numbers.)

79. $x^{2/3}(4x^{1/3} - x^{4/3})$ **80.** $x^{-2/5}(3x^{-7/5} - 4x^{7/5} + 3x^{2/5})$

81. $(2a^{1/2} - b^{-1/2})(2a^{1/2} + b^{-1/2})$ **82.** $(x^{-3/2} - 5y^{3/2})(x^{-3/2} + 5y^{3/2})$

83. $(x^{-3/2} + 5)^2$ **84.** $(x^{5/2} + y^{-1/2})^2$

85. $(x^{1/3} + y^{1/3})(x^{2/3} - x^{1/3}y^{1/3} + y^{2/3})$ **86.** $(x^{1/3} - 2)(x^{2/3} + 2x^{1/3} + 4)$

87. $(3x^{-1/4} + 2)(5x^{-1/4} - 7)$ **88.** $(x^{1/4} - 2)(x^{1/4} + 2)(x^{1/2} + 4)$

In problems 89–94, perform each addition or subtraction and simplify. (Assume that variables are restricted to values for which all expressions represent real numbers.)

89. $4x^{-1/2} + x^{1/2}$

90. $3y^{-1/2} - y^{3/2}$

91. $(x - 1)^{-1/2} - \frac{1}{2}x(x - 1)^{-3/2}$

92. $3(2x + 1)^{-1/2} - (3x - 4)(2x + 1)^{-3/2}$

93. $\dfrac{-x^2}{(x^2 + 1)^{1/2}} + (x^2 + 1)^{1/2}$

94. $\dfrac{1}{(x^2 + 4)^{1/2}} - \dfrac{x^2}{(x^2 + 4)^{3/2}}$

95. Use a calculator with a y^x key to evaluate each expression. Round off your answer to four significant digits.

(a) $4.18^{0.325}$ (b) $0.6^{3.41}$ (c) $7^{4/5}$

(d) $5.61^{\sqrt{2}}$ (e) $7.63^{\sqrt{3}}$ (f) $2^{\sqrt{5}}$

96. (a) Factor $5(x + 2)^{1/2}$ from $15(x + 2)^{3/2} - 10(x + 2)^{1/2}$.

(b) Factor $3(x + 7)^{2/5}$ from $21x^2(x + 7)^{4/5} + 9x(x + 7)^{2/5}$.

(c) Factor $7(x^2 + 5)^{-2/7}$ from $35x^2(x^2 + 5)^{-4/7} - 14x(x^2 + 5)^{-2/7}$.

(d) Complete the following:

$$x^{2/5} - x^{1/5} - 6 = (x^{1/5} - \underline{\quad})(\underline{\quad} + 2).$$

97. The formula $A = \sqrt{s(s - a)(s - b)(s - c)}$ gives the area of a triangle, where a, b, and c represent the lengths of the sides and

$$s = \frac{a + b + c}{2}. \qquad \text{(the semiperimeter)}$$

Find the area of the triangle whose sides are 42.3, 28.7, and 37.1 meters.

98. In statistics, the following formula is used:

$$\sigma = \sqrt{\frac{(x_1 - \bar{x})^2 + (x_2 - \bar{x})^2 + (x_3 - \bar{x})^2}{3}},$$

where x_1, x_2, and x_3 represent the values of 3 observations, and

$$\bar{x} = \frac{x_1 + x_2 + x_3}{3}. \qquad (\bar{x} \text{ is the average})$$

Find σ when $x_1 = 3.4$, $x_2 = 5.7$, and $x_3 = 6.8$.

99. Suppose that a population P of a certain culture of bacteria is given by the equation

$$P = 100(t^5 + 10t^2 + 9)^{1/4},$$

where t is the time in hours since the culture was started. Find the population P when $t = 2.5$ hours.

100. Kepler's third law [named after the German mathematician and astronomer Johann Kepler (1571–1630)] of planetary motion states that the time T in years that it takes a planet to revolve about the sun is given by the equation $T = d^{3/2}$, where d is the average distance in astronomical units between the planet and the sun. Find T when $d = 9.53$.

5.3 Radicals

Radical expressions can often be simplified by using certain properties. Consider the following example:

$$\sqrt{9} \cdot \sqrt{25} = 3 \cdot 5 = 15.$$

We can also see that

$$\sqrt{9 \cdot 25} = \sqrt{225} = 15.$$

Thus, we conclude that

$$\sqrt{9 \cdot 25} = \sqrt{9} \cdot \sqrt{25}.$$

Now, notice that

$$\sqrt[3]{\frac{27}{8}} = \frac{3}{2}, \quad \text{and that} \quad \frac{\sqrt[3]{27}}{\sqrt[3]{8}} = \frac{3}{2}.$$

We conclude that

$$\sqrt[3]{\frac{27}{8}} = \frac{\sqrt[3]{27}}{\sqrt[3]{8}}.$$

These examples imply the following properties of radicals:

Properties of Radicals

Let a and b be real numbers and let n be a positive integer. Then, provided that all expressions represent real numbers,

(i) $\sqrt[n]{ab} = \sqrt[n]{a} \cdot \sqrt[n]{b}$ (ii) $\sqrt[n]{\dfrac{a}{b}} = \dfrac{\sqrt[n]{a}}{\sqrt[n]{b}}$

Because $a^{1/n} = \sqrt[n]{a}$, the properties of radicals can be proved by using the properties of rational exponents. For instance, we verify Property (i) as follows:

$$\sqrt[n]{ab} = (ab)^{1/n} = a^{1/n} \cdot b^{1/n} = \sqrt[n]{a} \cdot \sqrt[n]{b}.$$

Similarly, Property (ii) is verified as follows:

$$\sqrt[n]{\frac{a}{b}} = \left(\frac{a}{b}\right)^{1/n} = \frac{a^{1/n}}{b^{1/n}} = \frac{\sqrt[n]{a}}{\sqrt[n]{b}}.$$

These properties of radicals allow us to express radical expressions in different forms and to simplify these expressions without changing their values.

To *simplify* a radical expression, we use the preceding properties and we write the expression in a form that satisfies the following conditions:

1. The power of any factor under the radical is less than the index of the radical, that is, in $\sqrt[n]{a^m}$, $m < n$. The radical $\sqrt[3]{x^4}$ violates this condition.

2. The exponents of factors under the radical and the index of the radical have no common factors, that is, in $\sqrt[n]{a^m}$, m and n have no common factors. The radical $\sqrt[6]{x^4}$ violates this condition.

3. The radicand contains no fractions. The radical $\sqrt{3/5}$ violates this condition.

4. There are no radicals in the denominator of a fraction. The fraction $3/\sqrt{7}$ violates this condition.

It should be noticed that special problems are encountered when we work with expressions like $\sqrt{a}$ and $\sqrt[4]{b}$. Since even roots of negative numbers do not exist as real numbers, we cannot let a and b represent negative numbers. Therefore, in this chapter, we impose restrictions on variables so that each radical expression represents a real number.

EXAMPLE 1   Use the properties of radicals to simplify each expression:

(a) $\sqrt{72}$ (b) $\sqrt[3]{-108}$ (c) $\sqrt[4]{\dfrac{7}{16}}$ (d) $\sqrt[5]{\dfrac{3}{-32}}$

SOLUTION   (a) First, we look for the largest perfect square factor of 72. This factor is 36, so we have

$$\sqrt{72} = \sqrt{36 \cdot 2}$$
$$= \sqrt{36} \cdot \sqrt{2} = 6\sqrt{2}. \qquad \text{[by Property (i)]}$$

(b) $\sqrt[3]{-108} = \sqrt[3]{(-27)(4)}$
$$= \sqrt[3]{-27} \cdot \sqrt[3]{4} \qquad \text{[by Property (i)]}$$
$$= -3\sqrt[3]{4}$$

(c) $\sqrt[4]{\dfrac{7}{16}} = \dfrac{\sqrt[4]{7}}{\sqrt[4]{16}} \qquad \text{[by Property (ii)]}$

$$= \dfrac{\sqrt[4]{7}}{2}$$

(d) $\sqrt[5]{\dfrac{3}{-32}} = \dfrac{\sqrt[5]{3}}{\sqrt[5]{-32}}$ [by Property (ii)]

$$= \dfrac{\sqrt[5]{3}}{-2} = -\dfrac{\sqrt[5]{3}}{2}.$$

Note that

$$\sqrt{9} = 3, \quad \text{since} \quad 3^2 = 9, \quad \text{so that} \quad (\sqrt{9})^2 = 9.$$

Also,

$$\sqrt{a} = b, \quad \text{since} \quad b^2 = a, \quad \text{so that} \quad (\sqrt{a})^2 = a.$$

Similarly,

$$\sqrt[3]{a} = b, \quad \text{since} \quad b^3 = a, \quad \text{so that} \quad (\sqrt[3]{a})^3 = a.$$

In general, if $\sqrt[n]{a}$ represents a real number, we have the following property:

Property (iii)

$$\boxed{(\sqrt[n]{a})^n = a.}$$

For example,

$$(\sqrt{7})^2 = 7, \quad (\sqrt[6]{3})^6 = 3, \quad (\sqrt[3]{-4})^3 = -4, \quad \text{and} \quad (\sqrt[5]{y})^5 = y.$$

Now consider the expression $\sqrt[3]{2^3}$. Since $2^3 = 8$, we have $\sqrt[3]{2^3} = \sqrt[3]{8} = 2$. Similarly, $\sqrt[3]{(-3)^3} = \sqrt[3]{-27} = -3$, and $\sqrt[5]{2^5} = \sqrt[5]{32} = 2$. In general, if a is a real number, and n is an odd positive integer, then we have the following property:

Property (iv)

$$\boxed{\sqrt[n]{a^n} = a.}$$

For instance, by using Property (iv), we find $\sqrt[3]{x^9}$ as follows:

$$\sqrt[3]{x^9} = \sqrt[3]{(x^3)^3} = x^3.$$

We can also find $\sqrt[3]{x^9}$ by dividing the exponent 9 of x by the index 3 of the radical. That is,

$$\sqrt[3]{x^9} = x^{9/3} = x^3.$$

Property (iv) also holds if a is a positive real number and n is an even positive integer. Thus, if all variables that appear in a radicand with an even index represent positive numbers, then

$$\sqrt{y^6} = \sqrt{(y^3)^2} = y^3, \qquad \sqrt[4]{t^{12}} = \sqrt[4]{(t^3)^4} = t^3,$$

and

$$\sqrt[6]{a^{30}} = \sqrt[6]{(a^5)^6} = a^5.$$

Radical expressions such as

$$\sqrt[3]{\sqrt{64}} \qquad \text{and} \qquad \sqrt{\sqrt[3]{64}}$$

can be simplified by observing that

$$\sqrt[3]{\sqrt{64}} = \sqrt[3]{8} = 2, \qquad \sqrt{\sqrt[3]{64}} = \sqrt{4} = 2 \qquad \text{and that} \qquad \sqrt[6]{64} = 2.$$

This suggests that

$$\sqrt[3]{\sqrt{64}} = \sqrt{\sqrt[3]{64}} = \sqrt[6]{64}.$$

Property (v)

> If m and n are positive integers and a and b are real numbers, then provided that all expressions are defined,
>
> $$\sqrt[n]{\sqrt[m]{a}} = \sqrt[m]{\sqrt[n]{a}} = \sqrt[nm]{a}.$$

For example,

$$\sqrt[4]{\sqrt[3]{x^{24}}} = \sqrt[3]{\sqrt[4]{x^{24}}} = \sqrt[12]{x^{24}} = x^2.$$

In Examples 2 and 3, use the properties of radicals to simplify each expression. Assume that variables are restricted to values for which all expressions represent real numbers.

EXAMPLE 2   (a) $\sqrt[3]{x^5}$ (b) $\sqrt{9y^3}$ (c) $\sqrt[3]{\dfrac{4a}{27c^3}}$ (d) $\sqrt[5]{\dfrac{7}{-32w^5}}$

(e) $(\sqrt[7]{t})^7$ (f) $\sqrt[3]{\sqrt[5]{-r^{15}}}$ (g) $\sqrt[7]{(a+b)^7}$ (h) $\sqrt[9]{-u^9}$

SOLUTION   (a) $\sqrt[3]{x^5} = \sqrt[3]{x^3 \cdot x^2} = \sqrt[3]{x^3} \cdot \sqrt[3]{x^2}$ [by Property (i)]
$$= x\sqrt[3]{x^2}$$

(b) $\sqrt{9y^3} = \sqrt{(9y^2)(y)} = \sqrt{9y^2} \cdot \sqrt{y}$ [by Property (i)]
$$= 3y\sqrt{y}$$

(c) $\sqrt[3]{\dfrac{4a}{27c^3}} = \dfrac{\sqrt[3]{4a}}{\sqrt[3]{27c^3}}$ [by Property (ii)]
$$= \frac{\sqrt[3]{4a}}{3c}$$

(d) $\sqrt[5]{\dfrac{7}{-32w^5}} = \dfrac{\sqrt[5]{7}}{\sqrt[5]{-32w^5}}$ [by Property (ii)]
$$= \frac{\sqrt[5]{7}}{-2w}$$

(e) $(\sqrt[7]{t})^7 = t$ [by Property (iii)]

(f) $\sqrt[3]{\sqrt[5]{-r^{15}}} = \sqrt[15]{-r^{15}}$ [by Property (v)]
$$= \sqrt[15]{(-r)^{15}}$$ (why?)
$$= -r$$

(g) $\sqrt[7]{(a+b)^7} = a + b$ [by Property (iv)]

(h) $\sqrt[9]{-u^9} = -\sqrt[9]{u^9}$ (why?)
$$= -u$$

EXAMPLE 3   (a) $\sqrt[3]{-125x^8y^{10}}$ (b) $\sqrt{3p^2q^3} \cdot \sqrt{6p^5q}$ (c) $\dfrac{\sqrt{324a^7b} \cdot \sqrt{9a^5}}{\sqrt{36a^4b^3}}$

SOLUTION   (a) $\sqrt[3]{-125x^8y^{10}} = \sqrt[3]{(-125x^6y^9)(x^2y)}$
$$= \sqrt[3]{-125x^6y^9} \cdot \sqrt[3]{x^2y}$$
$$= -5x^2y^3\sqrt[3]{x^2y}$$

(b) $\sqrt{3p^2q^3} \cdot \sqrt{6p^5q} = \sqrt{(3p^2q^3)(6p^5q)}$
$$= \sqrt{18p^7q^4} = \sqrt{(9p^6q^4)(2p)}$$
$$= \sqrt{9p^6q^4} \cdot \sqrt{2p} = 3p^3q^2\sqrt{2p}$$

(c) $\dfrac{\sqrt{324a^7b} \cdot \sqrt{9a^5}}{\sqrt{36a^4b^3}} = \dfrac{\sqrt{(324a^7b)(9a^5)}}{\sqrt{36a^4b^3}} = \sqrt{\dfrac{324(9)a^{12}b}{36a^4b^3}}$
$$= \sqrt{\frac{81a^8}{b^2}} = \frac{9a^4}{b}$$

In order to multiply or divide radicals with different indexes, we begin by building their indexes to a common index. The following property is a key to this operation.

Property (vi)

> If m and n are positive integers, and if each root exists, then
>
> $$\sqrt[cn]{a^{cm}} = \sqrt[n]{a^m}, \text{ where } c \text{ is a positive integer.}$$

EXAMPLE 4   Write $\sqrt[4]{x^3} \cdot \sqrt[3]{x^2}$ as a single radical, and then simplify the result. Assume that x is restricted to values for which all expressions are defined.

SOLUTION   We use the least common multiple (LCM) of the individual indexes as the common index. The LCM of indexes 4 and 3 is 12, so that

$$\sqrt[4]{x^3} = \sqrt[4 \cdot 3]{x^{3 \cdot 3}} \quad \text{[by Property (vi)]}$$
$$= \sqrt[12]{x^9},$$

and

$$\sqrt[3]{x^2} = \sqrt[3 \cdot 4]{x^{2 \cdot 4}} \quad \text{[by Property (vi)]}$$
$$= \sqrt[12]{x^8}.$$

Hence,

$$\sqrt[4]{x^3} \cdot \sqrt[3]{x^2} = \sqrt[12]{x^9} \cdot \sqrt[12]{x^8} = \sqrt[12]{x^9 \cdot x^8}$$
$$= \sqrt[12]{x^{17}} = \sqrt[12]{x^{12} \cdot x^5} = \sqrt[12]{x^{12}} \cdot \sqrt[12]{x^5}$$
$$= x\sqrt[12]{x^5}.$$

Let us go back and examine Property (iv), which deals with the expression $\sqrt[n]{a^n}$. It is tempting to say that in general, $\sqrt[n]{a^n} = a$. However, if a is a negative number and n is an even positive integer, this is not true. For example,

$$\sqrt{(-3)^2} = \sqrt{+9} = 3, \quad \text{not} \quad \sqrt{(-3)^2} = -3,$$

as the property would imply.

Therefore, regardless of the sign of a, if n is an even positive integer, a^n is always nonnegative. Its root always exists and it is also nonnegative.

In general, if n is an even positive integer, we have the following property:

Property (vii)

> $$\sqrt[n]{a^n} = |a|.$$

For example,

$$\sqrt{(-5)^2} = |-5| = 5, \quad \text{and} \quad \sqrt[4]{(-2)^4} = |-2| = 2.$$

PROBLEM SET 5.3

In problems 1–72, use the properties of radicals to simplify each expression. (Assume that variables are restricted to values for which all expressions represent real numbers.)

1. $\sqrt{27}$
2. $\sqrt{162}$
3. $\sqrt[3]{-54}$
4. $\sqrt[3]{-250}$

5. $\sqrt{288}$
6. $\sqrt[5]{64}$
7. $\sqrt{48x}$
8. $\sqrt[3]{24y^2}$

9. $\sqrt[3]{162c^2}$
10. $\sqrt[3]{54t^2}$
11. $\sqrt{20y^3}$
12. $\sqrt{32c^5}$

13. $\sqrt[3]{8x^4}$
14. $\sqrt[5]{64x^6}$
15. $\sqrt[3]{-16y^{10}}$
16. $\sqrt[7]{-128p^{15}}$

17. $\sqrt{98p^3}$
18. $\sqrt[5]{-96x^{11}}$
19. $\sqrt{\dfrac{7}{4}}$
20. $\sqrt{\dfrac{3}{25}}$

21. $\sqrt[3]{\dfrac{-5}{8}}$
22. $\sqrt[3]{\dfrac{11}{-125}}$
23. $\sqrt[3]{\dfrac{-x^2}{64}}$
24. $\sqrt[5]{\dfrac{x}{y^{10}}}$

25. $\sqrt{\dfrac{3w^3}{4w^5}}$
26. $\sqrt[4]{\dfrac{m}{625n^8}}$
27. $\sqrt[3]{\dfrac{a^2}{-8}}$
28. $\sqrt[7]{\dfrac{-3}{a^{14}}}$

29. $\sqrt{\dfrac{3}{25x^2}}$
30. $\sqrt{\dfrac{2x}{y^2}}$
31. $\sqrt{\dfrac{17}{y^4}}$
32. $\sqrt[3]{\dfrac{250x^3}{y^3}}$

33. $\sqrt{\dfrac{3}{9x^4}}$
34. $\sqrt[5]{\dfrac{7}{-32x^5}}$
35. $\sqrt[4]{\dfrac{486}{625y^4}}$
36. $\sqrt[7]{\dfrac{y^2}{-128}}$

37. $\sqrt[3]{-y^9}$
38. $\sqrt[4]{c^{12}}$
39. $\sqrt[5]{-32x^{10}}$
40. $\sqrt[3]{(2x+1)^6}$

41. $\sqrt{(a+b)^4}$
42. $\sqrt[5]{-x^{15}}$
43. $\sqrt[6]{(-x)^6}$
44. $\sqrt[3]{(-5x)^3}$

45. $\sqrt[5]{-32y^{20}z^{10}}$
46. $\sqrt[3]{27x^6y^9}$
47. $\sqrt[4]{64a^4b^8}$
48. $\sqrt[7]{-128y^{14}z^7}$

49. $(\sqrt[4]{5})^4$
50. $(5\sqrt[3]{x})^3$
51. $(\sqrt[7]{-p})^7$
52. $(\sqrt[5]{-t})^5$

53. $(\sqrt[11]{x})^{22}$
54. $\sqrt[5]{\sqrt{1024}}$
55. $\sqrt[5]{\sqrt[3]{y^{15}}}$
56. $\sqrt{\sqrt[4]{p^{32}}}$

57. $\sqrt[3]{\sqrt[10]{m^{30}}}$
58. $\sqrt{5x}\cdot\sqrt{10xy^2}$
59. $\sqrt[4]{(3x+1)^4y^6}\cdot\sqrt[4]{32y^6}$

60. $\sqrt[4]{x^2y^3}\cdot\sqrt[4]{x^3y^3}$
61. $\sqrt{18a^3b^7}\cdot\sqrt{2ab^3}$
62. $\sqrt[3]{25y^2}\cdot\sqrt[3]{5y^4}$

63. $\sqrt[3]{4xy^2}\cdot\sqrt[3]{2x^2y^3}$
64. $\sqrt[5]{-w^{15}t^4}\cdot\sqrt[5]{w^{10}t^6}$
65. $\sqrt[8]{x^{12}}\cdot\sqrt[8]{x^5y^{-8}}\cdot\sqrt[8]{x^2y^9}$

66. $\sqrt[4]{5p^2q^3}\cdot\sqrt[4]{5p^2q^3}$
67. $\dfrac{\sqrt[3]{u^{11}}\cdot\sqrt[3]{u^5}}{\sqrt[3]{u}}$
68. $\dfrac{\sqrt{m^2n}\cdot\sqrt{mn^4}}{\sqrt{mn^3}}$

69. $\sqrt[4]{\dfrac{a^4b^3}{a^3b}}\cdot\dfrac{\sqrt[4]{a^5b}}{\sqrt[4]{ab^{-1}}}$
70. $\dfrac{\sqrt{324x^5y}\cdot\sqrt{9x^2}}{\sqrt{25x^2y}}$
71. $\dfrac{\sqrt[3]{p^2q^3}\cdot\sqrt[3]{125p^3q^2}}{\sqrt[3]{8p^3q^4}}$
72. $\dfrac{\sqrt[3]{a^2b^4}\cdot\sqrt[3]{a^4b}\cdot\sqrt[3]{a^3b^4}}{\sqrt[3]{ab^2}\cdot\sqrt[3]{a^2b^7}}$

In problems 73–78, write each expression as a single radical, then simplify. (Assume that variables are restricted to values for which all expressions represent real numbers.)

73. $\sqrt[3]{x}\cdot\sqrt{x}$
74. $\sqrt[3]{p^2}\cdot\sqrt[4]{p^3}$
75. $\sqrt[3]{y^2}\cdot\sqrt[5]{y^4}$

76. $\sqrt[3]{p^2}\cdot\sqrt[7]{p^5}$
77. $\sqrt[5]{ab^2}\cdot\sqrt[4]{a^2b^3}$
78. $\sqrt[5]{c^2d^4}\cdot\sqrt[7]{c^3d^3}$

In problems 79 and 80, substitute the given numbers into the expression

$$\sqrt{b^2-4ac}$$

and then simplify.

79. $a = 1$, $b = 6$, $c = -3$ **80.** $a = 3$, $b = -8$, $c = -5$

81. Simplify the expression $\sqrt{x^2 + 8x + 16}$ for all real numbers x.

82. From geometry, the radius r of a sphere with volume V is given by

$$r = \sqrt[3]{\frac{3V}{4\pi}}.$$

Find r when $V = 44$ cubic centimeters. (Use $\frac{22}{7}$ for π.)

83. A biologist estimates that the population P of a certain culture of bacteria is given by

$$P = 1000\sqrt[4]{t^5 + 10t^2 + 9},$$

where t is the time in hours since the culture was started. Find the population of the bacteria when $t = 2$ hours.

84. Let x be a real number. Simplify the expression

$$\sqrt{x^2 + 6x + 9}.$$

$\boxed{\text{C}}$ In problems 85–88, evaluate the expression

$$\sqrt{(x_2 - x_1)^2 + (y_2 - y_1)^2}.$$

85. $x_1 = 2.4$, $x_2 = 4.3$, $y_1 = 4.2$, and $y_2 = -5.7$

86. $x_1 = 3.5$, $x_2 = -2.8$, $y_1 = -7.3$, and $y_2 = -9.8$

87. $x_1 = -2.53$, $x_2 = 9.11$, $y_1 = -13.21$, and $y_2 = -17.41$

88. $x_1 = -18.45$, $x_2 = -25.21$, $y_1 = -31.72$, and $y_2 = -43.58$

$\boxed{\text{C}}$ **89.** In electronics, the resonant frequency f (in hertz) for a tuned circuit is given by the equation

$$f = \frac{1}{2\pi\sqrt{LC}},$$

where L is the inductance and C is the capacitance. Find the frequency f for a tuned circuit in which $L = 3.57 \times 10^{-8}$ and $C = 121 \times 10^{-12}$.

5.4 Addition and Subtraction of Radical Expressions

In Section 2.2, we showed how to add and subtract *like* (or *similar*) terms when adding (or subtracting) polynomial expressions. For example, to add or subtract the similar terms $7x^2$ and $4x^2$, we have

$$7x^2 + 4x^2 = (7 + 4)x^2 = 11x^2 \quad \text{and} \quad 7x^2 - 4x^2 = (7 - 4)x^2 = 3x^2.$$

In this section, we add and subtract radical terms in much the same way as we add or subtract terms in a polynomial. Expressions such as

$$\sqrt{2} + \sqrt{3}, \qquad \sqrt[3]{x} - 3\sqrt[5]{x}, \qquad \text{and} \qquad y\sqrt[3]{2y + 1}$$

are radical expressions. Two or more radical terms are *like* (or *similar*) if they contain the same index and the same radicand. For example,

$$3\sqrt{6}, \qquad 5\sqrt{6}, \qquad \text{and} \qquad x\sqrt{6}$$

are like radicals, because they have the same index, 2, and the same radicand, 6; whereas

$$\sqrt[3]{2x + 5y} \qquad \text{and} \qquad \sqrt[5]{2x + 5y}$$

are not like terms, because the indices are not the same.

To add or subtract like radical terms, we apply the distributive property. For example,

$$3\sqrt{5} + 4\sqrt{5} = (3 + 4)\sqrt{5} = 7\sqrt{5} \qquad \text{and} \qquad 7\sqrt[3]{x} - 2\sqrt[3]{x} = (7 - 2)\sqrt[3]{x} = 5\sqrt[3]{x}.$$

Note that you can perform addition (or subtraction) only with like radical forms. Adding unlike radicals, as in this example:

$$\sqrt{9} + \sqrt{16} = \sqrt{9 + 16} = \sqrt{25},$$

is one of the most common errors made by students in algebra. Of course, you can easily verify that

$$\sqrt{9} + \sqrt{16} = 3 + 4 = 7$$

and

$$\sqrt{9 + 16} = \sqrt{25} = 5.$$

In Examples 1 and 2, combine like terms.

EXAMPLE 1 $2\sqrt{3} + 5\sqrt{3} - \sqrt{3}$

SOLUTION   All the radicals are similar. Thus, we apply the distributive property and combine:

$$2\sqrt{3} + 5\sqrt{3} - \sqrt{3} = (2 + 5 - 1)\sqrt{3} = 6\sqrt{3}.$$

EXAMPLE 2 $7\sqrt[3]{2} - 3\sqrt[3]{2} + 6\sqrt[3]{2}$

SOLUTION   All three radical terms are similar, so we apply the distributive property and combine:

$$7\sqrt[3]{2} - 3\sqrt[3]{2} + 6\sqrt[3]{2} = (7 - 3 + 6)\sqrt[3]{2} = 10\sqrt[3]{2}.$$

Each radical expression in Examples 1 and 2 is in simplified form. Occasionally two or more of the terms containing radical expressions do not appear to be similar but contain like terms when they are simplified. When this occurs, we write each expression in a simplified form, and then apply the distributive property to combine like terms. For example, to perform the addition

$$5\sqrt{8} + 11\sqrt{18}$$

we notice that the radicals $\sqrt{8}$ and $\sqrt{18}$ do not seem to be similar. However, each radical can be simplified as follows:

$$\sqrt{8} = \sqrt{4 \cdot 2} = \sqrt{4} \cdot \sqrt{2} = 2\sqrt{2},$$

and

$$\sqrt{18} = \sqrt{9 \cdot 2} = \sqrt{9} \cdot \sqrt{2} = 3\sqrt{2}.$$

Now we apply the distributive property and combine:

$$\begin{aligned}
5\sqrt{8} + 11\sqrt{18} &= 5(2\sqrt{2}) + 11(3\sqrt{2}) \\
&= 10\sqrt{2} + 33\sqrt{2} \\
&= (10 + 33)\sqrt{2} \\
&= 43\sqrt{2}.
\end{aligned}$$

In Examples 3–7, combine like terms.

EXAMPLE 3 $\sqrt{8} + \sqrt{32}$

SOLUTION   First we write the terms in a simplified form:

$$\begin{aligned}
\sqrt{8} + \sqrt{32} &= \sqrt{4 \cdot 2} + \sqrt{16 \cdot 2} \\
&= \sqrt{4} \cdot \sqrt{2} + \sqrt{16} \cdot \sqrt{2} \qquad \text{[by Property (i) of radicals on page 242]} \\
&= 2\sqrt{2} + 4\sqrt{2} = (2 + 4)\sqrt{2} \\
&= 6\sqrt{2}.
\end{aligned}$$

EXAMPLE 4 $2\sqrt[3]{54} - 2\sqrt[3]{16}$

SOLUTION

$$\begin{aligned}
2\sqrt[3]{54} - 2\sqrt[3]{16} &= 2\sqrt[3]{27 \cdot 2} - 2\sqrt[3]{8 \cdot 2} \\
&= 2\sqrt[3]{27} \cdot \sqrt[3]{2} - 2\sqrt[3]{8} \cdot \sqrt[3]{2} \\
&= 2(3)\sqrt[3]{2} - 2(2)\sqrt[3]{2} \\
&= 6\sqrt[3]{2} - 4\sqrt[3]{2} = (6 - 4)\sqrt[3]{2} \\
&= 2\sqrt[3]{2}.
\end{aligned}$$

EXAMPLE 5 $4\sqrt{12} + 5\sqrt{8} - \sqrt{50}$

SOLUTION

$$
\begin{aligned}
4\sqrt{12} + 5\sqrt{8} - \sqrt{50} &= 4\sqrt{4 \cdot 3} + 5\sqrt{4 \cdot 2} - \sqrt{25 \cdot 2} \\
&= 4\sqrt{4} \cdot \sqrt{3} + 5\sqrt{4} \cdot \sqrt{2} - \sqrt{25} \cdot \sqrt{2} \\
&= 4 \cdot 2\sqrt{3} + 5 \cdot 2\sqrt{2} - 5\sqrt{2} \\
&= 8\sqrt{3} + 10\sqrt{2} - 5\sqrt{2} \\
&= 8\sqrt{3} + (10 - 5)\sqrt{2} \\
&= 8\sqrt{3} + 5\sqrt{2}.
\end{aligned}
$$

EXAMPLE 6 $7\sqrt{4x} - 5\sqrt{9x}$, for $x > 0$.

SOLUTION

$$
\begin{aligned}
7\sqrt{4x} - 5\sqrt{9x} &= 7\sqrt{4} \cdot \sqrt{x} - 5\sqrt{9} \cdot \sqrt{x} \\
&= 7(2\sqrt{x}) - 5(3\sqrt{x}) \\
&= 14\sqrt{x} - 15\sqrt{x} = (14 - 15)\sqrt{x} \\
&= -\sqrt{x}.
\end{aligned}
$$

EXAMPLE 7 $5x\sqrt{xy^3} - 2y\sqrt{x^3y} + \sqrt{x^3y^3}$, $x > 0$, $y > 0$.

SOLUTION

$$
\begin{aligned}
5x\sqrt{xy^3} - 2y\sqrt{x^3y} + \sqrt{x^3y^3} &= 5x\sqrt{y^2 \cdot xy} - 2y\sqrt{x^2 \cdot xy} + \sqrt{x^2y^2 \cdot xy} \\
&= 5x\sqrt{y^2} \cdot \sqrt{xy} - 2y\sqrt{x^2} \cdot \sqrt{xy} + \sqrt{x^2y^2} \cdot \sqrt{xy} \\
&= 5xy\sqrt{xy} - 2yx\sqrt{xy} + xy\sqrt{xy} \\
&= (5xy - 2xy + xy)\sqrt{xy} \\
&= 4xy\sqrt{xy}.
\end{aligned}
$$

PROBLEM SET 5.4

In problems 1–42, simplify each expression by combining like terms. (Assume that variables are restricted to values for which all radical expressions represent real numbers.)

1. $5\sqrt{7} + 3\sqrt{7}$

2. $8\sqrt{3} + 2\sqrt{3}$

3. $7\sqrt{5} + 3\sqrt{5} - 2\sqrt{5}$

4. $9\sqrt{11} + 8\sqrt{11} - 3\sqrt{11}$

5. $8\sqrt[3]{4} - 3\sqrt[3]{4} + 2\sqrt[3]{4}$

6. $8\sqrt[5]{2} - 4\sqrt[5]{2} + 3\sqrt[5]{2}$

7. $\sqrt{18} + \sqrt{8}$

8. $3\sqrt{18} + 4\sqrt{2}$

9. $\sqrt{72} - 2\sqrt{8} + \sqrt{2}$

10. $2\sqrt{50} - 3\sqrt{128} + 4\sqrt{2}$

11. $5\sqrt{3} + 2\sqrt{12} - 2\sqrt{27}$

12. $4\sqrt{12} + 2\sqrt{27} - \sqrt{48}$

13. $4\sqrt{20} - 2\sqrt{45} + \sqrt{80}$

14. $\sqrt[4]{162} + \sqrt[4]{32} - 2\sqrt[4]{2}$

15. $5\sqrt[3]{81} - 3\sqrt[3]{24} + \sqrt[3]{192}$

16. $3\sqrt[3]{192} + 4\sqrt[3]{24} - 2\sqrt[3]{3}$

17. $\sqrt{75x} - \sqrt{3x} - \sqrt{12x}$

18. $2\sqrt{108y} - \sqrt{27y} + \sqrt{363y}$

19. $\sqrt{p^3} + \sqrt{25p^3} + \sqrt{9p}$

20. $\sqrt{m^3} - 2m\sqrt{m^5} + 3m\sqrt{m^7}$

21. $\sqrt{18p} + \sqrt{50p} - \sqrt{2p}$

22. $10\sqrt{3m} - 2\sqrt{75m} + 3\sqrt{243m}$

23. $\sqrt{12x} - \sqrt{3x} + \sqrt{108x}$

24. $a\sqrt{a^3b} + a^2\sqrt{32ab} - \sqrt{162a^5b}$

25. $\dfrac{3\sqrt[4]{m^9}}{m} - 5\sqrt[4]{m^5} + \sqrt[8]{m^{10}}$

26. $\sqrt[5]{32xy^{13}} - 8\sqrt[10]{x^2y^6} + \sqrt[5]{x^{11}y^3}$

27. $\dfrac{\sqrt{18x^3}}{3y} - \dfrac{x\sqrt{32x}}{2y} + \dfrac{6x\sqrt{2x}}{y}$

28. $\dfrac{5\sqrt{2x}}{x} + \dfrac{7\sqrt{2x}}{x} - \dfrac{\sqrt{2}\cdot\sqrt{x}}{x}$

29. $2y\sqrt{y} - 7\sqrt{y^3} + \dfrac{1}{7y}\sqrt{4y^3}$

30. $x\sqrt{25xy^3} - y\sqrt{4x^3y} - \sqrt{81x^3y^3}$

31. $x\sqrt[3]{81x^5y} - 5x\sqrt[3]{24x^2y^4}$

32. $11\sqrt[3]{x^4y^3c^6} + 7xy\sqrt[3]{xy^3}$

33. $5p\sqrt[3]{p^4q} - 7q\sqrt[3]{-pq^4} + 5\sqrt[3]{p^7q}$

34. $x\sqrt[3]{27x} - 17\sqrt[3]{-x^4} + \dfrac{8\sqrt[3]{x^7}}{x}$

35. $\sqrt{48x} - \sqrt[3]{16x} + \sqrt{108x} + \sqrt[3]{54x}$

36. $2y^2\sqrt{48x^7y^6} - 5x\sqrt{27x^5y^{10}}$

37. $7\sqrt[3]{m^5} + 3m\sqrt[3]{m^2} + 2m\sqrt[3]{8m^2}$

38. $7\sqrt[5]{ab} + 4\sqrt[5]{ab} - 5\sqrt{a} + 3\sqrt{a}$

39. $\sqrt{9y-9} - \sqrt{4y-4} + \sqrt{36y-36}$

40. $\sqrt{a^2+2b^2} + \sqrt{4a^2+8b^2} - \sqrt{16a^2+32b^2}$

41. $\sqrt{9x^2(x-2y)} - \sqrt{36y^2(x-2y)} + 2\sqrt{(x-2y)^3}$

42. $\sqrt{3(x+y)^3} - \sqrt[3]{3(x+y)^4} + \sqrt[4]{9(x+y)^6}$

C **43.** Find a four-decimal approximation:
(a) $\sqrt{5} + \sqrt{13}$ and (b) $\sqrt{18}$. (c) Is $\sqrt{5} + \sqrt{13} = \sqrt{18}$?

C **44.** Find a four-decimal approximation:
(a) $\sqrt{7} + \sqrt{19}$ and (b) $\sqrt{26}$. (c) Is $\sqrt{7} + \sqrt{19} = \sqrt{26}$?

5.5 Multiplication and Division of Radical Expressions

In this section, we study multiplication and division of expressions that contain radicals. We shall see that multiplication of expressions involving radicals is similar to the multiplication of polynomial expressions. The division of expressions involving radicals follows the division property [Property (ii)] in Section 5.3.

To multiply expressions involving radicals, we use Property (i) of Section 5.3, written in the form

$$\sqrt[n]{a} \cdot \sqrt[n]{b} = \sqrt[n]{ab}$$

along with the commutative and associative properties of multiplication. For example,

$$\sqrt{2} \cdot \sqrt{3} = \sqrt{2 \cdot 3} = \sqrt{6}$$

and

$$\sqrt{2x} \cdot \sqrt{5y} = \sqrt{2x \cdot 5y} = \sqrt{10xy}.$$

To multiply one radical term by a radical expression containing more than one term, we use the distributive property. For example,

$$\sqrt{3}(\sqrt{7} + \sqrt{2}) = \sqrt{3} \cdot \sqrt{7} + \sqrt{3} \cdot \sqrt{2} \qquad \text{(distributive property)}$$
$$= \sqrt{3 \cdot 7} + \sqrt{3 \cdot 2}$$
$$= \sqrt{21} + \sqrt{6}.$$

In Examples 1–8, perform each multiplication, and simplify the result when possible.

EXAMPLE 1 $\sqrt{8} \cdot \sqrt{2}$

SOLUTION

$$\sqrt{8} \cdot \sqrt{2} = \sqrt{8 \cdot 2} \qquad \text{[by Property (i)]}$$
$$= \sqrt{16} = 4.$$

EXAMPLE 2 $(5\sqrt{3})(4\sqrt{5})$

SOLUTION $(5\sqrt{3})(4\sqrt{5}) = (5 \cdot 4)(\sqrt{3} \cdot \sqrt{5}) \qquad \text{(by the commutative and associative properties)}$
$$= (5 \cdot 4)\sqrt{3 \cdot 5} \qquad \text{[by Property (i)]}$$
$$= 20\sqrt{15}.$$

EXAMPLE 3 $\sqrt{5}(3\sqrt{7} - 2\sqrt{5})$

SOLUTION $\sqrt{5}(3\sqrt{7} - 2\sqrt{5}) = \sqrt{5} \cdot 3\sqrt{7} - \sqrt{5} \cdot 2\sqrt{5} \qquad \text{(by the distributive property)}$
$$= 3\sqrt{5 \cdot 7} - 2(\sqrt{5})^2$$
$$= 3\sqrt{35} - 2(5)$$
$$= 3\sqrt{35} - 10.$$

EXAMPLE 4 $\sqrt{5}(\sqrt{15} + \sqrt{25})$

SOLUTION

$$\sqrt{5}(\sqrt{15} + \sqrt{25}) = \sqrt{5} \cdot \sqrt{15} + \sqrt{5} \cdot \sqrt{25}$$
$$= \sqrt{75} + 5\sqrt{5}$$
$$= \sqrt{25 \cdot 3} + 5\sqrt{5} = \sqrt{25} \cdot \sqrt{3} + 5\sqrt{5}$$
$$= 5\sqrt{3} + 5\sqrt{5}.$$

EXAMPLE 5 $(\sqrt{3} - \sqrt{2})(2\sqrt{3} + \sqrt{2})$

SOLUTION $(\sqrt{3} - \sqrt{2})(2\sqrt{3} + \sqrt{2}) = 2\sqrt{3} \cdot \sqrt{3} + \sqrt{3} \cdot \sqrt{2} - 2\sqrt{3} \cdot \sqrt{2} - \sqrt{2} \cdot \sqrt{2}$
$$= 2(3) + \sqrt{6} - 2\sqrt{6} - 2$$
$$= 6 - \sqrt{6} - 2 = 4 - \sqrt{6}.$$

EXAMPLE 6 $(\sqrt{x} + 2\sqrt{y})^2$

SOLUTION   Using the special product $(a + b)^2 = a^2 + 2ab + b^2$, we have

$$(\sqrt{x} + 2\sqrt{y})^2 = (\sqrt{x})^2 + 2\sqrt{x}(2\sqrt{y}) + (2\sqrt{y})^2$$
$$= x + 4\sqrt{x} \cdot \sqrt{y} + 4y$$
$$= x + 4\sqrt{xy} + 4y.$$

EXAMPLE 7 $(\sqrt{10} + \sqrt{2})(\sqrt{10} - \sqrt{2})$

SOLUTION   Using the special product $(a + b)(a - b) = a^2 - b^2$, we have

$$(\sqrt{10} + \sqrt{2})(\sqrt{10} - \sqrt{2}) = (\sqrt{10})^2 - (\sqrt{2})^2$$
$$= 10 - 2$$
$$= 8.$$

EXAMPLE 8 $(3\sqrt{x} - 5\sqrt{y})(3\sqrt{x} + 5\sqrt{y})$

SOLUTION

$$(3\sqrt{x} - 5\sqrt{y})(3\sqrt{x} + 5\sqrt{y}) = (3\sqrt{x})^2 - (5\sqrt{y})^2$$
$$= 9x - 25y.$$

Division of Radicals

We can divide radical expressions by using Property (ii) of Section 5.3 written in the form

$$\frac{\sqrt[n]{a}}{\sqrt[n]{b}} = \sqrt[n]{\frac{a}{b}}$$

Thus,

$$\frac{\sqrt{18}}{\sqrt{2}} = \sqrt{\frac{18}{2}} = \sqrt{9} = 3,$$

$$\frac{\sqrt{108y^3}}{\sqrt{3y}} = \sqrt{\frac{108y^3}{3y}} = \sqrt{36y^2} = 6y \qquad \text{for } y > 0,$$

and

$$\frac{\sqrt{3}}{\sqrt{21}} = \sqrt{\frac{3}{21}} = \sqrt{\frac{1}{7}} = \frac{1}{\sqrt{7}}.$$

Radicals may appear in the denominator of a fraction, such as $1/\sqrt{7}$. It is sometimes easier to work with fractions if their denominators do not contain radicals. To rewrite a fraction so that there are no radicals in the denominator, we multiply the numerator and the denominator by a **rationalizing factor** for the denominator. Whenever the product of two radical expressions is free of radicals, we say that the two expressions are rationalizing factors of each other. For instance, $\sqrt{7}$ is a rationalizing factor of $\sqrt{7}$, because $\sqrt{7} \cdot \sqrt{7} = 7$. If we multiply the numerator and the denominator of $1/\sqrt{7}$ by the rationalizing factor $\sqrt{7}$, we have

$$\frac{1}{\sqrt{7}} = \frac{1 \cdot \sqrt{7}}{\sqrt{7} \cdot \sqrt{7}} = \frac{\sqrt{7}}{7}.$$

Thus, we have a fraction whose denominator is free of radicals. Note that when we multiply

$$(\sqrt{5} - \sqrt{3})(\sqrt{5} + \sqrt{3}) = (\sqrt{5})^2 - (\sqrt{3})^2 = 5 - 3 = 2,$$

the expressions $(\sqrt{5} - \sqrt{3})$ and $(\sqrt{5} + \sqrt{3})$ are rationalizing factors for each other. If we multiply the numerator and the denominator of

$$\frac{2 + \sqrt{3}}{\sqrt{5} - \sqrt{3}}$$

by the rationalizing factor $\sqrt{5} + \sqrt{3}$, we have

$$\begin{aligned}
\frac{2 + \sqrt{3}}{\sqrt{5} - \sqrt{3}} &= \frac{(2 + \sqrt{3})(\sqrt{5} + \sqrt{3})}{(\sqrt{5} - \sqrt{3})(\sqrt{5} + \sqrt{3})} \\
&= \frac{2\sqrt{5} + 2\sqrt{3} + \sqrt{3} \cdot \sqrt{5} + \sqrt{3} \cdot \sqrt{3}}{(\sqrt{5})^2 - (\sqrt{3})^2} \\
&= \frac{2\sqrt{5} + 2\sqrt{3} + \sqrt{15} + 3}{5 - 3} = \frac{2\sqrt{5} + 2\sqrt{3} + \sqrt{15} + 3}{2}.
\end{aligned}$$

The resulting denominator is free of radicals. The process of writing fractions so that there are no radicals in the denominator is called **rationalizing the denominator.** In the previous example, the expression $(\sqrt{5} + \sqrt{3})$ is a rationalizing factor of the expression $(\sqrt{5} - \sqrt{3})$. In this context, each of the two factors $(\sqrt{5} + \sqrt{3})$ and $(\sqrt{5} - \sqrt{3})$ is called the **conjugate** of the other.

In Examples 9–15, rationalize the denominator of each fraction and simplify the result. Assume that all variables are positive.

EXAMPLE 9 $\dfrac{7}{\sqrt{3}}$

SOLUTION   Here we use $\sqrt{3}$ as the rationalizing factor. Thus, we multiply the numerator and the denominator of the fraction by $\sqrt{3}$:

$$\frac{7}{\sqrt{3}} = \frac{7 \cdot \sqrt{3}}{\sqrt{3} \cdot \sqrt{3}} = \frac{7\sqrt{3}}{3}.$$

EXAMPLE 10 $\dfrac{\sqrt{5}}{\sqrt{6}}$

SOLUTION   The rationalizing factor is $\sqrt{6}$:

$$\frac{\sqrt{5}}{\sqrt{6}} = \frac{\sqrt{5} \cdot \sqrt{6}}{\sqrt{6} \cdot \sqrt{6}} = \frac{\sqrt{5 \cdot 6}}{6} = \frac{\sqrt{30}}{6}.$$

EXAMPLE 11 $\dfrac{3}{5\sqrt{2x}} + \sqrt{2x}$

SOLUTION   We begin by writing the first term in simplified form:

$$\frac{3}{5\sqrt{2x}} + \sqrt{2x} = \frac{3}{5\sqrt{2x}} \cdot \frac{\sqrt{2x}}{\sqrt{2x}} + \sqrt{2x} \qquad \text{(We rationalize the denominator.)}$$

$$= \frac{3\sqrt{2x}}{5(2x)} + \sqrt{2x}$$

$$= \frac{3\sqrt{2x}}{10x} + \sqrt{2x} = \frac{3\sqrt{2x} + 10x\sqrt{2x}}{10x}$$

$$= \frac{(3 + 10x)\sqrt{2x}}{10x}.$$

EXAMPLE 12 $\dfrac{2}{\sqrt[3]{5}}$

SOLUTION   Since $\sqrt[3]{5^3} = 5$ and $\sqrt[3]{5^3} = \sqrt[3]{5} \cdot \sqrt[3]{5} \cdot \sqrt[3]{5}$, we multiply the numerator and the denominator by the rationalizing factor $\sqrt[3]{5} \cdot \sqrt[3]{5}$,

$$\frac{2}{\sqrt[3]{5}} = \frac{2\sqrt[3]{5} \cdot \sqrt[3]{5}}{\sqrt[3]{5} \cdot \sqrt[3]{5} \cdot \sqrt[3]{5}} = \frac{2\sqrt[3]{5 \cdot 5}}{5} = \frac{2\sqrt[3]{25}}{5}.$$

EXAMPLE 13 $\dfrac{5}{\sqrt{3} - 1}$

SOLUTION   We multiply the numerator and denominator by the rationalizing factor $\sqrt{3} + 1$:

$$\frac{5}{\sqrt{3} - 1} = \frac{5(\sqrt{3} + 1)}{(\sqrt{3} - 1)(\sqrt{3} + 1)}$$
$$= \frac{5\sqrt{3} + 5}{(\sqrt{3})^2 - 1^2} = \frac{5\sqrt{3} + 5}{3 - 1} = \frac{5\sqrt{3} + 5}{2}.$$

EXAMPLE 14 $\dfrac{3\sqrt{5} + 7\sqrt{2}}{6\sqrt{5} - 3\sqrt{2}}$

SOLUTION   We multiply the numerator and denominator by the rationalizing factor $6\sqrt{5} + 3\sqrt{2}$, so that

$$\frac{3\sqrt{5} + 7\sqrt{2}}{6\sqrt{5} - 3\sqrt{2}} = \frac{(3\sqrt{5} + 7\sqrt{2})}{(6\sqrt{5} - 3\sqrt{2})} \cdot \frac{(6\sqrt{5} + 3\sqrt{2})}{(6\sqrt{5} + 3\sqrt{2})}$$
$$= \frac{3\sqrt{5} \cdot 6\sqrt{5} + 3\sqrt{5} \cdot 3\sqrt{2} + 7\sqrt{2} \cdot 6\sqrt{5} + 7\sqrt{2} \cdot 3\sqrt{2}}{(6\sqrt{5})^2 - (3\sqrt{2})^2}$$
$$= \frac{18(5) + 9\sqrt{10} + 42\sqrt{10} + 21(2)}{36(5) - 9(2)}$$
$$= \frac{90 + 51\sqrt{10} + 42}{180 - 18}$$
$$= \frac{132 + 51\sqrt{10}}{162} = \frac{\cancel{3}(44 + 17\sqrt{10})}{\cancel{3}(54)}$$
$$= \frac{44 + 17\sqrt{10}}{54}.$$

EXAMPLE 15 $\dfrac{\sqrt{x} + \sqrt{y}}{\sqrt{x} - \sqrt{y}}$

SOLUTION   To rationalize the denominator, we multiply the numerator and denominator by the rationalizing factor $\sqrt{x} + \sqrt{y}$:

$$\frac{\sqrt{x} + \sqrt{y}}{\sqrt{x} - \sqrt{y}} = \frac{(\sqrt{x} + \sqrt{y})(\sqrt{x} + \sqrt{y})}{(\sqrt{x} - \sqrt{y})(\sqrt{x} + \sqrt{y})}$$
$$= \frac{(\sqrt{x})^2 + 2\sqrt{x} \cdot \sqrt{y} + (\sqrt{y})^2}{(\sqrt{x})^2 - (\sqrt{y})^2}$$
$$= \frac{x + 2\sqrt{xy} + y}{x - y}.$$

PROBLEM SET 5.5

In problems 1–56, find each product, and simplify. (Assume that variables are restricted to values for which all radical expressions represent real numbers.)

1. $\sqrt{12} \cdot \sqrt{3}$

2. $\sqrt{15} \cdot \sqrt{135}$

3. $\sqrt{3}(\sqrt{18} - \sqrt{2})$

4. $(5\sqrt{3})(2\sqrt{7})$

5. $(2\sqrt{6})(3\sqrt{7})$

6. $(3\sqrt{7})(4\sqrt{3})$

7. $(-3\sqrt{6})(-5\sqrt{2})$

8. $(-4\sqrt{5})(-2\sqrt{3})$

9. $(7\sqrt{3})(-11\sqrt{6})$

10. $(4\sqrt{x})(6\sqrt{y})$

11. $(2\sqrt{t})(-7\sqrt{u})$

12. $(4\sqrt{xy})(7\sqrt{ab})$

13. $\sqrt[3]{10} \cdot \sqrt[3]{75}$

14. $\sqrt[5]{27} \cdot \sqrt[5]{18}$

15. $(2\sqrt[3]{25})(3\sqrt[3]{5})$

16. $\sqrt{x-3} \cdot \sqrt{x-3}$

17. $\sqrt{x+1} \cdot \sqrt{x+3}$

18. $\sqrt{5x}\sqrt{5x-5}$

19. $\sqrt{a-3} \cdot \sqrt{a^2-9}$

20. $\sqrt{6} \cdot \sqrt[6]{32}$

21. $\sqrt{2}(2 - \sqrt{2})$

22. $\sqrt{2}(\sqrt{3} - \sqrt{2})$

23. $\sqrt{3}(4 - \sqrt{2})$

24. $\sqrt{6}(\sqrt{10} - \sqrt{15})$

25. $\sqrt{3}(\sqrt{5x} - \sqrt{10y})$

26. $\sqrt{ab}(\sqrt{a} - \sqrt{b})$

27. $\sqrt{5}(4\sqrt{5} - 3\sqrt{2})$

28. $\sqrt{8}(2\sqrt{6} - 3\sqrt{18})$

29. $\sqrt{11}(2\sqrt{3} - 4\sqrt{11})$

30. $\sqrt{2x}(\sqrt{3x} - \sqrt{y})$

31. $\sqrt{x-5}(\sqrt{x-5} - \sqrt{x+1})$

32. $\sqrt{7x}(3\sqrt{2x-2} - 7\sqrt{3x-3})$

33. $(\sqrt{3} - \sqrt{2})(2\sqrt{3} + \sqrt{2})$

34. $(3\sqrt{5} - 1)(5\sqrt{5} + 2)$

35. $(\sqrt{x} + 7)(2\sqrt{x} + 1)$

36. $(\sqrt{x} - 1)(3\sqrt{x} + 7)$

37. $(4\sqrt{2} - \sqrt{3})(5\sqrt{2} + 2\sqrt{3})$

38. $(2\sqrt{x} - \sqrt{y})(3\sqrt{x} + 2\sqrt{y})$

39. $(3\sqrt{6x} + 4\sqrt{2y})(\sqrt{6x} - \sqrt{2y})$

40. $(\sqrt{7y} - 3\sqrt{3})(2\sqrt{7y} + 2\sqrt{3})$

41. $(\sqrt{5} - \sqrt{3})^2$

42. $(\sqrt{3} - 2)^2$

43. $(2\sqrt{x} + \sqrt{y})^2$

44. $(x + 3\sqrt{y})^2$

45. $(2\sqrt{x} - \sqrt{y})^2$

46. $(\sqrt{3x} + 5\sqrt{2y})^2$

47. $(\sqrt{x-1} + 3)^2$

48. $(\sqrt{y+2} - 7)^2$

49. $(2\sqrt{3} - 1)(2\sqrt{3} + 1)$

50. $(5\sqrt{7} + \sqrt{2})(5\sqrt{7} - \sqrt{2})$

51. $(3\sqrt{5} - \sqrt{3})(3\sqrt{5} + \sqrt{3})$

52. $(2\sqrt{x} + 7)(2\sqrt{x} - 7)$

53. $(3\sqrt{x} - 11)(3\sqrt{x} + 11)$

54. $(4\sqrt{x} - \sqrt{y})(4\sqrt{x} + \sqrt{y})$

55. $(\sqrt{x+3} - 2)(\sqrt{x+3} + 2)$

56. $(\sqrt{3y-1} + 5)(\sqrt{3y-1} - 5)$

In problems 57–86, rationalize the denominator of each expression, and simplify the result. (Assume that all variables are restricted to values for which all radical expressions represent real numbers.)

57. $\dfrac{2}{\sqrt{3}}$

58. $\dfrac{9}{\sqrt{21}}$

59. $\dfrac{8}{7\sqrt{11x}}$

60. $\dfrac{10x}{3\sqrt{5x}}$

61. $\sqrt{\dfrac{1}{7y}}$

62. $\sqrt{\dfrac{1}{6t}}$

63. $\dfrac{5}{\sqrt[3]{7}}$

64. $\dfrac{8}{\sqrt[3]{36}}$

65. $\dfrac{5}{\sqrt[3]{9x}}$

66. $\dfrac{7}{\sqrt[3]{81y}}$

67. $\dfrac{3}{\sqrt[4]{3}}$

68. $\dfrac{5}{\sqrt[5]{4}}$

69. $\dfrac{7}{\sqrt[6]{2}}$

70. $\dfrac{4}{\sqrt[4]{3x}}$

71. $\dfrac{5}{\sqrt{x+2}}$

72. $\dfrac{3}{\sqrt{x-3}}$

73. $\dfrac{3}{\sqrt{5} + \sqrt{2}}$

74. $\dfrac{\sqrt{6}}{\sqrt{6} - \sqrt{3}}$

75. $\dfrac{10}{\sqrt{5} - 1}$

76. $\dfrac{36}{\sqrt{3} + 1}$

77. $\dfrac{\sqrt{2}}{1 + \sqrt{2}}$

78. $\dfrac{\sqrt{2} + 1}{\sqrt{3} + \sqrt{2}}$

79. $\dfrac{2 - 2\sqrt{3}}{\sqrt{7} - \sqrt{5}}$

80. $\dfrac{8}{6\sqrt{5} - 5\sqrt{3}}$

81. $\dfrac{\sqrt{y}}{3\sqrt{x} - 2\sqrt{y}}$

82. $\dfrac{3\sqrt{2} - \sqrt{3}}{2\sqrt{3} - 7\sqrt{2}}$

83. $\dfrac{\sqrt{x} - \sqrt{y}}{\sqrt{x} + \sqrt{y}}$

84. $\dfrac{\sqrt{a} + \sqrt{a - 4}}{\sqrt{a} - \sqrt{a - 4}}$

85. $\dfrac{4\sqrt{2} + 3\sqrt{5}}{7\sqrt{5} - 3\sqrt{2}}$

86. $\dfrac{3\sqrt{y + z}}{3 + \sqrt{y + z}}$

In problems 87 and 88, rationalize the denominator. (Here, it is necessary to multiply by a rationalizing factor twice to rationalize the denominator.)

87. $\dfrac{1}{\sqrt{5} + \sqrt{3} + \sqrt{2}}$

88. $\dfrac{1}{\sqrt{7} - \sqrt{5} + 2}$

89. Use special product $(a - b)(a^2 + ab + b^2) = a^3 - b^3$ to show that
$$(\sqrt[3]{5} - \sqrt[3]{2})(\sqrt[3]{25} + \sqrt[3]{10} + \sqrt[3]{4}) = 3.$$

90. Use special product $(a + b)(a^2 - ab + b^2) = a^3 + b^3$ to show that
$$(\sqrt[3]{t} + 5)(\sqrt[3]{t^2} - \sqrt[3]{5t} + 25) = t + 125.$$

91. Use problem 89 to rationalize the denominator.
$$\frac{1}{\sqrt[3]{5} - \sqrt[3]{2}}.$$

92. Evaluate $4x^2 - 12x + 7$ for $x = (3 + \sqrt{2})/2$.

93. Evaluate $3x^2 + 12x - 5$ for $x = (-6 - \sqrt{51})/3$.

94. Rationalize the numerator by multiplying the numerator and denominator by the rationalizing factor of the numerator:
$$\frac{\sqrt{7} + \sqrt{3}}{5}.$$

In problems 95–98, perform the indicated operation and simplify.

95. $\sqrt{5} + \dfrac{6}{\sqrt{5}}$

96. $\dfrac{2}{\sqrt{2x}} - \dfrac{5\sqrt{2x}}{x} + \sqrt{\dfrac{2}{x}}, \qquad x > 0$

97. $\sqrt{\dfrac{5}{3}} - \dfrac{15}{\sqrt{15}} + \dfrac{7\sqrt{15}}{3}$

98. $3\sqrt[3]{49} - \dfrac{11}{\sqrt[3]{7}} + 4\sqrt[3]{\dfrac{1}{7}}$

5.6 Complex Numbers

In 1545 the Italian physician and mathematician Geronimo Cardano (1501–1576) discovered that solutions to cubic equations involved square roots of negative numbers. Gradually, other mathematicians began to accept the idea of calculating with square roots of negative numbers. In 1637 the French philosopher and mathematician René Descartes (1596–1650) introduced the terms "real" and "imaginary." Later, in 1748, the Swiss mathematician Leonhard Euler (1707–1783) used the symbol "i" for $\sqrt{-1}$. Finally, in 1832 the great German mathematician Carl Friedrich Gauss (1777–1855) introduced the term "complex number." These numbers have been used in a wide variety of useful applications in electronics, engineering, and physics. In particular, Charles Steinmetz (1865–1923) used them to explain the behavior of electric circuits.

In our study, we work with expressions involving radicals in the form $\sqrt{-x}$, where $x > 0$. The square of a number represented by such a symbol must be negative:

$$(\sqrt{-x})^2 = [(-x)^{1/2}]^2 = (-x)^1 = -x.$$

However, the square of any real number (positive, negative, or zero) must be a positive number or zero. Accordingly, we must consider a new set of numbers, the *complex numbers* described above.

We define $\sqrt{-x}$, $x > 0$ to be the number such that

$$\sqrt{-x} \cdot \sqrt{-x} = -x.$$

In particular, if $x = 1$, we have

$$\sqrt{-1} \cdot \sqrt{-1} = -1.$$

We designate the number $\sqrt{-1}$ by the symbol i:

$$i = \sqrt{-1}.$$

We square both sides of this equation, and we have

$$i^2 = -1.$$

Remember, $\sqrt{-1} = i$, *not* $-i$, even though both i^2 and $(-i)^2 = -1$.

Using i, we can write the square root of a negative number, $-n$, $n > 0$, as

$$\sqrt{-n} = \sqrt{n(-1)} = \sqrt{n} \cdot \sqrt{-1} = \sqrt{n}\,i.$$

EXAMPLE 1   Write each expression in terms of i and simplify.

(a) $\sqrt{-4}$ (b) $-\sqrt{-25}$ (c) $\sqrt{-16} - \sqrt{-9}$ (d) $\sqrt{-16} \cdot \sqrt{-9}$

SOLUTION   (a) $\sqrt{-4} = \sqrt{4}\,i = 2i$

(b) $-\sqrt{-25} = -\sqrt{25}\,i = -5i$

(c) $\sqrt{-16} - \sqrt{-9} = \sqrt{16}i - \sqrt{9}i$
$= 4i - 3i = i$

(d) $\sqrt{-16} \cdot \sqrt{-9} = 4i \cdot 3i = (4 \cdot 3)(i \cdot i)$
$= 12i^2 = 12(-1) = -12$

Numbers such as $2i$, $5i$, and $3i$ are called **pure imaginary numbers.** In general, a number of the form bi, where b is a nonzero real number and $i = \sqrt{-1}$, is called a pure imaginary number. This leads to the following definition.

DEFINITION 1   **Complex Number**

> Any number, real or imaginary, of the form
> $$a + bi,$$
> where a and b are real numbers and $i = \sqrt{-1}$, is called a **complex number.**

For example,

$$4 + 3i, \qquad 2 - 7i, \qquad 3 + \sqrt{5}i, \qquad \text{and} \qquad \frac{3}{4} - 2\sqrt{7}i$$

are complex numbers.

The form $a + bi$ is called the **standard form** for complex numbers. The number a is called the **real part** of the complex number. The number b is called the **imaginary part** of the complex number.

$$\underset{\text{real part}}{a} \quad + \quad \underset{\text{imaginary part}}{bi}$$

Every real number a is a complex number, because the real number a can be written as

$$a = a + 0i.$$

Similarly, if b is a real number, then

$$bi = 0 + bi.$$

Two complex numbers $a + bi$ and $c + di$ are equal if and only if their real parts are equal and their imaginary parts are equal. That is,

$$a + bi = c + di \qquad \text{if and only if} \qquad a = c \qquad \text{and} \qquad b = d.$$

EXAMPLE 2   Find x and y, so that
$$5x - 20i = 15 + 4yi.$$

SOLUTION   Since the two complex numbers are equal, their real parts are equal and their imaginary parts are equal:

$$5x = 15 \quad \text{and} \quad -20 = 4y$$
$$x = 3 \quad \text{and} \quad y = -5.$$

We add and subtract complex numbers by combining their real parts and their imaginary parts separately, according to the following definition:

DEFINITION 2   **Addition and Subtraction of Complex Numbers**

> Let a, b, c, and d be real numbers. Then
>
> 1. **Addition**
> $$(a + bi) + (c + di) = a + bi + c + di = a + c + bi + di$$
> $$= (a + c) + (b + d)i.$$
>
> 2. **Subtraction**
> $$(a + bi) - (c + di) = a + bi - c - di = a - c + bi - di$$
> $$= (a - c) + (b - d)i.$$

EXAMPLE 3   Perform each operation.

(a) $(5 + 6i) + (9 + 3i)$ (b) $(4 - 2i) - (-3 + i)$

SOLUTION   (a) $(5 + 6i) + (9 + 3i) = 5 + 6i + 9 + 3i = 5 + 9 + 6i + 3i$
$$= (5 + 9) + (6 + 3)i = 14 + 9i$$

(b) $(4 - 2i) - (-3 + i) = 4 - 2i + 3 - i = 4 + 3 - 2i - i$
$$= (4 + 3) + (-2 - 1)i = 7 - 3i$$

Since complex numbers have the same form as binomials, we can treat a complex number as if it were a binomial, and multiply two complex numbers in the same way as we multiply binomials. That is, if a, b, c, and d are real numbers, then

$$(a + bi)(c + di) = ac + adi + bic + bidi = ac + bdi^2 + adi + bci$$
$$= ac + bd(-1) + (ad + bc)i = (ac - bd) + (ad + bc)i.$$

Thus, we have the following definition for multiplying complex numbers.

DEFINITION 3   **Multiplication of Complex Numbers**

$$(a + bi)(c + di) = (ac - bd) + (ad + bc)i.$$

It is probably not very useful to memorize Definition 3 at this point. It is easier to multiply two complex numbers as binomials, substitute -1 for i^2, and finally combine the real parts and imaginary parts to express each result in the form $a + bi$.

EXAMPLE 4   Perform each of the following operations.

(a) $4i(3 + 7i)$ (b) $(4 + 3i)(2 - 4i)$

(c) $(1 + 2i)(1 - 2i)$ (d) $(3 + 4i)^2$

SOLUTION   (a) By applying the distributive property, we have

$$4i(3 + 7i) = 4i(3) + 4i(7i) = 12i + 28i^2$$
$$= 12i + 28(-1) = -28 + 12i.$$

(b) We multiply each term in the second complex number by each term in the first:

$$(4 + 3i)(2 - 4i) = 4 \cdot 2 - 4 \cdot 4i + 2 \cdot 3i - 3 \cdot 4i^2$$
$$= 8 - 16i + 6i - 12(-1) \qquad \text{(Use } i^2 = -1.\text{)}$$
$$= 8 - 16i + 6i + 12$$
$$= 20 - 10i.$$

(c) This product has the form of the special product $(a - b)(a + b) = a^2 - b^2$, so that

$$(1 + 2i)(1 - 2i) = 1^2 - (2i)^2$$
$$= 1 - 4i^2$$
$$= 1 - 4(-1) = 5.$$

(d) This product has the form of the special product $(a + b)^2 = a^2 + 2ab + b^2$, so that

$$(3 + 4i)^2 = 3^2 + 2(3)(4i) + (4i)^2$$
$$= 9 + 24i + 16i^2$$
$$= 9 + 24i - 16$$
$$= -7 + 24i.$$

Notice in Example 4(c) that the product of the two complex numbers $1 + 2i$ and $1 - 2i$ is the real number 5. The two complex numbers $1 + 2i$

and $1 - 2i$ are called **complex conjugates.** In general, the complex numbers $a + bi$ and $a - bi$ are called the **complex conjugates of each other,** and

$$(a + bi)(a - bi) = a^2 - (bi)^2$$
$$= a^2 - b^2i^2$$
$$= a^2 + b^2.$$

If $a + bi$ is a complex number, then its complex conjugate is denoted by $\overline{a + bi}$. Thus,

$$\overline{a + bi} = a - bi.$$

For example,

$$\overline{4 + 3i} = 4 - 3i,$$

$$\overline{1 - 4i} = 1 + 4i,$$

and $\quad \overline{-5 - 7i} = -5 + 7i.$

To perform the division

$$\frac{3i}{4 - 3i},$$

we look for a complex number in standard form that is equivalent to $3i/(4 - 3i)$. To do this, we need to eliminate i from the denominator, so we multiply the numerator and the denominator by the complex conjugate of the denominator, which is $4 + 3i$.

$$\frac{3i}{4 - 3i} = \frac{3i(4 + 3i)}{(4 - 3i)(4 + 3i)} = \frac{12i + 9i^2}{16 - 9i^2} = \frac{12i - 9}{16 + 9} = \frac{-9 + 12i}{25}$$

$$= -\frac{9}{25} + \frac{12}{25} i.$$

In general,

$$\frac{a + bi}{c + di} = \frac{(a + bi)(c - di)}{(c + di)(c - di)} = \frac{(ac + bd) + (bc - ad)i}{c^2 + d^2} = \frac{ac + bd}{c^2 + d^2} + \frac{(bc - ad)}{c^2 + d^2} i.$$

In Examples 5–7, divide each and express the answer in the form $a + bi$.

EXAMPLE 5
$$\frac{1}{3 - 2i}$$

SOLUTION
We multiply the numerator and the denominator by the complex conjugate $3 + 2i$:

$$\frac{1}{3 - 2i} = \frac{1}{3 - 2i} \cdot \frac{3 + 2i}{3 + 2i} = \frac{3 + 2i}{3^2 + 2^2} = \frac{3 + 2i}{9 + 4} = \frac{3 + 2i}{13} = \frac{3}{13} + \frac{2}{13} i.$$

EXAMPLE 6 $\dfrac{1 + i}{1 - i}$

SOLUTION   The complex conjugate of the denominator is $1 + i$. Therefore, we have

$$\frac{1 + i}{1 - i} = \frac{1 + i}{1 - i} \cdot \frac{1 + i}{1 + i} = \frac{1 + 2i + i^2}{1^2 + 1^2} = \frac{2i}{2} = i.$$

EXAMPLE 7 $\dfrac{4 - 3i}{5 + 7i}$

SOLUTION   The complex conjugate of the denominator is $5 - 7i$:

$$\frac{4 - 3i}{5 + 7i} = \frac{4 - 3i}{5 + 7i} \cdot \frac{5 - 7i}{5 - 7i} = \frac{20 - 28i - 15i + 21i^2}{5^2 + 7^2}$$

$$= \frac{(20 - 21) + (-28 - 15)i}{25 + 49}$$

$$= \frac{-1 - 43i}{74} = -\frac{1}{74} - \frac{43}{74}i.$$

Positive integer exponents have the same meaning, in terms of repeated multiplication, for both complex numbers and real numbers. Therefore, we can extend the definition of positive integer exponents to include complex numbers. In particular, for powers of i, we have

$$\begin{aligned}
i^1 &= i \\
i^2 &= -1 \\
i^3 &= i^2 \cdot i = -1(i) = -i \\
i^4 &= i^2 \cdot i^2 = (-1)(-1) = 1.
\end{aligned}$$

If we continue the list, we repeat the same sequence of answers because we can replace every factor of i^4 by 1. Thus,

$$\begin{aligned}
i^5 &= i^4 \cdot i = 1(i) = i \\
i^6 &= i^4 \cdot i^2 = 1(-1) = -1 \\
i^7 &= i^4 \cdot i^3 = 1(-i) = -i \\
i^8 &= i^4 \cdot i^4 = 1(1) = 1 \\
i^9 &= i^8 \cdot i = 1(i) = i,
\end{aligned}$$

and so on.

EXAMPLE 8   Write each expression as i, 1, $-i$, or -1.

(a) i^{18} (b) i^{27} (c) i^{105} (d) i^{204}

SOLUTION   (a) $i^{18} = i^{16} \cdot i^2 = (i^4)^4 \cdot i^2 = 1^4(-1) = -1$

(b) $i^{27} = i^{24} \cdot i^3 = (i^4)^6 \cdot i^3 = 1^6(-i) = -i$

(c) $i^{105} = i^{104} \cdot i = (i^4)^{26} \cdot i = 1^{26}(i) = i$

(d) $i^{204} = (i^4)^{51} = 1^{51} = 1$

The next example illustrates how to perform operations with a negative power of i.

EXAMPLE 9   Express i^{-3} as the indicated product of a real number and i.

SOLUTION   We note that $i^{-3} = 1/i^3$. By multiplying the numerator and denominator by i, we obtain a real number in the denominator, so that

$$i^{-3} = \frac{1}{i^3} = \frac{1}{i^3} \cdot \frac{i}{i} = \frac{i}{i^4} = \frac{i}{1} = i.$$

PROBLEM SET 5.6

In problems 1–12, write each expression in the form $a + bi$, where a and b are real numbers. (If $a = 0$, write the number as bi.)

1. $\sqrt{-81}$ 2. $-\sqrt{\dfrac{-9}{16}}$ 3. $-\sqrt{\dfrac{-25}{4}}$ 4. $-\sqrt{-8}$

5. $-\sqrt{-x^4}$ 6. $\sqrt{-9y^8}$ 7. $3\sqrt{-25y^4}$ 8. $-5\sqrt{-8x^4}$

9. $5 + \sqrt{-81}$ 10. $-8 + \sqrt{-72}$ 11. $2 - \sqrt{-4x^4}$ 12. $6 - \sqrt{-16y^4}$

In problems 13–16, simplify each expression.

13. $\sqrt{-4} \cdot \sqrt{-9}$ 14. $\sqrt{-4} \cdot \sqrt{-25}$ 15. $\sqrt{-8} \cdot \sqrt{-2}$ 16. $\sqrt{-8} \cdot \sqrt{-18}$

In problems 17 and 18, find x and y such that each statement is true.

17. $4x + 8i = 20 + 2yi$

18. $(4x - 3) + 21i = 5 + (2y - 3)i$

In problems 19–54, perform each operation. Express the answer in the form $a + bi$.

19. $(-3 + 6i) + (2 + 3i)$ 20. $(-2 + 5i) + (-5 + i)$ 21. $(-5 + 3i) + (5i - 1)$

22. $(10 - 24i) + (3 + 7i)$ 23. $(6 + 8i) + (6 - 8i)$ 24. $(3 + 2i) + (-7 + 2i)$

25. $(-2 - 3i) - (-3 - 2i)$ 26. $(7 + 24i) - (-3 - 4i)$ 27. $(10 - 8i) - (10 + 8i)$

28. $(5 - 7i) - (5 - 13i)$

29. $(6 - 8i) - (5 + 3i)$

30. $(4 - 5i) - (2 - 6i)$

31. $[(4 + 3i) - (3 + 6i)] + (2 - 3i)$

32. $(8 - 5i) + [(3 + 7i) - (2 - 5i)]$

33. $4i(4 - 5i)$

34. $-7i(-2 + 4i)$

35. $3i(2 + 3i)$

36. $-6i(3 + 7i)$

37. $(2 - 4i)(3 + 2i)$

38. $(1 + 2i)(4 - i)$

39. $(7 + 2i)(3 - 2i)$

40. $(2 + 3i)(3 - 5i)$

41. $(-3 - 2i)(2 + 5i)$

42. $(2 + 3i)(1 - 3i)$

43. $(3 - 7i)(2 + 3i)$

44. $(1 - 6i)(2 + 5i)$

45. $(3 + 2i)^2$

46. $(-7 - 2i)^2$

47. $(1 + i)^2$

48. $(-2 + 7i)(-2 - 7i)$

49. $(5 - 4i)(5 + 4i)$

50. $(3 - 2i)(3 + 2i)$

51. $(7 - 11i)(7 + 11i)$

52. $(5 - 12i)(5 + 12i)$

53. $(2 - 7i)(2 + 7i)$

54. $(2 + i^{11})(2 - i^{11})$

In problems 55–58, find (a) $\bar{z}$ and (b) $z\bar{z}$.

55. $z = 2 + 4i$ **56.** $z = 3 - 4i$ **57.** $z = 2i$ **58.** $z = -3i$

In problems 59–76, perform each division and express the answer in the form $a + bi$.

59. $\dfrac{1}{4 + 3i}$

60. $\dfrac{1}{3 - 5i}$

61. $\dfrac{7}{5 - 4i}$

62. $\dfrac{-3}{3 + 4i}$

63. $\dfrac{2 + i}{2 + 3i}$

64. $\dfrac{1 - i}{-1 + i}$

65. $\dfrac{3 + 4i}{-2 + 5i}$

66. $\dfrac{7 + 2i}{1 - 5i}$

67. $\dfrac{4 - i^2}{2 - 7i}$

68. $\dfrac{5 - i^2}{11 + 2i}$

69. $\dfrac{i}{-1 - i}$

70. $\dfrac{1 + i}{2 - 3i}$

71. $\dfrac{5 + 2i}{i}$

72. $\dfrac{3 - 2i}{-i}$

73. $\dfrac{3 - 2i}{5 - 2i}$

74. $\dfrac{7 + 4i}{1 - 2i}$

75. $\dfrac{3 - \sqrt{-9}}{2 + \sqrt{-25}}$

76. $\dfrac{2 + \sqrt{-36}}{1 - \sqrt{-4}}$

In problems 77–88, write each expression as i, -1, $-i$, or 1.

77. i^{29} **78.** i^{37} **79.** i^{54} **80.** i^{49} **81.** i^{65} **82.** i^{74}

83. i^{108} **84.** i^{119} **85.** i^{-7} **86.** i^{-18} **87.** i^{-5} **88.** i^{-14}

89. Find the value of the expression $2x^2 + 3x + 2$ for

$$x = \frac{-3 - \sqrt{7}i}{4}.$$

90. Simplify the expression $(7 - 3i)^{-2}$ and express the answer in the form $a + bi$.

REVIEW PROBLEM SET

In problems 1–14, rewrite each expression so that it contains only positive exponents, and simplify. (Assume that variables are restricted to values for which all expressions represent real numbers.)

1. 3^{-2} **2.** $(\frac{1}{5})^{-2}$ **3.** y^{-4} **4.** $(3u)^{-3}$ **5.** $t^{-4}s^{-2}$

6. $4x^{-3}y^{-7}$

7. $\left(\dfrac{x^2}{y^3}\right)^{-1}$

8. $\dfrac{u^{-7}}{v^{-5}}$

9. $(3x^4)^0$

10. $(7y^3 + 1)^0$

11. $\dfrac{1 + 3^{-2}}{2^{-2}}$

12. $\dfrac{1}{2^{-3} + 3^{-2}}$

13. $\dfrac{2}{x^{-3} + y^{-2}}$

14. $\dfrac{w^{-3}}{z^3} + \dfrac{z^{-3}}{w^3}$

In problems 15–32, use the properties of exponents to rewrite each expression so that it contains only positive exponents, and simplify.

15. $3^{-11} \cdot 3^{15}$

16. $19^{-3} \cdot 19^5$

17. $x^{-8} \cdot x^{15}$

18. $w^{12} \cdot w^{-7}$

19. $(2^{-3})^2$

20. $(z^4)^{-3}$

21. $(t^{-3})^{-4}$

22. $(3u^{-2})^{-2}$

23. $(4^{-1}x^{-3})^{-2}$

24. $(3x^{-3}y^{-4})^{-3}$

25. $\left(\dfrac{2}{u^{-2}}\right)^{-3}$

26. $\left(\dfrac{m^3}{n^{-4}}\right)^{-2}$

27. $\dfrac{5^{-4}}{5^{-6}}$

28. $\dfrac{11^{-3}}{11^{-2}}$

29. $\dfrac{r^{-5}}{r^{-2}}$

30. $\dfrac{(xy)^2}{(xy)^{-4}}$

31. $\dfrac{8y^2}{2y^{-3}}$

32. $\dfrac{(ab)^{-6}}{(ab)^{-11}}$

In problems 33–38, rewrite each expression without negative exponents, and simplify.

33. $\left(\dfrac{x^{-3}y^2}{x^{-2}y^{-4}}\right)^{-2}$

34. $\left(\dfrac{w^3z^{-4}}{w^{-4}z^2}\right)^{-1}$

35. $\left(\dfrac{u^4}{2v^{-3}}\right)^{-1}\left(\dfrac{u^{-1}v^2}{4^{-1}}\right)^{-2}$

36. $\left(\dfrac{m^{-4}n^{-3}}{m^{-7}n^{-1}}\right)^{-2}\left(\dfrac{m^4n}{m^{-3}n^4}\right)^{-1}$

37. $\left[\dfrac{r^3s^{-2}t^{-4}}{(rst)^{-1}}\right]^{-2}$

38. $\left[\dfrac{x^{-5}y^3z^{-4}}{(xy)^{-2}(yz)^{-3}}\right]^{-1}$

In problems 39 and 40, simplify each expression and write each answer in scientific notation.

C **39.** $\dfrac{(8.34 \times 10^6) \cdot (2.3 \times 10^{-2})}{(7.4 \times 10^{-3}) \cdot (1.9 \times 10^5)}$

C **40.** $\dfrac{(2.38 \times 10^3)^2 \cdot (5.3 \times 10^{-4})^3}{(4.8 \times 10^{-2})^2 \cdot (6.31 \times 10^5)}$

In problems 41–46, find the principal root (if it is defined as a real number).

41. $\sqrt{81}$

42. $\sqrt[3]{-64}$

43. $\sqrt[3]{-216}$

44. $\sqrt[4]{-16}$

45. $\sqrt{-9}$

46. $\sqrt{289}$

In problems 47–50, rewrite each expression in terms of rational exponents.

47. $\sqrt{15}$

48. $\sqrt[3]{-7}$

49. $\sqrt[4]{(a + b)^3}$

50. $\sqrt[6]{9x^5}$

In problems 51–54, rewrite each expression in terms of radicals.

51. $13^{2/3}$

52. $19^{4/5}$

53. $(2x + 3y)^{2/7}$

54. $(-3u^2v^4)^{4/9}$

In problems 55–62, find the value of each expression.

55. $(-32)^{3/5}$

56. $(-27)^{-4/3}$

57. $16^{-3/4}$

58. $-(-8)^{1/3}$

59. $-4^{3/2}$

60. $-25^{-1/2}$

61. $(-8)^{-2/3}$

62. $36^{-5/2}$

In problems 63–86, rewrite each expression so that it contains only positive exponents, and simplify. (Assume that variables are restricted to values for which all expressions represent real numbers.)

63. $7^{-2/3} \cdot 7^{5/3}$

64. $9^{-4/7} \cdot 9^{11/7}$

65. $x^{3/2} \cdot x^{-5/4}$

66. $y^{5/11} \cdot y^{-16/11}$

67. $(2^{-1/3})^9$

68. $(8^2)^{-5/12}$

69. $(x^{-11/3})^{3/22}$

70. $(y^{-51})^{3/17}$

71. $(-8t^6)^{1/3}$

72. $(16u^4)^{-1/2}$

73. $(-32m^5)^{2/5}$

74. $(4x^2y^8)^{3/2}$

75. $\left(\dfrac{8}{v^9}\right)^{-2/3}$

76. $\left(\dfrac{128}{p^7}\right)^{-2/7}$

77. $\left(\dfrac{c^{-5/2}}{d^{-2/5}}\right)^{-30}$

78. $\left(\dfrac{16z^4}{w^8}\right)^{-3/4}$

79. $\left(\dfrac{-x^5y^{10}}{32z^{15}}\right)^{-3/5}$

80. $\left(\dfrac{8m^{-6}n^{-30}}{27n^{-12}}\right)^{-1/3}$

81. $\dfrac{a^{-2/3}}{a^{-5/7}}$

82. $\dfrac{(y^2)^{-3/2}}{y^{7/2}}$

83. $\dfrac{(r^{-1}y^{-2/3})^{-3}}{(r^{-1}y^{-2/3})^5}$

84. $\dfrac{(w^{-7/3})^{2/7}}{w^{-3}}$

85. $\left(\dfrac{x^{3/2}y^{-1/3}}{w^{-2}}\right)^{-6}\left(\dfrac{x^{1/3}y^{-2/3}}{w^{-2}}\right)^6$

86. $\left[\dfrac{(u^{2/3}v^{-1/2})(u^{1/2}v^{-3})}{u^{5/6}v^{-2}}\right]^{-1}$

In problems 87–90, perform each multiplication. Write each answer with positive exponents. (Assume that variables are restricted to values for which all expressions represent real numbers.)

87. $x^{1/4}(2x^{3/4} - 3x^{7/4})$

88. $(x^{-1/2} + 2y^{2/3})(x^{-1/2} - 2y^{2/3})$

89. $(z^{-2/3} + 3)^2$

90. $(x^{1/3} + 2^{1/3})(x^{2/3} - 2^{1/3}x^{1/3} + 2^{2/3})$

In problems 91–94, perform each addition or subtraction and simplify. (Assume that variables are restricted to values for which all expressions represent real numbers.)

91. $3x^{-2/3} + x^{1/3}$

92. $x^{4/5} - 7x^{-1/5}$

93. $(x^2 + 2)^{1/2} - 5(x^2 + 2)^{-1/2}$

94. $\dfrac{2}{(x^2 + 1)^{1/2}} - \dfrac{x^2 - 1}{(x^2 + 1)^{3/2}}$

95. Factor $3(y - 1)^{1/3}$ from $12(y - 1)^{4/3} + 16(y - 1)^{1/3}$.

96. Factor $4(z^2 + 1)^{-3/5}$ from $20(z^2 + 1)^{-1/5}$ to $8(z^2 + 1)^{-3/5}$.

|C| In problems 97–100, use a calculator with a y^x key to evaluate each expression. Round off your answer to four significant digits.

97. $3.12^{4.3}$

98. $0.91^{2.71}$

99. $9^{\sqrt{2}}$

100. $\sqrt{5}^{\sqrt{3}}$

In problems 101–122, use the properties of radicals to simplify each expression. (Assume that variables are restricted to values for which all expressions represent real numbers.)

101. $\sqrt{125}$

102. $\sqrt[5]{-64}$

103. $\sqrt[4]{32t^5}$

104. $\sqrt[3]{54w^7}$

105. $\sqrt[3]{-24x^4y^{11}}$

106. $\sqrt{250z^3y^{10}}$

107. $\sqrt{\dfrac{4}{25}}$

108. $\sqrt[3]{-\dfrac{8}{27}}$

109. $\sqrt[5]{-\dfrac{t^{10}}{32}}$

110. $\sqrt[4]{\dfrac{16x^8y^{12}}{81}}$

111. $\sqrt[3]{\dfrac{-5}{c^6d^9}}$

112. $\sqrt{\dfrac{11}{w^4z^8}}$

113. $\dfrac{\sqrt{125u^7}}{\sqrt{5u}}$

114. $\dfrac{\sqrt[3]{-24m^{11}}}{\sqrt[3]{3m^2}}$

115. $\sqrt{\sqrt{16x^8}}$

116. $\sqrt[3]{\sqrt[3]{t^{18}}}$

117. $\sqrt[3]{\sqrt[5]{v^{15}}}$

118. $\sqrt[4]{\sqrt[6]{x^{24}}}$

119. $\sqrt{8u^3v^4} \cdot \sqrt{2uv^2}$

120. $\sqrt[3]{4m^2} \cdot \sqrt[3]{2m}$

121. $\sqrt[3]{\dfrac{x^3\sqrt{x^5}}{x^4}}$

122. $\sqrt[6]{y\sqrt[3]{z\sqrt{w}}}$

In problems 123–126, write each expression as a single radical, and then simplify the result. (Assume that variables are restricted to values for which all expressions represent real numbers.)

123. $\sqrt[3]{z}\sqrt[4]{z}$

124. $\sqrt[5]{x^2}\sqrt[3]{x}$

125. $\sqrt{2u}\sqrt[3]{3u^2}$

126. $\sqrt[4]{a^2b^3}\sqrt[5]{a^4b^2}$

Ⓒ In problems 127 and 128, use a calculator to verify each equation.

127. $\sqrt{11}\sqrt{3}=\sqrt{33}$

128. $\sqrt{19}\sqrt{3}=\sqrt{57}$

In problems 129–154, perform the indicated operations and simplify the result. (Assume that variables are restricted to values for which all expressions represent real numbers.)

129. $7\sqrt{2}-3\sqrt{2}+4\sqrt{2}$

130. $\sqrt{5}-6\sqrt{5}+2\sqrt{5}$

131. $4\sqrt{x}+7\sqrt{x}-5\sqrt{x}$

132. $3\sqrt[3]{t}-\sqrt[3]{t}+2\sqrt[3]{t}$

133. $\sqrt{128}+\sqrt{8}$

134. $\sqrt{48}-\sqrt{12}$

135. $\sqrt{108}-\sqrt{27}$

136. $\sqrt{45}+\sqrt{80}$

137. $\sqrt{32z}+\sqrt{72z}$

138. $\sqrt{27x}-\sqrt{48x}$

139. $\sqrt{63u}+2\sqrt{112u}-\sqrt{252u}$

140. $\sqrt[3]{16y^2}-\sqrt[3]{54y^2}+\sqrt[3]{250y^2}$

141. $\sqrt{3}(\sqrt{2}+\sqrt{5})$

142. $\sqrt{5}(\sqrt{7}-\sqrt{2})$

143. $\sqrt{3}(\sqrt{6}-\sqrt{8})$

144. $\sqrt{5}(\sqrt{15}+\sqrt{10})$

145. $(\sqrt{10}+\sqrt{2})^2$

146. $(2\sqrt{6}-3\sqrt{2})^2$

147. $(\sqrt{8}-\sqrt{2})(\sqrt{8}+\sqrt{2})$

148. $(\sqrt{a}-\sqrt{b})(\sqrt{a}+\sqrt{b})$

149. $(\sqrt{3t}+\sqrt{5})(\sqrt{3t}-\sqrt{5})$

150. $(\sqrt{xy}-\sqrt{z})(\sqrt{xy}+\sqrt{z})$

151. $(\sqrt{3}+\sqrt{6})(2\sqrt{3}-\sqrt{6})$

152. $(\sqrt{6}+1)(\sqrt{8}+\sqrt{18})$

153. $(\sqrt{x}-\sqrt{y})(2\sqrt{x}+\sqrt{y})$

154. $(3\sqrt{u}-v)(5\sqrt{u}+2v)$

In problems 155–168, rationalize the denominator of each expression and simplify the result. (Assume that all variables are restricted to values for which all radical expressions represent real numbers.)

155. $\dfrac{4}{\sqrt{3}}$

156. $\dfrac{7}{\sqrt{14}}$

157. $\dfrac{5}{\sqrt{32}}$

158. $\dfrac{8}{\sqrt{18w}}$

159. $\dfrac{6t}{\sqrt{27ts}}$

160. $\dfrac{2}{\sqrt{7}-1}$

161. $\dfrac{5}{\sqrt{11}+5}$

162. $\dfrac{6}{\sqrt{2}+3\sqrt{5}}$

163. $\dfrac{10}{\sqrt{x}-2}$

164. $\dfrac{\sqrt{m}}{3+\sqrt{m}}$

165. $\dfrac{\sqrt{7}-\sqrt{6}}{\sqrt{7}+\sqrt{6}}$

166. $\dfrac{2\sqrt{x}-\sqrt{y}}{\sqrt{x}+\sqrt{y}}$

167. $\dfrac{u-\sqrt{v}}{u-2\sqrt{v}}$

168. $\dfrac{2}{1+\sqrt{2}-\sqrt{3}}$

In problems 169 and 170, perform the indicated operation and simplify.

169. $\sqrt{3}-\dfrac{2}{3\sqrt{3}}$

170. $5\sqrt[3]{16}-\dfrac{7}{\sqrt[3]{4}}+\sqrt[3]{\dfrac{1}{4}}$

In problems 171–186, write each expression in the form of $a+bi$, where a and b are real numbers. (If $a=0$, write the number as bi.)

171. $4\sqrt{-9}$

172. $-2\sqrt{-25}$

173. $3+\sqrt{-t^2},\,t>0$

174. $-7-\sqrt{-4y^2},\,y>0$

175. $(-4-i)+(3+17i)$

176. $(3-42i)+(17+4i)$

177. $(-7+6i)+(3-2i)$

178. $(12+6i)+(12-6i)$

179. $(-24+i)-(3-9i)$

180. $(65 - 4i) - (43 - 2i)$

181. $(15 - 7i) - (-8 + 3i)$

182. $(-12 + 3i) - (-11 - 5i)$

183. $(6 - 2i)(1 + 3i)$

184. $(2 + 3i)(1 - 7i)$

185. $(3 + 8i)(2 - 5i)$

186. $(3 - 2i\sqrt{7})(3 + 2i\sqrt{7})$

In problems 187–190, find the conjugate of each complex number.

187. $3 + 5i$

188. $-6i$

189. $-4 - 7i$

190. $-11 + 3i$

In problems 191–196, perform each division and express the answer in the form $a + bi$.

191. $\dfrac{3 + 2i}{4 + i}$

192. $\dfrac{1 - i}{2 - 3i}$

193. $\dfrac{7 + 3i}{3 - 2i}$

194. $\dfrac{2 + 7i}{2 + 2i}$

195. $\dfrac{2 + i}{-5 - i}$

196. $\dfrac{3 + \sqrt{-16}}{2 - \sqrt{-9}}$

In problems 197–202, simplify.

197. i^{33}

198. i^{44}

199. i^{55}

200. i^{69}

201. i^{402}

202. i^{570}

CHAPTER 5 TEST

1. Rewrite each expression so that it contains only positive exponents.

(a) z^{-3} (b) $(8^0 p)^{-1}$ (c) $\left(\dfrac{x^3}{y^4}\right)$

2. Use the properties of exponents to rewrite each expression so that it contains only positive exponents, and simplify.

(a) $x^{-11} \cdot x^7$ (b) $(2x^2 y^{-3})^{-2}$ (c) $\dfrac{(xy)^{-3}}{(xy)^{-2}}$

3. Rewrite each expression without negative exponents, and simplify.

(a) $\left(\dfrac{w^{-3} z^{-2}}{wz^{-4}}\right)^{-2}$ (b) $\left(\dfrac{x^{-3} y^{-2}}{x^{-5} y^{-1}}\right)^{-1} \cdot \left(\dfrac{x^3 y^2}{x^{-2} y}\right)^{-1}$

$\boxed{\text{C}}$ **4.** Simplify the given expression and write the answer in scientific notation:

$$\frac{(3.74 \times 10^4) \cdot (4.31 \times 10^{-3})}{(6.11 \times 10^{-2}) \cdot (1.78 \times 10^2)}.$$

5. In each of the following, find the principal root (if it is defined as a real number.)

(a) $\sqrt[4]{16}$ (b) $\sqrt[3]{-125}$ (c) $\sqrt{-25}$

6. Rewrite $\sqrt[5]{(x + 2)^3}$ in terms of rational exponents.

7. Rewrite $(3w - z)^{4/7}$ in terms of radicals.

8. Find the value of each expression.

(a) $16^{3/4}$ (b) $(-27)^{2/3}$ (c) $(-8)^{-4/3}$

9. Rewrite each expression so that it contains only positive exponents, and simplify.

   (a) $y^{-2/3} \cdot y^{7/3}$ (b) $x^{-3/5} \div x^{-1/2}$ (c) $\left(\dfrac{x^{-4}}{4y^{-2/3}}\right)^{-1/2}$

10. Perform each multiplication and write answers with positive exponents.

    (a) $y^{1/3}(2y^{5/3} - 4y^{-4/3})$ (b) $(x^{1/2} - y^{-1/3})(x^{1/2} + y^{-1/3})$

11. Use the properties of radicals to simplify each expression.

    (a) $\sqrt{75x^5y^3}$ (b) $\sqrt[3]{-81u^4v^{14}}$ (c) $\sqrt{\dfrac{32w^3}{25x^2y^4}}$ (d) $\sqrt{8uv^3} \cdot \sqrt{2v^5}$

12. Write the expression $\sqrt[3]{x^2} \cdot \sqrt[5]{x^4}$ as a single radical in simplified form.

13. Perform the indicated operations, and simplify the result.

    (a) $\sqrt{48x^3} + \sqrt{12x}$ (b) $\sqrt{3}(\sqrt{15} + \sqrt{6})$ (c) $(\sqrt{7x} - \sqrt{3y})(\sqrt{7x} + \sqrt{3y})$

14. Rationalize the denominator of each expression, and simplify the result.

    (a) $\dfrac{7}{\sqrt{21}}$ (b) $\dfrac{3\sqrt{x} + 1}{\sqrt{x} + 2}$

15. Write each expression in the form of $a + bi$, where a and b are real numbers.

    (a) $3 + \sqrt{-9}$ (b) $(2 + 3i) + (-7 - i)$ (c) $(-11 - 3i) - (4 - 6i)$

    (d) $(2 + 5i)(1 - 2i)$ (e) $\dfrac{1 + i}{3 - 2i}$

6 Nonlinear Equations and Inequalities

Nonlinear equations and inequalities are used frequently in the physical sciences and engineering. We examine some nonlinear equations and inequalities, including quadratic (second-degree) equations and inequalities, radical equations, and equations containing rational exponents. We then use these equations and inequalities to solve practical problems.

A **quadratic equation** in one variable is an equation that can be written in the form

$$ax^2 + bx + c = 0,$$

in which a, b, and c are constants with $a \neq 0$. When a quadratic equation is written this way, it is said to be in **standard form.**

In the next three sections we discuss various methods for solving quadratic equations. Then we apply these methods to solving equations that are quadratic in form as well as to quadratic inequalities.

6.1 Solving Quadratic Equations by Factoring

To solve quadratic equations of the form

$$ax^2 + bx + c = 0$$

by factoring, we use the methods of factoring developed in Sections 2.5 through 2.7, *and* the following special property:

Zero Product Property

> Assume that a and b are real numbers.
>
> If $ab = 0$, then $a = 0$ or $b = 0$.

For instance, to solve the equation $2x^2 + x - 6 = 0$, we factor the left side of the equation:

$$(2x - 3)(x + 2) = 0.$$

Now set each factor equal to zero and solve the resulting first-degree equations:

$$2x - 3 = 0 \quad \Big| \quad x + 2 = 0$$
$$2x = 3 \quad \Big| \quad x = -2$$
$$x = \frac{3}{2} \quad \Big|$$

To check these solutions, we replace x by $\frac{3}{2}$ in the original equation:

$$2\left(\frac{3}{2}\right)^2 + \frac{3}{2} - 6 = 2\left(\frac{9}{4}\right) + \frac{3}{2} - 6 = \frac{9}{2} + \frac{3}{2} - 6 = \frac{12}{2} - 6 = 0.$$

For $x = -2$, we have

$$2(-2)^2 + (-2) - 6 = 2(4) + (-2) - 6 = 8 - 2 - 6 = 0.$$

To solve a quadratic equation by factoring, we suggest the following procedure as a guideline:

Step 1. Write the quadratic equation in standard form.

Step 2. Factor the expression on the left side of the equation.

Step 3. Use the zero property to set each factor equal to zero.

Step 4. Solve the resulting linear equations.

Step 5. Check the solutions.

In Examples 1–5, use the factoring method to solve each equation.

EXAMPLE 1 $x^2 + 4x = 21$

SOLUTION

Step 1. Write the equation in standard form:

$$x^2 + 4x - 21 = 0.$$

Step 2. Factor the left side of the equation:

$$(x - 3)(x + 7) = 0.$$

Steps 3 and 4. Set each factor equal to zero and solve the resulting first-degree equations:

$$x - 3 = 0 \quad \Big| \quad x + 7 = 0$$
$$x = 3 \quad \Big| \quad x = -7$$

Step 5. Check for $x = 3$:

$$3^2 + 4(3) - 21 = 9 + 12 - 21 = 0.$$

For $x = -7$:

$$(-7)^2 + 4(-7) - 21 = 49 - 28 - 21 = 0.$$

Therefore, the solutions are 3 and -7.

In the following examples, we omit showing the check. However, we encourage you to check the solutions whenever the check is omitted.

EXAMPLE 2 $2y^2 - y = 1$

SOLUTION   First we rewrite the equation in standard form:

$$2y^2 - y - 1 = 0.$$

We factor the left side of the equation:

$$(2y + 1)(y - 1) = 0.$$

We set each factor equal to zero and solve the resulting equations:

$$
\begin{array}{c|c}
2y + 1 = 0 & y - 1 = 0 \\
2y = -1 & y = 1 \\
y = -\dfrac{1}{2} &
\end{array}
$$

Therefore, the solutions are $-\frac{1}{2}$ and 1.

EXAMPLE 3 $5m^2 = -15m$

SOLUTION   First we rewrite the equation in standard form:

$$5m^2 + 15m = 0.$$

Next we factor the left side of the equation:

$$5m(m + 3) = 0.$$

We set each factor equal to zero and solve the resulting equations:

$$
\begin{array}{c|c}
5m = 0 & m + 3 = 0 \\
m = 0 & m = -3
\end{array}
$$

Therefore, the solutions are 0 and -3.

EXAMPLE 4 $9r^2 = 7(6r - 7)$

SOLUTION   First we rewrite the equation in standard form:

$$9r^2 - 42r + 49 = 0.$$

We factor the left side of the equation:

$$(3r - 7)^2 = 0.$$

We set each factor equal to zero:

$$3r - 7 = 0 \qquad\Big|\qquad 3r - 7 = 0$$
$$3r = 7 \qquad\Big|\qquad 3r = 7$$
$$r = \frac{7}{3} \qquad\Big|\qquad r = \frac{7}{3}$$

Therefore, the only solution is $\frac{7}{3}$.

EXAMPLE 5 $(t + 6)(t - 2) = -7$

SOLUTION   To solve this equation for t, it is incorrect to set each of the factors of the left side of the equation equal to -7. Because the product of the two expressions is -7 does not imply that either factor must be -7. So we begin by multiplying the two factors on the left side of the equation; then we write the equation in standard form. That is,

$$(t + 6)(t - 2) = -7$$
$$t^2 + 4t - 12 = -7$$
$$t^2 + 4t - 5 = 0 \qquad \text{(We write the equation in standard form.)}$$
$$(t + 5)(t - 1) = 0 \qquad \text{(We factor the left side.)}$$

$$t + 5 = 0 \qquad\Big|\qquad t - 1 = 0$$
$$t = -5 \qquad\Big|\qquad t = 1$$

Therefore, the solutions are -5 and 1.

EXAMPLE 6   Solve the equation

$$15x^2 + 11xb - 14b^2 = 0$$

for x.

SOLUTION   First we factor the left side of the equation:

$$15x^2 + 11xb - 14b^2 = (3x - 2b)(5x + 7b) = 0.$$

By setting each factor equal to zero, we have

$$3x - 2b = 0 \qquad \bigg| \qquad 5x + 7b = 0$$

$$3x = 2b \qquad \bigg| \qquad 5x = -7b$$

$$x = \frac{2b}{3} \qquad \bigg| \qquad x = \frac{-7b}{5}$$

We often encounter equations that contain fractions. Sometimes, when this occurs, a quadratic equation may be obtained by multiplying each side of the given equation by the LCD of the fractions. This possibility is illustrated in Example 7 below:

EXAMPLE 7   Solve the equation

$$\frac{5}{t + 4} - \frac{3}{t - 2} = 4.$$

SOLUTION   First we multiply both sides of the equation by $(t + 4)(t - 2)$, the LCD of the fractions. (Remember, any value of the variable that makes the denominator zero is not allowed. Thus, $t = -4$ or $t = 2$ cannot be solutions to the original equation.) This gives us an equation that contains no fractions:

$$(t + 4)(t - 2)\left(\frac{5}{t + 4} - \frac{3}{t - 2}\right) = 4(t + 4)(t - 2)$$

$$5(t - 2) - 3(t + 4) = 4(t^2 + 2t - 8)$$

$$5t - 10 - 3t - 12 = 4t^2 + 8t - 32$$

$$2t - 22 = 4t^2 + 8t - 32.$$

We write this equation in standard form:

$$4t^2 + 6t - 10 = 0$$

or

$$2t^2 + 3t - 5 = 0. \qquad \text{(We divided both sides by 2.)}$$

We factor the left side of the equation:

$$(2t + 5)(t - 1) = 0.$$

We set each factor equal to zero:

$$2t + 5 = 0 \qquad \bigg| \qquad t - 1 = 0$$

$$2t = -5 \qquad \bigg| \qquad t = 1$$

$$t = -\frac{5}{2} \qquad \bigg|$$

Check For $t = 1$,

$$\frac{5}{1+4} - \frac{3}{1-2} = \frac{5}{5} - \frac{3}{-1}$$
$$= 1 + 3$$
$$= 4.$$

For $t = -\frac{5}{2}$,

$$\frac{5}{-\frac{5}{2}+4} - \frac{3}{-\frac{5}{2}-2} = \frac{5}{\frac{3}{2}} - \frac{3}{-\frac{9}{2}}$$
$$= 5\left(\frac{2}{3}\right) - 3\left(-\frac{2}{9}\right)$$
$$= \frac{10}{3} + \frac{2}{3} = 4.$$

Therefore, the solutions are 1 and $-\frac{5}{2}$.

PROBLEM SET 6.1

In problems 1–48, use the factoring method to solve each equation. Check the solutions in problems 39–44.

1. $x^2 - 3x + 2 = 0$ **2.** $y^2 - 7y - 8 = 0$ **3.** $c^2 - 6c + 8 = 0$

4. $u^2 + 5u - 66 = 0$ **5.** $t^2 - t = 20$ **6.** $z^2 + 3z = 10$

7. $y^2 + 2y = 35$ **8.** $b^2 + 9b = 10$ **9.** $u^2 - u = 12$

10. $y^2 + y = 12$ **11.** $3x^2 - 2x - 5 = 0$ **12.** $15x^2 - 19x + 6 = 0$

13. $10y^2 + y - 2 = 0$ **14.** $6z^2 + 17z - 3 = 0$ **15.** $9r^2 + 6r - 8 = 0$

16. $5w^2 + 34w - 7 = 0$ **17.** $-10t^2 + 11t - 3 = 0$ **18.** $3y^2 - y - 14 = 0$

19. $4z^2 + 20 = 21z$ **20.** $12c^2 + 5c = 2$ **21.** $6u^2 + 7u = 20$

22. $4x^2 = 27x + 7$ **23.** $10y^2 - 31y = 14$ **24.** $18w^2 + 61w = 7$

25. $x^2 = 7x$ **26.** $x^2 = -8x$ **27.** $y^2 + 7y = 0$

28. $3m^2 - 7m = 0$ **29.** $49z^2 - 14z + 1 = 0$ **30.** $25y^2 - 20y + 4 = 0$

31. $9y^2 - (y + 2)^2 = 0$ **32.** $4x^2 - (x + 1)^2 = 0$ **33.** $z(3z + 11) = 20$

34. $m(4m + 25) = -6$ **35.** $n(5n + 2) = 3$ **36.** $x(x - 2) = 9 - 2x$

37. $r(2r - 19) = 33$ **38.** $(x + 3)(2x + 3) = 11(x + 3)$ **39.** $(x + 4)(x - 1) = 24$

40. $(x + 7)(x - 4) = 26$ **41.** $(2x + 1)(3x - 2) = 10$ **42.** $(2x - 1)(x + 2) = -3$

43. $\dfrac{12}{y} - 7 = \dfrac{12}{1 - y}$ **44.** $\dfrac{5}{4(t + 4)} - 1 = \dfrac{3}{4(t - 2)}$ **45.** $\dfrac{15}{(x - 2)^2} + \dfrac{2}{x - 2} = 1$

46. $\dfrac{3 - 2m}{4m} - 4 = \dfrac{3}{4m - 3}$ **47.** $\dfrac{2u - 5}{2u + 1} = \dfrac{7}{4} - \dfrac{6}{2u - 3}$ **48.** $\dfrac{6}{t^2 - 1} = \dfrac{1}{2} + \dfrac{1}{1 - t}$

In problems 49–56, solve each equation for the indicated unknown.

49. $x^2 - 2ax - 15a^2 = 0$, for x. **50.** $12y^2 - 10my = 12m^2$, for y. **51.** $w^2 + 2qw = p^2 - q^2$, for w.
52. $(at - bt)^2 = t(b - a)$, for t. **53.** $6m^2 + mb = 2b^2$, for m. **54.** $7y^2 - 11hy = 6h^2$, for y.
55. $z^2 + 4d^2 - z + 2d = 4dz$, for z. **56.** $b^2t^2 - 4bt + 3 = 4t^2 - 4t$, for t.

6.2 Solving Quadratic Equations by Roots Extraction and Completing the Square

It is clear that factoring is an efficient method for solving quadratic equations in standard form. But it is not always possible to factor such equations. In this section, we develop other methods for solving quadratic equations.

Roots Extraction

In Section 6.1, we solved equations such as

$$x^2 - 4 = 0$$

by factoring to get $(x - 2)(x + 2) = 0$, and found that $x = 2$ and $x = -2$. We can also solve this equation by determining the numbers whose squares are 4. That is, we determine the two square roots of 4. Thus, if $x^2 = 4$, then

$$x = -\sqrt{4} = -2 \qquad \text{or} \qquad x = \sqrt{4} = 2,$$

since

$$(-2)^2 = 4 \qquad \text{and} \qquad 2^2 = 4.$$

This example suggests the following property:

If $au^2 = p$, which is equivalent to $u^2 = p/a$, then

$$u = -\sqrt{\frac{p}{a}} \qquad \text{or} \qquad u = \sqrt{\frac{p}{a}}.$$

This method is called **roots extraction.**

The two equations $u = -\sqrt{p/a}$ and $u = \sqrt{p/a}$ are often written as the single equation $u = \pm\sqrt{p/a}$. The symbol "$\pm$" is read "plus or minus."

In Examples 1–4, solve each equation by the roots extraction method.

EXAMPLE 1 $x^2 = 81$

SOLUTION   If $x^2 = 81$, then $x = \pm\sqrt{81} = \pm 9$. So that

$$x = -9 \quad \text{or} \quad x = 9.$$

The solutions are -9 and 9.

EXAMPLE 2 $2t^2 = 24$

SOLUTION   We divide both sides by 2:

$$t^2 = 12,$$

so that

$$t = \pm\sqrt{12}.$$

$$
\begin{array}{c|c}
t = -\sqrt{12} & t = \sqrt{12} \\
t = -\sqrt{4(3)} & t = \sqrt{4(3)} \\
t = -2\sqrt{3} & t = 2\sqrt{3}
\end{array}
$$

Therefore, the solutions are $-2\sqrt{3}$ and $2\sqrt{3}$.

EXAMPLE 3 $(2y + 3)^2 = 36$

SOLUTION   The equation is of the form

$$u^2 = 36 \quad \text{with} \quad u = 2y + 3,$$

so that

$$u = \pm\sqrt{36} = \pm 6.$$

$$
\begin{array}{c|c}
2y + 3 = -6 & 2y + 3 = 6 \\
2y = -9 & 2y = 3 \\
y = -\dfrac{9}{2} & y = \dfrac{3}{2}
\end{array}
$$

Therefore, the solutions are $-\frac{9}{2}$ and $\frac{3}{2}$.

EXAMPLE 4 $w^2 + 25 = 0$

SOLUTION   First we rewrite the equation as

$$w^2 = -25,$$

so that

$$w = \pm\sqrt{-25}.$$

$w = -\sqrt{-25}$	$w = \sqrt{-25}$
$w = -i\sqrt{25}$	$w = i\sqrt{25}$
$w = -5i$	$w = 5i$

Check For $w = -5i$,

$$(-5i)^2 + 25 = 25i^2 + 25 = -25 + 25 = 0.$$

For $w = 5i$,

$$(5i)^2 + 25 = 25i^2 + 25 = -25 + 25 = 0.$$

Therefore, the solutions are $-5i$ and $5i$.

Completing the Square

Consider the following quadratic equations:

$$x^2 - 4x - 8 = 0 \quad \text{and} \quad (x - 2)^2 = 12.$$

Although these two equations appear to be different, they are simply two forms of the same equation, as you will see when you put the second equation into standard form:

$$(x - 2)^2 = 12$$
$$x^2 - 4x + 4 = 12$$
$$x^2 - 4x - 8 = 0 \quad \text{(We subtracted 12 from both sides.)}$$

Since the first equation $x^2 - 4x - 8 = 0$ cannot be solved by factoring, we ask: How can we convert this first equation into the second equation $(x - 2)^2 = 12$? To do this, we first separate the variable terms from the constant term; that is, we write the equation as

$$x^2 - 4x = 8.$$

Next, we find a constant term, which when added to both sides of the equation makes the left side of the equation factorable into a perfect square.

By working backward from

$$x^2 - 4x + \quad \text{to} \quad (x - \quad)^2,$$

we see that the missing constant term is 4. (Why 4?) Thus the left side of the equation becomes

$$x^2 - 4x + 4,$$

a perfect square, which can be written in factored form as $(x - 2)^2$. Therefore, adding 4 to both sides of the equation

$$x^2 - 4x = 8,$$

the equation becomes

$$x^2 - 4x + 4 = 8 + 4$$

or

$$(x - 2)^2 = 12$$

The method just used is called **completing the square,** since we complete the square on the left side of the original equation by adding the appropriate constant term.

The method of completing the square is based on the fact that

$$x^2 + 2kx + k^2 = (x + k)^2.$$

Thus, if we start with an expression of the form $x^2 + 2kx$, we see from the preceding equation that we can "complete the square" by adding k^2. We obtain the term k^2 by squaring one-half the coefficient of the first-degree term x. That is, we complete the square by adding

$$\left[\frac{1}{2}(2k)\right]^2 = k^2,$$

so that

$$x^2 + 2kx + \left[\frac{1}{2}(2k)\right]^2 = x^2 + 2kx + k^2 = (x + k)^2.$$

coefficient of x term

For example, given $x^2 + 8x$, we can complete the square as follows:

$$x^2 + 8x + \left[\frac{1}{2}(8)\right]^2 = x^2 + 8x + 16 = (x + 4)^2.$$

coefficient of x term

EXAMPLE 5   Find the term necessary to complete the square of the given expression and write the result as the square of a binomial.

(a) $x^2 + 4x$ (b) $y^2 - 18y$ (c) $w^2 - 5w$ (d) $t^2 + 13t$

SOLUTION   In each case, we complete the square by adding the square of one-half the coefficient of the first-degree term.

(a) $x^2 + 4x + [\frac{1}{2}(4)]^2 = x^2 + 4x + 4 = (x + 2)^2$

(b) $y^2 - 18y + [\frac{1}{2}(-18)]^2 = y^2 - 18y + 81 = (y - 9)^2$

(c) $w^2 - 5w + [\frac{1}{2}(-5)]^2 = w^2 - 5w + \frac{25}{4} = (w - \frac{5}{2})^2$

(d) $t^2 + 13t + [\frac{1}{2}(13)]^2 = t^2 + 13t + \frac{169}{4} = (t + \frac{13}{2})^2$

Now we can combine the method of roots extraction and the procedure of completing the square to solve quadratic equations. For instance, to solve the equation

$$x^2 + 8x - 4 = 0,$$

we proceed as follows:

$$x^2 + 8x = 4 \qquad \text{(We added 4 to both sides.)}$$

$$x^2 + 8x + \left[\frac{1}{2}(8)\right]^2 = 4 + \left[\frac{1}{2}(8)\right]^2 \qquad \text{(We completed the square on the left side and added the same quantity to the right side.)}$$

$$x^2 + 8x + 16 = 4 + 16$$

$$(x + 4)^2 = 20.$$

By using the method of roots extraction, we have $x + 4 = \pm\sqrt{20}$, so that

$$
\begin{array}{c|c}
x + 4 = -\sqrt{20} & x + 4 = \sqrt{20} \\
x + 4 = -2\sqrt{5} & x + 4 = 2\sqrt{5} \\
x = -4 - 2\sqrt{5} & x = -4 + 2\sqrt{5}
\end{array}
$$

Therefore, the solutions are $-4 - 2\sqrt{5}$ and $-4 + 2\sqrt{5}$.

The following procedure summarizes the process of solving a quadratic equation $ax^2 + bx + c = 0$ by completing the square:

Procedure for Solving a Quadratic Equation by Completing the Square

Step 1. Write the equation in the equivalent form $ax^2 + bx = -c$.

Step 2. Divide both sides of the equation by the coefficient a of the x^2 term. (If $a = 1$, then proceed to step 3.)

Step 3. Complete the square on the left side of the equation by adding the square of one-half the coefficient b/a of the x term to both sides.

Step 4. Write the left side of the equation as the square of a binomial expression and solve the resulting equation by roots extraction.

In Examples 6–9, solve each equation by completing the square.

EXAMPLE 6 $x^2 + 4x - 2 = 0$

SOLUTION   Step 1. We add 2 to both sides of the equation:

$$x^2 + 4x = 2.$$

Step 2. We proceed to the next step, because the coefficient of x^2 is 1.

Step 3. We complete the square on the left side of the equation and we add the same quantity to the right side of the equation:

$$x^2 + 4x + \left[\frac{1}{2}(4)\right]^2 = 2 + \left[\frac{1}{2}(4)\right]^2$$

$$x^2 + 4x + 4 = 2 + 4$$

Step 4. We rewrite the equation as

$$(x + 2)^2 = 6,$$

and we solve the equation by the method of roots extraction, so that

$$x + 2 = \pm\sqrt{6}.$$

$$\begin{array}{c|c} x + 2 = -\sqrt{6} & x + 2 = \sqrt{6} \\ x = -2 - \sqrt{6} & x = -2 + \sqrt{6} \end{array}$$

Therefore, the solutions are $-2 - \sqrt{6}$ and $-2 + \sqrt{6}$.

Check For $x = -2 - \sqrt{6}$

$$(-2 - \sqrt{6})^2 + 4(-2 - \sqrt{6}) - 2 = 4 + 4\sqrt{6} + 6 - 8 - 4\sqrt{6} - 2 = 0.$$

For $x = -2 + \sqrt{6}$

$$(-2 + \sqrt{6})^2 + 4(-2 + \sqrt{6}) - 2 = 4 - 4\sqrt{6} + 6 - 8 + 4\sqrt{6} - 2 = 0.$$

EXAMPLE 7 $4y^2 - 2y - 3 = 0$

SOLUTION   Step 1. We add 3 to each side of the equation:

$$4y^2 - 2y = 3.$$

Step 2. We divide both sides of the equation by 4:

$$y^2 - \frac{2}{4}y = \frac{3}{4}$$

$$y^2 - \frac{1}{2}y = \frac{3}{4}.$$

Step 3. We complete the square on the left side of the equation and we add the same quantity to the right side of the equation:

$$y^2 - \frac{1}{2}y + \left[\frac{1}{2}\left(-\frac{1}{2}\right)\right]^2 = \frac{3}{4} + \left[\frac{1}{2}\left(-\frac{1}{2}\right)\right]^2$$

$$y^2 - \frac{1}{2}y + \frac{1}{16} = \frac{3}{4} + \frac{1}{16} = \frac{13}{16}.$$

Step 4. We rewrite the equation as

$$\left(y - \frac{1}{4}\right)^2 = \frac{13}{16},$$

and we solve the equation by the method of roots extraction, so that

$$y - \frac{1}{4} = \pm\sqrt{\frac{13}{16}}.$$

$$y - \frac{1}{4} = -\sqrt{\frac{13}{16}} \qquad \bigg| \qquad y - \frac{1}{4} = \sqrt{\frac{13}{16}}$$

$$y - \frac{1}{4} = \frac{-\sqrt{13}}{4} \qquad \bigg| \qquad y - \frac{1}{4} = \frac{\sqrt{13}}{4}$$

$$y = \frac{1}{4} - \frac{\sqrt{13}}{4} \qquad \bigg| \qquad y = \frac{1}{4} + \frac{\sqrt{13}}{4}$$

$$y = \frac{1 - \sqrt{13}}{4} \qquad \bigg| \qquad y = \frac{1 + \sqrt{13}}{4}$$

Therefore, the solutions are

$$\frac{1 - \sqrt{13}}{4} \quad \text{and} \quad \frac{1 + \sqrt{13}}{4}.$$

Check For $x = \dfrac{1 - \sqrt{13}}{4}$

$$4\left(\frac{1 - \sqrt{13}}{4}\right)^2 - 2\left(\frac{1 - \sqrt{13}}{4}\right) - 3 = \frac{4}{16}(1 - 2\sqrt{13} + 13) - \frac{2}{4}(1 - \sqrt{13}) - 3$$

$$= \frac{1}{4}(1 - 2\sqrt{13} + 13) - \frac{1}{2}(1 - \sqrt{13}) - 3$$

$$= \frac{1 - 2\sqrt{13} + 13 - 2(1 - \sqrt{13}) - 3(4)}{4}$$

$$= \frac{1 - 2\sqrt{13} + 13 - 2 + 2\sqrt{13} - 12}{4}$$

$$= 0$$

For $x = \dfrac{1 + \sqrt{13}}{4}$

$$4\left(\frac{1+\sqrt{13}}{4}\right)^2 - 2\left(\frac{1+\sqrt{13}}{4}\right) - 3 = \frac{4}{16}(1 + 2\sqrt{13} + 13) - \frac{2}{4}(1 + \sqrt{13}) - 3$$

$$= \frac{1}{4}(1 + 2\sqrt{13} + 13) - \frac{1}{2}(1 + \sqrt{13}) - 3$$

$$= \frac{1 + 2\sqrt{13} + 13 - 2(1 + \sqrt{13}) - 3(4)}{4}$$

$$= \frac{1 + 2\sqrt{13} + 13 - 2 - 2\sqrt{13} - 12}{4}$$

$$= 0.$$

EXAMPLE 8 $7w^2 - 4w - 1 = 0$

SOLUTION   Step 1. $7w^2 - 4w = 1$

Step 2. $w^2 - \frac{4}{7}w = \frac{1}{7}$

Step 3. $w^2 - \frac{4}{7}w + \left[\frac{1}{2}\left(-\frac{4}{7}\right)\right]^2 = \frac{1}{7} + \left[\frac{1}{2}\left(-\frac{4}{7}\right)\right]^2$

$\qquad\quad w^2 - \frac{4}{7}w + \frac{4}{49} = \frac{1}{7} + \frac{4}{49}$

$\qquad\quad w^2 - \frac{4}{7}w + \frac{4}{49} = \frac{11}{49}$

Step 4. $(w - \frac{2}{7})^2 = \frac{11}{49}$, so that $w - \frac{2}{7} = \pm\sqrt{\frac{11}{49}}$

$$w - \frac{2}{7} = -\sqrt{\frac{11}{49}} \qquad\qquad w - \frac{2}{7} = \sqrt{\frac{11}{49}}$$

$$w - \frac{2}{7} = \frac{-\sqrt{11}}{7} \qquad\qquad w - \frac{2}{7} = \frac{\sqrt{11}}{7}$$

$$w = \frac{2}{7} - \frac{\sqrt{11}}{7} \qquad\qquad w = \frac{2}{7} + \frac{\sqrt{11}}{7}$$

$$w = \frac{2 - \sqrt{11}}{7} \qquad\qquad w = \frac{2 + \sqrt{11}}{7}$$

Therefore, the solutions are

$$\frac{2 - \sqrt{11}}{7} \quad \text{and} \quad \frac{2 + \sqrt{11}}{7}.$$

EXAMPLE 9 $2r^2 + 3r + 2 = 0$

SOLUTION   Step 1. $2r^2 + 3r = -2$

Step 2. $r^2 + \frac{3}{2}r = -1$

Step 3. $r^2 + \frac{3}{2}r + [\frac{1}{2}(\frac{3}{2})]^2 = -1 + [\frac{1}{2}(\frac{3}{2})]^2$

$\qquad r^2 + \frac{3}{2}r + \frac{9}{16} = -1 + \frac{9}{16} = -\frac{7}{16}$

Step 4. $(r + \frac{3}{4})^2 = -\frac{7}{16}$, so that $r + \frac{3}{4} = \pm\sqrt{-\frac{7}{16}}$

$$r + \frac{3}{4} = -\sqrt{\frac{-7}{16}} \qquad\qquad r + \frac{3}{4} = \sqrt{\frac{-7}{16}}$$

$$r + \frac{3}{4} = -\frac{\sqrt{-7}}{4} \qquad\qquad r + \frac{3}{4} = \frac{\sqrt{-7}}{4}$$

$$r = -\frac{3}{4} - \frac{\sqrt{-7}}{4} \qquad\qquad r = -\frac{3}{4} + \frac{\sqrt{-7}}{4}$$

$$r = -\frac{3}{4} - \frac{\sqrt{7}}{4}i \qquad\qquad r = -\frac{3}{4} + \frac{\sqrt{7}}{4}i$$

Therefore, the solutions are

$$-\frac{3}{4} - \frac{\sqrt{7}}{4}i \qquad \text{and} \qquad -\frac{3}{4} + \frac{\sqrt{7}}{4}i.$$

Both solutions are complex, but we check each solution in the usual manner. Here is a check of the first solution. The check of the second solution is left for you to do (see problem 57).

Check For $r = -\dfrac{3}{4} - \dfrac{\sqrt{7}}{4}i$

$$2\left(-\frac{3}{4} - \frac{\sqrt{7}}{4}i\right)^2 + 3\left(-\frac{3}{4} - \frac{\sqrt{7}}{4}i\right) + 2$$

$$= \frac{2}{16}(-3 - \sqrt{7}i)^2 + \frac{3}{4}(-3 - \sqrt{7}i) + 2$$

$$= \frac{1}{8}(-3 - \sqrt{7}i)^2 + \frac{3}{4}(-3 - \sqrt{7}i) + 2$$

$$= \frac{9 + 6\sqrt{7}i + 7i^2 + 6(-3 - \sqrt{7}i) + 2(8)}{8}$$

$$= \frac{9 + 6\sqrt{7}i - 7 - 18 - 6\sqrt{7}i + 16}{8}$$

$$= 0.$$

EXAMPLE 10   A device for measuring the speed of the current in a stream (or river) is an open-ended, L-shaped tube as shown in Figure 1. **Torricelli's law** in physics tells us that the height h (in feet) that the water is pushed up into the tube above the surface is related to the water's speed s (in feet per second) by the equation

$$s^2 = 64h.$$

Figure 1

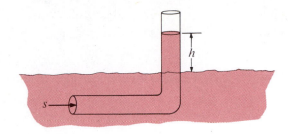

Find the speed of the current when the height is 0.5 feet.

SOLUTION   Substituting $h = 0.5$ into the equation

$$s^2 = 64h$$

we have

$$s^2 = 64(0.5) = 32$$
$$s = \pm\sqrt{32} = \pm 4\sqrt{2}.$$

Since the speed cannot be negative, we reject $-4\sqrt{2}$. Therefore the speed of the current is $4\sqrt{2}$ or approximately 5.7 feet per second.

PROBLEM SET 6.2

In problems 1–20, solve each equation by the roots extraction method.

1. $x^2 = 64$ **2.** $9x^2 = 4$ **3.** $2y^2 = 18$ **4.** $3t^2 = 75$

5. $4z^2 = 60$ **6.** $5u^2 = 75$ **7.** $(x-1)^2 = 9$ **8.** $(y+2)^2 = 25$

9. $(2m+1)^2 = 36$ **10.** $(2b-5)^2 = 11$ **11.** $(6x-5)^2 - 4 = 0$ **12.** $(8x-3)^2 - 49 = 0$

13. $(4y-3)^2 - 16 = 0$ **14.** $(3x-2)^2 - 81 = 0$ **15.** $w^2 + 81 = 0$ **16.** $4y^2 + 49 = 0$

17. $(u-6)^2 + 25 = 0$ **18.** $(t-3)^2 + 8 = 0$ **19.** $(3x-2)^2 + 49 = 0$ **20.** $(3y-7)^2 + 4 = 0$

In problems 21–30, find the term necessary to complete the square for each expression and write the result as the square of a binomial.

21. $x^2 + 6x$ **22.** $t^2 + 8t$ **23.** $y^2 - 10y$ **24.** $z^2 + 12z$ **25.** $m^2 + 20m$

26. $b^2 + 18b$ **27.** $x^2 + 7x$ **28.** $t^2 + 11t$ **29.** $c^2 + 17c$ **30.** $y^2 + 9y$

In problems 31–50, solve each equation by completing the square.

31. $x^2 + 2x - 3 = 0$ **32.** $x^2 + 4x - 7 = 0$ **33.** $y^2 - 12y - 17 = 0$

34. $t^2 - 6t - 11 = 0$ **35.** $3m^2 - 12m - 3 = 0$ **36.** $3u^2 - 18u - 7 = 0$

37. $5x^2 - 10x - 1 = 0$ **38.** $4y^2 - 8y + 3 = 0$ **39.** $25y^2 - 25y - 14 = 0$

40. $5z^2 + 10z - 3 = 0$ **41.** $5m^2 - 8m + 17 = 0$ **42.** $9x^2 - 27x + 14 = 0$

43. $4y^2 + 7y + 5 = 0$

44. $2p^2 + 7p + 3 = 0$

45. $3t^2 - 8t + 4 = 0$

46. $7m^2 + 5m + 1 = 0$

47. $9x^2 - 6x - 1 = 0$

48. $25x^2 - 50x + 21 = 0$

49. $16y^2 - 24y = -5$

50. $9t^2 - 30t = -21$

In problems 51–54, solve each equation for the indicated unknown.

51. $ax^2 + bx + 3b = 0$, for x.

52. $a^2y^2 - 3by + c^2 = 0$, for y.

53. $-gt^2 + vt = s$, for t.

54. $b^2t^2 - 7bt - 3a^2 = 0$, for t.

55. Check the solution $x = -3 + 2\sqrt{3}$ in the equation $x^2 + 6x - 3 = 0$.

56. Check the solution $x = (-5 - \sqrt{37})/2$ in the equation $x^2 + 5x - 12 = 0$.

57. Check the solution $r = -3/4 + (\sqrt{7}/4)i$ in the equation $2r^2 + 3r + 2 = 0$.

58. Check the solution $x = 4/3 - (\sqrt{5}/3)i$ in the equation $3x^2 - 8x + 7 = 0$.

59. Check the solution $x = 1/3 + (\sqrt{2}/3)i$ in the equation $3x^2 - 2x + 1 = 0$.

60. Check the solutions $y = -3/10 \pm (\sqrt{131}/10)i$ in the equation $5y^2 + 3y + 7 = 0$.

61. Find the speed of the current in Example 10 if $h = 0.75$ foot.

62. The pressure p in pounds per square foot of wind blowing at v miles per hour is given by the equation $p = 0.003v^2$. Find v if $p = 14.5$ pounds per square foot.

63. If P dollars is invested at r percent compounded annually, at the end of two years it will grow to the sum S, where S is given by the equation $S = P(1 + r)^2$. At what interest rate will \$1,000 grow to \$1,421 in 2 years?

6.3 Using the Quadratic Formula to Solve Quadratic Equations

The method of completing the square can be used to solve any quadratic equation. However, it is often more efficient to use what is called the **quadratic formula.** To derive this formula, we use the method of completing the square to solve, for x, the general quadratic equation

$$ax^2 + bx + c = 0$$

in which a, b, and c are constants and $a \neq 0$. We follow the procedure outlined in Section 6.2.

Step 1. We subtract c from each side of the equation:

$$ax^2 + bx = -c.$$

Step 2. We divide both sides by a:

$$x^2 + \frac{b}{a}x = -\frac{c}{a}.$$

Step 3. We complete the square of the left side by adding

$$\left[\frac{1}{2}\left(\frac{b}{a}\right)\right]^2 = \frac{b^2}{4a^2}$$

to both sides:

$$x^2 + \frac{b}{a}x + \frac{b^2}{4a^2} = \frac{b^2}{4a^2} - \frac{c}{a} = \frac{b^2 - 4ac}{4a^2}.$$

Step 4. We rewrite the equation as

$$\left(x + \frac{b}{2a}\right)^2 = \frac{b^2 - 4ac}{4a^2}.$$

We solve the equation by roots extraction:

$$x + \frac{b}{2a} = \sqrt{\frac{b^2 - 4ac}{4a^2}} \qquad\qquad x + \frac{b}{2a} = -\sqrt{\frac{b^2 - 4ac}{4a^2}}$$

$$x + \frac{b}{2a} = \frac{\sqrt{b^2 - 4ac}}{2a} \qquad\qquad x + \frac{b}{2a} = -\frac{\sqrt{b^2 - 4ac}}{2a}$$

$$x = -\frac{b}{2a} + \frac{\sqrt{b^2 - 4ac}}{2a} \qquad\qquad x = -\frac{b}{2a} - \frac{\sqrt{b^2 - 4ac}}{2a}$$

$$x = \frac{-b + \sqrt{b^2 - 4ac}}{2a} \qquad\qquad x = \frac{-b - \sqrt{b^2 - 4ac}}{2a}$$

These two solutions are usually written in the compact form shown in the following statement of the quadratic formula.

The Quadratic Formula

If $ax^2 + bx + c = 0$ and $a \neq 0$, then

$$x = \frac{-b \pm \sqrt{b^2 - 4ac}}{2a}.$$

Note that if $b^2 - 4ac$ is negative, the quadratic equation has no real roots—its roots are two complex numbers.

In Examples 1–4, use the quadratic formula to solve each equation.

EXAMPLE 1 $3x^2 + 4x - 4 = 0$

SOLUTION $a = 3$, $b = 4$, and $c = -4$. We substitute these values in the quadratic formula

$$x = \frac{-b \pm \sqrt{b^2 - 4ac}}{2a}$$

and we obtain

$$x = \frac{-4 \pm \sqrt{4^2 - 4(3)(-4)}}{2(3)} = \frac{-4 \pm \sqrt{64}}{6}.$$

Thus,

$$x = \frac{-4 + 8}{6} \qquad x = \frac{-4 - 8}{6}$$

$$x = \frac{4}{6} = \frac{2}{3} \qquad x = \frac{-12}{6} = -2$$

Therefore, the solutions are $\frac{2}{3}$ and -2.

EXAMPLE 2 $y^2 - 4y + 3 = -y^2 + 2y$

SOLUTION   First, we write the equation in standard form:

$$2y^2 - 6y + 3 = 0.$$

Now we substitute $a = 2$, $b = -6$, and $c = 3$ in the quadratic formula

$$y = \frac{-b \pm \sqrt{b^2 - 4ac}}{2a},$$

so that

$$y = \frac{-(-6) \pm \sqrt{(-6)^2 - 4(2)(3)}}{2(2)} = \frac{6 \pm \sqrt{12}}{4}$$

$$= \frac{6 \pm 2\sqrt{3}}{4} = \frac{2(3 \pm \sqrt{3})}{2(2)} = \frac{3 \pm \sqrt{3}}{2}.$$

Therefore, the solutions are

$$\frac{3 + \sqrt{3}}{2} \qquad \text{and} \qquad \frac{3 - \sqrt{3}}{2}.$$

EXAMPLE 3 $(t - 1)(t + 3) = -5$

SOLUTION   First, we write the equation in standard form:

$$(t - 1)(t + 3) = -5$$
$$t^2 + 2t - 3 = -5$$
$$t^2 + 2t + 2 = 0.$$

Now, we substitute $a = 1$, $b = 2$, and $c = 2$ in the quadratic formula

$$t = \frac{-b \pm \sqrt{b^2 - 4ac}}{2a},$$

so that

$$t = \frac{-2 \pm \sqrt{2^2 - 4(1)(2)}}{2(1)}$$

$$= \frac{-2 \pm \sqrt{-4}}{2}$$

$$= \frac{-2 \pm 2i}{2} \qquad \text{(since } \sqrt{-4} = \sqrt{4}i = 2i\text{)}$$

$$= \frac{2(-1 \pm i)}{2}$$

$$= -1 \pm i.$$

Therefore, the solutions are $-1 + i$ and $-1 - i$.

EXAMPLE 4   C Use the quadratic formula and a calculator to find approximate solutions of the equation $2.41x^2 - 12.3x - 8.39 = 0$. Round off the results to two decimal places.

SOLUTION $a = 2.41$, $b = -12.3$, and $c = -8.39$. We substitute these values in the quadratic formula:

$$x = \frac{-(-12.3) \pm \sqrt{(-12.3)^2 - 4(2.41)(-8.39)}}{2(2.41)}$$

$$= \frac{12.3 \pm \sqrt{232.1696}}{4.82}$$

$$= \frac{12.3 \pm 15.24}{4.82}.$$

Thus,

$$x = \frac{12.3 + 15.24}{4.82} \qquad \qquad x = \frac{12.3 - 15.24}{4.82}$$

$$x = 5.71 \qquad \qquad \qquad \quad x = -0.61$$

Therefore, the approximate solutions are 5.71 and -0.61.

The Quadratic Discriminant

The expression $b^2 - 4ac$, which appears under the radical sign in the quadratic formula

$$x = \frac{-b \pm \sqrt{b^2 - 4ac}}{2a},$$

is called the **discriminant** of the quadratic equation $ax^2 + bx + c = 0$. We can use the algebraic sign of the discriminant to determine the number and the kind of solutions to a quadratic equation:

> Case 1. If $b^2 - 4ac > 0$, the quadratic equation has two real and unequal solutions.
>
> Case 2. If $b^2 - 4ac = 0$, the quadratic equation has only one real solution, a *double* solution.
>
> Case 3. If $b^2 - 4ac < 0$, the quadratic equation has two complex solutions.

In Examples 5–7, use the discriminant to determine the number and the kind of solutions to each quadratic equation.

EXAMPLE 5 $2x^2 - 4x + 1 = 0$

SOLUTION $a = 2$, $b = -4$, and $c = 1$. Thus,

$$b^2 - 4ac = (-4)^2 - 4(2)(1)$$
$$= 8 > 0.$$

Therefore, the equation has two unequal real solutions.

EXAMPLE 6 $4x^2 - 28x + 49 = 0$

SOLUTION $a = 4$, $b = -28$, and $c = 49$. Thus,

$$b^2 - 4ac = (-28)^2 - 4(4)(49)$$
$$= 0.$$

Therefore, the equation has just one solution—a double solution—and this solution is a real number.

EXAMPLE 7 $2x^2 + 3x + 2 = 0$

SOLUTION $a = 2$, $b = 3$, and $c = 2$. Thus,

$$b^2 - 4ac = 3^2 - 4(2)(2) = 9 - 16 = -7 < 0.$$

Therefore, the equation has two complex solutions.

PROBLEM SET 6.3

In problems 1–32, use the quadratic formula to solve each equation. In problems 27–32, round off the answers to two decimal places.

1. $x^2 - 5x + 4 = 0$
2. $x^2 - 2x - 3 = 0$
3. $6y^2 - y - 1 = 0$

4. $4t^2 - 8t + 3 = 0$
5. $2u^2 - 5u - 3 = 0$
6. $4z^2 - 4z - 9 = 0$

7. $6y^2 - 7y - 5 = 0$
8. $6p^2 + 3p - 5 = 0$
9. $6t^2 - 8t - 3 = 0$

10. $6z^2 - 7z - 2 = 0$
11. $2x^2 - 5x + 1 = 0$
12. $6y^2 + 5y + 2 = 0$

13. $5y^2 - 2y + 7 = 0$
14. $9m^2 - 2m + 11 = 0$
15. $7x^2 - 8x + 3 = 0$

16. $12p^2 - 4p + 3 = 0$
17. $15y^2 + 2y - 8 = 0$
18. $4x^2 + 11x - 3 = 0$

19. $3x^2 - x + 7 = 0$
20. $6x^2 + 17x - 14 = 0$
21. $4y(y - 1) = 19$

22. $t(t + 8) + 8(t - 8) = 0$
23. $(t - 2)(t - 3) = 5$
24. $(3y - 1)(2y + 5) = 3$

25. $(y - 1)(y - 5) = 9$
26. $(t + 1)(2t - 1) = 4$
© **27.** $0.8w^2 - 0.16w - 1.91 = 0$

© **28.** $0.17u^2 - 0.55u - 3.87 = 0$
© **29.** $1.47y^2 - 3.82y - 5.71 = 0$
© **30.** $2.81t^2 - 7.14t - 31.72 = 0$

© **31.** $1.32x^2 + 2.78x - 9.31 = 0$
© **32.** $8.84x^2 - 71.41x - 94.03 = 0$

In problems 33–36, use the quadratic formula to solve each equation for the indicated unknown.

33. $nx^2 + mnx - m^2 = 0$, for x.
34. $2ay^2 - 7ay - 4ab^2 = 0$, for y.

35. $LI^2 + RI + \dfrac{1}{C} = 0$, for I.
36. $Mx^2 + 2Rx + K = 0$, for x.

In problems 37–46, use the discriminant to determine the number and the kind of solutions to each equation.

37. $x^2 + 6x - 7 = 0$
38. $6x^2 - x + 1 = 0$
39. $4y^2 + 12y + 9 = 0$
40. $16t^2 - 40t + 25 = 0$

41. $t^2 + 3t + 5 = 0$
42. $2u^2 + u + 5 = 0$
43. $3x^2 - 7x + 4 = 0$
44. $6y^2 + y - 2 = 0$

45. $9z^2 + 30z + 25 = 0$
46. $20x^2 - 11x + 3 = 0$

47. Find the values of k so that the equation $-3x^2 + 4x + k = 0$ will have two different real roots. (*Hint:* Let $b^2 - 4ac > 0$.)

48. Find the values of k so that the equation $kx^2 - 7x - 3 = 0$ will have two different real roots.

49. Find the values of k so that the equation $2x^2 + 7x + 3k = 0$ will have two complex roots. (*Hint:* Let $b^2 - 4ac < 0$.)

50. Find the values of k so that the equation $2kx^2 + 5x - 7 = 0$ will have two complex roots.

6.4 **Applications of Quadratic Equations**

We saw in Chapter 4 that many word problems can be expressed mathematically in the form of a linear equation, and hence be solved. In this section, we examine real-world applications that require solving quadratic equations. When the solutions to real-world problems involve quadratic equations, it is important to check both solutions in terms of the original problem to determine whether we must reject either solution.

EXAMPLE 1   Find two consecutive odd positive integers whose product is 195.

SOLUTION   Let

$$x = \text{the first odd integer.}$$

Then

$$x + 2 = \text{the next consecutive odd integer.}$$

The product of the two numbers is 195. Therefore, the equation is

$$x(x + 2) = 195$$
$$x^2 + 2x = 195 \qquad \text{or} \qquad x^2 + 2x - 195 = 0.$$

We solve this equation by factoring:

$$(x - 13)(x + 15) = 0.$$

$$
\begin{array}{c|c}
x - 13 = 0 & x + 15 = 0 \\
x = 13 & x = -15
\end{array}
$$

We reject the solution -15 because it is not a positive integer. Therefore, the numbers are 13 and $13 + 2 = 15$.

Check The numbers 13 and 15 are consecutive odd positive integers and $13 \cdot 15 = 195$.

EXAMPLE 2   A rectangular garden 20 meters wide and 60 meters long is surrounded by a walkway of uniform width. If the total area of the walkway is 516 square meters, how wide is the walkway?

SOLUTION   Since the walkway has a uniform width around the garden (Figure 1), we let the width be x meters. The length of the outer rectangle is $x + 60 + x = 2x + 60$, and the width of the outer rectangle is $x + 20 + x = 2x + 20$.

Figure 1

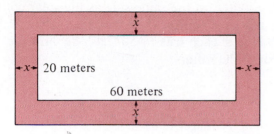

20 meters

60 meters

Therefore, the total area of the outer rectangle is given by

$$(2x + 60)(2x + 20).$$

The area of the inner rectangle is $(60)(20) = 1200$ square meters. Therefore, the total area of the outer rectangle is $516 + 1200 = 1716$ square meters. Thus,

$$(2x + 60)(2x + 20) = 1716$$

$$4x^2 + 160x + 1200 = 1716$$

$$4x^2 + 160x - 516 = 0$$

$$x^2 + 40x - 129 = 0.$$

We solve this equation by factoring:

$$(x - 3)(x + 43) = 0.$$

$$x - 3 = 0 \quad | \quad x + 43 = 0$$

$$x = 3 \quad | \quad x = -43$$

Since the width must be positive, 3 is the only solution. Therefore, the width of the walkway is 3 meters.

Check $(6 + 60)(6 + 20) = (66)(26) = 1716.$

We learned in geometry that the Pythagorean theorem expresses the following property for a right triangle (Figure 2):

$$c^2 = a^2 + b^2.$$

Figure 2

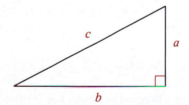

Use this theorem to solve the following examples.

EXAMPLE 3   In order to support a solar collector at the correct angle, the roof trusses of a house are designed as right triangles. Rafters form the right angle, and

Figure 3

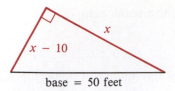

base = 50 feet

the base of the truss is the hypotenuse (Figure 3). If the rafter on the same side as the solar collector is 10 feet shorter than the other rafter, and if the base of each truss is 50 feet long, how long is each of the rafters?

SOLUTION   Let

$$x = \text{the length of the longer rafter in feet.}$$

Then

$$x - 10 = \text{the length of the rafter on the side of the collector.}$$

We use the Pythagorean theorem:

$$x^2 + (x - 10)^2 = 50^2$$
$$x^2 + x^2 - 20x + 100 = 2{,}500$$
$$2x^2 - 20x - 2{,}400 = 0$$
$$x^2 - 10x - 1{,}200 = 0$$
$$(x + 30)(x - 40) = 0.$$

$x + 30 = 0$	$x - 40 = 0$
$x = -30$	$x = 40$

Because the length must be positive, the lengths of the rafters are 40 feet and $40 - 10 = 30$ feet.

EXAMPLE 4   A rocket is fired straight up from the ground at a time $t = 0$ with an initial speed of 560 feet per second. The rocket's distance of h feet above the ground t seconds later (Figure 4) is given by the equation

$$h = 560t - 16t^2.$$

At what time will the rocket be 3,136 feet above the ground?

SOLUTION   When the rocket is 3,136 feet above the ground, $h = 3{,}136$, and the equation becomes

$$3{,}136 = 560t - 16t^2$$

or

$$16t^2 - 560t + 3{,}136 = 0$$
$$t^2 - 35t + 196 = 0.$$

Figure 4

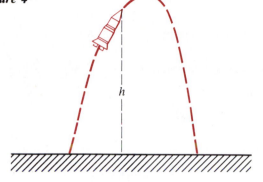

We solve this equation by factoring:

$$(t - 28)(t - 7) = 0.$$

$t - 28 = 0$	$t - 7 = 0$
$t = 28$	$t = 7$

Both solutions make sense, because the rocket can be at a given height at two different times: once on its way up and once on its way down. Therefore, the rocket will be 3,136 feet above the ground after 7 seconds and after 28 seconds.

Check At $t = 7$,

$$560(7) - 16(7)^2 = 3,136.$$

At $t = 28$,

$$560(28) - 16(28)^2 = 3,136.$$

EXAMPLE 5   Carlos and Pedro both ran 26 miles in the Boston Marathon. Carlos averaged 2 miles per hour less than Pedro. If Carlos finished 26 minutes behind Pedro, what was Pedro's average speed?

SOLUTION   Let

$$x = \text{Pedro's average speed in miles per hour.}$$

Then

$$x - 2 = \text{Carlos's average speed in miles per hour.}$$

We use the formula $d = rt$ (distance = rate × time) and solve for t:

$$t = \frac{d}{r}.$$

The following table summarizes the given information:

	Average Speed r (in miles per hour)	Distance d (in miles)	Time t (in hours)
Pedro	x	26	$\dfrac{26}{x}$
Carlos	$x - 2$	26	$\dfrac{26}{x - 2}$

We are told that

$$\text{Carlos's time} - \text{Pedro's time} = 26 \text{ minutes} = \frac{26}{60} \text{ of an hour.}$$

The following equation expresses this relationship:

$$\frac{26}{x - 2} - \frac{26}{x} = \frac{26}{60}$$

or

$$\frac{1}{x - 2} - \frac{1}{x} = \frac{1}{60}.$$

We multiply both sides of the equation by $60x(x - 2)$, the LCD, and we solve the resulting equation by factoring:

$$60x(x - 2)\left(\frac{1}{x - 2} - \frac{1}{x}\right) = 60x(x - 2)\left(\frac{1}{60}\right)$$

$$60x - 60(x - 2) = x(x - 2)$$

$$x^2 - 2x - 120 = 0$$

$$(x - 12)(x + 10) = 0.$$

$$x - 12 = 0 \quad \bigg| \quad x + 10 = 0$$

$$x = 12 \quad \bigg| \quad x = -10$$

We reject $x = -10$ because speed cannot be negative. Therefore, Pedro's average speed is 12 miles per hour.

Check

$$\text{Pedro's time} = \frac{d}{r} = \frac{26}{12} = \frac{13}{6} \text{ hours}$$

$$\text{Carlos's time} = \frac{d}{r} = \frac{26}{10} = \frac{13}{5} \text{ hours}$$

Carlos finishes $\frac{13}{5} - \frac{13}{6} = \frac{13}{30} = \frac{26}{60}$ hour behind Pedro.

EXAMPLE 6   It takes a crew $1\frac{1}{2}$ hours to complete a round trip, rowing 10 kilometers with the current and 10 kilometers against the current. If the rate of the current is 5 kilometers per hour, find the rate at which the crew can row in still water.

SOLUTION   Let x be the rate (in kilometers per hour) at which the crew can row in still water. Then $x + 5$ is the rate (in kilometers per hour) going with the current, and $x - 5$ is the rate (in kilometers per hour) against the current. We use the formula $d = rt$ (distance = rate × time). The following table summarizes the given information:

	r	d	$t = \dfrac{d}{r}$
With the current	$x + 5$	10	$\dfrac{10}{x + 5}$
Against the current	$x - 5$	10	$\dfrac{10}{x - 5}$

Since the total time is $1\frac{1}{2}$ hours, we have the following relationship:

$$\frac{10}{x + 5} + \frac{10}{x - 5} = \frac{3}{2}$$

We multiply both sides of the equation by $2(x + 5)(x - 5)$, the LCD, and solve the resulting equation by factoring:

$$20(x - 5) + 20(x + 5) = 3(x - 5)(x + 5)$$
$$20x - 100 + 20x + 100 = 3x^2 - 75$$
$$3x^2 - 40x - 75 = 0$$
$$(x - 15)(3x + 5) = 0.$$

$$x - 15 = 0 \qquad\qquad 3x + 5 = 0$$
$$x = 15 \qquad\qquad 3x = -5$$
$$x = -\frac{5}{3}$$

We reject $x = -\frac{5}{3}$ because the rate cannot be negative. Therefore, the crew can row in still water at the rate of 15 kilometers per hour.

Check The time it takes the crew to row with the current is

$$t = \frac{d}{r} = \frac{10}{15 + 5} = \frac{10}{20} = \frac{1}{2} \text{ hour.}$$

The time it takes the crew to row against the current is

$$t = \frac{d}{r} = \frac{10}{15 - 5} = \frac{10}{10} = 1 \text{ hour.}$$

PROBLEM SET 6.4

In many of the following problems a calculator may be useful to speed up the arithmetic.

1. Find two consecutive positive integers whose product is 30.
2. Find two consecutive negative integers whose product is 56.

3. Two brothers were born in consecutive years. The product of their present ages is 156. How old are they now?

4. Maria throws two dice. The difference between the face values of the two dice is 2, and the product of their values is 24. What numbers did Maria throw?

5. If five times David's age plus 6 years is the same as the square of his age, how old is David?

6. Tom is 4 years older than Kim, and the product of their ages is three times what the product of their ages was 4 years ago. Find their present ages.

7. What are the dimensions of a rectangle if the area of the rectangle is 60 square feet and the perimeter is 32 feet?

8. A candy manufacturer makes rectangular chocolate bars in such a way that the length of each bar is four times the width. If the length of each bar is increased by 3 centimeters and the width is increased by 1 centimeter, the new area of each bar becomes 33 square centimeters. What were the original dimensions of the candy bar?

9. The area of a rectangular garden is 1,320 square feet. If the perimeter is 148 feet, what are the dimensions of the garden?

10. If increasing the lengths of the sides of a square by 8 feet results in a square whose area is 25 times the area of the original square, what was the length of a side of the original square?

11. A skating rink is 100 feet long and 70 feet wide. If we want to increase the area of the rink to 13,000 square feet by adding rectangular strips of equal width to one side and one end, maintaining the rectangular shape of the rink, how wide should these strips be?

12. A rectangular garden that is 4 meters wide and 6 meters long is surrounded by a path of uniform width. If the area of the path equals the area of the garden, what is the width of the path?

13. A lawn 30 feet long and 20 feet wide has a flower border of uniform width around it. If the area of the lawn and the area of the border are the same, find the width of the border.

14. A box without a top is to be constructed from a square piece of tin by removing a 3-inch square from each corner of the square, and bending up the sides. If the box is to have a volume of 300 cubic inches, what is the length of the side of the original tin square?

In problems 15 and 16, find the dimensions of each right triangle in Figures 5 and 6.

15. *Figure 5*

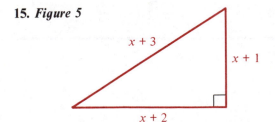

16. *Figure 6*

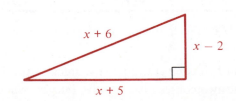

17. In order to support a solar collector at the correct angle, the roof trusses of a building are designed as right triangles. Rafters form the right angle, and the base of the truss is the hypotenuse (see Figure 3 on page 298). If the rafter on the same side as the solar collector is 7 meters shorter than the other rafter, and if the base of each truss is 13 meters long, how long is each of the rafters?

18. Two bicyclists, each moving at a uniform speed, leave the same point and travel at right angles to each other. One travels 5 kilometers per hour faster than the other. In 2 hours the bicyclists are 50 kilometers apart. How fast does each bicyclist travel?

19. Two joggers leave the same point and travel at right angles to each other, each jogging at a uniform speed. One jogs 7 miles per hour faster than the other. After 1 hour they are 13 miles apart. Find the speed of each jogger.

20. Two planes leave the same point and fly at right angles to each other. After 2 hours they are 1,000 miles apart. If one plane flies 100 miles per hour faster than the other, and each is moving at a uniform speed, how fast are the planes flying?

21. A ball thrown straight upward from ground level with an initial velocity of 128 feet per second will be at a height $h = 128t - 16t^2$, t seconds later (neglecting air resistance). How long will it take the ball to reach a height of 112 feet?

22. A rocket is fired upward from the ground at an initial velocity of 120 meters per second. The rocket's height h in meters above the ground t seconds later is given by the equation $h = 120t - 4.9t^2$. In how many seconds will the rocket reach a height of 176 meters?

23. A rocket is fired upward from the ground at an initial velocity of 480 feet per second. Its height h in feet above the ground t seconds later is $h = 480t - 16t^2$. In how many seconds will the rocket reach a height of 2,000 feet ?

24. Howard leaves a ski lodge and accelerates as he skis downhill. After t seconds he has traveled a distance of h meters—h is given by the equation $h = 10t^2 + 10t$. How long does it take him to go 560 meters downhill?

25. In a 300-mile car race, Mario averages 20 miles per hour less than his competitor Andy. If Mario finishes 30 minutes behind Andy, what is Andy's average speed?

26. A carpenter, working alone, can finish a job in 2 hours less time than his assistant. They can do the job in 7 hours working together. Find the time that each requires to do the job alone.

27. Susan and June, working together, can type a chapter of a manuscript in 8 hours. Typing alone, it would take June 12 hours longer than it would take Susan to type the chapter. How long would it take Susan to type the chapter by herself?

28. A motorist travels for 20 miles at one speed, and then increases her speed by 20 miles per hour for the next 30 miles. How fast was she traveling originally if the total trip took 1 hour?

29. A company determines that its total monthly sales revenue (in dollars) is given by the equation $R = 200x - \frac{1}{3}x^2$, in which x is the price (in dollars) of each unit sold. At what price per unit will the revenue for the month be $10,800, if the price per unit must be greater than $60?

30. A cable television company plans to begin operations in a small town. The company foresees that 1,000 people will subscribe to the service if the price per subscriber is $3 per month, but that for each 10-cent decrease in the monthly subscription price 50 more people will subscribe. The company begins operations, and its total revenue for the first month is $3,125. How much did the company charge each subscriber?

31. A crew can row 9 kilometers downstream and then back upstream to their starting point in 10 hours. If the rate of the stream is 4 kilometers per hour, find the rate at which the crew can row in still water.

6.5 Equations in Quadratic Form

An equation that is not quadratic may have the following characteristics: If you substitute a letter (for example, u) for an appropriate expression containing the unknown, the resulting equation is a quadratic equation in u. The original equation is called *an equation that is quadratic in form.* Examples of such equations are:

$$3x^4 + 2x^2 - 1 = 0 \quad (u = x^2),$$

$$(x^2 - 1)^2 + 2(x^2 - 1) - 3 = 0 \quad (u = x^2 - 1),$$

and

$$t^2 + t + \frac{12}{t^2 + t} = 8 \quad (u = t^2 + t).$$

To solve an equation that is quadratic in form, we make a substitution that transforms the given equation into a quadratic equation in another variable, such as u.

In Examples 1–4, solve each equation.

EXAMPLE 1 $x^4 - 5x^2 + 4 = 0$

SOLUTION   First we write $x^4 - 5x^2 + 4 = 0$ as

$$(x^2)^2 - 5x^2 + 4 = 0.$$

If we let $u = x^2$, we have

$$u^2 - 5u + 4 = 0.$$

We solve this equation by factoring:

$$(u - 1)(u - 4) = 0.$$

$$u - 1 = 0 \quad | \quad u - 4 = 0$$
$$u = 1 \quad | \quad u = 4$$

If $u = 1$, then $x^2 = 1$; therefore,

$$x = -\sqrt{1} \quad | \quad x = \sqrt{1}$$
$$x = -1 \quad | \quad x = 1$$

If $u = 4$, then $x^2 = 4$; therefore,

$$x = -\sqrt{4} \quad | \quad x = \sqrt{4}$$
$$x = -2 \quad | \quad x = 2$$

Thus, -1, 1, -2, and 2 are the solutions.

EXAMPLE 2 $t^{-2} + t^{-1} - 6 = 0$

SOLUTION   First we write $t^{-2} + t^{-1} - 6 = 0$ as

$$(t^{-1})^2 + t^{-1} - 6 = 0.$$

If we let $u = t^{-1}$, we have

$$u^2 + u - 6 = 0.$$

We solve this equation by factoring:

$$(u + 3)(u - 2) = 0.$$

$$u + 3 = 0 \quad | \quad u - 2 = 0$$
$$u = -3 \quad | \quad u = 2$$

If $u = -3$, then $t^{-1} = -3$; therefore,

$$\frac{1}{t} = -3 \quad \text{or} \quad t = -\frac{1}{3}.$$

If $u = 2$, then $t^{-1} = 2$; therefore,

$$\frac{1}{t} = 2 \quad \text{or} \quad t = \frac{1}{2}.$$

Thus, $-\frac{1}{3}$ and $\frac{1}{2}$ are the solutions.

EXAMPLE 3 $3(y + 4)^2 - (y + 4) - 14 = 0$

SOLUTION   If we substitute u for $y + 4$, the equation becomes

$$3u^2 - u - 14 = 0.$$

We solve this equation by factoring:

$$(3u - 7)(u + 2) = 0.$$

$$
\begin{array}{c|c}
3u - 7 = 0 & u + 2 = 0 \\
3u = 7 & u = -2 \\
u = \dfrac{7}{3} &
\end{array}
$$

If $u = \frac{7}{3}$, then $y + 4 = \frac{7}{3}$ and $y = -\frac{5}{3}$. If $u = -2$, then $y + 4 = -2$ and $y = -6$. Thus, $-\frac{5}{3}$ and -6 are the solutions.

EXAMPLE 4 $\left(r - \dfrac{8}{r} \right)^2 + \left(r - \dfrac{8}{r} \right) = 42$

SOLUTION   If we substitute u for $r - 8/r$, the equation becomes

$$u^2 + u = 42 \qquad \text{or} \qquad u^2 + u - 42 = 0.$$

We solve this equation by factoring:

$$(u + 7)(u - 6) = 0.$$

$$
\begin{array}{c|c}
u + 7 = 0 & u - 6 = 0 \\
u = -7 & u = 6
\end{array}
$$

If $u = -7$, then

$$r - \dfrac{8}{r} = -7$$

$$r^2 - 8 = -7r$$

$$r^2 + 7r - 8 = 0.$$

We solve this equation by factoring:

$$(r + 8)(r - 1) = 0.$$

$$
\begin{array}{c|c}
r + 8 = 0 & r - 1 = 0 \\
r = -8 & r = 1
\end{array}
$$

If $u = 6$, then

$$r - \dfrac{8}{r} = 6$$

$$r^2 - 6r - 8 = 0.$$

We solve this equation by using the quadratic formula:

$$r = \frac{6 \pm \sqrt{36 - 4(1)(-8)}}{2} = \frac{6 \pm \sqrt{68}}{2} = 3 \pm \sqrt{17}.$$

Thus, -8, 1, $3 + \sqrt{17}$, and $3 - \sqrt{17}$ are the solutions.

Since the original equation contains a variable in a denominator, you should check the four proposed solutions. Following is a check for one of these solutions.

Check For $r = 3 + \sqrt{17}$, we have the expression

$$r - \frac{8}{r} = 3 + \sqrt{17} - \frac{8}{3 + \sqrt{17}}$$

$$= \frac{(3 + \sqrt{17})^2 - 8}{3 + \sqrt{17}}$$

$$= \frac{9 + 6\sqrt{17} + 17 - 8}{3 + \sqrt{17}}$$

$$= \frac{18 + 6\sqrt{17}}{3 + \sqrt{17}} = \frac{6(3 + \sqrt{17})}{3 + \sqrt{17}} = 6.$$

Now, substituting this value for $r - 8/r$ in the equation

$$\left(r - \frac{8}{r}\right)^2 + \left(r - \frac{8}{r}\right) = 42,$$

we have

$$6^2 + 6 = 42$$

$$36 + 6 = 42$$

$$42 = 42.$$

Thus, $3 + \sqrt{17}$ is a solution.

PROBLEM SET 6.5

In problems 1–32, reduce each equation to a quadratic form by using an appropriate substitution, and solve the equation for the original variable.

1. $x^4 - 13x^2 + 36 = 0$, $u = x^2$

2. $y^4 - 17y^2 + 16 = 0$, $u = y^2$

3. $t^4 - 3t^2 - 4 = 0$

4. $z^4 - 10z^2 + 9 = 0$

5. $y^4 - 29y^2 + 100 = 0$

6. $9t^4 - 226t^2 + 25 = 0$

7. $t^{-2} - 2t^{-1} - 8 = 0$, $u = t^{-1}$

8. $6x^{-2} + 5x^{-1} - 4 = 0$, $u = x^{-1}$

9. $6y^{-2} + 13y^{-1} - 5 = 0$

10. $28p^{-2} - 17p^{-1} - 3 = 0$

11. $x^{-4} - 9x^{-2} + 20 = 0$, $u = x^{-2}$

12. $y^{-4} + 63y^{-2} - 64 = 0$, $u = y^{-2}$

13. $t^{-4} - 20t^{-2} + 64 = 0$

14. $w^{-4} - 34w^{-2} + 225 = 0$

15. $(x^2 + 1)^2 - 3(x^2 + 1) + 2 = 0$, $u = x^2 + 1$

16. $(2t^2 + 7t)^2 - 3(2t^2 + 7t) = 10$, $u = 2t^2 + 7t$

17. $(y^2 + 2y)^2 - 2(y^2 + 2y) = 3$

18. $(w^2 - w)^2 + 12 = 8(w^2 - w)$

19. $(w^2 + 2w)^2 - 14(w^2 + 2w) = 15$

20. $(2y^2 - y)^2 - 16(2y^2 - y) + 60 = 0$

21. $(m^2 + 2m)^2 + m^2 + 2m - 12 = 0$

22. $3(x + 1)^2 + 2(x + 1) = 2$

23. $(2p^2 + p + 4)^2 - 4(2p^2 + p + 4) = 5$

24. $(2w^2 - 3w)^2 - 2(2w^2 - 3w) = 3$

25. $\left(3x - \dfrac{2}{x}\right)^2 + 6\left(3x - \dfrac{2}{x}\right) + 5 = 0$, $u = 3x - \dfrac{2}{x}$

26. $\left(y - \dfrac{5}{y}\right)^2 - 2y + \dfrac{10}{y} = 8$, $u = y - \dfrac{5}{y}$

27. $t^2 + t + \dfrac{36}{t^2 + t} = 15$

28. $\dfrac{w^2 + 1}{w} + \dfrac{4w}{w^2 + 1} - 4 = 0$

29. $\dfrac{m + 1}{m} + 2 = \dfrac{3m}{m + 1}$

30. $\dfrac{p^2}{p + 1} + \dfrac{2(p + 1)}{p^2} = 3$

31. $\left(2x + \dfrac{1}{x}\right)^2 + 5\left(2x + \dfrac{1}{x}\right) + 6 = 0$

32. $\left(y + \dfrac{3}{y}\right)^2 - 2y - \dfrac{6}{y} = 15$

6.6 Equations Involving Radicals

The following equations are examples of **radical equations:**

$$\sqrt{x} = 3, \qquad \sqrt{5t + 1} = 4, \qquad \sqrt[3]{5y + 2} = 3, \qquad \text{and} \qquad u - \sqrt{u - 2} = 4.$$

To solve a radical equation, we use the following procedure:

Procedure for Solving a Radical Equation

> **Step 1.** Isolate one of the radical expressions containing the variable on one side of the equation.
>
> **Step 2.** Eliminate the radical by raising both sides of the equation to a power equal to the index of the radical. [Recall that $(\sqrt[n]{a})^n = a$.] You may have to repeat this technique in order to eliminate all radicals. When the equation is free of radicals, simplify and solve the equation.
>
> **Step 3.** Check all solutions in the original equation whenever a radical with an *even* index is involved.

It is important to understand that step 3 of this procedure is not an optional step. Wherever an even index is involved, it is necessary to determine which of the solutions actually satisfies the original radical equation. We never lose solutions to a radical equation by applying the property $(\sqrt[n]{a})^n = a$, but when n is even, we may introduce extraneous solutions. Extraneous solutions satisfy the equation obtained in step 2 but do not satisfy the original equation.

Extraneous solutions are introduced when we raise opposites to an even power. For example, -2 and 2 are opposites and $-2 \neq 2$, but $(-2)^2 = 2^2$. This situation does not occur when we raise opposites to an odd power. For example, $(-2)^3 \neq (2)^3$.

In Examples 1–6, solve each equation.

EXAMPLE 1 $\sqrt{2x + 5} - 3 = 0$

SOLUTION

Step 1. We isolate the radical expression on one side of the equation by adding 3 to both sides:

$$\sqrt{2x + 5} = 3.$$

Step 2. We eliminate the radical by squaring both sides of the equation:

$$(\sqrt{2x + 5})^2 = 3^2$$
$$2x + 5 = 9$$
$$2x = 4$$
$$x = 2$$

Step 3. Since the index is even, we must check $x = 2$:

$$\sqrt{2(2) + 5} - 3 = \sqrt{4 + 5} - 3$$
$$= \sqrt{9} - 3$$
$$= 0.$$

Hence, 2 is the solution.

EXAMPLE 2 $\sqrt[3]{3t - 1} - 2 = 0$

SOLUTION

Step 1. We isolate the radical expression by adding 2 to both sides of the equation:

$$\sqrt[3]{3t - 1} = 2.$$

Step 2. We eliminate the radical by raising both sides of the equation to the third power:

$$(\sqrt[3]{3t - 1})^3 = 2^3$$
$$3t - 1 = 8$$
$$3t = 9$$
$$t = 3.$$

Step 3. It is not necessary to check for extraneous solutions because the index is odd. Hence, 3 is the solution.

EXAMPLE 3 $\sqrt[4]{y^2 - 5y + 6} = \sqrt{y - 2}$

SOLUTION   We raise both sides of the equation to the fourth power:

$$(\sqrt[4]{y^2 - 5y + 6})^4 = (\sqrt{y - 2})^4$$
$$y^2 - 5y + 6 = (y - 2)^2$$
$$y^2 - 5y + 6 = y^2 - 4y + 4$$
$$-5y + 4y = 4 - 6$$
$$-y = -2.$$
$$y = 2.$$

Check For $y = 2$,

$$\sqrt[4]{4 - 10 + 6} = \sqrt{2 - 2}$$
$$\sqrt[4]{0} = \sqrt{0}$$
$$0 = 0.$$

Therefore, 2 is the solution.

EXAMPLE 4 $\sqrt{4w^2 - 3} = 2w + 1$

SOLUTION   We square both sides of the equation:

$$(\sqrt{4w^2 - 3})^2 = (2w + 1)^2$$
$$4w^2 - 3 = 4w^2 + 4w + 1$$
$$-4w = 1 + 3$$
$$-4w = 4$$
$$w = -1.$$

Check For $w = -1$,

$$\sqrt{4(-1)^2 - 3} = 2(-1) + 1$$
$$\sqrt{4 - 3} = -2 + 1$$
$$1 \neq -1.$$

The solution, $w = -1$, does not satisfy the original equation. Therefore, -1 is an **extraneous** solution. The original equation has no solutions.

This example illustrates how extraneous solutions are introduced. Note that for $w = -1$, $\sqrt{4w^2 - 3} = 1$ and $2w + 1 = -1$. Hence, when we squared both sides of the equation, we were in effect saying $(+1)^2 = (-1)^2$, which is true.

EXAMPLE 5 $\sqrt{z - 2} + 4 = z$

SOLUTION   We follow the procedure on page 308:

$$\sqrt{z - 2} = z - 4$$
$$(\sqrt{z - 2})^2 = (z - 4)^2$$
$$z - 2 = z^2 - 8z + 16$$
$$z^2 - 9z + 18 = 0$$
$$(z - 3)(z - 6) = 0.$$

$$z - 3 = 0 \quad | \quad z - 6 = 0$$
$$z = 3 \quad | \quad z = 6$$

Check For $z = 3$,

$$\sqrt{3 - 2} + 4 = 1 + 4 = 5 \neq 3.$$

For $z = 6$,

$$\sqrt{6 - 2} + 4 = \sqrt{4} + 4 = 2 + 4 = 6.$$

Therefore, 3 is an extraneous solution. The only solution that satisfies the original equation is 6.

EXAMPLE 6 $\sqrt{1 - 5x} + \sqrt{1 - x} = 2$

SOLUTION   We add $-\sqrt{1 - x}$ to both sides of the equation to isolate $\sqrt{1 - 5x}$ on the left side:

$$\sqrt{1 - 5x} = 2 - \sqrt{1 - x}.$$

We square both sides of the equation:

$$(\sqrt{1-5x})^2 = (2 - \sqrt{1-x})^2$$
$$1 - 5x = 4 - 4\sqrt{1-x} + 1 - x.$$

The equation still contains a radical, so we simplify and isolate this radical:

$$-4 - 4x = -4\sqrt{1-x}$$
$$1 + x = \sqrt{1-x}. \qquad \text{(We divided both sides by } -4.)$$

Again, we square both sides of the equation:

$$(1+x)^2 = (\sqrt{1-x})^2$$
$$1 + 2x + x^2 = 1 - x$$
$$x^2 + 3x = 0$$
$$x(x+3) = 0.$$
$$x = 0 \quad \Big| \quad x + 3 = 0$$
$$x = -3$$

Check For $x = -3$,

$$\sqrt{1 - 5(-3)} + \sqrt{1 - (-3)} = \sqrt{16} + \sqrt{4}$$
$$= 4 + 2 = 6 \neq 2.$$

For $x = 0$,

$$\sqrt{1 - 0} + \sqrt{1 - 0} = 1 + 1 = 2.$$

Therefore, -3 is an extraneous solution and 0 is the only solution.

PROBLEM SET 6.6

In problems 1–26, solve each equation (find only the real solutions) and check the solution whenever the equation has a radical with an even index.

1. $\sqrt{x} - 2 = 3$ **2.** $\sqrt{x} + 1 = 4$ **3.** $\sqrt{t+1} - 2 = 0$

4. $\sqrt{y-3} - 5 = 0$ **5.** $\sqrt{2w+5} - 4 = 0$ **6.** $\sqrt{6t-3} - 27 = 0$

7. $8 - \sqrt{y-1} = 6$ **8.** $2 + \sqrt{7m-5} = 6$ **9.** $\sqrt{11-x} = \sqrt{x+6}$

10. $\sqrt{3p+1} = \sqrt{p+1}$ **11.** $\sqrt{q+14} = \sqrt{5q}$ **12.** $\sqrt{y+5} - \sqrt{y} = 0$

13. $\sqrt{x^2+3x} = x+1$ **14.** $\sqrt{9x^2-7} - 3x = 2$ **15.** $\sqrt[3]{y-3} = 3$

16. $\sqrt[3]{t+2} = -2$ **17.** $\sqrt[3]{3b-4} = 2$ **18.** $\sqrt[4]{y-1} = 2$

19. $\sqrt[4]{y^2-7y+1} = \sqrt{y-5}$ **20.** $\sqrt[4]{t^2+1} = \sqrt{t+1}$ **21.** $5 + \sqrt[4]{x-5} = 0$

22. $\sqrt[4]{2x-1} + 3 = 0$ **23.** $\sqrt[4]{t+8} = \sqrt{3t}$ **24.** $\sqrt[3]{x^2+2x-6} = \sqrt[3]{x^2}$

25. $2 + \sqrt[4]{7x-5} = 6$ **26.** $8 + \sqrt[4]{x-1} = 3$

In problems 27–44, solve each equation and check the solution.

27. $\sqrt{t} = \sqrt{t + 16} - 2$ 28. $\sqrt{2y^2 + 4} + 2 = 2y$ 29. $\sqrt{p + 12} = 2 + \sqrt{p}$

30. $\sqrt{m^2 + 6m} = m + \sqrt{2m}$ 31. $\sqrt{y + 5} = \sqrt{y} + 1$ 32. $\sqrt{x + 7} = 5 + \sqrt{x - 2}$

33. $\sqrt{5t + 1} = 1 + \sqrt{3t}$ 34. $\sqrt{2w + 1} = 1 + 2\sqrt{w}$ 35. $\sqrt{2x - 5} = \sqrt{x - 2} + 2$

36. $\sqrt{3 - t} - \sqrt{2 + t} = 3$ 37. $\sqrt{t + 2} + \sqrt{t - 3} = 5$ 38. $\sqrt{3y + 1} - 1 = \sqrt{3y - 8}$

39. $\sqrt{m + 4} + 1 = \sqrt{m + 11}$ 40. $\sqrt{6z + 7} = \sqrt{3z + 3} + 1$ 41. $2\sqrt{1 - 3y} = 2 - \sqrt{2 - 4y}$

42. $\sqrt{7x - 6} = \sqrt{7x + 22} - 2$ 43. $\sqrt{7 - 4t} - \sqrt{3 - 2t} = 1$ 44. $\sqrt{2t} + \sqrt{2t + 12} = 2\sqrt{4t + 1}$

In problems 45–48, solve each formula for the indicated unknown.

45. $r = \sqrt{\dfrac{V}{\pi h}}$, for V 46. $r = \sqrt[3]{\dfrac{3V}{4\pi}}$, for V

47. $C = 331\sqrt{1 + \dfrac{T}{273}}$, for T 48. $I = \sqrt{\dfrac{P}{R}}$, for P

6.7 Equations Involving Rational Exponents

The following equations contain rational exponents:

$$x^{1/6} = 2, \qquad y^{2/3} = 4, \qquad (t - 3)^{2/5} = 1, \qquad \text{and} \qquad z^{2/3} - z^{1/3} - 12 = 0.$$

This type of equation can be solved using slight variations of the methods we have already discussed. Keep in mind that an equation containing rational exponents (in their lowest terms) with an even denominator is equivalent to a radical equation with even indexes. Therefore, in this case, you *must* check your solution.

In Examples 1–6, solve each equation.

EXAMPLE 1 $x^{1/6} = 2$

SOLUTION   To solve the equation $x^{1/6} = 2$, we raise each side of the equation to the sixth power. That is,

$$(x^{1/6})^6 = 2^6$$
$$x^{6/6} = 2^6$$
$$x^1 = 64$$
$$x = 64.$$

ALTERNATIVE
SOLUTION

The equation can be rewritten as

$$\sqrt[6]{x} = 2.$$

We eliminate the radical by raising both sides of the equation to the sixth power:

$$(\sqrt[6]{x})^6 = 2^6$$

$$x = 64.$$

Because the original equation contains a rational exponent (in its lowest terms) with an even denominator, we must check the solution.

Check For $x = 64$,

$$\sqrt[6]{64} = 2.$$

Hence, 64 is the solution.

EXAMPLE 2 $y^{2/3} = 4$

SOLUTION

We raise each side of the equation to the third power. That is,

$$(y^{2/3})^3 = 4^3$$

$$y^{(2/3) \cdot 3} = (2^2)^3$$

$$y^2 = 2^6$$

$$y = \pm 2^3$$

$$y = \pm 8.$$

ALTERNATIVE
SOLUTION

The equation can be rewritten as

$$\sqrt[3]{y^2} = 4.$$

We raise both sides of the equation to the third power:

$$(\sqrt[3]{y^2})^3 = 4^3$$

$$y^2 = 64$$

$$y = \pm\sqrt{64}$$

$$y = \pm 8.$$

Therefore, -8 and 8 are the solutions.

EXAMPLE 3 $(t - 3)^{2/5} = 1$

SOLUTION We raise each side of the equation to the fifth power. That is,

$$[(t-3)^{2/5}]^5 = 1^5$$
$$(t-3)^{(2/5) \cdot 5} = 1$$
$$(t-3)^2 = 1$$
$$t-3 = \pm\sqrt{1}$$
$$t-3 = \pm 1.$$

$t-3 = -1$	$t-3 = 1$
$t = 2$	$t = 4$

ALTERNATIVE SOLUTION The equation can be rewritten as

$$\sqrt[5]{(t-3)^2} = 1.$$

We raise both sides of the equation to the fifth power:

$$[\sqrt[5]{(t-3)^2}]^5 = 1^5$$
$$(t-3)^2 = 1$$
$$t-3 = \pm\sqrt{1}$$
$$t-3 = \pm 1.$$

$t-3 = -1$	$t-3 = 1$
$t = 2$	$t = 4$

Hence, 2 and 4 are the solutions.

EXAMPLE 4 $2(w+2)^{-2/7} = 1$

SOLUTION We begin by dividing both sides of the equation by 2, so that $(w+2)^{-2/7} = \frac{1}{2}$. We raise each side of the equation to the negative seventh power. That is,

$$[(w+2)^{-2/7}]^{-7} = (\tfrac{1}{2})^{-7}$$
$$(w+2)^{-(2/7) \cdot (-7)} = (\tfrac{1}{2})^{-7}$$
$$(w+2)^2 = 2^7$$
$$w+2 = \pm\sqrt{2^7}$$
$$w+2 = \pm 2^3\sqrt{2}$$
$$w+2 = \pm 8\sqrt{2}.$$

$w+2 = -8\sqrt{2}$	$w+2 = 8\sqrt{2}$
$w = -2 - 8\sqrt{2}$	$w = -2 + 8\sqrt{2}.$

ALTERNATIVE
SOLUTION

We begin by rewriting the equation as

$$\frac{2}{(w+2)^{2/7}} = 1 \quad \text{or} \quad (w+2)^{2/7} = 2.$$

The last equation can also be written as

$$\sqrt[7]{(w+2)^2} = 2.$$

We raise both sides of the equation to the seventh power:

$$[\sqrt[7]{(w+2)^2}]^7 = 2^7$$
$$(w+2)^2 = 128$$
$$w + 2 = \pm\sqrt{128}.$$

$$
\begin{array}{c|c}
w + 2 = -\sqrt{128} & w + 2 = \sqrt{128} \\
w + 2 = -8\sqrt{2} & w + 2 = 8\sqrt{2} \\
w = -2 - 8\sqrt{2} & w = -2 + 8\sqrt{2}
\end{array}
$$

Hence, $-2 - 8\sqrt{2}$ and $-2 + 8\sqrt{2}$ are the solutions.

EXAMPLE 5 $z^{2/3} - z^{1/3} - 12 = 0$

SOLUTION   Let $u = z^{1/3}$, so that $u^2 = (z^{1/3})^2 = z^{2/3}$. Then the equation becomes

$$u^2 - u - 12 = 0$$
$$(u - 4)(u + 3) = 0.$$

$$
\begin{array}{c|c}
u - 4 = 0 & u + 3 = 0 \\
u = 4 & u = -3
\end{array}
$$

Since $u = z^{1/3}$, we have

$$
\begin{array}{c|c}
z^{1/3} = 4 & z^{1/3} = -3 \\
\text{or} \quad \sqrt[3]{z} = 4 & \sqrt[3]{z} = -3 \\
\text{so that} \quad z = 4^3 & z = (-3)^3 \\
z = 64 & z = -27
\end{array}
$$

Hence, 64 and -27 are the solutions.

EXAMPLE 6 $x + 2 + (x + 2)^{1/2} - 2 = 0$

SOLUTION   Let $u = (x + 2)^{1/2}$, so that $u^2 = [(x + 2)^{1/2}]^2 = x + 2$. The equation becomes

$$u^2 + u - 2 = 0$$

$$(u + 2)(u - 1) = 0.$$

| $u + 2 = 0$ | $u - 1 = 0$ |
| $u = -2$ | $u = 1$ |

Since $u = (x + 2)^{1/2} = \sqrt{x + 2}$, we have

$\sqrt{x + 2} = -2$	$\sqrt{x + 2} = 1$
$x + 2 = (-2)^2$	$x + 2 = 1^2$
$x + 2 = 4$	$x + 2 = 1$
$x = 2$	$x = -1$

Here we must check the solutions. (Why?)

Check For $x = -1$,

$$[(-1) + 2] + \sqrt{(-1) + 2} - 2 = 1 + \sqrt{1} - 2 = 2 - 2 = 0.$$

Thus, $x = -1$ is a solution.
     For $x = 2$,

$$(2 + 2) + \sqrt{2 + 2} - 2 = 4 + 2 - 2 \neq 0.$$

Thus, $x = 2$ is *not* a solution. The only solution is -1.

PROBLEM SET 6.7

In problems 1–32, find the real solutions of each equation. Be sure to check the solutions whenever the equation has a rational exponent (in its lowest terms) with an even denominator.

1. $y^{1/5} = -3$ **2.** $x^{1/6} = \frac{1}{2}$ **3.** $t^{1/7} = -2$ **4.** $y^{1/9} = -1$

5. $x^{2/3} = 16$ **6.** $m^{3/5} = 8$ **7.** $u^{4/7} = 16$ **8.** $t^{3/7} = -8$

9. $(x + 1)^{3/5} = 1$ **10.** $(t - 1)^{2/3} = 4$ **11.** $(z - 1)^{5/2} = 32$ **12.** $(x - 3)^{3/2} = 8$

13. $(3y - 7)^{4/3} = 1$ **14.** $(5y - 7)^{4/3} = 16$ **15.** $(2t - 1)^{2/7} = 4$ **16.** $(3m + 1)^{5/3} = -32$

17. $(2x - 1)^{-1/3} = 8$ **18.** $(3y - 7)^{-4/3} = 1$ **19.** $(3x + 1)^{-5/3} = \frac{1}{32}$ **20.** $(7t - 1)^{-3/5} = \frac{1}{27}$

21. $y^{2/3} + 2y^{1/3} = 8, u = y^{1/3}$ **22.** $x^{1/2} + 2x^{1/4} = 3, u = x^{1/4}$

23. $x^{1/3} - 1 - 12x^{-1/3} = 0$ **24.** $z^5 - 33z^{5/2} + 32 = 0$

25. $x^3 - 9x^{3/2} + 8 = 0$ **26.** $8t^{2/3} + 7t^{1/3} = 1$

27. $x + 7 - (x + 7)^{1/2} - 2 = 0$ **28.** $y^2 + 3y + (y^2 + 3y - 2)^{1/2} = 22$

29. $(m + 20)^{1/2} - 4(m + 20)^{1/4} + 3 = 0$ **30.** $2(1 - y)^{1/3} + 3(1 - y)^{1/6} = 2$

31. $2t^2 + t - 4(2t^2 + t + 4)^{1/2} = 1$ **32.** $x^2 + 6x - 6(x^2 + 6x - 2)^{1/2} + 3 = 0$

6.8 Nonlinear Inequalities

Up to this point, we have dealt only with inequalities containing first-degree polynomials in one unknown (variable). In this section, we study quadratic and rational inequalities in one unknown.

Quadratic Inequalities

An inequality such as $2x^2 - x + 1 < x^2 + 4x - 5$ is called a **quadratic inequality.** Before solving such inequalities, we rewrite them in **standard form** with zero on the right side and a quadratic polynomial on the left side.

Suppose we have to solve the quadratic inequality

$$2x^2 - x + 1 < x^2 + 4x - 5.$$

We begin by subtracting $x^2 + 4x - 5$ from both sides in order to obtain the standard form

$$x^2 - 5x + 6 < 0.$$

Now, let us consider the possible values of this quadratic polynomial as we vary the values of the unknown. As the values we assign to x move along the number line, the values of the quantity $x^2 - 5x + 6$ are sometimes positive, sometimes negative, and sometimes zero.

For example, when we let $x = -1$, the expression has the numerical value $1 + 5 + 6 = 12$, and $12 > 0$. When $x = \frac{5}{2}$, the value of the expression is $\frac{25}{4} - \frac{25}{2} + 6 = -\frac{1}{4}$, and $-\frac{1}{4} < 0$. When $x = 4$, the value of the expression becomes $16 - 20 + 6 = 2$, and $2 > 0$.

To solve the inequality, we must find the values of x for which the expression $x^2 - 5x + 6$ is negative. Quadratic inequalities have a characteristic that will help us do this. As we move along the number line, substituting values for x, we find that the parts of the line where $x^2 - 5x + 6$ is positive are separated from the parts where $x^2 - 5x + 6$ is negative by the values of x for which $x^2 - 5x + 6$ is zero. To locate these points, we begin by solving the equation

$$x^2 - 5x + 6 = 0.$$

Factoring, we obtain

$$(x - 2)(x - 3) = 0,$$

which has the solutions 2 and 3. These solutions divide the number line into

Figure 1

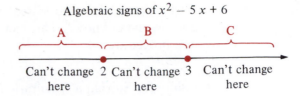

Algebraic signs of $x^2 - 5x + 6$

A   B   C

Can't change 2 Can't change 3 Can't change
here   here   here

three parts, which we label A, B, and C (Figure 1). In part A, $x < 2$; in part B, $2 < x < 3$; and in part C, $x > 3$. The quantity $x^2 - 5x + 6$ will have a constant algebraic sign over each of these parts.

To find out whether $x^2 - 5x + 6$ is positive or negative over A (in which $x < 2$), we select any convenient test number in this part of the number line—say, $x = 0$—and substitute it into $x^2 - 5x + 6$:

$$0^2 - 5(0) + 6 = 6 > 0.$$

Because $x^2 - 5x + 6$ is positive at one number in part A, and cannot change its algebraic sign over this part, we conclude that $x^2 - 5x + 6 > 0$ for all values of x in A.

Similarly, to find the algebraic sign of $x^2 - 5x + 6$ over part B, we substitute a test number—say, $x = \frac{5}{2}$—in the expression to obtain

$$\left(\frac{5}{2}\right)^2 - 5\left(\frac{5}{2}\right) + 6 = \frac{25}{4} - \frac{25}{2} + 6$$

$$= -\frac{1}{4} < 0.$$

Thus we know that the expression is negative for all values of x in part B.

Finally, we choose $x = 4$ in part C:

$$4^2 - 5(4) + 6 = 16 - 20 + 6$$
$$= 2 > 0.$$

Therefore, the expression $x^2 - 5x + 6$ is positive for all values of x in part C.

The information we now have about the algebraic sign of $x^2 - 5x + 6$ in each part of the number line is summarized in Figure 2. Therefore, the solution set of $x^2 - 5x + 6 < 0$ is $\{x \mid 2 < x < 3\}$.

Figure 2

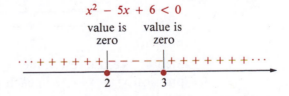

$x^2 - 5x + 6 < 0$

value is   value is
zero   zero

$\cdots + + + + + + \vert - - - - \vert + + + + + + + \cdots$

2   3

The method just illustrated can be used to solve any quadratic inequality in one unknown. This method is summarized in the following step-by-step procedure:

Procedure for Solving a Quadratic Inequality

> Step 1. Rewrite the inequality in standard form.
>
> Step 2. Set the quadratic expression equal to zero and find all real solutions of the resulting equation.
>
> Step 3. Arrange the solutions obtained in step 2 in increasing order on a number line. These solutions will divide the number line into at most three parts (say, A, B, and C).
>
> Step 4. Determine the algebraic sign for each part of the number line by selecting a test number from that part and substituting it for the unknown in the quadratic expression. The algebraic sign of the resulting value is the sign of the quadratic expression over the entire part.
>
> Step 5. Using the information obtained in step 4, draw a figure showing the algebraic signs of the expression over the various parts of the number line. The solution set of the inequality can be read from this figure.

In Examples 1–4, solve each quadratic inequality.

EXAMPLE 1 $x^2 + x - 12 \leq 0$

SOLUTION   We carry out the steps in the procedure above.

Step 1. The inequality is given in standard form.

Step 2. Set $x^2 + x - 12 = (x + 4)(x - 3) = 0$. The solutions of this equation are -4 and 3.

Step 3. The solutions -4 and 3 (of the equation in step 2) divide the number line into three parts, A, B, and C (Figure 3).

Figure 3

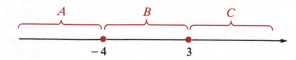

Step 4. We use test values to determine the algebraic sign of the expression $x^2 + x - 12$ in each part. The following table summarizes the results:

Part	Test Number	Test Value of $x^2 + x - 12$	Sign of $x^2 + x - 12$
A: $x < -4$	-5	$(-5)^2 + (-5) - 12 = 8$	$+$
B: $-4 < x < 3$	0	$0^2 + 0 - 12 = -12$	$-$
C: $x > 3$	4	$4^2 + 4 - 12 = 8$	$+$

Figure 4

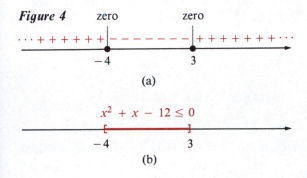

(a)

(b)

Step 5. The information obtained in step 4 is illustrated in Figure 4a. We see that

$$x^2 + x - 12 \leq 0$$

if $-4 \leq x \leq 3$. Thus, the solution set is $\{x \mid -4 \leq x \leq 3\}$ (Figure 4b).

EXAMPLE 2 $x^2 + x - 2 > 0$

SOLUTION   Step 1. The inequality is already in standard form.

Step 2. Set $x^2 + x - 2 = (x + 2)(x - 1) = 0$. The solutions of this equation are -2 and 1.

Step 3. The numbers -2 and 1 divide the number line into three parts, A, B, and C (Figure 5).

Figure 5

A B C

-2 1

Step 4. We use test values to determine the algebraic sign of the expression $x^2 + x - 2$ in each part. The following table summarizes the results:

Part	Test Number	Test Value of $x^2 + x - 2$	Sign of $x^2 + x - 2$
A: $x < -2$	-3	$(-3)^2 + (-3) - 2 = 4$	$+$
B: $-2 < x < 1$	0	$0^2 + 0 - 2 = -2$	$-$
C: $x > 1$	2	$2^2 + 2 - 2 = 4$	$+$

Figure 6

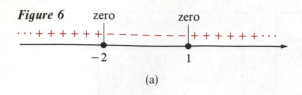

(a)

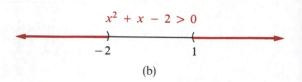

(b)

Step 5. The information obtained in step 4 is illustrated in Figure 6a. We see that $x^2 + x - 2 > 0$ if $x < -2$ or $x > 1$. Therefore, the solution set is $\{x|x < -2\} \cup \{x|x > 1\}$ (Figure 6b).

EXAMPLE 3 $3x^2 - 10x + 5 \geq -6x^2 + 14x - 11$

SOLUTION Step 1. First we write the inequality in standard form:

Figure 7

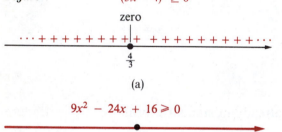

(a)

(b)

$$9x^2 - 24x + 16 \geq 0.$$

The inequality $9x^2 - 24x + 16 \geq 0$ is equivalent to $(3x - 4)^2 \geq 0$. This inequality is true for all real values of x, because the square of any real expression is always nonnegative (Figure 7a). Therefore, the solution set consists of all real numbers $\mathbb{R}$ (Figure 7b).

EXAMPLE 4 $4x^2 + 20x + 25 < 0$

SOLUTION The inequality $4x^2 + 20x + 25 < 0$ is equivalent to $(2x + 5)^2 < 0$. The inequality is not true for any real values of x, because the square of the expression $2x + 5$ is *never* negative. Therefore, the solution set is the empty set.

Rational Inequalities

An inequality such as

$$\frac{2x + 5}{x + 1} < 0 \qquad \text{or} \qquad \frac{x + 3}{5x - 7} \geq 0$$

that involves rational expressions is called a **rational inequality.** Rational inequalities, such as

$$\frac{3x + 1}{x - 1} < 2$$

can be written in **standard form** by subtracting the expression on the right side from both sides and then simplifying the left side. Thus, the inequality

$$\frac{3x + 1}{x - 1} < 2$$

can be written in standard form as follows:

$$\frac{3x + 1}{x - 1} - 2 < 0$$

$$\frac{3x + 1 - 2(x - 1)}{x - 1} < 0$$

$$\frac{x + 3}{x - 1} < 0$$

When we have written a rational inequality in standard form, we can solve it by considering how the algebraic sign of the expression on the left side changes as the values of the unknown are varied. Notice that a fraction can change its algebraic sign only if its numerator or denominator changes its algebraic sign. Hence, the parts where a rational expression is positive are separated from parts where it is negative by values of the variable for which the numerator or denominator is zero. It is thus possible to solve rational inequalities by a simple modification of the step-by-step procedure for solving quadratic inequalities.

For example, to solve the rational inequality

$$\frac{x + 3}{x - 1} < 0,$$

we first determine that the numerator of

$$\frac{x + 3}{x - 1}$$

is zero when $x = -3$, and the denominator is zero when $x = 1$. These values of x will divide the number line into three parts. We can use test values in each of the three parts as we did for the solution in the quadratic inequalities.

In Examples 5 and 6, solve each rational inequality.

EXAMPLE 5 $$\frac{3x - 12}{x + 2} < 0$$

SOLUTION   Step 1. The given rational inequality is in standard form.

Step 2. The numerator of

$$\frac{3x - 12}{x + 2}$$

equals zero if

$$3x - 12 = 0 \quad \text{or} \quad x = 4.$$

The denominator is zero when

$$x + 2 = 0 \quad \text{or} \quad x = -2.$$

Notice that -2 is not a solution of

$$\frac{3x - 12}{x + 2} < 0$$

since the expression is undefined for $x = -2$. Also, 4 is not a solution of

$$\frac{3x - 12}{x + 2} < 0$$

since

$$\frac{3(4) - 12}{4 + 2} = 0.$$

Step 3. The numbers -2 and 4 divide the number line into three parts, A, B, and C (Figure 8).

Figure 8

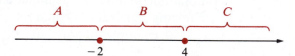

Step 4. We use test values to determine the algebraic sign of

$$\frac{3x - 12}{x + 2}$$

in each part:

Part	Test Number	Test Value of $\dfrac{3x - 12}{x + 2}$	Sign of $\dfrac{3x - 12}{x + 2}$
A: $x < -2$	-3	$\dfrac{-9 - 12}{-3 + 2} = 21$	$+$
B: $-2 < x < 4$	0	$\dfrac{0 - 12}{0 + 2} = -6$	$-$
C: $x > 4$	5	$\dfrac{15 - 12}{5 + 2} = \dfrac{3}{7}$	$+$

Figure 9

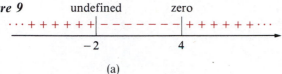

(a)

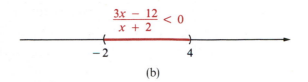

(b)

Step 5. The information obtained in step 4 is illustrated in Figure 9a. We see that

$$\frac{3x - 12}{x + 2} < 0$$

if $-2 < x < 4$. Hence, the solution set is $\{x \mid -2 < x < 4\}$ (Figure 9b).

EXAMPLE 6 $\dfrac{2x + 1}{x - 2} \geq 1$

SOLUTION   Step 1. We write the inequality in standard form:

$$\frac{2x + 1}{x - 2} \geq 1$$

$$\frac{2x + 1}{x - 2} - 1 \geq 0$$

$$\frac{2x + 1 - x + 2}{x - 2} \geq 0$$

$$\frac{x + 3}{x - 2} \geq 0.$$

Step 2. The numerator of

$$\frac{x + 3}{x - 2}$$

equals zero if

$$x + 3 = 0 \qquad \text{or} \qquad x = -3.$$

The denominator is zero if

$$x - 2 = 0 \qquad \text{or} \qquad x = 2.$$

So the fraction

$$\frac{x + 3}{x - 2}$$

is undefined when $x = 2$.

Step 3. The numbers -3 and 2 divide the number line into three parts, A, B, and C (Figure 10).

Figure 10

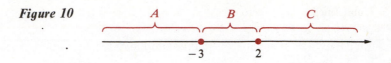

Step 4. We use test values to determine the algebraic sign of

$$\frac{x+3}{x-2}$$

in each part:

Part	Test Number	Test Value of $\dfrac{x+3}{x-2}$	Sign of $\dfrac{x+3}{x-2}$
A: $x < -3$	-4	$\dfrac{-4+3}{-4-2} = \dfrac{1}{6}$	$+$
B: $-3 < x < 2$	0	$\dfrac{0+3}{0-2} = -\dfrac{3}{2}$	$-$
C: $x > 2$	3	$\dfrac{3+3}{3-2} = 6$	$+$

Therefore,

$$\frac{x+3}{x-2} > 0$$

for all values of *x* for which $x < -3$ or $x > 2$.

Step 5. The information obtained in step 4 is illustrated in Figure 11a. From this figure we see that

$$\frac{x+3}{x-2} \geq 0$$

if $x \leq -3$ or $x > 2$. (Recall that 2 isn't included in the solution because the inequality is undefined at that point.) Therefore, the solution set is $\{x \mid x \leq -3\} \cup \{x \mid x > 2\}$ (Figure 11b).

Figure 11

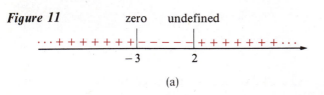

(a)

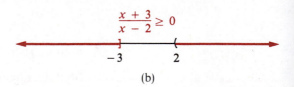

(b)

PROBLEM SET 6.8

In problems 1–22, solve each quadratic inequality and illustrate the solution set on a number line.

1. $x^2 + x - 2 < 0$ **2.** $x^2 - x - 6 < 0$ **3.** $x^2 - 3x + 2 \leq 0$ **4.** $x^2 - 6x + 8 \leq 0$

5. $x^2 + 2x - 3 > 0$ **6.** $x^2 + 5x + 6 > 0$ **7.** $x^2 \geq 4x + 12$ **8.** $x^2 + 3x \geq 10$

9. $x^2 - x \leq 20$ **10.** $x^2 + 4x \leq 21$ **11.** $2x^2 + x - 1 \leq 0$ **12.** $2x^2 + 9x - 5 \geq 0$

13. $3x^2 - 2x - 5 > 0$ **14.** $40 - 3x - x^2 < 0$ **15.** $4x^2 \geq 27x + 7$ **16.** $10x^2 - 31x \leq 14$

17. $2x^2 - 11x + 13 \geq -2x^2 + 9x - 12$ **18.** $12x^2 + 7x + 1 \geq -4x^2 - x$

19. $5x^2 + 6x + 1 > -4x^2$ **20.** $13x^2 - 19x + 5 < -12x^2 + 11x - 4$

21. $16x^2 - 24x < -9$ **22.** $49x^2 + 14x \leq -1$

In problems 23–36, solve each rational inequality and illustrate the solution set on a number line.

23. $\dfrac{x-1}{x-4} < 0$ **24.** $\dfrac{x+3}{x-3} > 0$ **25.** $\dfrac{x+1}{x-2} > 0$ **26.** $\dfrac{2x+3}{4x+8} < 0$ **27.** $\dfrac{x-4}{5x+2} \leq 0$

28. $\dfrac{x-2}{3x-1} \geq 0$ **29.** $\dfrac{5x-1}{3x-2} \geq 0$ **30.** $\dfrac{2x-1}{5x+1} \leq 0$ **31.** $\dfrac{1-x}{3-x} \geq 0$ **32.** $\dfrac{7x}{3x+1} > 0$

33. $\dfrac{x+1}{x-3} \leq 1$ **34.** $\dfrac{x+2}{x-1} \geq 2$ **35.** $\dfrac{1-x}{x} < 1$ **36.** $\dfrac{x}{x+2} \leq 3$

37. For what values of x will the expression

$$\sqrt{x^2 - 9x - 10}$$

be a real number?

38. For what value of x will the expression

$$\sqrt{\frac{x-2}{x+4}}$$

be a real number?

39. If an object is projected straight upward from ground level with an initial speed of 96 feet per second, then its height h in feet t seconds later is given by $h = 96t - 16t^2$. For what time interval is the object more than 112 feet above the ground?

40. Wildlife biologists predict that the population N of a certain endangered species after t years will be given by the equation

$$N = 50t^2 + 200t + 250.$$

How many years will it take for the population N to be at least 5050?

REVIEW PROBLEM SET

In problems 1–14, use the factoring method to solve each equation.

1. $x^2 - 14x - 15 = 0$ **2.** $z^2 - 5z - 84 = 0$ **3.** $t^2 + t - 12 = 0$ **4.** $6y^2 - 15y = 0$

5. $5w^2 - 10w = 0$ **6.** $5 - 14x - 3x^2 = 0$ **7.** $8 - 2y - y^2 = 0$ **8.** $16t^2 - 24t + 9 = 0$

9. $4 + 5x = 9x^2$ **10.** $5x^2 = 2x + 24$ **11.** $z(6z - 7) = 20$ **12.** $y(4y - 31) = 8$

13. $\dfrac{3}{4x^2} + \dfrac{7}{8x} - \dfrac{5}{2} = 0$ **14.** $\dfrac{2}{w - 1} - \dfrac{3}{2w + 5} = \dfrac{5}{3}$

In problems 15 and 16, solve each equation for the indicated variable.

15. $x^2 + bx - 6b^2 = 0$, for x.   **16.** $6y^2 - 7cy - 3c^2 = 0$, for y.

In problems 17–24, solve each equation by the roots extraction method.

17. $t^2 = 144$ **18.** $5u^2 - 80 = 0$ **19.** $3x^2 + 48 = 0$ **20.** $(v - 1)^2 + 4 = 0$

21. $(y - \frac{3}{5})^2 = \frac{16}{25}$ **22.** $(x + \frac{5}{2})^2 = \frac{7}{4}$ **23.** $(3m - 1)^2 - 256 = 0$ **24.** $(2z + 3)^2 + 25 = 0$

In problems 25–28, complete the square for each expression and write the result as the square of a binomial.

25. $u^2 - 8u$ **26.** $z^2 + 13z$ **27.** $x^2 + 5x$ **28.** $y^2 - 14y$

In problems 29–40, solve each equation by completing the square.

29. $x^2 - 6x + 7 = 0$ **30.** $y^2 + 4y - 21 = 0$ **31.** $t^2 - 8t + 9 = 0$ **32.** $u^2 - 4u + 5 = 0$

33. $9y^2 - 6y + 2 = 0$ **34.** $9t^2 - 12t + 1 = 0$ **35.** $4z^2 + 13 = 12z$ **36.** $16y^2 - 24y + 5 = 0$

37. $4x^2 + 4x - 3 = 0$ **38.** $2w^2 + 3 = 8w$ **39.** $y^2 + cy - 6c^2 = 0$, $c > 0$, solve for y

40. $3x^2 - dx + d^2 = 0$, $d < 0$, solve for x.

In problems 41–52, solve each equation by using the quadratic formula. In problems 51 and 52, round off your answers to two decimal places.

41. $x^2 + 10x - 13 = 0$ **42.** $5 - 8t + t^2 = 0$ **43.** $y^2 + y + 3 = 0$

44. $3 - z - z^2 = 0$ **45.** $4u^2 - u - 1 = 0$ **46.** $3x^2 - 8x + 1 = 0$

47. $2t^2 + 5t - 17 = 0$ **48.** $16 - 16y - 3y^2 = 0$ **49.** $(2x - 1)(x + 2) = 5$

50. $(w - 1)(3w + 5) = 2$ © **51.** $0.9x^2 - 4.1x - 3.14 = 0$ © **52.** $1.7y^2 + 3.8y - 15.6 = 0$

In problems 53–58, use the discriminant to determine the number and the kind of solutions to each equation.

53. $t^2 - 8t + 9 = 0$ **54.** $2y^2 - y - 1 = 0$ **55.** $5x^2 - 2x + 1 = 0$

56. $4m^2 - 12m + 13 = 0$ **57.** $4y^2 - 12y + 9 = 0$ **58.** $25z^2 - 20z + 4 = 0$

In problems 59–68, reduce each equation to the quadratic form by using an appropriate substitution and solve the equation for the original variable.

59. $x^4 - 5x^2 + 4 = 0$ **60.** $y^4 - 8y^2 + 16 = 0$ **61.** $16t^4 - 17t^2 + 1 = 0$

62. $v^8 - 2v^4 + 1 = 0$ **63.** $4z^{-4} - 11z^{-2} - 3 = 0$ **64.** $1 - 2x^{-2} - 3x^{-4} = 0$

65. $(x^2 + x)^2 - 8(x^2 + x) + 12 = 0$

66. $(y^2 + 4y)^2 - 17(y^2 + 4y) = 60$

67. $v^2 + 2v + \dfrac{3}{v^2 + 2v} = 4$

68. $2t^2 - t - \dfrac{6}{2t^2 - t} + 1 = 0$

In problems 69–84, find the real solutions of each equation. Check the solutions whenever the equation has a radical with even index.

69. $\sqrt{x - 1} = 4$

70. $\sqrt{2u + 5} - 7 = -4$

71. $\sqrt{t + 17} + 1 = 5$

72. $1 + \sqrt{x - 3} = 4$

73. $2\sqrt{2y - 3} + 4 = 1$

74. $\sqrt{z - 7} = 2\sqrt{z}$

75. $\sqrt[3]{3x - 2} = 4$

76. $1 - \sqrt[3]{t - 3} = 3$

77. $\sqrt[4]{5x - 7} = \sqrt[4]{x}$

78. $\sqrt[4]{4x - 3} = \sqrt[4]{5x - 7}$

79. $\sqrt{w - 3} - \sqrt{w} = -1$

80. $\sqrt{2x + 1} + \sqrt{8 - x} = 5$

81. $\sqrt{2t + 1} + \sqrt{t} = 1$

82. $\sqrt{5z - 4} - \sqrt{2z + 1} = 1$

83. $\sqrt{5x + 5} + \sqrt{x - 4} = 5$

84. $\sqrt{2x + \sqrt{3 + x}} = 2$

In problems 85–94, solve each equation. Check the solution whenever the equation has a rational exponent (in its lowest terms) with an even denominator.

85. $t^{1/3} = -2$

86. $x^{1/4} = 3$

87. $y^{3/4} = 8$

88. $z^{2/5} = 4$

89. $(x - 2)^{2/3} = 9$

90. $(2 - 3w)^{5/3} = 32$

91. $(3y + 2)^{-1/3} = 2$

92. $(2x - 1)^{-4/3} = 81$

93. $z^{2/3} + 2z^{1/3} - 3 = 0$

94. $x^{-3/2} - 28x^{-3/4} + 27 = 0$

In problems 95–102, solve each inequality and illustrate the solution set on a number line.

95. $x^2 + 2x - 15 < 0$

96. $u^2 - 2u - 3 \geq 0$

97. $4t^2 - 3t \geq 10 + 3t^2$

98. $6y^2 + 7y + 1 < 6 + 3y^2 + 9y$

99. $\dfrac{x - 2}{x + 3} > 0$

100. $\dfrac{3w + 1}{2w - 5} \leq 0$

101. $\dfrac{3y - 1}{5y - 7} \leq 0$

102. $\dfrac{7x - 3}{9x + 1} > 0$

103. Find three consecutive positive integers such that the sum of their squares is 149.

104. Find a number such that the sum of the number and its reciprocal is 4.

105. A boy has mowed a strip of uniform width around a rectangular lawn that is 80 feet by 60 feet. He still has half the lawn to mow. How wide is the strip?

106. A swimming pool 25 feet by 15 feet is bordered by a concrete walk of uniform width. The area of the walk is 329 square feet. How wide is it?

107. If the distance s in feet that a bomb falls in t seconds is given by the formula $s = 16t^2/(1 + 0.06t)$, how many seconds are required for a bomb released at 20,000 feet to reach its target?

108. A group of students chartered a bus for $60. When four students withdrew from the group, the share of each of the other students was increased by $2.50. How many students were in the group originally?

109. Two joggers leave the same point and travel at right angles to each other, each moving at a uniform speed. One jogs 1 mile per hour faster than the other. After 3 hours, they are 15 miles apart. Find the speed of each jogger.

110. A girl asked her father: "How old are you?" The father answered: "I was 30 years old when you were born and the product of our present ages is 736." How old is the father?

CHAPTER 6 TEST

1. Use the factoring method to solve each equation.

(a) $2x^2 - 6x = 0$ (b) $y^2 - 2y - 15 = 0$ (c) $6x(x - 1) = 12$

2. Solve each equation by the roots extraction method.

(a) $z^2 = 36$ (b) $(3w - 2)^2 + 9 = 0$

3. Solve each equation by completing the square.

(a) $x^2 - 4x + 2 = 0$ (b) $3y^2 + 12y + 1 = 0$

4. Solve each equation by using the quadratic formula.

(a) $3x^2 - 5x + 4 = 0$ (b) $2y(y - 1) = 3$

5. Use the discriminant to determine the number and the kind of solutions to each equation.

(a) $4x^2 + 7x + 2 = 0$ (b) $8t^2 - 3t + 5 = 0$

6. Solve the equation $x^4 - 3x^2 - 4 = 0$.

7. Find the real solutions of each equation.

(a) $\sqrt{2x - 1} = 5$ (b) $\sqrt{2z + 1} + \sqrt{z} = 1$

8. Solve each of the following equations.

(a) $(3x + 1)^{-1/3} = 3$ (b) $w^{2/3} + 4w^{1/3} - 5 = 0$

9. Solve each inequality and illustrate the solution set on a number line.

(a) $2x^2 + x - 1 > 0$ (b) $\dfrac{2x + 3}{x - 5} \leq 0$

10. Find the dimensions of a rectangular plot if its area is 96 square feet and its perimeter is 40 feet.

7 Graphing Linear Equations and Inequalities in Two Variables

In Chapter 4, we discussed a procedure for solving linear equations and inequalities in one variable. Now, we turn our attention to the discussion of linear equations and inequalities in two variables. We begin this chapter by extending the techniques of graphing to include points in a plane. This enables us to graph equations and inequalities in two variables.

7.1 The Cartesian Coordinate System

In Chapter 1, Section 1.2, we saw that a point P on a number line can be specified by a real number x, called the *coordinate* of that point. Similarly, if we use a **Cartesian coordinate system,** named in honor of the French mathematician René Descartes (1596–1650), a point P in a plane can be specified by two real numbers, also called *coordinates*.

A Cartesian coordinate system consists of two perpendicular number lines, called the **coordinate axes.** The point at which the two lines intersect is called the **origin** (Figure 1a). Ordinarily, one of the number lines is horizontal and is called the **x axis,** and the other is vertical and is called the **y axis.**

We normally use the same scale on the two axes (Figure 1b), although in some figures space considerations make it convenient to use different

Figure 1

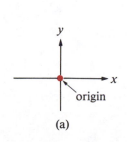

(a)

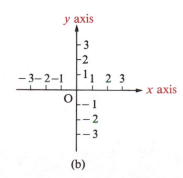

(b)

scales. By convention, numerical coordinates increase to the right along the x axis and upward along the y axis. The positive portion of the x axis is to the right of the origin, and the negative portion is to the left of the origin. The positive portion of the y axis is above the origin, and the negative portion is below the origin.

If P is a point in the Cartesian plane, the coordinates of P are the coordinates x and y of the points where the perpendiculars to the axes from P meet the two axes (Figure 2). The coordinates of P are traditionally written as (x, y). The notation (x, y) represents an **ordered pair,** which is a pair of numbers or symbols in which the order of listing is important. For example,

$$(5, 6) \neq (6, 5).$$

The first number, x, the x coordinate of an ordered pair, is called the **abscissa** of P. The second number, y, the y coordinate of the ordered pair, is called the **ordinate** of P. To **plot,** or **graph,** the point P with coordinates (x, y) means to draw Cartesian coordinate axes and to place a dot representing P at the point with abscissa x and ordinate y.

For example, to locate the point $(3, 2)$, start at the origin and move three units to the right along the x axis; then move two units up from the x axis. Similarly, to locate $(-4, 3)$, move four units to the left of 0 and three units up from the x axis; and to locate $(0, -\frac{5}{2})$, move no units along the x axis and $\frac{5}{2}$ units down from the x axis along the y axis (Figure 3).

You can think of the ordered pair (x, y) as the numerical "address" of P. We show the correspondence between P and (x, y) by identifying the point P with its "address" (x, y) and write $P = (x, y)$.

We relate each ordered pair of real numbers (x, y) to a **point,** and we refer to the set of all such ordered pairs as the **Cartesian plane** or the **xy plane.**

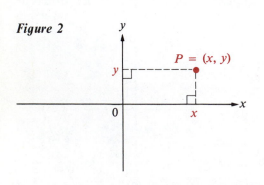

Figure 2

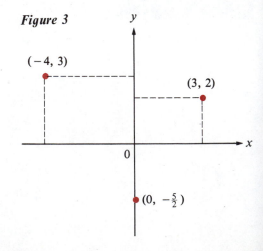

Figure 3

The x and y axes divide the plane into four regions called **quadrants** I, II, III, and IV (Figure 4).

Figure 4

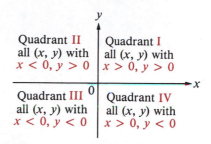

EXAMPLE 1 Plot each point and indicate which quadrant or coordinate axis contains the point.

(a) $(4, 1)$ (b) $(-4, 2)$ (c) $(-2, -3)$ (d) $(2, -5)$

(e) $(2, 0)$ (f) $(-\frac{3}{2}, 0)$ (g) $(0, 5)$ (h) $(0, 0)$

SOLUTION The points are plotted in Figure 5.

Figure 5

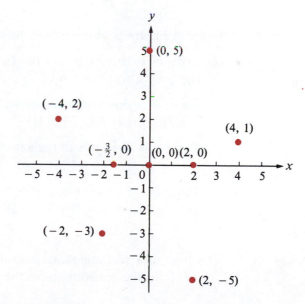

(a) $(4, 1)$ lies in quadrant I.   (b) $(-4, 2)$ lies in quadrant II.

(c) $(-2, -3)$ lies in quadrant III.   (d) $(2, -5)$ lies in quadrant IV.

(e) $(2, 0)$ lies on the positive x axis.   (f) $(-\frac{3}{2}, 0)$ lies on the negative x axis.

(g) $(0, 5)$ lies on the positive y axis.   (h) $(0, 0)$, the origin, lies on both axes.

EXAMPLE 2   Find the coordinates of each point on the Cartesian coordinate system in Figure 6.

Figure 6

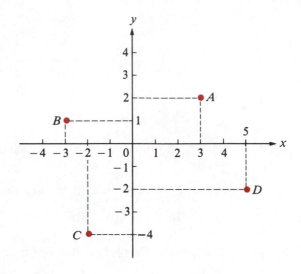

SOLUTION   Look at Figure 6. We see that the coordinates of each point are:

(a) Point *A* is three units to the right of the *y* axis and two units up from the *x* axis, so that $A = (3, 2)$.

(b) Point *B* is three units to the left of the *y* axis and one unit up from the *x* axis, so that $B = (-3, 1)$.

(c) Point *C* is two units to the left of the *y* axis and four units down from the *x* axis, so that $C = (-2, -4)$.

(d) Point *D* is five units to the right of the *y* axis and two units down from the *x* axis, so that $D = (5, -2)$.

EXAMPLE 3   The base of a rectangle has endpoints $(1, 0)$ and $(8, 0)$, and $(1, 4)$ is a third vertex. Find the coordinates of the fourth vertex.

SOLUTION   We plot the three vertices $(1, 0)$, $(8, 0)$, and $(1, 4)$ and connect them (Figure 7). Since the opposite sides of a rectangle are parallel and have the same length, and since the fourth vertex is in quadrant I, we start at $(1, 4)$ and count 7 units horizontally to the right to locate the fourth vertex at the point $(8, 4)$.

Figure 7

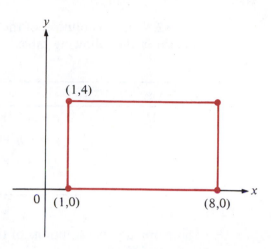

Solutions of Linear Equations in Two Variables

Equations such as

$$2x + 3y = 6, \qquad y = 5 - \frac{1}{2}x, \qquad \text{and} \qquad \sqrt{3}x + 5y - 2 = 0$$

are called **linear equations** (or **first-degree equations**) in two variables, x and y. An ordered pair (a, b) is said to **satisfy** a linear equation if, when a is substituted for x and b is substituted for y, the resulting statement is true. A **solution** of a linear equation in two variables, x and y, is an ordered pair that satisfies the equation. An equation in two variables can have an infinite number of solutions.

EXAMPLE 4   Find five ordered pairs that satisfy the equation $3x - y = -2$, given the following values of x: $-2, -1, 0, 1, 2$.

SOLUTION   Find we solve for y in terms of x:

$$3x - y = -2$$
$$-y = -3x - 2$$
$$y = 3x + 2.$$

Now we substitute the numbers $-2, -1, 0, 1, 2$ for x to obtain the corresponding values for y. Substituting -2 for x, we have

$$y = 3(-2) + 2 = -6 + 2 = -4.$$

So $(-2, -4)$ is a solution of the equation. The other four solutions are organized in the following table.

x	$y = 3x + 2$	Solution
-1	$y = 3(-1) + 2 = -3 + 2 = -1$	$(-1, -1)$
0	$y = 3(0) + 2 = 0 + 2 = 2$	$(0, 2)$
1	$y = 3(1) + 2 = 3 + 2 = 5$	$(1, 5)$
2	$y = 3(2) + 2 = 6 + 2 = 8$	$(2, 8)$

Therefore, the five solutions of the equation are $(-2, -4)$, $(-1, -1)$, $(0, 2)$, $(1, 5)$, and $(2, 8)$.

EXAMPLE 5   Plot the five solutions of the equation obtained in Example 4.

SOLUTION   The points are plotted in Figure 8.

Figure 8

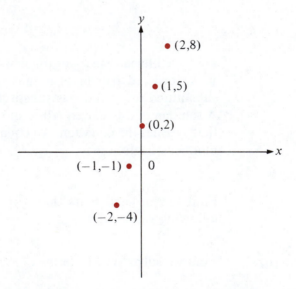

PROBLEM SET 7.1

1. Plot each point and indicate which quadrant or coordinate axis contains the point.

(a) $(1, 2)$ (b) $(-1, 1)$ (c) $(-1, -2)$ (d) $(3, -2)$

(e) $(0, 1)$ (f) $(-6, 0)$ (g) $(3, 0)$ (h) $(0, -4)$

2. Plot each point and indicate which quadrant or coordinate axis contains the point.

(a) $(3, \sqrt{2})$ (b) $(-\sqrt{3}, 1)$ (c) $(-\sqrt{3}, -\sqrt{2})$ (d) $(0, -\frac{3}{4})$

(e) $(\frac{2}{3}, 0)$ (f) $(-\pi, 0)$ (g) $(-\frac{1}{2}, \frac{1}{2})$ (h) $(\frac{2}{3}, -\frac{3}{4})$

3. Find the coordinates of each point on the Cartesian coordinate system in Figure 9.

Figure 9

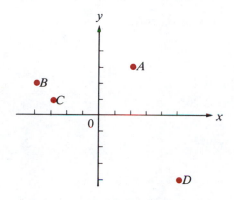

4. On a Cartesian coordinate grid, an aircraft carrier is detected by radar at point $P = (3, 2)$. It then moved to the points Q, R, and finally S. Locate the coordinates of Q, R, and S on the grid if:

(a) the line segment $\overline{PQ}$ is perpendicular to the x axis and is bisected by it.

(b) the line segment $\overline{PR}$ is perpendicular to the y axis and is bisected by it.

(c) the line segment $\overline{PS}$ is bisected by the origin.

5. The base of a rectangle has endpoints $(2, 0)$ and $(10, 0)$, and $(2, 6)$ is a third vertex. Find the coordinates of the fourth vertex.

6. The base of a parallelogram has endpoints $(0, 0)$ and $(6, 0)$, and $(2, 4)$ is a third vertex. If the fourth vertex is in quadrant I, find its coordinates.

7. The center of a square, whose four sides are parallel to the coordinate axes, is at the origin. If one vertex is at $(2, 2)$, find the coordinates of the other three vertices.

8. The center of a rectangle is at the origin. If the four sides are parallel to the coordinate axes and if the coordinates of two vertices are $(-4, 3)$ and $(4, 3)$, find the coordinates of the other vertices.

In problems 9–22, find five ordered pairs that satisfy each equation, given the following values of x: -2, -1, 0, 1, 2. Also plot the five ordered pairs.

9. $3x - 2y = 6$ 10. $x - 3y = 6$ 11. $y = -3x + 2$ 12. $3x + y = 4$

13. $y = \frac{1}{2}x + 2$ 14. $y = -\frac{1}{3}x + 1$ 15. $y = -\frac{2}{3}x + 1$ 16. $y = -\frac{4}{5}x + 2$

17. $2y = 4x + 7$ 18. $y = \frac{3}{7}x + 4$ 19. $5x - 3y = 15$ 20. $2x + 3y = 12$

21. $3x + 5y = 15$ 22. $x - 3y = 9$

In problems 23–30, complete each ordered pair to make it a solution to the given equation. Also plot the ordered pairs.

23. $x + y = 3$, $(0,\ )$, $(\ , 0)$, $(-1,\ )$, $(\ , 2)$

24. $4x - 5y = 20$, $(0,\ )$, $(\ , 0)$, $(5,\ )$, $(\ , 4)$

25. $y = x$, $(0,\ )$, $(\ , 0)$, $(1,\ )$, $(\ , -1)$

26. $y = -x$, $(0,\ )$, $(\ , 0)$, $(-2,\ )$, $(\ , -2)$

27. $y = 2$, $(0,\ )$, $(1,\ )$, $(-1,\ )$, $(2,\ )$

28. $y = -4$, $(0,\ )$, $(-1,\ )$, $(1,\ )$, $(-2,\ )$

29. $x = -5$, $(\ , 0)$, $(\ , -1)$, $(\ , -3)$, $(\ , 4)$

30. $x = 3$, $(\ , 0)$, $(\ , -2)$, $(\ , 3)$, $(\ , 4)$

In problems 31–38, write an equation that describes each statement.

31. The y value is 3 more than twice the x value.

32. The y value is 5 less than 4 times the x value.

33. The sum of the y value and 3 times the x value is 4.

34. The sum of the x value and 5 times the y value is 20.

35. The x value is equal to the y value.

36. The y value is equal to the negative of the x value.

37. The x value is -3.

38. The y value is 5.

In problems 39–42, let $c > 0$ and $d < 0$; state the quadrant in which each point lies.

39. (c, d) **40.** $(c, -d)$ **41.** $(d, -c)$ **42.** $(-c, -d)$

7.2 Distance Between Two Points

One useful feature of the Cartesian coordinate system is that there is a formula that gives the distance between two points in the xy plane. We represent the distance between two points P_1 and P_2 by

$$d = |\overline{P_1 P_2}|.$$

1. If $P_1 = P_2$, then $|\overline{P_1 P_2}| = 0$.

2. If $P_1 = (x_1, y_1)$ and $P_2 = (x_2, y_2)$ lie on the same horizontal line, that is, if $y_1 = y_2$, then $d = |\overline{P_1 P_2}| = |x_2 - x_1|$ (Figure 1a).

3. If $P_1 = (x_1, y_1)$ and $P_2 = (x_2, y_2)$ lie on the same vertical line, that is, if $x_1 = x_2$, then $d = |\overline{P_1 P_2}| = |y_2 - y_1|$ (Figure 1b).

Because $|x_2 - x_1| = |x_1 - x_2|$, the formula in (2) is true whether P_1 lies to the left of P_2 or to the right of P_2. Moreover, it does not matter in which quadrants the points lie.

Figure 1

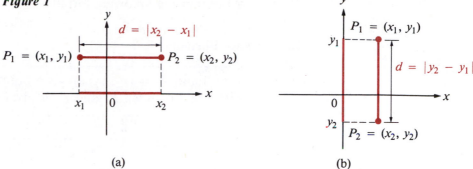

(a)   (b)

Finally, let us consider the general case in which the points $P_1 = (x_1, y_1)$ and $P_2 = (x_2, y_2)$ do not lie on the same horizontal or vertical line, that is, the case in which the line segment $\overline{P_1 P_2}$ is neither horizontal nor vertical. To determine the length of $\overline{P_1 P_2}$, we can draw a right triangle with $\overline{P_1 P_2}$ as the hypotenuse and with one leg parallel to the x axis and the other leg parallel to the y axis (Figure 2). The two legs intersect at some point P_3. Point P_3 has the same y coordinate as P_1 and the same x coordinate as P_2. Therefore, we can represent P_3 by $P_3 = (x_2, y_1)$ (Figure 2). Then $|\overline{P_1 P_3}| = |x_2 - x_1|$ and $|\overline{P_3 P_2}| = |y_2 - y_1|$.

Figure 2

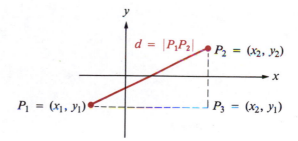

We can use the Pythagorean theorem to determine the length of the hypotenuse $\overline{P_1 P_2}$ of the right triangle $P_1 P_2 P_3$:

$$d^2 = |\overline{P_1 P_2}|^2 = |\overline{P_1 P_3}|^2 + |\overline{P_3 P_2}|^2,$$

or

$$d^2 = |\overline{P_1 P_2}|^2 = |x_2 - x_1|^2 + |y_2 - y_1|^2.$$

Since the square of each expression is automatically nonnegative, we write

$$|x_2 - x_1|^2 = (x_2 - x_1)^2 \quad \text{and} \quad |y_2 - y_1|^2 = (y_2 - y_1)^2.$$

We can now obtain the following formula:

The Distance Formula

If $P_1 = (x_1, y_1)$ and $P_2 = (x_2, y_2)$ are two points in the Cartesian plane, then the **distance** d between P_1 and P_2 is given by

$$d = |\overline{P_1 P_2}| = \sqrt{(x_2 - x_1)^2 + (y_2 - y_1)^2}.$$

Note that it doesn't make any difference which point you call P_1 or P_2, since $(c - d)^2 = (d - c)^2$, so the distance formula can also be written as

$$d = \sqrt{(x_1 - x_2)^2 + (y_1 - y_2)^2}.$$

EXAMPLE 1   Plot the points $(2, -4)$ and $(-2, -1)$ and find the distance d between them.

SOLUTION   The points $P_1 = (2, -4)$ and $P_2 = (-2, -1)$ are plotted in Figure 3 and the distance d is given by

$$
\begin{aligned}
d &= \sqrt{[-2 - 2]^2 + [-1 - (-4)]^2} \\
&= \sqrt{(-4)^2 + 3^2} \\
&= \sqrt{16 + 9} \\
&= \sqrt{25} \\
&= 5.
\end{aligned}
$$

Figure 3

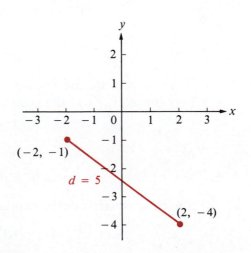

EXAMPLE 2   Plot the points $(-1, -2)$ and $(3, 4)$ and find the distance d between them.

SOLUTION   The points $P_1 = (-1, -2)$ and $P_2 = (3, 4)$ are plotted in Figure 4, and the distance d is given by

$$d = \sqrt{[3 - (-1)]^2 + [4 - (-2)]^2}$$
$$= \sqrt{4^2 + 6^2}$$
$$= \sqrt{16 + 36}$$
$$= \sqrt{52}$$
$$= 2\sqrt{13}.$$

Figure 4

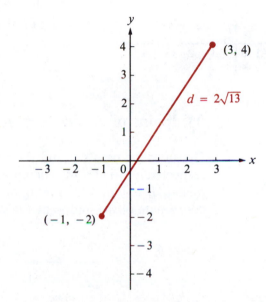

EXAMPLE 3   © Let $P_1 = (75.39, -14.06)$ and $P_2 = (18.64, 19.37)$. Find the distance d between P_1 and P_2. Round off the answer to four significant digits.

SOLUTION   The distance d between P_1 and P_2 is given by

$$d = \sqrt{(18.64 - 75.39)^2 + [19.37 - (-14.06)]^2}$$
$$= \sqrt{3220.5625 + 1117.5649}$$
$$= \sqrt{4338.1274} = 65.86.$$

EXAMPLE 4   Let $A = (-1, 3)$, $B = (2, 6)$, and $C = (3, 2)$.

(a) Plot the points A, B, and C, and draw the triangle ABC.

(b) Find the distances $|\overline{AB}|$, $|\overline{AC}|$, and $|\overline{BC}|$.

(c) Show that ABC is an isosceles triangle.

SOLUTION   (a) The points A, B, and C are plotted and the triangle ABC is drawn in Figure 5.

Figure 5

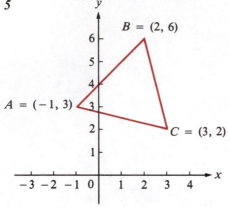

$A = (-1, 3)$
$B = (2, 6)$
$C = (3, 2)$

(b)
$$|\overline{AB}| = \sqrt{[2 - (-1)]^2 + (6 - 3)^2}$$
$$= \sqrt{9 + 9} = \sqrt{18} = 3\sqrt{2}$$
$$|\overline{AC}| = \sqrt{[3 - (-1)]^2 + (2 - 3)^2}$$
$$= \sqrt{16 + 1} = \sqrt{17}$$
$$|\overline{BC}| = \sqrt{(3 - 2)^2 + (2 - 6)^2}$$
$$= \sqrt{1 + 16} = \sqrt{17}.$$

(c) Because $|\overline{AC}| = |\overline{BC}|$, we conclude that the triangle ABC is isosceles.

Midpoint of a Line Segment Joining Two Points

Consider the points A and B on a number line with coordinates x_1 and x_2, respectively (Figure 6). The coordinate of the midpoint M of the line segment $\overline{AB}$ is found by averaging the coordinates of A and B. That is, the coordinate of the midpoint M is $(x_1 + x_2)/2$.

Figure 6

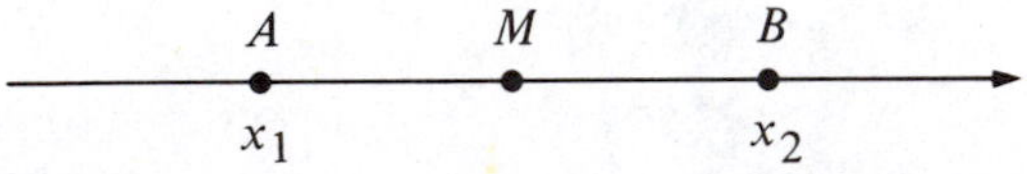

A M B
x_1 x_2

Now let $\overline{PQ}$ represent any segment in the xy plane, and let M be the midpoint (Figure 7). Then the coordinates of the midpoint are obtained by averaging

Figure 7

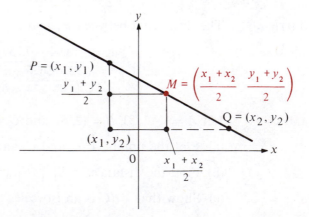

$P = (x_1, y_1)$

$\dfrac{y_1 + y_2}{2}$

$M = \left(\dfrac{x_1 + x_2}{2}, \dfrac{y_1 + y_2}{2} \right)$

$Q = (x_2, y_2)$

(x_1, y_2)

$\dfrac{x_1 + x_2}{2}$

the x values and averaging the y values of P and Q. More formally, we have the following formula:

Midpoint Formula

> Let $P = (x_1, y_1)$ and $Q = (x_2, y_2)$ be two points in the xy plane. Then the coordinates of the **midpoint** M of the line segment $\overline{PQ}$ are given by
>
> $$M = \left(\frac{x_1 + x_2}{2}, \frac{y_1 + y_2}{2}\right).$$

EXAMPLE 5 Determine the coordinates of the midpoint of the line segment containing the points $(3, 6)$ and $(5, 8)$.

SOLUTION Here $(x_1, y_1) = (3, 6)$ and $(x_2, y_2) = (5, 8)$, so that the coordinates of the midpoint are given by

$$\left(\frac{x_1 + x_2}{2}, \frac{y_1 + y_2}{2}\right) = \left(\frac{3 + 5}{2}, \frac{6 + 8}{2}\right)$$
$$= \left(\frac{8}{2}, \frac{14}{2}\right) = (4, 7).$$

The Circle

Now we use the distance formula to derive an equation of a circle. By definition, a **circle** consists of all points P whose distance from a fixed point C is constant. The fixed point C is called the **center** of the circle, and the constant distance from the center to the circle is called the **radius,** r. The **graph** of an equation in two variables x and y is defined to be the set of all points $P = (x, y)$ in the Cartesian plane whose coordinates x and y satisfy the equation.

In accordance with the definition of a circle, r is the distance from the center $C = (h, k)$ to any point $P = (x, y)$ on the circle. Hence, by the distance formula, we can write an equation for this circle as

$$\sqrt{(x - h)^2 + (y - k)^2} = r$$

or equivalently,

$$(x - h)^2 + (y - k)^2 = r^2.$$

This last equation holds if and only if the point $P = (x, y)$ is r units from the point $C = (h, k)$. Therefore, the graph of the equation consists of all points that lie on a circle of radius r with center at (h, k) (Figure 8).

Figure 8

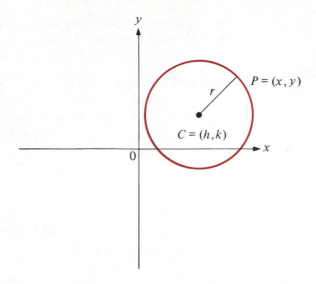

The equation

$$(x - h)^2 + (y - k)^2 = r^2$$

is called the **standard form** for an equation of the circle with radius r and center (h, k).

EXAMPLE 6   Find an equation of a circle with center C and radius r if:

(a) $C = (0, 0)$ and $r = 4$ (b) $C = (-3, -4)$ and $r = 5$

SOLUTION   (a) Here $r = 4$ and $(h, k) = (0, 0)$, so, in standard form, an equation of the circle is

$$(x - 0)^2 + (y - 0)^2 = 4^2$$

or

$$x^2 + y^2 = 16.$$

(b) We note that $r = 5$ and $(h, k) = (-3, -4)$, so, in standard form, an equation of the circle is

$$[x - (-3)]^2 + [y - (-4)]^2 = 5^2$$

or

$$(x + 3)^2 + (y + 4)^2 = 25.$$

If we are given an equation of a circle in standard form, we can easily identify the radius and the center. However, if an equation is not in standard form, then we use the technique of completing the square to put it in that form.

EXAMPLE 7   Find the center and radius of each circle. Also graph each circle.

(a) $(x + 2)^2 + (y - 3)^2 = 16$ (b) $x^2 + y^2 + 6x - 2y - 15 = 0$

SOLUTION   (a) We compare this equation with the standard form of an equation of a circle.

$$(x - h)^2 + (y - k)^2 = r^2$$

with $h = -2$, $k = 3$, and $r = 4$. Therefore, the graph is a circle of radius $r = 4$ and center at $(-2, 3)$ (Figure 9a).

(b) First, we change the equation into standard form. To do this, we rewrite the equation in the form

$$(x^2 + 6x \quad) + (y^2 - 2y \quad) = 15.$$

Now, we complete the square in each of the parentheses by adding 9 to the terms in x, and 1 to the terms in y. Since we are adding 9 and 1 to the left side of the equation, we must also add 9 and 1 to the right side.

Figure 9

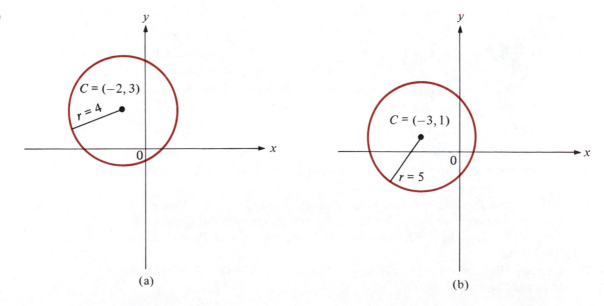

(a)   (b)

The equation now has the form

$$(x^2 + 6x + 9) + (y^2 - 2y + 1) = 15 + 9 + 1$$

or

$$(x + 3)^2 + (y - 1)^2 = 25.$$

The last equation has the form

$$(x - h)^2 + (y - k)^2 = r^2$$

with $h = -3$, $k = 1$, and $r = 5$. Therefore, the graph is a circle of radius $r = 5$ with the center at $(-3, 1)$ (Figure 9b).

PROBLEM SET 7.2

In problems 1–16, find the distance between the two points with the given coordinates.

1. $(7, 10)$ and $(1, 2)$
2. $(-3, -4)$ and $(-5, -7)$
3. $(1, 1)$ and $(-3, 2)$
4. $(-2, 5)$ and $(3, -1)$
5. $(4, -3)$ and $(6, 2)$
6. $(1, 5)$ and $(4, 9)$
7. $(-5, 0)$ and $(-2, -4)$
8. $(-4, 7)$ and $(0, -8)$
9. $(7, -3)$ and $(3, -3)$
10. $(6, 2)$ and $(6, -2)$
11. $(-4, -3)$ and $(0, 0)$
12. $(0, 3)$ and $(-4, 0)$
13. $(7, -1)$ and $(7, 3)$
14. $(2, -b)$ and $(4, b)$
15. $(-\frac{1}{2}, \frac{1}{3})$ and $(2, 3)$
16. $(t, 8)$ and $(t, -2)$

In problems 17–22, use the distance formula to determine whether or not the triangle ABC is isosceles.

17. $A = (-5, 1)$, $B = (-6, 5)$, $C = (-2, 4)$
18. $A = (3, 1)$, $B = (4, 3)$, $C = (6, 2)$
19. $A = (-2, -3)$, $B = (3, -1)$, $C = (1, 4)$
20. $A = (-4, 2)$, $B = (1, 4)$, $C = (3, -1)$
21. $A = (5, -3)$, $B = (2, 4)$, $C = (2, 5)$
22. $A = (-6, 3)$, $B = (2, 1)$, $C = (-2, -3)$

In problems 23–26, use the distance formula and the Pythagorean theorem to show that the triangle ABC is a right triangle.

23. $A = (1, 1)$, $B = (5, 7)$, $C = (5, 1)$
24. $A = (1, 2)$, $B = (5, 5)$, $C = (5, 2)$
25. $A = (-3, 1)$, $B = (3, 10)$, $C = (3, 1)$
26. $A = (-2, -2)$, $B = (3, -3)$, $C = (0, 0)$

27. Find all values of u so that the distance between the points $(-2, 3)$ and (u, u) is five units.

28. If P_1, P_2, and P_3 are points in a plane, then P_1, P_2, and P_3 are **collinear** if P_2 lies on the line segment $\overline{P_1 P_3}$, that is, P_1, P_2, and P_3 are collinear if and only if $|\overline{P_1 P_3}| = |\overline{P_1 P_2}| + |\overline{P_2 P_3}|$. Illustrate this fact for $P_1 = (-3, -2)$, $P_2 = (1, 2)$, and $P_3 = (3, 4)$.

29. Show that the points $A = (-2, 2)$, $B = (1, 4)$, $C = (3, 1)$, and $D = (0, -1)$ are the vertices of the square $ABCD$.

30. A median of a triangle is a line segment from a vertex to the midpoint of the opposite side. Find the length of each median of the triangle with the vertices $A = (2, 3)$, $B = (3, -3)$, and $C = (-1, -1)$.

In problems 31–34, find the coordinates of the midpoint M of the line segment $\overline{PQ}$.

31. $P = (5, 6)$ and $Q = (-7, 8)$
32. $P = (-4, 7)$ and $Q = (-3, 0)$
33. $P = (-2, 3)$ and $Q = (4, -2)$
34. $P = (2, -5)$ and $Q = (-1, -3)$

35. If one end of a line segement is the point $(-4, 2)$ and the midpoint is $(3, -1)$, find the coordinates of the other end of the line segment.

36. Find the coordinates of a point P on the line joining the two points $A = (-3, 4)$ and $B = (2, 5)$ such that $|\overline{AP}| = 2|\overline{PB}|$.

C In problems 37–40, find the distance between the two points. Round off the answers to four significant digits.

37. $(81.31, 74.01)$ and $(93.72, 61.37)$
38. $(39.14, 61.78)$ and $(49.07, 32.14)$
39. $(17.81, 13.75)$ and $(45.03, 21.82)$
40. $(19.41, 27.72)$ and $(34.16, 12.14)$

In problems 41–48, find the center and radius of each circle. Also graph each circle.

41. $(x + 1)^2 + (y - 2)^2 = 4$
42. $(x - 2)^2 + (y + 1)^2 = 9$
43. $x^2 + y^2 - 6x - 8y + 9 = 0$
44. $x^2 + y^2 - 10x - 10y + 25 = 0$
45. $2x^2 + 2y^2 + 12x - 8y + 13 = 0$
46. $2x^2 + 2y^2 - 2x + 2y - 7 = 0$
47. $x^2 + y^2 + 7x - 5y - 44 = 0$
48. $3x^2 + 3y^2 + 4y - 7 = 0$

7.3 Graphs of Linear Equations

We can graphically display the solutions of an equation in two variables by using the Cartesian coordinate system.

For example, suppose we want to graph the equation

$$2x + 3y = 6.$$

This equation can be rewritten in the form

$$y = 2 - \frac{2}{3}x.$$

Now we choose some values of x, say $-1, 0, 3$, and 6, and then we determine the corresponding values for y. This procedure will yield some ordered pairs

(x, y), which represent points on the graph. We show these points in the following table:

x	$y = 2 - \frac{2}{3}x$	Point (x, y)
-1	$y = 2 - \frac{2}{3}(-1) = \frac{8}{3}$	$(-1, \frac{8}{3})$
0	$y = 2 - \frac{2}{3}(0) = 2$	$(0, 2)$
3	$y = 2 - \frac{2}{3}(3) = 0$	$(3, 0)$
6	$y = 2 - \frac{2}{3}(6) = -2$	$(6, -2)$

The points $(-1, \frac{8}{3})$, $(0, 2)$, $(3, 0)$, and $(6, -2)$ are plotted in Figure 1a. These points appear to lie on a straight line. It can be shown that every solution of the equation $2x + 3y = 6$ corresponds to a point on a single line. It can also be shown that the coordinates of each point on this line satisfy the equation $2x + 3y = 6$. Hence, the line is the graph of the equation $2x + 3y = 6$, and we can **sketch the graph** (Figure 1b). This sketch is understood to be a portion of the complete graph, because the line does not terminate, but continues on in both directions. In general, the graph of any first-degree equation in the two variables x and y of the form

$$Ax + By = C,$$

where A, B, and C are constants, and where A and B cannot both equal 0, is a **straight line.** For this reason, such equations are called **linear equations.**

Figure 1

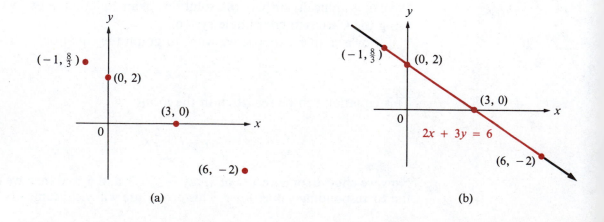

(a)   (b)

A straight line is completely determined by any two different points of the line. Therefore, we can graph a linear equation by finding two points whose coordinates are solutions of the equation, and drawing a straight line through these points. However, to avoid mistakes, we suggest that you plot a third point as a check. In general, we perform the following steps:

Procedure for Graphing a Linear Equation

> Step 1. Find three points on the graph of the given equation. (Choose three different values of x and calculate the corresponding values of y.)
>
> Step 2. Plot the three points on a Cartesian coordinate system.
>
> Step 3. Draw a straight line through the three points.

In Examples 1–4, sketch the graph of each equation.

EXAMPLE 1 $y = 3x$

SOLUTION   We choose three values of x and then determine the corresponding values for y. This will yield three points of the form (x, y). Let us choose the values $-2, 0,$ and 1 for x, and find the corresponding values for y. We show the resulting points in the following table:

x	$y = 3x$	Point (x, y)
-2	$y = 3(-2) = -6$	$(-2, -6)$
0	$y = 3(0) = 0$	$(0, 0)$
1	$y = 3(1) = 3$	$(1, 3)$

We plot these points and draw a straight line through them (Figure 2).

Figure 2

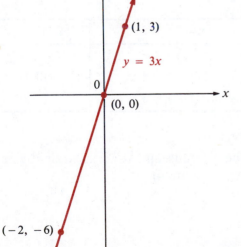

EXAMPLE 2 $y = -2x$

SOLUTION We choose three values of x and then determine the corresponding values for y to obtain three points of the form (x, y). Let us choose the values -2, 0, and 1 for x, and find the corresponding values for y. The resulting points are shown in the following table:

Figure 3

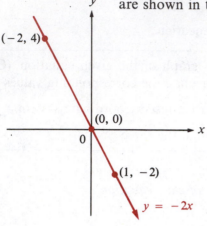

x	$y = -2x$	Point (x, y)
-2	$y = -2(-2) = 4$	$(-2, 4)$
0	$y = -2(0) = 0$	$(0, 0)$
1	$y = -2(1) = -2$	$(1, -2)$

We plot these points and draw a straight line through them (Figure 3).

EXAMPLE 3 $y = 2x + 1$

SOLUTION We choose three values of x and then determine the corresponding values for y to obtain three points of the form (x, y). Let us choose the values -2, 0, and 1 for x, and find the corresponding values for y. The resulting points are shown in the following table:

Figure 4

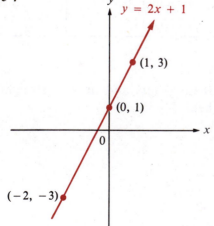

x	$y = 2x + 1$	Point (x, y)
-2	$y = 2(-2) + 1 = -3$	$(-2, -3)$
0	$y = 2(0) + 1 = 1$	$(0, 1)$
1	$y = 2(1) + 1 = 3$	$(1, 3)$

We plot these points and draw a straight line through them (Figure 4).

EXAMPLE 4 $y = -\frac{2}{3}x + 3$

SOLUTION   We choose the values -3, 0, and $\frac{9}{2}$ for x, and find the corresponding values for y. The resulting points are shown in the following table:

Figure 5

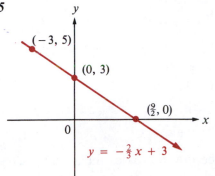

(−3, 5)

(0, 3)

$(\frac{9}{2}, 0)$

0

$y = -\frac{2}{3}x + 3$

x	$y = -\frac{2}{3}x + 3$	Point (x, y)
-3	$y = -\frac{2}{3}(-3) + 3 = 5$	$(-3, 5)$
0	$y = -\frac{2}{3}(0) + 3 = 3$	$(0, 3)$
$\frac{9}{2}$	$y = -\frac{2}{3}(\frac{9}{2}) + 3 = 0$	$(\frac{9}{2}, 0)$

We plot these points and draw a straight line through them (Figure 5).

The two points $(\frac{9}{2}, 0)$ and $(0, 3)$ used to sketch the graph in Example 4 are the points of intersection of the given line $y = -\frac{2}{3}x + 3$ with the x and y axes. Thus the x intercept is $\frac{9}{2}$ and the y intercept is 3. In general, suppose that L is any nonvertical line. If $(a, 0)$ is the only point of intersection of the line L with the x axis, then a is called the **x intercept** of L. Similarly, if $(0, b)$ is the only point of intersection of L with the y axis, then b is called the **y intercept** of L (Figure 6).

Figure 6

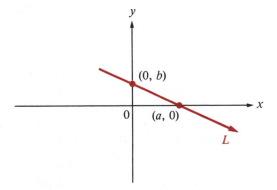

(0, b)

0   (a, 0)

L

To determine the x intercept of line L, set $y = 0$ in any equation of L and solve for x; to determine the y intercept, set $x = 0$ and solve for y.

EXAMPLE 5   Find the intercepts of the line $3x - 4y - 12 = 0$. Also sketch the graph.

SOLUTION   To locate the x intercept, set $y = 0$ and solve for x:

$$3x - 0 - 12 = 0$$
$$3x = 12$$
$$x = 4.$$

Therefore, the x intercept is 4. To locate the y intercept, set $x = 0$ and solve for y:

$$0 - 4y - 12 = 0$$
$$-4y = 12$$
$$y = -3.$$

Thus, the y intercept is -3. So, the graph crosses the x and y axes at the points $(4, 0)$ and $(0, -3)$, respectively (Figure 7).

Figure 7

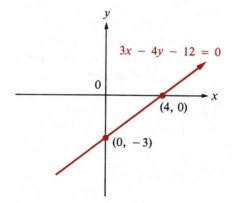

Lines that are parallel to the x or y axes have only one intercept. These lines have equations that are special cases of the form $Ax + By = C$. If $B = 0$, then $Ax = C$ or $x = C/A$ for all values of y. Similarly, if $A = 0$, then $By = C$, or $y = C/B$ for all values of x. These special cases give rise to **vertical** and **horizontal** lines as the following examples show.

In Examples 6 and 7, sketch the graph of each line.

EXAMPLE 6 $x - 2 = 0$.

SOLUTION   The line $x - 2 = 0$ or $x = 2$ is the set of all points whose x coordinate is 2. The variable y does not appear in the equation, so the y coordinate can be

Figure 8

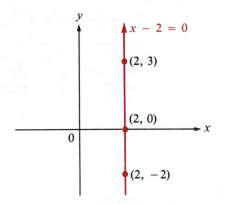

any number. The graph is a vertical line parallel to the y axis. The x intercept is 2, but there is no y intercept (Figure 8).

EXAMPLE 7 $2y = -4$

SOLUTION   The line $2y = -4$ or $y = -2$ is the set of all points whose y coordinates are -2. The variable x does not appear in the equation, so the x coordinate can be any number. The graph is a horizontal line parallel to the x axis. The y intercept is -2, but there is no x intercept (Figure 9).

Figure 9

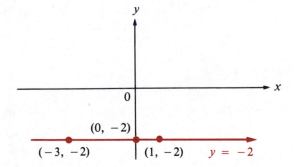

Many applications of mathematics involve the use of linear equations. The following example illustrates one of these applications.

EXAMPLE 8   The formula for changing temperature readings from Celsius C to Fahrenheit F is given by the linear equation

$$F = \frac{9}{5}C + 32.$$

(a) If the temperature is 15°C, what is the Fahrenheit temperature?

(b) If the temperature is 85°C, what is the Fahrenheit temperature?

(c) Sketch the graph of the equation by representing C on the horizontal axis and F on the vertical axis.

SOLUTION (a) When $C = 15$,

$$F = \frac{9}{5}(15) + 32$$
$$= 27 + 32$$
$$= 59.$$

(b) When $C = 85$,

$$F = \frac{9}{5}(85) + 32$$
$$= 153 + 32$$
$$= 185.$$

(c) We obtain the graph by drawing a straight line through the points (15, 59) and (85, 185) (Figure 10).

Figure 10

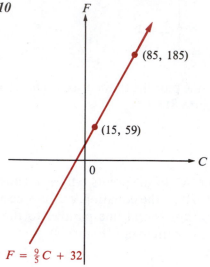

$F = \frac{9}{5}C + 32$

PROBLEM SET 7.3

In problems 1–16, sketch the graph of each equation by following the procedure on page 349.

1. $y = 2x$ **2.** $y = -5x$ **3.** $y = -\frac{2}{3}x$ **4.** $y = 7x$

5. $y = -4x$ **6.** $y = -3x$ **7.** $y = \frac{3}{7}x$ **8.** $y = \frac{4}{5}x$

9. $y = 3x + 1$ **10.** $y = 5x + 2$ **11.** $y = \frac{2}{3}x - 4$ **12.** $y = \frac{3}{4}x - 1$

13. $y = -2x + 5$ **14.** $y = -3x + 2$ **15.** $y = -\frac{2}{5}x + 1$ **16.** $y = -\frac{3}{4}x - 5$

In problems 17–36, find the x intercept and the y intercept, and use the intercepts to sketch the graph of each equation.

17. $2x + 3y = -6$ **18.** $4x + 3y = 12$ **19.** $x - 3y = 5$ **20.** $4x - 3y - 12 = 0$

21. $-x + 5y = 10$ **22.** $-3x + 5y = 4$ **23.** $-2x - 4y + 1 = 0$ **24.** $-x - 2y + 4 = 0$

25. $-3x + 7y - 4 = 0$ **26.** $2x - 5y - 3 = 0$ **27.** $3y = 6$ **28.** $4y = 7$

29. $-2y = 5$ **30.** $-3y = 8$ **31.** $y - 6 = 0$ **32.** $2y - 3 = 0$

33. $3x - 2 = 0$ **34.** $5x + 1 = 0$ **35.** $-4x + 1 = 0$ **36.** $-7x + 2 = 0$

37. Chemists have found that the volume V (in cubic centimeters) of a liquid in a test tube changes as the temperature T (in Celsius) changes. The formula

$$V = \frac{1}{4}T + 32$$

describes this relationship. Sketch the graph of this equation by representing T on the horizontal axis and V on the vertical axis. Also find the volume when the temperature is 5°C.

38. A produce store makes a profit P (in cents) from selling x melons, and

$$P = 15x - 180.$$

(a) Sketch the graph of this equation by representing x on the horizontal axis and P on the vertical axis.

(b) How many melons must be sold for the store to break even?

(c) What is the store's profit if 100 melons are sold?

39. A manufacturer can produce x small calculators at a total cost of c dollars, where c is given by the equation

$$c = 8x + 350.$$

(a) Sketch the graph of this equation by representing x on the horizontal axis and c on the vertical axis for $2 \le x \le 150$.

(b) What is the total cost of producing 130 calculators?

40. In physics it is known from **Hooke's law** that the relationship between the stretch s (in inches) of a spring and the weight w of an object (in pounds) is given by the equation

$$s = \frac{1}{10}w.$$

(a) Sketch the graph of this equation by representing w on the horizontal axis and s on the vertical axis for $0 \le w \le 50$.

(b) What is the stretch of the spring for a 40-pound weight?

41. Suppose that the value V of an automobile depreciates according to the equation

$$V = 10{,}000 - 2{,}000t,$$

where t is the number of years after it was bought.

(a) Sketch the graph of this equation by representing t on the horizontal axis and V on the vertical axis for $0 \le t \le 5$.

(b) What is the value V of the automobile after 3 years?

7.4 The Slope of a Line

Figure 1

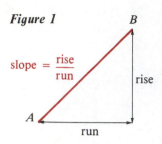

slope = $\dfrac{\text{rise}}{\text{run}}$

We commonly use the word "slope" to refer to a steepness of an incline, or some deviation from the horizontal. For instance, we speak of a ski slope or the slope of a roof. In mathematics, the word slope has a similar meaning. Consider the line segment $\overline{AB}$ in Figure 1. The horizontal distance between A and B is called the **run.** The vertical distance between A and B is called the **rise.** The ratio of the rise to the run is called the **slope** of the line segment

A ski slope in the Swiss Alps.

$\overline{AB}$, and is represented by the letter m:

$$m = \frac{\text{rise}}{\text{run}}.$$

If the line segment $\overline{AB}$ is horizontal, its rise is zero, and its slope $m = \text{rise/run} = 0$ (Figure 2a). If $\overline{AB}$ slants upward to the right, its rise is considered to be positive, and the slope $m = \text{rise/run}$ is positive (Figure 2b). If $\overline{AB}$ slants downward to the right, its rise is considered to be negative; hence, its slope $m = \text{rise/run}$ is negative (Figure 2c). The run is always considered to be nonnegative.

Now let's consider the line segment $\overline{AB}$ in which $A = (x_1, y_1)$ and $B = (x_2, y_2)$ (Figure 3). Note in Figure 3 that B is above and to the right of A.

Figure 2

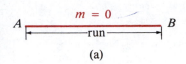

(a)

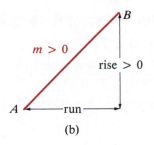

(b)

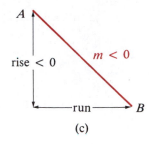

(c)

Figure 3

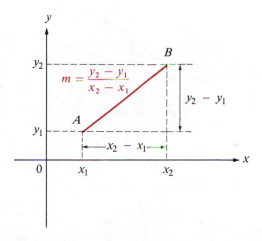

The line segment $\overline{AB}$ has rise $= y_2 - y_1$ and run $= x_2 - x_1$. Thus the slope is given by the formula

$$m = \frac{y_2 - y_1}{x_2 - x_1}.$$

provided that $x_1 \neq x_2$. Notice that

$$\frac{y_2 - y_1}{x_2 - x_1} = \frac{y_1 - y_2}{x_1 - x_2}.$$

Therefore, the slope of a line segment is the same, no matter which endpoint is called (x_1, y_1) and which is called (x_2, y_2). Notice also that the slope m of a vertical line segment is undefined because the denominator (the run) is zero, that is,

$$m = \frac{y_2 - y_1}{0}$$

is undefined.

EXAMPLE 1   Sketch each line segment $\overline{AB}$ and find its slope m.

(a) $A = (4, 1)$ and $B = (7, 6)$ (b) $A = (3, 9)$ and $B = (7, 4)$

(c) $A = (-1, 4)$ and $B = (3, 4)$ (d) $A = (2, -3)$ and $B = (2, 2)$

SOLUTION   The line segments are sketched in Figure 4.

(a) $m = \dfrac{y_2 - y_1}{x_2 - x_1} = \dfrac{6 - 1}{7 - 4} = \dfrac{5}{3}$

(b) $m = \dfrac{y_2 - y_1}{x_2 - x_1} = \dfrac{4 - 9}{7 - 3} = -\dfrac{5}{4}$

(c) $m = \dfrac{y_2 - y_1}{x_2 - x_1} = \dfrac{4 - 4}{3 - (-1)} = \dfrac{0}{4} = 0$

(d) m is undefined because $x_2 - x_1 = 2 - 2 = 0$.

Figure 4

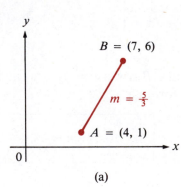

(a)

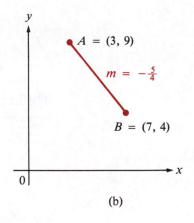

(b)

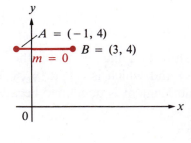

(c)

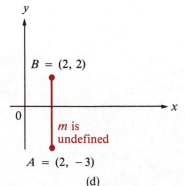

(d)

Figure 5

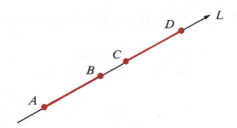

If two line segments $\overline{AB}$ and $\overline{CD}$ lie on the same line L, then they have the same slope (Figure 5). In fact, the slope of a line L is the same no matter which two points of the line are considered. The common slope of all line segments lying on a line L is called the **slope** of L.

EXAMPLE 2   Sketch the line L that contains the point $P = (-2, 3)$ and that has slope (a) $m = \frac{3}{5}$ and (b) $m = -\frac{3}{5}$.

SOLUTION   (a) The condition $m = \frac{3}{5}$ means that for every five units we move to the right from a point on L, we must move up three units to get back to L. If we start at the point $P = (-2, 3)$ on L and move five units to the right and three units up, we arrive at another point on line L, the point $Q = (-2 + 5, 3 + 3) = (3, 6)$. Because any two points on a line determine the line, we simply plot $P = (-2, 3)$ and $Q = (3, 6)$, and draw the line L (Figure 6a).

(b) The condition $m = -\frac{3}{5}$ means that for every five units we move to the right from a point on L, we must move down three units to get back to L. If we start at the point $P = (-2, 3)$ on L and move five units to the right and three units down, we arrive at the point $Q = (-2 + 5, 3 - 3) = (3, 0)$ on L. Thus, we plot $P = (-2, 3)$ and $Q = (3, 0)$ and draw the line L (Figure 6b).

Figure 6

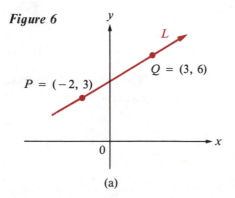

(a)

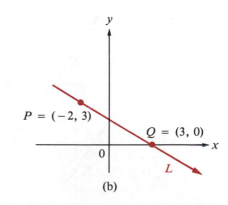

(b)

Figure 7

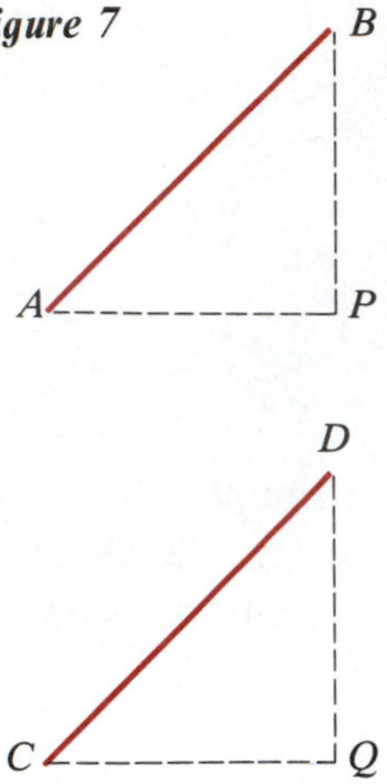

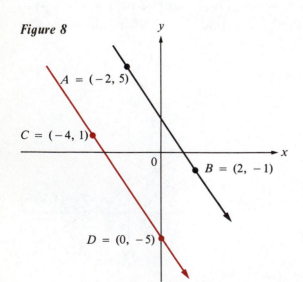

Parallel and Perpendicular Lines

Consider the similar triangles APB and CQD in Figure 7. The slopes of the two parallel line segments $\overline{AB}$ and $\overline{CD}$ are

$$\frac{|BP|}{|AP|} \quad \text{and} \quad \frac{|DQ|}{|CQ|},$$

respectively. These slopes are equal because they are the ratios of the corresponding sides of similar triangles:

$$\frac{|BP|}{|AP|} = \frac{|DQ|}{|CQ|}.$$

It follows that two parallel lines have the same slope. We can also use geometry to show that two different lines with the same slope are parallel. Therefore, we have the following condition:

The Parallelism Condition

> Two different nonvertical lines in the Cartesian plane with slopes m_1 and m_2 are **parallel** if and only if $m_1 = m_2$.

EXAMPLE 3   Determine whether or not the line containing the points A and B is parallel to the line containing the points C and D if $A = (-2, 5)$, $B = (2, -1)$, $C = (-4, 1)$, and $D = (0, -5)$. Sketch the lines.

SOLUTION   The lines are sketched in Figure 8. We compute the slopes m_1 of $\overline{AB}$ and m_2 of $\overline{CD}$:

Figure 8

$$m_1 = \frac{y_2 - y_1}{x_2 - x_1}$$

$$= \frac{-1 - 5}{2 - (-2)} = -\frac{6}{4} = -\frac{3}{2}$$

$$m_2 = \frac{-5 - 1}{0 - (-4)} = -\frac{6}{4} = -\frac{3}{2}.$$

The two lines are parallel because the slopes are equal.

EXAMPLE 4

Find a constant real number k such that the line containing points A and B is parallel to the line containing points C and D, if $A = (-1, 1)$, $B = (1, k)$, $C = (3, 3)$, and $D = (5, 6)$.

SOLUTION

We compute the slopes m_1 of $\overline{AB}$ and m_2 of $\overline{CD}$:

$$m_1 = \frac{y_2 - y_1}{x_2 - x_1} = \frac{k - 1}{1 - (-1)} = \frac{k - 1}{2}$$

$$m_2 = \frac{6 - 3}{5 - 3} = \frac{3}{2}.$$

Because the two lines are to be parallel, we must have

$$m_1 = m_2.$$

We substitute the computed values for the slopes:

$$\frac{k - 1}{2} = \frac{3}{2},$$

or

$$k - 1 = 3$$

$$k = 4.$$

If we have two perpendicular lines with slopes m_1 and m_2, we can show that the product of their slopes $m_1 m_2$ is -1. Conversely, if we have two lines with slopes m_1 and m_2 such that the product of their slopes $m_1 m_2$ is -1, we can show that the lines are perpendicular.

The Perpendicularity Condition

> Two nonvertical lines in the Cartesian plane with slopes m_1 and m_2 are **perpendicular** if and only if $m_1 m_2 = -1$, or, equivalently, $m_2 = -1/m_1$.

EXAMPLE 5

Determine whether or not the line containing the points A and B is perpendicular to the line containing the points C and D, if $A = (-1, 1)$, $B = (1, 5)$, $C = (-2, -3)$, and $D = (2, -5)$.

SOLUTION

The lines are sketched in Figure 9. We compute the slopes m_1 of $\overline{AB}$ and m_2 of $\overline{CD}$:

$$m_1 = \frac{y_2 - y_1}{x_2 - x_1} = \frac{5 - 1}{1 - (-1)} = \frac{4}{2} = 2$$

$$m_2 = \frac{-5 - (-3)}{2 - (-2)} = \frac{-2}{4} = -\frac{1}{2}.$$

The two lines are perpendicular because $m_1 m_2 = 2(-\tfrac{1}{2}) = -1$.

Figure 9

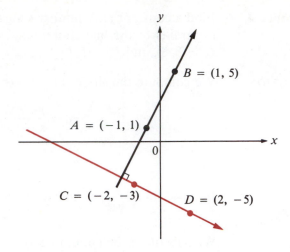

PROBLEM SET 7.4

In problems 1–10, sketch the line segment $\overline{AB}$ and find its slope m.

1. $A = (2, -1)$ and $B = (-3, 4)$ **2.** $A = (1, -4)$ and $B = (2, 3)$ **3.** $A = (1, 5)$ and $B = (-2, 3)$

4. $A = (4, 3)$ and $B = (-3, -4)$ **5.** $A = (6, -1)$ and $B = (0, 2)$ **6.** $A = (7, 1)$ and $B = (-8, 3)$

7. $A = (-2, 5)$ and $B = (4, 5)$ **8.** $A = (-1, 4)$ and $B = (3, 4)$ **9.** $A = (4, -1)$ and $B = (4, 3)$

10. $A = (3, 2)$ and $B = (3, -5)$

In problems 11–20, sketch the line L that contains the given point P and that has slope m.

11. $P = (1, 5)$ and $m = -3$ **12.** $P = (2, 5)$ and $m = 4$ **13.** $P = (2, -1)$ and $m = 2$

14. $P = (2, -3)$ and $m = -\frac{1}{2}$ **15.** $P = (7, -2)$ and $m = \frac{2}{3}$ **16.** $P = (-1, -6)$ and $m = \frac{3}{4}$

17. $P = (2, 3)$ and $m = -\frac{3}{7}$ **18.** $P = (1, 4)$ and $m = -\frac{4}{3}$ **19.** $P = (-1, -3)$ and $m = -\frac{4}{5}$

20. $P = (-1, -5)$ and $m = -\frac{3}{4}$

In problems 21–28, determine whether the line containing the points A and B is parallel or perpendicular (or neither) to the line containing the points C and D.

21. $A = (2, 4)$, $B = (3, 8)$, $C = (5, 1)$, and $D = (4, -3)$

22. $A = (2, -3)$, $B = (-4, 5)$, $C = (-1, 0)$, and $D = (-4, 4)$

23. $A = (1, 9)$, $B = (4, 0)$, $C = (0, 6)$, and $D = (5, 3)$

24. $A = (8, -1)$, $B = (2, 3)$, $C = (5, 1)$, and $D = (2, -7)$

25. $A = (2, 4)$, $B = (3, 8)$, $C = (8, -2)$, and $D = (-4, 1)$

26. $A = (-\frac{5}{3}, 0)$, $B = (0, 5)$, $C = (2, 1)$, and $D = (5, 0)$

27. $A = (8, -7)$, $B = (-7, 8)$, $C = (10, -7)$, and $D = (-4, 6)$

28. $A = (-2, 8)$, $B = (8, 2)$, $C = (-8, -2)$, and $D = (2, -8)$

In problems 29–34, find a constant real number k such that the lines containing the segments $\overline{AB}$ and $\overline{CD}$ are (a) parallel and (b) perpendicular.

29. $A = (2, 1)$, $B = (6, 3)$, $C = (4, k)$, and $D = (3, 1)$
30. $A = (k, 1)$, $B = (3, 2)$, $C = (7, 1)$, and $D = (6, 3)$
31. $A = (-10, k)$, $B = (-5, -5)$, $C = (4, 4)$, and $D = (10, 10)$
32. $A = (4, 9)$, $B = (10, k)$, $C = (-2, 11)$, and $D = (4, -2)$
33. $A = (2, 1)$, $B = (0, 4)$, $C = (2, k)$, and $D = (-3, 1)$
34. $A = (-9, -1)$, $B = (-6, 0)$, $C = (7, 1)$, and $D = (4, k)$
35. Show that the four points $A = (1, 9)$, $B = (4, 0)$, $C = (0, 6)$, and $D = (5, 3)$ are vertices of a parallelogram.
36. Show that the four points $A = (-4, -1)$, $B = (0, 2)$, $C = (-2, -1)$, and $D = (2, 2)$ are vertices of a parallelogram.

In problems 37–40, use the concept of slope to show that the triangle with vertices A, B, and C is a right triangle.

37. $A = (-4, 2)$, $B = (1, 4)$, and $C = (3, -1)$
38. $A = (2, 1)$, $B = (3, -1)$, and $C = (1, -2)$
39. $A = (8, 5)$, $B = (1, -2)$, and $C = (-3, 2)$
40. $A = (-6, 3)$, $B = (3, -5)$, and $C = (-1, 5)$

In problems 41–44, use the concept of slope to determine whether or not the points in each set are collinear. The points $P_1 = (x_1, y_1)$, $P_2 = (x_2, y_2)$, and $P_3 = (x_3, y_3)$ are **collinear** if the slope of the line between P_1 and P_2 is the same as the slope of the line between P_1 and P_3 (or between P_2 and P_3).

41. $(1, 1)$, $(2, 4)$, and $(3, 2)$
42. $(0, 3)$, $(1, 1)$, and $(2, -1)$
43. $(1, -3)$, $(-1, -11)$, and $(-2, -15)$
44. $(1, 5)$, $(-2, -1)$, and $(-3, -3)$
45. Show by means of slopes that the points $A = (-4, -1)$, $B = (3, \frac{8}{3})$, $C = (8, -4)$, and $D = (2, -9)$ are the vertices of a trapezoid.
46. Show by means of slopes that the diagonals of a rhombus, with vertices $A = (0, 0)$, $B = (5, 0)$, $C = (3, 4)$, and $D = (8, 4)$ are perpendicular.

7.5 Equations of Lines

Suppose that we are given a point on a line and the slope of the line. We ask: Can we find an equation to describe all the points on the line? To answer this question, consider a nonvertical line L having slope m and containing the point $P_1 = (x_1, y_1)$ (Figure 1). If $P = (x, y)$ is any other point on L, then the slope m is given by

$$m = \frac{y - y_1}{x - x_1}.$$

Figure 1

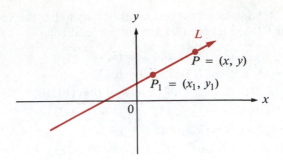

We multiply both sides of this equation by $x - x_1$:

$$y - y_1 = m(x - x_1).$$

The latter equation is known as the **point-slope form** of an equation for the line L, which contains the point $P_1 = (x_1, y_1)$ and has the slope m.

In Examples 1 and 2, find an equation for the line L in point-slope form.

EXAMPLE 1   The line L contains the point $(-2, 3)$ and has slope $m = 4$.

SOLUTION   We substitute $x_1 = -2$, $y_1 = 3$, and $m = 4$ in $y - y_1 = m(x - x_1)$:

$$y - 3 = 4[x - (-2)]$$

or

$$y - 3 = 4(x + 2).$$

EXAMPLE 2   The line L contains the points $(6, 1)$ and $(8, 7)$.

SOLUTION   The slope m is given by

$$m = \frac{y_2 - y_1}{x_2 - x_1} = \frac{7 - 1}{8 - 6} = \frac{6}{2} = 3.$$

We use $P_1 = (x_1, y_1) = (6, 1)$:

$$y - 1 = 3(x - 6).$$

We can find an equation of a line L in point-slope form if we know the slope m and a point on the line. If, for example, the point is $P_1 = (0, b)$, where b is the y intercept (Figure 2), then an equation of L is given by

$$y - b = m(x - 0).$$

Figure 2

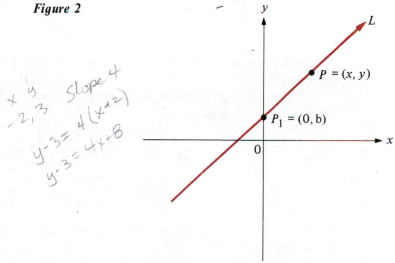

We can simplify this equation:

$$y = mx + b.$$

The latter equation is called the **slope-intercept** form of an equation for L. In the equation $y = mx + b$ *the coefficient of x is the slope and the constant term b is the y intercept.*

EXAMPLE 3   Find the slope-intercept form of an equation of a line L having slope $m = -\frac{2}{3}$ and y intercept $b = 5$.

SOLUTION   We substitute $m = -\frac{2}{3}$ and $b = 5$ in the equation $y = mx + b$ to obtain

$$y = -\frac{2}{3}x + 5.$$

EXAMPLE 4   Write the equation $3x + 2y - 6 = 0$ in slope-intercept form. Find the slope m, the y intercept b, the x intercept, and sketch the graph.

SOLUTION   We begin by solving the equation for y in terms of x:

$$3x + 2y - 6 = 0$$

$$2y = -3x + 6$$

$$y = -\frac{3}{2}x + 3.$$

This last equation of the line is in slope-intercept form with $m = -\frac{3}{2}$ and y intercept $b = 3$.

To find the x intercept we substitute $y = 0$ in the original equation $3x + 2y - 6 = 0$ and solve for x:

$$3x + 2(0) - 6 = 0$$

$$3x - 6 = 0$$

$$3x = 6$$

$$x = 2.$$

Thus, we obtain the graph by drawing the line through the points $(2, 0)$ and $(0, 3)$ (Figure 3).

A **horizontal line** has slope $m = 0$. Hence, in slope-intercept form, an equation for a horizontal line is (Figure 4)

$$y = 0(x) + b$$

or

$$y = b.$$

Figure 3

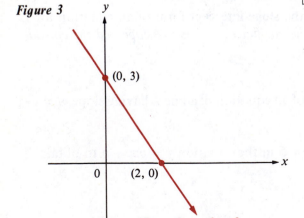

Figure 4

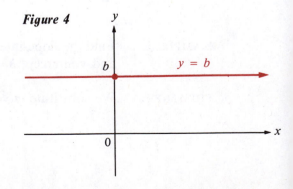

A **vertical line** has an undefined slope. Therefore, you cannot write an equation for a vertical line in slope-intercept form. However, because all points on a vertical line have the same abscissa or x coordinate—say, a—an equation of such a line is (Figure 5)

$$x = a.$$

Figure 5

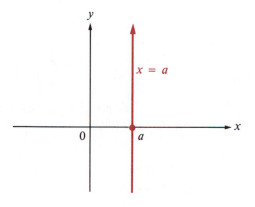

EXAMPLE 5	(a) Determine an equation for the horizontal line that contains the point $(1, 4)$.
	(b) Determine an equation for the vertical line that contains the point $(-4, 3)$.
SOLUTION	(a) The slope of a horizontal line is 0. The y intercept is the y coordinate of the given point, that is, $b = 4$. Therefore, an equation is

$$y = 4.$$

(b) The slope is undefined because the line is vertical. The vertical line consists of all points with the x coordinate -4. Therefore, an equation is

$$x = -4.$$

If A, B, and C are constants, and if A and B are not both zero, any equation of the form

$$Ax + By + C = 0$$

represents a line. This equation is called the **general form** of an equation of the line.

In Examples 6 and 7, find an equation of the line L in (a) point-slope form, (b) slope-intercept form, and (c) general form.

EXAMPLE 6 The line L contains the point $(3, -5)$ and is parallel to the line L_1, whose equation is $3x - 4y + 9 = 0$.

SOLUTION We obtain the slope m_1 of L_1 by solving the equation $3x - 4y + 9 = 0$ for y in terms of x:

$$3x - 4y + 9 = 0$$

$$-4y = -3x - 9$$

$$y = \frac{3}{4}x + \frac{9}{4}.$$

Thus, the slope of L_1 is $m_1 = \frac{3}{4}$. We use the parallelism condition to find the slope m of L:

$$m = m_1 = \frac{3}{4}.$$

(a) Because L has the slope $m = \frac{3}{4}$ and contains the point $(3, -5)$, its equation in point-slope form is

$$y - (-5) = \frac{3}{4}(x - 3) \qquad \text{or} \qquad y + 5 = \frac{3}{4}(x - 3).$$

(b) To obtain an equation of L in slope-intercept form, we solve the equation in part (a) for y in terms of x.

$$y + 5 = \frac{3}{4}(x - 3)$$

$$y + 5 = \frac{3}{4}x - \frac{9}{4}$$

$$y = \frac{3}{4}x - \frac{9}{4} - 5$$

$$y = \frac{3}{4}x - \frac{29}{4}.$$

(c) To obtain an equation of L in general form, we multiply both sides of the equation in part (b) by 4 and rearrange terms.

$$y = \frac{3}{4}x - \frac{29}{4}$$

$$4y = 3x - 29$$

$$3x - 4y - 29 = 0.$$

EXAMPLE 7 L contains the point $(-4, 7)$ and is perpendicular to the line L_1, whose equation is $5x + 7y - 11 = 0$.

SOLUTION   We obtain the slope m_1 of L_1 by solving the equation $5x + 7y - 11 = 0$ for y in terms of x:

$$5x + 7y - 11 = 0$$

$$7y = -5x + 11$$

$$y = -\frac{5}{7}x + \frac{11}{7}.$$

Thus, the slope of L_1 is $m_1 = -\frac{5}{7}$. We use the perpendicularity condition to find the slope m of L:

$$m = -\frac{1}{m_1} = \frac{7}{5}.$$

(a) Because L has slope $m = \frac{7}{5}$ and contains the point $(-4, 7)$, its equation in point-slope form is

$$y - 7 = \frac{7}{5}[x - (-4)],$$

or

$$y - 7 = \frac{7}{5}(x + 4).$$

(b) We solve the equation in part (a) for y in terms of x and obtain an equation of L in slope-intercept form:

$$y - 7 = \frac{7}{5}(x + 4)$$

$$y - 7 = \frac{7}{5}x + \frac{28}{5}$$

$$y = \frac{7}{5}x + \frac{63}{5}.$$

(c) We multiply both sides of the equation in part (b) by 5 and rearrange terms to obtain an equation of L in general form:

$$y = \frac{7}{5}x + \frac{63}{5}$$

$$5y = 7x + 63$$

$$7x - 5y + 63 = 0.$$

EXAMPLE 8   A book salesperson receives a monthly salary plus commission on sales. The relationship between income (salary plus commission) and the amount of sales is linear. The salesperson's income was $3,400 during a month when sales totaled $12,000 and was $2,900 in another month, when sales totaled $9,500.

(a) Write an equation that expresses the salesperson's monthly income I in terms of the total sales T for that month. What is the monthly salary?

(b) Find the salesperson's rate of commission.

(c) If sales were $14,000 in a particular month, what was the salesperson's income I for that month?

SOLUTION   (a) The information given can be expressed as the two ordered pairs

$$(T_1, I_1) = (9,500, 2,900) \quad \text{and} \quad (T_2, I_2) = (12,000, 3,400).$$

The slope of the line through these points is

$$m = \frac{I_2 - I_1}{T_2 - T_1} = \frac{3,400 - 2,900}{12,000 - 9,500}$$

$$= \frac{500}{2,500} = \frac{1}{5}.$$

Using the point (9,500, 2,900), we obtain the equation

$$I - 2,900 = \frac{1}{5}(T - 9,500).$$

This is equivalent to

$$I - 2,900 = \frac{1}{5}T - 1,900,$$

or

$$I = \frac{1}{5}T + 1,000.$$

This equation informs us that the monthly salary was $1,000, because if the total monthly sales T were zero, no commission would be earned and income I would equal salary, $1,000.

(b) The slope $m = \frac{1}{5}$ gives the salesperson's rate of commission. We can express this rate as a percentage:

$$\frac{1}{5} = 0.20$$

$$= 20\%.$$

(c) Here $T = 14,000$. Thus,

$$I = \frac{1}{5}(14,000) + 1,000$$
$$= 2,800 + 1,000$$
$$= 3,800.$$

Hence, the salesperson's income for a month in which sales were $14,000 was $3,800, made up of $2,800 in commission and $1,000 in salary.

PROBLEM SET 7.5

In problems 1–18, find an equation of the line L.

1. L contains $P = (-1, 2)$ and has slope $m = 5$.

2. L contains $P = (0, 3)$ and has slope $m = -7$.

3. L contains $P = (7, 3)$ and has slope $m = -3$.

4. L contains $P = (-1, 4)$ and has slope $m = \frac{2}{5}$.

5. L contains $P = (5, -1)$ and has slope $m = -\frac{3}{7}$.

6. L contains $P = (-4, 6)$ and has slope $m = -\frac{3}{4}$.

7. L contains $P = (0, 0)$ and has slope $m = \frac{3}{8}$.

8. L contains $P = (0, 0)$ and has slope $m = -\frac{4}{9}$.

9. L contains $P = (-1, -5)$ and has slope $m = 0$.

10. L contains $P = (-\frac{1}{2}, 4)$ and has slope $m = 0$.

11. L contains $P_1 = (-3, 2)$ and $P_2 = (3, 5)$.

12. L contains $P_1 = (-2, 4)$ and $P_2 = (0, 1)$.

13. L has slope $m = -3$ and y intercept $b = 5$.

14. L has slope $m = -7$ and y intercept $b = 2$.

15. L has slope $m = -\frac{3}{7}$ and y intercept $b = 0$.

16. L has slope $m = -\frac{5}{11}$ and y intercept $b = 0$.

17. L contains $P = (-3, 4)$ with undefined slope.

18. L contains $P = (2, -5)$ with undefined slope.

In problems 19–26, express each equation in slope-intercept form. Find the slope m, the x intercept, the y intercept, and sketch the graph.

19. $2x - 3y - 1 = 0$

20. $2x + 3y + 12 = 0$

21. $y - 1 = -2(x - 2)$

22. $5x - 7y - 8 = 0$

23. $4x - y + 5 = 0$

24. $3x - 4y - 5 = 0$

25. $-2x + y = 0$

26. $y - 3 = -4(x - 3)$

In problems 27–40, find equations of the line L in (a) point-slope form, (b) slope-intercept form, and (c) general form.

27. L contains $P = (3, -1)$ and is parallel to the line with equation $x + 2y + 7 = 0$.

28. L contains $P = (-2, 1)$ and is parallel to the line with equation $5x - 7y - 8 = 0$.

29. L contains $P = (-\frac{1}{2}, 5)$ and is parallel to the line containing $P_1 = (4, -3)$ and $P_2 = (5, 7)$.

30. L contains $P = (6, -1)$ and is parallel to the line containing $P_1 = (-2, 3)$ and $P_2 = (3, -4)$.

31. L contains $P = (1, 5)$ and is perpendicular to the line whose equation is given by $2x + 3y - 1 = 0$.

32. L contains $P = (-3, 2)$ and is perpendicular to the line whose equation is given by $x + 2y - 5 = 0$.

33. L contains $P = (0, 0)$ and is perpendicular to the line containing $P_1 = (3, 0)$ and $P_2 = (-2, 3)$.

34. L contains $P = (-1, 2)$ and is perpendicular to the line containing $P_1 = (5, 1)$ and $P_2 = (-2, 3)$.

35. L has y intercept 3 and is parallel to the line $x - 2y = 5$.

36. L has y intercept 5 and is perpendicular to the line $y = 3x$.

37. L has x intercept -1 and is perpendicular to the line $3x - 2y = 7$.

38. L contains $P = (7, -3)$ and is parallel to the x axis.

39. L contains $P = (-1, 6)$ and is perpendicular to the y axis.

40. L is the perpendicular bisector of the line segment $\overline{AB}$, where $A = (4, -5)$ and $B = (8, 5)$.

41. Find a value of k so that each of the following conditions will hold.

   (a) The line $3x + ky + 2 = 0$ is parallel to the line $6x - 5y + 3 = 0$.

   (b) The line $y = (2 - k)x + 2$ is perpendicular to the line $y = 3x - 1$.

42. Suppose that the line L has nonzero x and y intercepts a and b, respectively. Show that an equation of L can be written in the **intercept form**

$$\frac{x}{a} + \frac{y}{b} = 1.$$

In problems 43–48, use the result of problem 42 to find the equation of the line in intercept form.

43. x intercept 5 and y intercept 6

44. x intercept -2 and y intercept 7

45. x intercept -3 and y intercept -1

46. x intercept -1 and y intercept -5

47. $5x - 2y = 5$

48. $9x + 5y = 13$

49. A jogger's heartbeat N (in beats per minute) is related to her speed V (in feet per second) by a linear equation. The jogger's heartbeat is 75 beats per minute when her speed is 10 feet per second, and her heartbeat is 80 beats per minute when her speed is 12 feet per second.

   (a) Write an equation that expresses the jogger's heartbeat N in terms of her speed V.

   (b) If the jogger's heartbeat is 90 beats per minute, what is her speed?

50. The annual simple interest earned I is related to the amount P invested in a bank by a linear equation. If you invest \$650, you earn \$45.50 in 1 year, and if you invest \$1,375, you earn \$96.25 in a year.

   (a) Write an equation that expresses the relationship between the annual interest earned I and the amount P invested.

   (b) How much money should you invest to earn \$515.55 in one year?

7.6 Graphs of Linear Inequalities

Recall from Section 7.3 that the graph of a linear equation in two unknowns is a straight line. Now we graph linear inequalities such as

$$y < 3x + 6, \qquad 5x - 7 > y, \qquad \text{and} \qquad y \geq 5x + 3.$$

The **graph** of an inequality in two unknowns is defined to be the set of all points (x, y) in the plane whose coordinates satisfy the inequality. To study the graph of a linear inequality in x and y, we begin by considering the graph of the associated linear equation. This latter graph is obtained by temporarily replacing the inequality sign with an equals sign. The graph is a straight line that divides the xy plane into two regions called **half-planes,** one above the line and one below it (assuming the line is not vertical). In Figure 1, the graph of the line $y = 3x + 6$ divides the xy plane into an upper half-plane A and a lower half-plane C. The set of all points on the line is represented by B. The solution of the inequality $y \leq 3x + 6$ consists of all points in the half-plane C and includes all points on the line B (Figure 1). The procedure for sketching the graph of a linear inequality follows:

Figure 1

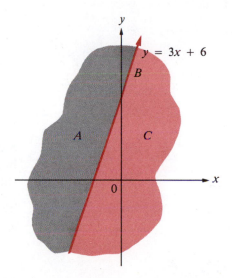

Procedure for Sketching Graphs of Linear Inequalities

Step 1. Sketch the graph of the linear equation obtained by replacing the inequality sign with an equals sign. If the inequality has the symbols $<$ or $>$, draw a dashed line; if the inequality has the symbols $\leq$ or $\geq$, draw a solid line.

Step 2. Determine which half-plane corresponds to the inequality. To do this, select any convenient test point (x, y) *not* on the line.
 (i) If the point satisfies the original inequality, shade the half-plane containing the test point.
(ii) If the point does not satisfy the inequality, shade the half-plane not containing the test point.

In Examples 1–4, sketch the graph of each inequality.

EXAMPLE 1 $y < x + 2$

SOLUTION   Step 1. Because the inequality contains the symbol $<$, we draw the graph of $y = x + 2$ as a dashed line (Figure 2).

Step 2. We select the test point $(1, 2)$ not on the line. We see that $2 < 1 + 2$ is true.

Therefore, we shade the lower half-plane, which contains the point $(1, 2)$ (Figure 2).

Figure 2

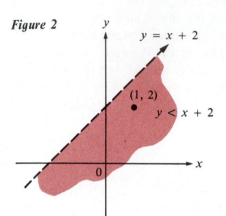

EXAMPLE 2 $2x - y \le 1$

SOLUTION   Step 1. Because the inequality contains the symbol $\le$, we draw the graph of $2x - y = 1$ as a solid line (Figure 3).

Step 2. We test the inequality at the point $(1, -3)$ and find that

$$2(1) - (-3) \le 1$$

or

$$5 \le 1$$

is *false*.

Therefore, we shade the upper half-plane, which does not contain the point $(1, -3)$ (Figure 3).

Figure 3

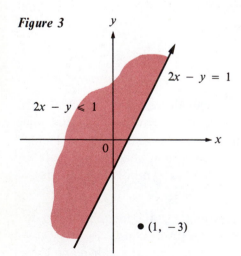

EXAMPLE 3 $y \geq 3$

SOLUTION

Step 1. We draw the graph of $y = 3$ as a solid line (Figure 4).

Step 2. We test the inequality at the point $(-2, 1)$ and find that

$$1 \geq 3$$

is *false*.

Therefore, we shade the upper half-plane, which does not contain the point $(-2, 1)$ (Figure 4).

Figure 4

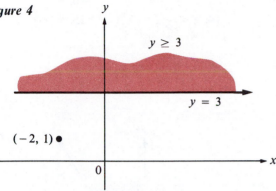

EXAMPLE 4 $x < -1$

SOLUTION

Step 1. We draw the graph of $x = -1$ as a dashed line (Figure 5).

Step 2. We test the inequality at $(0, 3)$ and find that

$$0 < -1$$

is *false*.

Therefore, we shade the plane to the left of the line $x = -1$, which does not contain the point $(0, 3)$.

Figure 5

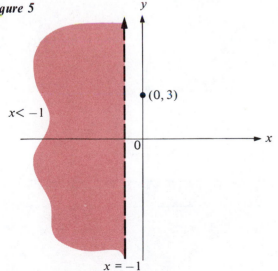

PROBLEM SET 7.6

In problems 1–18, sketch the graph of each inequality.

1. $y \leq 2x + 5$ **2.** $y \geq 3x + 4$ **3.** $y > 3x$ **4.** $y \leq -4x$ **5.** $y > -2x + 3$

6. $2y > 3x + 7$ **7.** $2x + y \leq 3$ **8.** $3x + 2y > 4$ **9.** $3y \leq -4$ **10.** $5y \geq 7$

11. $y \geq 2$ **12.** $y \leq -1$ **13.** $3x - 4y > 12$ **14.** $x < -2$ **15.** $x \geq 3$

16. $2y - 6 < 0$ **17.** $3x + 9 < 0$ **18.** $y + 3x \leq 6$

In problems 19–24, match the shaded region with the appropriate inequality.

(a) $x \geq 2$ (b) $y > 2x - 3$ (c) $y < 3$

(d) $x + y \geq 4$ (e) $10y > -x + 5$ (f) $y > x$

19.

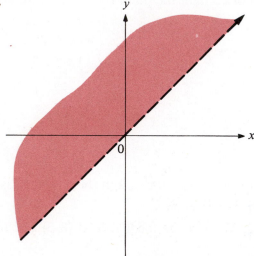

20.

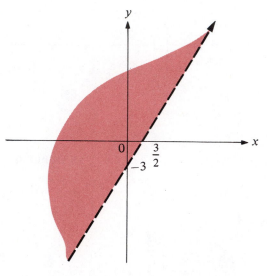

21.

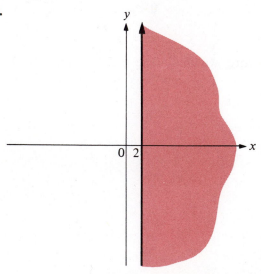

22.

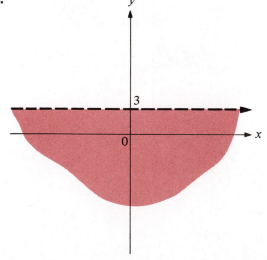

23.

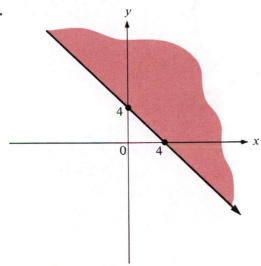

24.

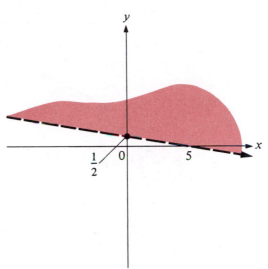

In problems 25–30, write an inequality whose solution is the given graph.

25.

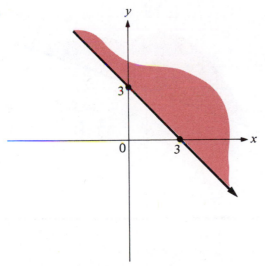

26.

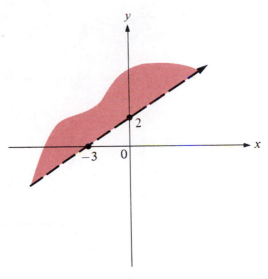

27.

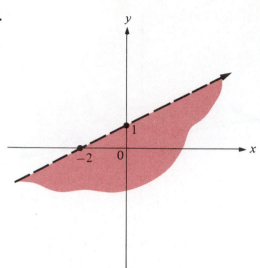

28.

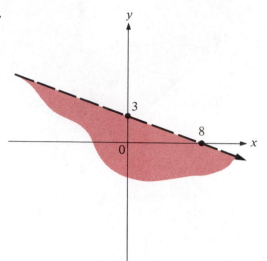

29.

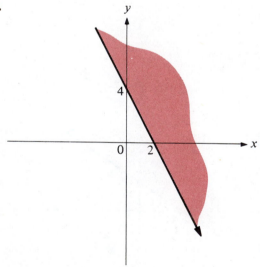

30.

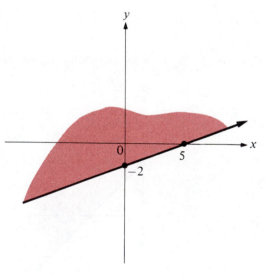

REVIEW PROBLEM SET

In problems 1–8, plot each point and indicate which quadrant or coordinate axis
contains the point.

1. (3, 2)   **2.** (−2, −3)   **3.** (−4, 3)   **4.** (1, −1)
5. (0, 2)   **6.** (3, 0)   **7.** (2, −3)   **8.** (−2, 2)

In problems 9 and 10, find five ordered pairs in the solution of each equation corresponding to $x = -2, -1, 0, 1, 2$. Also, plot the five ordered pairs.

9. $x + 2y = 6$

10. $3x - 4y = 5$

In problems 11–14, find the distance between the two points.

11. $(3, 1)$ and $(15, 6)$
12. $(4, -3)$ and $(-1, 7)$
13. $(-2, 1)$ and $(-5, 5)$
14. $(a, -b)$ and $(-a, b)$

In problems 15 and 16, use the distance formula to show that the triangle ABC is isosceles.

15. $A = (2, 1)$, $B = (9, 3)$, and $C = (4, -6)$

16. $A = (0, 2)$, $B = (-1, 4)$, and $C = (-3, 3)$

In problems 17 and 18, find the coordinates of the midpoint M of the line segment $\overline{PQ}$.

17. $P = (2, 3)$ and $Q = (4, 5)$

18. $P = (-3, 4)$ and $Q = (5, -6)$

In problems 19–22, find the center and radius of each circle. Also, graph each circle.

19. $(x - 3)^2 + (y + 2)^2 = 36$

20. $(x + 2)^2 + (y - 7)^2 = 11$

21. $x^2 + y^2 + 4x - 6y + 4 = 0$

22. $2x^2 + 2y^2 - 4x + 8y - 13 = 0$

In problems 23–26, sketch the graph of each equation by following the procedure on page 349.

23. $y = \frac{3}{4}x$

24. $y = -\frac{5}{3}x$

25. $y = 2x - 3$

26. $y = -2x + 1$

In problems 27–30, find the x and y intercepts and use them to sketch the graph of each equation.

27. $2x - 5y = 10$

28. $-3x + 6y + 5 = 0$

29. $4x + 12 = 0$

30. $7 + 3y = 0$

In problems 31–34, sketch the line segment $\overline{AB}$ and find its slope m.

31. $A = (7, 3)$ and $B = (2, -2)$
32. $A = (11, 12)$ and $B = (7, 4)$
33. $A = (-3, 5)$ and $B = (1, 3)$
34. $A = (3, 0)$ and $B = (0, -2)$

In problems 35–38, sketch the line L that contains the given point P and has slope m.

35. $P = (1, 2)$ and $m = \frac{3}{4}$
36. $P = (4, -5)$ and $m = -\frac{1}{2}$
37. $P = (-8, 6)$ and $m = -\frac{5}{2}$
38. $P = (-3, -2)$ and $m = \frac{4}{7}$

In problems 39–42, determine whether the line containing the points A and B is parallel or perpendicular (or neither) to the line containing the points C and D.

39. $A = (1, 3)$, $B = (-1, -1)$, $C = (2, 1)$, and $D = (-3, -9)$
40. $A = (4, 6)$, $B = (-2, -3)$, $C = (3, 0)$, and $D = (6, -2)$
41. $A = (8, -6)$, $B = (-4, 3)$, $C = (3, 3)$, and $D = (9, 11)$
42. $A = (5, 3)$, $B = (0, 1)$, $C = (2, 3)$, and $D = (-4, -12)$

In problems 43–50, find an equation of the line L.

43. L contains $P = (1, 1)$ and has slope $m = 3$.

44. L contains $P = (2, -3)$ and has slope $m = -2$.

45. L contains $P = (-3, -2)$ and has slope $m = 0$.

46. L contains $P_1 = (4, 1)$ and $P_2 = (2, 3)$.

47. L has slope $m = 2$ and y intercept $b = 4$.

48. L has slope $m = -3$ and y intercept $b = -2$.

49. L contains $P = (3, 5)$ with undefined slope.

50. L contains $P = (-4, 7)$ with undefined slope.

In problems 51–54, express each equation in slope-intercept form. Find the slope m, the x intercept, the y intercept, and sketch the graph.

51. $2x + y = 4$

52. $x + y - 2 = 0$

53. $-3x + 4y - 12 = 0$

54. $\dfrac{x}{2} + \dfrac{y}{3} = 1$

In problems 55–58, find equations of the line L in (a) point-slope form, (b) slope-intercept form, and (c) general form.

55. L contains $P = (2, 3)$ and is parallel to the line $3x - 2y + 5 = 0$.

56. L contains $P = (-3, 4)$ and is parallel to the line whose equation is $2x - 5y = 7$.

57. L contains $P = (1, -2)$ and is perpendicular to the line containing $P_1 = (3, 4)$ and $P_2 = (-5, 6)$.

58. L contains $P = (-7, 5)$ and is perpendicular to the line whose equation is $y = -\frac{3}{2}x + 6$.

In problems 59–64, sketch the graph of each inequality.

59. $y < -2x + 4$

60. $y \geq 3x - 7$

61. $y - 3x \geq 1$

62. $4x - 2y > 3$

63. $y \leq -2$

64. $4y - 12 > 0$

CHAPTER 7 TEST

1. Indicate which quadrant or coordinate axis contains the given point.

(a) $(-3, 7)$ (b) $(0, 5)$ (c) $(7, -2)$

(d) $(-4, -1)$ (e) $(3, 6)$ (f) $(-4, 0)$

2. Find the distance between the points $(-3, 4)$ and $(2, -8)$.

3. Find the coordinates of the midpoint of the line segment joining the points $(-1, 9)$ and $(3, 5)$.

4. Find the center and radius of each circle.

(a) $(x - 3)^2 + (y + 2)^2 = 16$ (b) $x^2 + y^2 - 4x + 6y - 12 = 0$

5. Find the ordered pairs that satisfy each equation, given the following values of x: $-2, 0, 1$. Use these results to sketch the graph of each equation.

(a) $y = 4x$ (b) $y = -2x + 3$

6. Find the x and y intercepts and use them to sketch the graph of each equation.

(a) $3x - 2y - 6 = 0$ (b) $3x - 9 = 0$ (c) $2y + 5 = 0$

7. Find the slope of the line containing the points $(-3, 2)$ and $(-1, 7)$.

8. Find an equation of the line L such that
   (a) L contains $P = (-3, 4)$ and has slope $m = \frac{2}{3}$.
   (b) L contains $P = (2, 5)$ and $Q = (-3, -5)$.
   (c) L contains $P = (-4, 2)$ with undefined slope.
   (d) L has slope $m = \frac{2}{3}$ and y intercept $b = -5$.
9. Find equations of the line L in (i) point-slope form, (ii) slope-intercept form, and (iii) general form.
   (a) L contains $(3, -1)$ and is parallel to the line $4x + 2y = 7$.
   (b) L contains $(-2, 4)$ and is perpendicular to the line $y = -\frac{3}{5}x + 7$.
10. Sketch the graph of each inequality.
   (a) $y < 2x - 4$ (b) $2y - 3x \geq 4$

8 Systems of Linear Equations and Inequalities

In Chapter 7, we studied linear equations and inequalities in two variables. Now, we turn our attention to systems of linear equations and inequalities containing two or more variables. Because applications of mathematics and other disciplines frequently involve many unknown quantities, it is important to be able to deal with systems of linear equations and inequalities. In this chapter, we study algebraic methods of solving these systems. Determinants and their use in solving systems of linear equations are also included in the chapter.

8.1 Systems of Linear Equations in Two Variables

Two linear equations in two variables may be associated in what is called a **system** of equations. The equations in such a system are usually written in a column with a brace on the left. For instance,

$$\begin{cases} 3x - y = 7 \\ 2x + y = 8 \end{cases}$$

is a system of two linear equations in two variables x and y. If each of the two equations in a system is a true statement when we substitute particular numbers for the variables, we say that the pair of numbers substituted is a **solution** to the system. For example, a solution to the system

$$\begin{cases} 2x + 3y = 12 \\ 6x - 2y = 14 \end{cases}$$

is $x = 3$ and $y = 2$. We usually write this solution as the ordered pair (3, 2). We assume that 3 is substituted for x and 2 for y.

The graph of each linear equation in the preceding system is a straight line. Therefore, the solution (3, 2) can be interpreted graphically as the point where the graph of $2x + 3y = 12$ *intersects* the graph of $6x - 2y = 14$, assuming, of course, that the graphs are drawn on the same coordinate system (Figure 1).

Figure 1

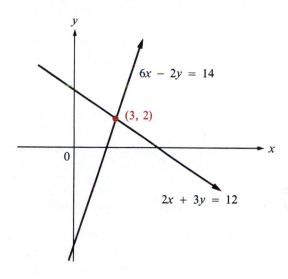

The graph of a system of linear equations yields considerable information about the solution to the system. If we graph the two lines of a system of linear equations on the same coordinate axes, one of the following cases will occur:

Case 1. The two lines intersect at exactly one point. Therefore, there is exactly one solution. In this case, we say that the system is **independent.**

Case 2. The two lines are parallel, and therefore do not intersect. In this case, there is no solution, and we say that the system is **inconsistent.**

Case 3. The two lines coincide. In this case, every point on the common line corresponds to a solution, and we say that the system is **dependent.**

EXAMPLE 1　Use graphs to determine whether each system is independent, inconsistent, or dependent. Indicate the solutions (if any).

(a) $\begin{cases} 2x - 3y = -7 \\ x + 2y = 7 \end{cases}$　(b) $\begin{cases} 2x - 3y = -7 \\ 4x - 6y = 8 \end{cases}$　(c) $\begin{cases} 2x - 3y = -7 \\ 4x - 6y = -14 \end{cases}$

SOLUTION　The graphs of equations in systems (a), (b), and (c) are shown in Figure 2. Graphs of both equations in each system are drawn on the same coordinate system.

Figure 2

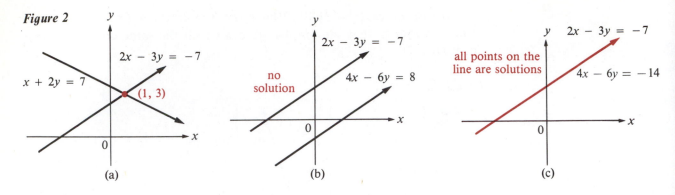

(a)   (b)   (c)

(a) In Figure 2a, the graphs intersect at one point. Therefore, there is one solution, and the system is independent. We see that the solution is (1, 3).

(b) In Figure 2b, the two lines do not appear to intersect. In fact, they are parallel. Therefore, there is no solution, and the system is inconsistent.

(c) In Figure 2c, we see that the two equations have the same graph. Therefore, there are infinitely many solutions (one for each point on the graph of the line), and the system is dependent.

The Substitution Method

The graphic method for solving systems of linear equations is not always practical when we are interested in decimal-place accuracy. We can, however, use algebraic methods to obtain an exact solution of any independent system. The procedure for the algebraic technique, known as the **substitution method,** follows:

Procedure for Solving a Linear System by the Substitution Method

Step 1. Choose one of the equations and solve it for one of the variables in terms of the remaining variable.

Step 2. Substitute the resulting expression for the variable in the other equation to obtain an equation in one variable.

Step 3. Solve the equation resulting from step 2 for the variable.

Step 4. Substitute the value from step 3 in any of the equations involving both variables to obtain an equation in the remaining variable. Solve this equation.

In Examples 2–4, use the substitution method to solve each system.

EXAMPLE 2
$$\begin{cases} 3x + y = 7 \\ 2x - 3y = 1 \end{cases}$$

SOLUTION We solve the equation $3x + y = 7$ for y in terms of x and obtain

$$y = -3x + 7.$$

We then substitute $-3x + 7$ for y in the equation $2x - 3y = 1$:

$$2x - 3(-3x + 7) = 1$$
$$2x + 9x - 21 = 1$$
$$11x = 22$$
$$x = 2.$$

Finally, we find y by substituting $x = 2$ in the equation $y = -3x + 7$:

$$y = -3(2) + 7$$
$$= -6 + 7$$
$$= 1.$$

Check To check the solution, we replace x by 2 and y by 1 in the original system:

$$\begin{cases} 3(2) + 1 = 7 \\ 2(2) - 3(1) = 1. \end{cases}$$

Therefore, the solution is $x = 2$ and $y = 1$, or $(2, 1)$.

EXAMPLE 3
$$\begin{cases} 7x + 5y = 62 \\ x - 9y = -30 \end{cases}$$

SOLUTION We solve the equation $x - 9y = -30$ for x in terms of y and obtain

$$x = 9y - 30.$$

We then substitute $9y - 30$ for x in the equation $7x + 5y = 62$:

$$7(9y - 30) + 5y = 62$$
$$63y - 210 + 5y = 62$$
$$68y = 272$$
$$y = 4.$$

To obtain the corresponding value of x, we substitute $y = 4$ in the equation $x = 9y - 30$:

$$x = 9(4) - 30 = 36 - 30 = 6.$$

Therefore, the solution is $x = 6$ and $y = 4$, or $(6, 4)$.

EXAMPLE 4 $\begin{cases} 2x - 3y = 16 \\ -5x + 7y = -39 \end{cases}$

SOLUTION   We solve the equation $2x - 3y = 16$ for x:

$$x = \frac{3}{2}y + 8.$$

We then substitute $\frac{3}{2}y + 8$ for x in the equation $-5x + 7y = -39$:

$$-5\left(\frac{3}{2}y + 8\right) + 7y = -39$$

$$-\frac{15}{2}y - 40 + 7y = -39$$

$$-15y - 80 + 14y = -78$$

$$-y = 2$$

$$y = -2.$$

To obtain the corresponding value for x, we substitute $y = -2$ in the equation $x = \frac{3}{2}y + 8$:

$$x = \frac{3}{2}(-2) + 8 = -3 + 8 = 5.$$

Therefore, the solution is $x = 5$ and $y = -2$, or $(5, -2)$.

The Elimination Method

Another algebraic method can be used to obtain an exact solution of an independent system of linear equations in two variables. We eliminate one of the variables by adding or subtracting the two equations in the system. This is known as the **elimination method,** or the **addition or subtraction method.** The procedure is as follows:

Procedure for Solving a Linear System by the Elimination Method

Step 1. Multiply the terms of one or both equations by numbers that will make the corresponding coefficients of one of the variables the same or negatives of each other. (This step may not be necessary.)

Step 2. Add or subtract the two equations resulting from step 1 to produce a single equation in a single variable.

Step 3. Solve the equation resulting from step 2 for that variable.

Step 4. Substitute the solution from step 3 in one of the equations in the original system, and solve the resulting equation to obtain the value of the other variable.

In Examples 5–7, use the elimination method to solve each system.

EXAMPLE 5

$$\begin{cases} x - y = 1 \\ 2x + y = 5 \end{cases}$$

SOLUTION We notice that the y terms in the equations are negatives of each other. We add the two equations to eliminate the y term:

$$\begin{cases} x - y = 1 \\ 2x + y = 5 \quad \text{add} \end{cases}$$
$$\overline{3x = 6}$$
$$x = 2.$$

We substitute 2 for x in the equation $x - y = 1$:

$$2 - y = 1$$
$$-y = -1$$
$$y = 1.$$

Check We replace x by 2 and y by 1 in the original system:

$$\begin{cases} 2 - 1 = 1 \\ 2(2) + 1 = 5. \end{cases}$$

Therefore, the solution is $x = 2$ and $y = 1$, or $(2, 1)$.

EXAMPLE 6

$$\begin{cases} 2r + 3s = 7 \\ 5r + s = -2 \end{cases}$$

SOLUTION   We eliminate s as follows:

$$\begin{cases} 2r + 3s = 7 \\ 5r + s = -2 \end{cases} \xrightarrow[\text{We multiply each side by 3.}]{} \begin{cases} 2r + 3s = 7 \\ 15r + 3s = -6 \end{cases}$$

We subtract the second equation from the first:

$$-13r = 13$$
$$r = -1.$$

We substitute $r = -1$ in the equation $5r + s = -2$:

$$5(-1) + s = -2$$
$$-5 + s = -2$$
$$s = 3.$$

Therefore, the solution is $r = -1$ and $s = 3$, or $(-1, 3)$.

EXAMPLE 7 $\begin{cases} 0.3p + 0.2q = 0.8 \\ 0.2p - 0.3q = 1.4 \end{cases}$

SOLUTION   We begin by rewriting the system so that the coefficients are free from decimals:

$$\begin{cases} 0.3p + 0.2q = 0.8 \\ 0.2p - 0.3q = 1.4 \end{cases} \xrightarrow[\text{We multiply each side by 10.}]{\text{We multiply each side by 10.}} \begin{cases} 3p + 2q = 8 \\ 2p - 3q = 14 \end{cases}$$

We eliminate q as follows:

$$\begin{cases} 3p + 2q = 8 \\ 2p - 3q = 14 \end{cases} \xrightarrow[\text{We multiply each side by 2.}]{\text{We multiply each side by 3.}} \begin{cases} 9p + 6q = 24 \\ 4p - 6q = 28 \end{cases}$$

Adding the two equations, we have

$$13p = 52$$
$$p = 4.$$

We substitute $p = 4$ in the equation $3p + 2q = 8$:

$$3(4) + 2q = 8$$
$$12 + 2q = 8$$

$$2q = -4$$

$$q = -2.$$

Therefore, the solution is $p = 4$ and $q = -2$, or $(4, -2)$.

Some systems of equations such as

$$\begin{cases} \dfrac{1}{p} - \dfrac{2}{q} = 8 \\ \dfrac{3}{p} + \dfrac{1}{q} = 3 \end{cases}$$

are not linear, but can be made linear and solved if we make the substitutions $x = 1/p$ and $y = 1/q$ as shown in the following example.

EXAMPLE 8   Solve the system

$$\begin{cases} \dfrac{1}{p} - \dfrac{2}{q} = 8 \\ \dfrac{3}{p} + \dfrac{1}{q} = 3. \end{cases}$$

SOLUTION   Substituting $x = 1/p$ and $y = 1/q$ in the preceding system, we have

$$\begin{cases} x - 2y = 8 \\ 3x + y = 3. \end{cases}$$

We eliminate y as follows:

$$\begin{cases} x - 2y = 8 \\ 3x + y = 3 \end{cases} \xrightarrow[\text{We multiply each side by 2.}]{} \begin{cases} x - 2y = 8 \\ 6x + 2y = 6 \end{cases}$$

Adding the two equations, we have

$$7x = 14$$

$$x = 2.$$

We substitute $x = 2$ in the equation $3x + y = 3$:

$$3(2) + y = 3$$

$$6 + y = 3$$

$$y = -3.$$

Since $x = 1/p$, then $p = 1/x = 1/2$, and since $y = 1/q$, then $q = 1/y = 1/(-3)$. Therefore, the solution is $p = \frac{1}{2}$ and $q = -\frac{1}{3}$ or $(\frac{1}{2}, -\frac{1}{3})$.

PROBLEM SET 8.1

In problems 1–6, sketch the graph of each system of linear equations. Use the graph to determine whether the system is dependent, inconsistent, or independent. If the system is independent, determine the coordinates of the solution graphically.

1. $\begin{cases} 3x - 2y = 1 \\ 6x - 8y = 2 \end{cases}$

2. $\begin{cases} y = 2x - 3 \\ 4x - 2y = 6 \end{cases}$

3. $\begin{cases} x + 3y = 6 \\ 2x + 6y = 8 \end{cases}$

4. $\begin{cases} 2x = y + 3 \\ 4x - 2y = 5 \end{cases}$

5. $\begin{cases} 2x - 3y = 1 \\ 5x + 2y = 12 \end{cases}$

6. $\begin{cases} 3x + 2y = 11 \\ -2x + y = 2 \end{cases}$

In problems 7–18, use the substitution method to solve each system.

7. $\begin{cases} 2x - y = 5 \\ x + 3y = 13 \end{cases}$

8. $\begin{cases} 3x + y = 4 \\ 7x - y = 6 \end{cases}$

9. $\begin{cases} 3x - y = -2 \\ x + y = 6 \end{cases}$

10. $\begin{cases} 5x + 7y = 4 \\ -x + 6y = 14 \end{cases}$

11. $\begin{cases} 5p - q = 13 \\ p + q = 1 \end{cases}$

12. $\begin{cases} 3r - 4s = -5 \\ 2r + 2s = 3 \end{cases}$

13. $\begin{cases} 13a + 11b = 21 \\ 7a + 6b = -3 \end{cases}$

14. $\begin{cases} 0.2m - 0.3n = 0.1 \\ 0.3m + 0.2n = 2.1 \end{cases}$

15. $\begin{cases} \frac{1}{2}x - \frac{3}{4}y = 1 \\ 3x + y = 1 \end{cases}$

16. $\begin{cases} \frac{1}{3}x + \frac{1}{2}y = 4 \\ \frac{1}{4}x - \frac{1}{3}y = -1 \end{cases}$

17. $\begin{cases} 0.5x - 1.2y = 0.3 \\ 0.7x + 1.5y = 3.6 \end{cases}$

18. $\begin{cases} 2u - v = 3 \\ 3u + v = 22 \end{cases}$

In problems 19–36, use the elimination method to solve each system.

19. $\begin{cases} 2x - y = 1 \\ x + y = 2 \end{cases}$

20. $\begin{cases} 2x + y = 10 \\ 3x - y = 5 \end{cases}$

21. $\begin{cases} u + 3v = 9 \\ u - v = 1 \end{cases}$

22. $\begin{cases} 5u + v = 14 \\ 2u + v = 5 \end{cases}$

23. $\begin{cases} 2r + 4s = 2 \\ -r + s = 8 \end{cases}$

24. $\begin{cases} 7p + q = 5 \\ -2p + q = 3 \end{cases}$

25. $\begin{cases} 3x + 2y = 4 \\ 5x + 3y = 7 \end{cases}$

26. $\begin{cases} 2x - 7y = 5 \\ x + y = -8 \end{cases}$

27. $\begin{cases} 7p + q = 3 \\ 5p + q = 6 \end{cases}$

28. $\begin{cases} 3y + 2z = 5 \\ 2y + 3z = 1 \end{cases}$

29. $\begin{cases} -3x + y = 3 \\ 4x + 2y = 10 \end{cases}$

30. $\begin{cases} 5x - 2y = 35 \\ x + 4y = 25 \end{cases}$

31. $\begin{cases} 4u - 3v = 1 \\ 3u + 4v = 6 \end{cases}$

32. $\begin{cases} 13y + 5z = 2 \\ 6y + 2z = 2 \end{cases}$

33. $\begin{cases} \frac{1}{2}x + \frac{1}{3}y = 13 \\ \frac{1}{5}x + \frac{1}{8}y = 5 \end{cases}$

34. $\begin{cases} \frac{1}{3}x - \frac{1}{4}y = 2 \\ \frac{1}{4}x - \frac{1}{2}y = 7 \end{cases}$

35. $\begin{cases} 3x + y = a \\ x - 3y = b \end{cases}$ a and b are constants

36. $\begin{cases} 3ax + 2by = 6 \\ 2ax - 5by = 7 \end{cases}$ a and b are constants

In problems 37–40, solve each system by using an appropriate method.

37. $\begin{cases} 0.5x + 0.2y = 0.9 \\ 0.3x - 0.4y = -0.5 \end{cases}$

38. $\begin{cases} 0.2p + 0.3q = 0.7 \\ 0.4p - 0.5q = 0.3 \end{cases}$

Ⓒ **39.** $\begin{cases} 1.41p - 4.70q = 12.69 \\ -1.04p + 3.40q = -9.60 \end{cases}$

Ⓒ **40.** $\begin{cases} 4.01u + 6.07v = 2.33 \\ 7.57u - 3.11v = 13.81 \end{cases}$

In problems 41–46, solve each system by using the substitution $u = 1/x$ and $v = 1/y$.

41. $\begin{cases} \dfrac{2}{x} - \dfrac{1}{y} = 9 \\ \dfrac{5}{x} - \dfrac{3}{y} = 14 \end{cases}$

42. $\begin{cases} \dfrac{5}{x} - \dfrac{2}{y} = 1 \\ \dfrac{8}{x} + 11 = \dfrac{5}{y} \end{cases}$

43. $\begin{cases} \dfrac{3}{x} + \dfrac{2}{y} = 2 \\ \dfrac{1}{x} - \dfrac{1}{y} = 9 \end{cases}$

44. $\begin{cases} \dfrac{4}{x} - \dfrac{3}{y} = 1 \\ \dfrac{3}{x} - \dfrac{4}{y} = 6 \end{cases}$

45. $\begin{cases} \dfrac{5}{x} + \dfrac{2}{y} = 1 \\ \dfrac{13}{x} + \dfrac{8}{y} = 11 \end{cases}$

46. $\begin{cases} \dfrac{4}{x} + \dfrac{1}{y} = 16 \\ \dfrac{3}{x} + \dfrac{1}{y} = 11 \end{cases}$

47. One car rental agency charges \$20 for the first day and \$18 for each additional day. If the number of additional days after the first is x, the equation that gives the total cost y, in dollars, for the rental of the car is given by $y = 18x + 20$. A second car rental agency charges \$26 for the first day and \$15 for each additional day, so the equation that gives the total cost, y, for this company's car is given by $y = 15x + 26$. For how many additional days must a car be rented for the total costs of the two agencies to be equal?

48. A manufacturing company produces x units per month of item A and y units per month of item B. The manufacturing process emits 1.5 cubic meters of carbon monoxide and 3 cubic meters of sulfur dioxide per unit of item A, and 2 cubic meters each of carbon monoxide and sulfur dioxide per unit of item B. The pollution standards allow the company to emit a maximum of 4,800 cubic meters of carbon monoxide and 5,550 cubic meters of sulfur dioxide per month. The equation that expresses the number of units of item A and of item B that will cause a total emission of 4,800 cubic meters of carbon monoxide is given by $1.5x + 2y = 4,800$.

(a) Write an equation to express the number of units of item A and of item B that will cause a total emission of 5,550 cubic meters of sulfur dioxide.

(b) How many units of items A and B can be produced at the maximum allowable standards?

8.2 Systems of Linear Equations in Three Variables

An equation such as

$$3x + 5y + z = 8$$

is an example of a **linear equation in three variables,** because each of the three unknowns is raised to the first power. One solution of this equation is $x = -1$, $y = 2$, and $z = 1$, because

$$3(-1) + 5(2) + 1 = -3 + 10 + 1$$
$$= 8.$$

This solution can be written as the **ordered triple** $(-1, 2, 1)$. Graphic methods for illustrating the solutions of systems of linear equations in three unknowns are impractical because such graphs may only be drawn in a three-dimensional coordinate system. Therefore, we consider only the algebraic methods of substitution and elimination to solve systems of linear equations in three unknowns. To use the method of substitution, we choose one of the equations and solve it for one of the variables in terms of the remaining two. Then we substitute this solution into the remaining equations. This produces a system involving two equations and two unknowns.

In Examples 1 and 2, use the substitution method to solve each system of equations.

EXAMPLE 1

$$\begin{cases} x + y + z = 6 \\ 2x - y - z = 0 \\ x - y + 2z = 7 \end{cases}$$

SOLUTION

We begin by solving the first equation, $x + y + z = 6$, for z in terms of x and y:

$$z = 6 - x - y.$$

Now we substitute $6 - x - y$ for z in the second and third equations:

$$2x - y - (6 - x - y) = 0$$

and

$$x - y + 2(6 - x - y) = 7.$$

Simplifying these two equations, we have the system

$$\begin{cases} 3x = 6 \\ -x - 3y = -5. \end{cases}$$

We can now solve this latter system by using the substitution procedure again:

$$3x = 6$$

$$x = 2.$$

We substitute $x = 2$ in the equation $-x - 3y = -5$:

$$-2 - 3y = -5$$
$$-3y = -3$$
$$y = 1.$$

We have found that $x = 2$ and $y = 1$. Therefore, we need only substitute these values in the equation $z = 6 - x - y$ to find a value for z:

$$z = 6 - 2 - 1$$
$$= 3.$$

Check We replace x by 2, y by 1, and z by 3 in the original system:

$$\begin{cases} 2 + 1 + 3 = 6 \\ 2(2) - 1 - 3 = 0 \\ 2 - 1 + 2(3) = 7. \end{cases}$$

Hence, our solution is $x = 2$, $y = 1$, and $z = 3$, or $(2, 1, 3)$.

EXAMPLE 2
$$\begin{cases} r + 2s + 4t = 12 \\ 2r - 3s + t = 10 \\ 3r - s - 2t = 1 \end{cases}$$

SOLUTION   We solve the first equation $r + 2s + 4t = 12$ for r in terms of s and t:

$$r = 12 - 2s - 4t.$$

Now we substitute $12 - 2s - 4t$ for r in the second and third equations:

$$2(12 - 2s - 4t) - 3s + t = 10$$

and

$$3(12 - 2s - 4t) - s - 2t = 1.$$

Simplifying these two equations gives us the system

$$\begin{cases} -7s - 7t = -14 \\ -7s - 14t = -35 \end{cases}$$

or

$$\begin{cases} s + t = 2 \\ s + 2t = 5. \end{cases}$$

We solve the equation $s + t = 2$ for t in terms of s:

$$t = 2 - s.$$

We substitute $2 - s$ for t in the equation $s + 2t = 5$:

$$s + 2(2 - s) = 5$$
$$s + 4 - 2s = 5$$
$$-s = 1$$
$$s = -1.$$

Now we substitute $s = -1$ in the equation $t = 2 - s$:

$$t = 2 - (-1) = 3.$$

Finally, we substitute $s = -1$ and $t = 3$ in the equation $r = 12 - 2s - 4t$:

$$r = 12 - 2(-1) - 4(3) = 2.$$

Therefore, the solution is $r = 2$, $s = -1$, and $t = 3$, or $(2, -1, 3)$.

The elimination method is sometimes more efficient than the method of substitution, and it has the additional advantage that it can be programmed on a computer.

In Examples 3 and 4, use the elimination method to solve each system.

EXAMPLE 3

$$\begin{cases} x + y + z = 2 \\ 2x + 3y - z = 3 \\ 3x + 5y + z = 8 \end{cases}$$

SOLUTION   First we eliminate z as follows:

$$\begin{cases} x + y + z = 2 \\ 2x + 3y - z = 3 \end{cases} \quad \text{add}$$
$$\overline{3x + 4y \quad\quad = 5}$$

and

$$\begin{cases} 2x + 3y - z = 3 \\ 3x + 5y + z = 8 \end{cases} \quad \text{add}$$
$$\overline{5x + 8y \quad\quad = 11.}$$

Next we eliminate y as follows:

$$\begin{cases} 3x + 4y = 5 \\ 5x + 8y = 11 \end{cases} \xrightarrow{\text{We multiply each side by 2.}} \begin{cases} 6x + 8y = 10 \\ 5x + 8y = 11 \quad \text{subtract} \end{cases}$$
$$\overline{x \quad\quad = -1.}$$

We substitute $x = -1$ in the equation $3x + 4y = 5$:

$$3(-1) + 4y = 5$$
$$4y = 8$$
$$y = 2.$$

We substitute $x = -1$ and $y = 2$ in the equation $x + y + z = 2$:

$$-1 + 2 + z = 2$$
$$z = 1.$$

Therefore, the solution is $x = -1$, $y = 2$, and $z = 1$, or $(-1, 2, 1)$.

EXAMPLE 4

$$\begin{cases} p + 2q + 5r = 4 \\ 4p + q + 3r = 9 \\ 6p + 9q + r = 21 \end{cases}$$

SOLUTION   First we eliminate p as follows:

$$\begin{cases} p + 2q + 5r = 4 \\ 4p + q + 3r = 9 \\ 6p + 9q + r = 21 \end{cases} \xrightarrow{\text{We multiply each side by 4.}} \begin{cases} 4p + 8q + 20r = 16 \\ 4p + \ q + \ 3r = 9 \quad \text{subtract} \\ \overline{7q + 17r = 7} \end{cases}$$

$$\begin{cases} p + 2q + 5r = 4 \\ 4p + q + 3r = 9 \\ 6p + 9q + r = 21 \end{cases} \xrightarrow{\text{We multiply each side by 6.}} \begin{cases} 6p + 12q + 30r = 24 \\ 6p + \ 9q + \ \ r = 21 \quad \text{subtract} \\ \overline{3q + 29r = 3.} \end{cases}$$

We now have a system of two equations in two unknowns. We can eliminate q as follows:

$$\begin{cases} 7q + 17r = 7 \\ 3q + 29r = 3 \end{cases} \begin{array}{l} \xrightarrow{\text{We multiply each side by 3.}} \\ \xrightarrow{\text{We multiply each side by 7.}} \end{array} \begin{cases} 21q + \ 51r = 21 \\ 21q + 203r = 21 \quad \text{subtract} \\ \overline{-152r = 0} \end{cases}$$

or

$$r = 0.$$

We substitute $r = 0$ in the equation $3q + 29r = 3$:

$$3q + 29(0) = 3$$
$$3q = 3$$
$$q = 1.$$

Finally, we substitute $r = 0$ and $q = 1$ in the equation $p + 2q + 5r = 4$:

$$p + 2(1) + 5(0) = 4$$

$$p = 2.$$

Therefore, the solution is $p = 2$, $q = 1$, and $r = 0$, or $(2, 1, 0)$.

PROBLEM SET 8.2

In problems 1–6, use the substitution method to solve each system.

1. $\begin{cases} x + y = 5 \\ x + z = 1 \\ y + z = 2 \end{cases}$

2. $\begin{cases} 2x + 3y = 28 \\ 3y + 4z = 46 \\ 4z + 5x = 53 \end{cases}$

3. $\begin{cases} x + y + 2z = 11 \\ x - y + z = 3 \\ 2x + y + 3z = 17 \end{cases}$

4. $\begin{cases} x - 3y = -11 \\ 2y - 5z = 26 \\ 7x - 3z = -2 \end{cases}$

5. $\begin{cases} 2p - q + r = 8 \\ p + 2q + 3r = 9 \\ 4p + q - 2r = 1 \end{cases}$

6. $\begin{cases} s + 3t - u = 4 \\ 3s - 2t + 4u = 11 \\ 2s + t + 3u = 13 \end{cases}$

In problems 7–20, use the elimination method to solve each system.

7. $\begin{cases} x + y + 2z = 4 \\ x + y - 2z = 0 \\ x - y = 0 \end{cases}$

8. $\begin{cases} x + y + z = 2 \\ x + 2y - z = 4 \\ 2x - y + z = 0 \end{cases}$

9. $\begin{cases} x + y + z = 6 \\ x - y + 2z = 12 \\ 2x + y + z = 1 \end{cases}$

10. $\begin{cases} x + y + 2z = 4 \\ x - 5y + z = 5 \\ 3x - 4y + 7z = 24 \end{cases}$

11. $\begin{cases} x + y = 4 \\ 3x - y + 3z = 7 \\ 5x - 7y + 2z = -2 \end{cases}$

12. $\begin{cases} 7x + y + 3z = -6 \\ 4x - 5y + 6z = -27 \\ x + 15y - 9z = 61 \end{cases}$

13. $\begin{cases} 2p + q - 3r = 9 \\ p - 2q + 4r = 5 \\ 3p + q - 2r = 15 \end{cases}$

14. $\begin{cases} u + 3v - w = -2 \\ 7u - 5v + 4w = 11 \\ 2u + v + 3w = 21 \end{cases}$

15. $\begin{cases} 2r + 3s + t = 6 \\ r - 2s + 3t = 3 \\ 3r + s - t = 8 \end{cases}$

16. $\begin{cases} 8s + 3t - 18u = -76 \\ 10s + 6t - 6u = -50 \\ 4s + 9t + 12u = 10 \end{cases}$

17. $\begin{cases} a - 5b + 4c = 8 \\ 3a + b - 2c = 4 \\ 9a - 3b + 6c = 6 \end{cases}$

18. $\begin{cases} 2p + 3q - 2r = 3 \\ 8p + q + r = 2 \\ 2p + 2q + r = 1 \end{cases}$

19. $\begin{cases} \dfrac{p}{2} + \dfrac{q}{3} - \dfrac{r}{4} = -1 \\[2mm] \dfrac{p}{3} + \dfrac{r}{2} = 8 \\[2mm] \dfrac{2p}{3} + \dfrac{q}{3} - \dfrac{3r}{4} = -6 \end{cases}$

20. $\begin{cases} 0.5u + 1.5v - 0.5w = 2 \\ -1.5u - 2.5v + 0.5w = -4 \\ -0.5v + 1.5w = 7 \end{cases}$

In problems 21 and 22, solve each system by using the substitutions $u = 1/x$, $v = 1/y$, and $w = 1/z$.

21.
$$\begin{cases} \dfrac{3}{x} - \dfrac{4}{y} + \dfrac{6}{z} = 1 \\[2mm] \dfrac{9}{x} + \dfrac{8}{y} - \dfrac{12}{z} = 3 \\[2mm] \dfrac{9}{x} - \dfrac{4}{y} + \dfrac{12}{z} = 4 \end{cases}$$

22.
$$\begin{cases} \dfrac{3}{x} + \dfrac{1}{y} - \dfrac{1}{z} = 5 \\[2mm] \dfrac{4}{x} - \dfrac{1}{y} + \dfrac{2}{z} = 13 \\[2mm] \dfrac{2}{x} + \dfrac{2}{y} + \dfrac{3}{z} = 22 \end{cases}$$

23. In the electric circuit shown in Figure 1, I_1, I_2, and I_3 represent the amount of current (in amperes) flowing across the 1-ohm, 2-ohm, and 2-ohm resistors. The system of linear equations used to find the currents I_1, I_2, and I_3 is given by

$$\begin{cases} I_1 + I_2 - I_3 = 0 \\ I_1 + 2I_3 = 12 \\ I_1 - 2I_2 = -4. \end{cases}$$

Solve the system for I_1, I_2, and I_3.

Figure 1

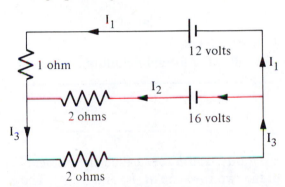

24. Three companies produce products A, B, and C. In one hour, the first company produces 3 units of A, 1 unit of B, and 2 units of C. The second company produces, in one hour, 2 units of A and 3 units of B. The hourly production for the third company is 1 unit of A, 1 unit of B, and 3 units of C. If the first company operates x hours, the second company operates y hours, and the third company operates z hours, then the number of hours that each company needs to operate to fill an order for 70 units of A, 65 units of B, and 50 units of C is given by the system

$$\begin{cases} 3x + 2y + z = 70 \\ x + 3y + z = 65 \\ 2x + 3z = 50. \end{cases}$$

Solve the system for x, y and z.

8.3 Determinants

Consider the following system of two linear equations in two variables:

$$\begin{cases} a_1x + b_1y = c_1 \\ a_2x + b_2y = c_2. \end{cases}$$

We can solve this system by using the substitution method or the elimination method (Section 8.1) to obtain the solution (x, y), where

$$x = \frac{b_2c_1 - b_1c_2}{a_1b_2 - a_2b_1} \quad \text{and} \quad y = \frac{a_1c_2 - a_2c_1}{a_1b_2 - a_2b_1},$$

provided that $a_1b_2 - a_2b_1 \neq 0$.

Any system of two linear equations in two variables can be arranged in the preceding form. Note that the solution appears as fractions with a common denominator. We can represent this denominator by the symbol

$$\begin{vmatrix} a_1 & b_1 \\ a_2 & b_2 \end{vmatrix}$$

which is called a **determinant.**

We have the following formal definition:

$$\begin{vmatrix} a_1 & b_1 \\ a_2 & b_2 \end{vmatrix} = a_1b_2 - a_2b_1.$$

The numbers a_1, b_1, a_2, and b_2 are called the **elements** of the determinant. The elements of the two **rows** are a_1, b_1 and a_2, b_2. The elements of the two **columns** are a_1, a_2 and b_1, b_2. Therefore, this is called a **two-by-two** (2×2) determinant, or a determinant of **order 2.** The expression $a_1b_2 - a_2b_1$ is referred to as the **expansion** of the determinant and is used to find the value of the determinant.

In Examples 1 and 2, find the value of the following 2 × 2 determinants.

EXAMPLE 1

$$\begin{vmatrix} 1 & -2 \\ 3 & 4 \end{vmatrix}$$

SOLUTION

Here we have $a_1 = 1$, $b_1 = -2$, $a_2 = 3$, and $b_2 = 4$. So by substitution in the expression $a_1b_2 - a_2b_1$, we have

$$\begin{vmatrix} 1 & -2 \\ 3 & 4 \end{vmatrix} = 1(4) - 3(-2) = 4 + 6 = 10.$$

EXAMPLE 2 $\begin{vmatrix} 1 & 0 \\ 2 & -1 \end{vmatrix}$

SOLUTION $\begin{vmatrix} 1 & 0 \\ 2 & -1 \end{vmatrix} = 1(-1) - 2(0) = -1 - 0 = -1$

The value of a 3×3 determinant is defined in terms of 2×2 determinants as follows:

$$\begin{vmatrix} a_1 & b_1 & c_1 \\ a_2 & b_2 & c_2 \\ a_3 & b_3 & c_3 \end{vmatrix} = a_1 \begin{vmatrix} b_2 & c_2 \\ b_3 & c_3 \end{vmatrix} - a_2 \begin{vmatrix} b_1 & c_1 \\ b_3 & c_3 \end{vmatrix} + a_3 \begin{vmatrix} b_1 & c_1 \\ b_2 & c_2 \end{vmatrix}.$$

We can use the definition of a 2×2 determinant to obtain:

$$\begin{vmatrix} a_1 & b_1 & c_1 \\ a_2 & b_2 & c_2 \\ a_3 & b_3 & c_3 \end{vmatrix} = a_1 b_2 c_3 - a_1 b_3 c_2 - a_2 b_1 c_3 + a_2 b_3 c_1 + a_3 b_1 c_2 - a_3 b_2 c_1.$$

The right-hand side of this equation is called the **expansion** of a 3×3 determinant. We can express the expansions of 2×2 determinants and 3×3 determinants in terms of sums and differences of the products of their elements.

In Examples 3 and 4, find the value of the 3×3 determinants.

EXAMPLE 3 $\begin{vmatrix} 1 & 0 & 2 \\ 4 & 6 & -1 \\ -1 & 0 & -1 \end{vmatrix}$

SOLUTION   By definition,

$$\begin{vmatrix} 1 & 0 & 2 \\ 4 & 6 & -1 \\ -1 & 0 & -1 \end{vmatrix} = 1 \begin{vmatrix} 6 & -1 \\ 0 & -1 \end{vmatrix} - 4 \begin{vmatrix} 0 & 2 \\ 0 & -1 \end{vmatrix} + (-1) \begin{vmatrix} 0 & 2 \\ 6 & -1 \end{vmatrix}$$

$$= 1[6(-1) - 0(-1)] - 4[0(-1) - 0(2)] - 1[0(-1) - 6(2)]$$

$$= 1(-6) - 4(0) - 1(-12) = -6 + 12 = 6.$$

EXAMPLE 4

$$\begin{vmatrix} 3 & 1 & -1 \\ 0 & 2 & 4 \\ -1 & 4 & 2 \end{vmatrix}$$

SOLUTION

$$\begin{vmatrix} 3 & 1 & -1 \\ 0 & 2 & 4 \\ -1 & 4 & 2 \end{vmatrix} = 3 \begin{vmatrix} 2 & 4 \\ 4 & 2 \end{vmatrix} - 0 \begin{vmatrix} 1 & -1 \\ 4 & 2 \end{vmatrix} + (-1) \begin{vmatrix} 1 & -1 \\ 2 & 4 \end{vmatrix}$$

$$= 3(-12) - 0 - 1(6) = -42$$

PROBLEM SET 8.3

In problems 1–10, evaluate each 2×2 determinant.

1. $\begin{vmatrix} 7 & 1 \\ -5 & 3 \end{vmatrix}$

2. $\begin{vmatrix} 0 & 1 \\ 1 & 0 \end{vmatrix}$

3. $\begin{vmatrix} 6 & 0 \\ -3 & 4 \end{vmatrix}$

4. $\begin{vmatrix} 1 & 2 \\ 3 & 5 \end{vmatrix}$

5. $\begin{vmatrix} -3 & -1 \\ -5 & \frac{1}{2} \end{vmatrix}$

6. $\begin{vmatrix} 2 & 1 \\ -10 & 4 \end{vmatrix}$

7. $\begin{vmatrix} 3 & -1 \\ 6 & -2 \end{vmatrix}$

8. $\begin{vmatrix} 3 & -2 \\ 3 & 2 \end{vmatrix}$

9. $\begin{vmatrix} 1 & 0 \\ 0 & 1 \end{vmatrix}$

10. $\begin{vmatrix} 7 & 14 \\ 3 & 6 \end{vmatrix}$

In problems 11–20, evaluate each 3×3 determinant.

11. $\begin{vmatrix} -3 & 1 & 7 \\ 0 & 2 & 6 \\ -4 & 5 & 1 \end{vmatrix}$

12. $\begin{vmatrix} 2 & 1 & 1 \\ 9 & 3 & 6 \\ 0 & 0 & 1 \end{vmatrix}$

13. $\begin{vmatrix} 1 & 0 & 0 \\ 0 & 1 & 0 \\ 0 & 0 & 1 \end{vmatrix}$

14. $\begin{vmatrix} -10 & -1 & 5 \\ -7 & 8 & 2 \\ 3 & -6 & 0 \end{vmatrix}$

15. $\begin{vmatrix} -1 & 3 & 5 \\ -7 & 4 & 2 \\ -6 & 2 & 0 \end{vmatrix}$

16. $\begin{vmatrix} 3 & -1 & 2 \\ 0 & 1 & -5 \\ 6 & 7 & 4 \end{vmatrix}$

17. $\begin{vmatrix} 2 & 3 & 5 \\ 9 & 4 & 2 \\ 11 & -6 & 2 \end{vmatrix}$

18. $\begin{vmatrix} 2 & 2 & 2 \\ 3 & 3 & 3 \\ 4 & 4 & 4 \end{vmatrix}$

19. $\begin{vmatrix} -2 & -1 & 3 \\ 7 & -7 & 4 \\ 8 & -6 & 2 \end{vmatrix}$

20. $\begin{vmatrix} \frac{1}{2} & 4 & 7 \\ 1 & -1 & 2 \\ 3 & 2 & 5 \end{vmatrix}$

In problems 21–25, solve each equation for x.

21. $\begin{vmatrix} x & x \\ 5 & 3 \end{vmatrix} = 2$

22. $\begin{vmatrix} x+1 & x \\ x & x-2 \end{vmatrix} = -6$

23. $\begin{vmatrix} x & 4 & 5 \\ 0 & 1 & x \\ 5 & 2 & 0 \end{vmatrix} = 7$

24. $\begin{vmatrix} x & 0 & 1 \\ 4x & 1 & 2 \\ 3x & 1 & 3 \end{vmatrix} = 4$

25. $\begin{vmatrix} x & 5 \\ 4 & 2-x \end{vmatrix} = -x^2 + 3$

26. Show that the equation

$$\begin{vmatrix} 0 & x-2 & x-3 \\ x+2 & 0 & x-4 \\ x+3 & x+4 & 0 \end{vmatrix} = 0 \qquad \text{has 0 as a root.}$$

27. For what values of x is it true that $\begin{vmatrix} x & 2 \\ 2 & x \end{vmatrix} > 0$?

28. What kind of solutions does $ax^2 + bx + c = 0$, $a \neq 0$ have if

(a) $\begin{vmatrix} b & 4a \\ c & b \end{vmatrix} > 0$ 　　(b) $\begin{vmatrix} b & 4a \\ c & b \end{vmatrix} = 0$ 　　(c) $\begin{vmatrix} b & 4a \\ c & b \end{vmatrix} < 0$

8.4 Cramer's Rule

As we mentioned in Section 8.3, the solution to the linear system

$$\begin{cases} a_1 x + b_1 y = c_1 \\ a_2 x + b_2 y = c_2 \end{cases}$$

is the ordered pair (x, y), where

$$x = \frac{b_2 c_1 - b_1 c_2}{a_1 b_2 - a_2 b_1} \qquad \text{and} \qquad y = \frac{a_1 c_2 - a_2 c_1}{a_1 b_2 - a_2 b_1}.$$

We can express the numerators and the denominators of these two equations by using determinants:

$$a_1 b_2 - a_2 b_1 = \begin{vmatrix} a_1 & b_1 \\ a_2 & b_2 \end{vmatrix}.$$

$$b_2 c_1 - b_1 c_2 = \begin{vmatrix} c_1 & b_1 \\ c_2 & b_2 \end{vmatrix} \qquad \text{and} \qquad a_1 c_2 - a_2 c_1 = \begin{vmatrix} a_1 & c_1 \\ a_2 & c_2 \end{vmatrix}.$$

We may therefore write the solution of the linear system

$$\begin{cases} a_1 x + b_1 y = c_1 \\ a_2 x + b_2 y = c_2 \end{cases}$$

in the form

$$x = \frac{\begin{vmatrix} c_1 & b_1 \\ c_2 & b_2 \end{vmatrix}}{\begin{vmatrix} a_1 & b_1 \\ a_2 & b_2 \end{vmatrix}} \qquad \text{and} \qquad y = \frac{\begin{vmatrix} a_1 & c_1 \\ a_2 & c_2 \end{vmatrix}}{\begin{vmatrix} a_1 & b_1 \\ a_2 & b_2 \end{vmatrix}}.$$

This is **Cramer's rule** for two linear equations in two unknowns, which is stated formally as follows:

Cramer's Rule for Solving
Linear Systems of Two Equations in Two Unknowns

Suppose that $D \neq 0$, and that D, D_x, and D_y are given by

$$D = \begin{vmatrix} a_1 & b_1 \\ a_2 & b_2 \end{vmatrix}, \qquad D_x = \begin{vmatrix} c_1 & b_1 \\ c_2 & b_2 \end{vmatrix}, \qquad D_y = \begin{vmatrix} a_1 & c_1 \\ a_2 & c_2 \end{vmatrix}.$$

Then the system

$$\begin{cases} a_1 x + b_1 y = c_1 \\ a_2 x + b_2 y = c_2 \end{cases}$$

has one and only one solution:

$$x = \frac{D_x}{D}, \qquad y = \frac{D_y}{D}.$$

In Cramer's rule, we call D the **coefficient determinant** because the elements of D are the coefficients of the unknowns in the system:

$$\begin{cases} a_1 x + b_1 y = c_1 \\ a_2 x + b_2 y = c_2 \end{cases}, \qquad D = \begin{vmatrix} a_1 & b_1 \\ a_2 & b_2 \end{vmatrix}.$$

You will notice that we obtain D_x by replacing the *first* column of D (the coefficients of x) with the constants on the right in the system of equations:

$$\begin{cases} a_1 x + b_1 y = c_1 \\ a_2 x + b_2 y = c_2 \end{cases}, \qquad D_x = \begin{vmatrix} c_1 & b_1 \\ c_2 & b_2 \end{vmatrix}, \qquad D_y = \begin{vmatrix} a_1 & c_1 \\ a_2 & c_2 \end{vmatrix}.$$

We obtain D_y by replacing the *second* column of D (the coefficients of y) with these same constants.

In Examples 1–3, solve the given systems by using Cramer's rule.

EXAMPLE 1
$$\begin{cases} 2x - 3y = 3 \\ x + 4y = 7 \end{cases}$$

SOLUTION

$$D = \begin{vmatrix} 2 & -3 \\ 1 & 4 \end{vmatrix} = 2(4) - 1(-3) = 11$$

$$D_x = \begin{vmatrix} 3 & -3 \\ 7 & 4 \end{vmatrix} = 3(4) - 7(-3) = 33$$

$$D_y = \begin{vmatrix} 2 & 3 \\ 1 & 7 \end{vmatrix} = 2(7) - 1(3) = 11.$$

Therefore,

$$x = \frac{D_x}{D} = \frac{33}{11} = 3 \quad \text{and} \quad y = \frac{D_y}{D} = \frac{11}{11} = 1.$$

EXAMPLE 2

$$\begin{cases} 3x = 4y - 1 \\ y = 2x + 2 \end{cases}$$

SOLUTION

First we write the system in the form

$$\begin{cases} 3x - 4y = -1 \\ -2x + y = 2. \end{cases}$$

Thus,

$$D = \begin{vmatrix} 3 & -4 \\ -2 & 1 \end{vmatrix} = 3(1) - (-2)(-4) = -5$$

$$D_x = \begin{vmatrix} -1 & -4 \\ 2 & 1 \end{vmatrix} = -1(1) - 2(-4) = 7$$

$$D_y = \begin{vmatrix} 3 & -1 \\ -2 & 2 \end{vmatrix} = 3(2) - (-1)(-2) = 4.$$

The solution is

$$x = \frac{D_x}{D} = -\frac{7}{5} \quad \text{and} \quad y = \frac{D_y}{D} = -\frac{4}{5}.$$

EXAMPLE 3

$$\begin{cases} 2x + y = 1 \\ 4x + 2y = 3 \end{cases}$$

SOLUTION

$$D = \begin{vmatrix} 2 & 1 \\ 4 & 2 \end{vmatrix} = 2(2) - 4(1) = 0$$

$$D_x = \begin{vmatrix} 1 & 1 \\ 3 & 2 \end{vmatrix} = 1(2) - 3(1) = -1$$

$$D_y = \begin{vmatrix} 2 & 1 \\ 4 & 3 \end{vmatrix} = 2(3) - 4(1) = 2.$$

The system is inconsistent because $x = D_x/D = -1/0$ and $y = D_y/D = 2/0$. Thus, there is no solution.

When we solve systems of linear equations by this method, we are using the **Cramer's rule method.** This method can be used to solve any system of n linear equations in n variables (n is a positive integer). Here, however, we only extend this rule to include systems of three equations in three variables. Consider the following system of linear equations:

$$\begin{cases} a_1x + b_1y + c_1z = d_1 \\ a_2x + b_2y + c_2z = d_2 \\ a_3x + b_3y + c_3z = d_3. \end{cases}$$

We can now state Cramer's rule for solving this system:

**Cramer's Rule for Solving
Linear Systems of Three Equations in Three Unknowns**

Suppose that $D \neq 0$ and that D, D_x, D_y, and D_z are given by

$$D = \begin{vmatrix} a_1 & b_1 & c_1 \\ a_2 & b_2 & c_2 \\ a_3 & b_3 & c_3 \end{vmatrix}, \qquad D_x = \begin{vmatrix} d_1 & b_1 & c_1 \\ d_2 & b_2 & c_2 \\ d_3 & b_3 & c_3 \end{vmatrix}, \qquad D_y = \begin{vmatrix} a_1 & d_1 & c_1 \\ a_2 & d_2 & c_2 \\ a_3 & d_3 & c_3 \end{vmatrix},$$

$$D_z = \begin{vmatrix} a_1 & b_1 & d_1 \\ a_2 & b_2 & d_2 \\ a_3 & b_3 & d_3 \end{vmatrix},$$

then the solution of the preceding system is given by

$$x = \frac{D_x}{D}, \qquad y = \frac{D_y}{D}, \qquad z = \frac{D_z}{D}.$$

If $D = 0$, the system is inconsistent (or dependent), and Cramer's rule is not applicable.

EXAMPLE 4   Use Cramer's rule to find the solution set of the given system:

$$\begin{cases} x + y + z = 2 \\ 2x - y + z = 0 \\ x + 2y - z = 4 \end{cases}$$

SOLUTION

$$D = \begin{vmatrix} 1 & 1 & 1 \\ 2 & -1 & 1 \\ 1 & 2 & -1 \end{vmatrix}$$

$$= 1\begin{vmatrix} -1 & 1 \\ 2 & -1 \end{vmatrix} - 2\begin{vmatrix} 1 & 1 \\ 2 & -1 \end{vmatrix} + 1\begin{vmatrix} 1 & 1 \\ -1 & 1 \end{vmatrix}$$

$$= 1(-1) - 2(-3) + 1(2)$$

$$= 7.$$

Since $D \neq 0$, we proceed to find D_x, D_y, and D_z as follows:

$$D_x = \begin{vmatrix} 2 & 1 & 1 \\ 0 & -1 & 1 \\ 4 & 2 & -1 \end{vmatrix}$$

$$= 2\begin{vmatrix} -1 & 1 \\ 2 & -1 \end{vmatrix} - 0\begin{vmatrix} 1 & 1 \\ 2 & -1 \end{vmatrix} + 4\begin{vmatrix} 1 & 1 \\ -1 & 1 \end{vmatrix}$$

$$= 2(-1) - 0(-3) + 4(2) = 6$$

$$D_y = \begin{vmatrix} 1 & 2 & 1 \\ 2 & 0 & 1 \\ 1 & 4 & -1 \end{vmatrix}$$

$$= 1\begin{vmatrix} 0 & 1 \\ 4 & -1 \end{vmatrix} - 2\begin{vmatrix} 2 & 1 \\ 4 & -1 \end{vmatrix} + 1\begin{vmatrix} 2 & 1 \\ 0 & 1 \end{vmatrix}$$

$$= 1(-4) - 2(-6) + 1(2) = 10$$

$$D_z = \begin{vmatrix} 1 & 1 & 2 \\ 2 & -1 & 0 \\ 1 & 2 & 4 \end{vmatrix}$$

$$= 1\begin{vmatrix} -1 & 0 \\ 2 & 4 \end{vmatrix} - 2\begin{vmatrix} 1 & 2 \\ 2 & 4 \end{vmatrix} + 1\begin{vmatrix} 1 & 2 \\ -1 & 0 \end{vmatrix}$$

$$= 1(-4) - 2(0) + 1(2) = -2.$$

Then we have

$$x = \frac{D_x}{D} = \frac{6}{7}, \qquad y = \frac{D_y}{D} = \frac{10}{7}, \qquad z = \frac{D_z}{D} = -\frac{2}{7}.$$

PROBLEM SET 8.4

In problems 1–20, use Cramer's rule to solve each system.

1. $\begin{cases} 2x - y = 0 \\ x + y = 1 \end{cases}$

2. $\begin{cases} -3x + y = 3 \\ -2x - y = -5 \end{cases}$

3. $\begin{cases} u + v = 0 \\ u - v = 0 \end{cases}$

4. $\begin{cases} 3t + s = 1 \\ 9t + 3s = -4 \end{cases}$

5. $\begin{cases} p + q = 30 \\ 2p - 2q = 25 \end{cases}$

6. $\begin{cases} 7x - 9y = 13 \\ 5x + 2y = 10 \end{cases}$

7. $\begin{cases} 3x + 7y = 16 \\ 2x + 5y = 13 \end{cases}$

8. $\begin{cases} 7m + 4n = 1 \\ 9m + 4n = 3 \end{cases}$

9. $\begin{cases} 8z - 2w = 52 \\ 3z - 5w = 45 \end{cases}$

10. $\begin{cases} 5x + 11y - 102 = 0 \\ x - 3y + 16 = 0 \end{cases}$

11. $\begin{cases} x + y + z = 6 \\ 3x - y + 2z = 7 \\ 2x + 3y - z = 5 \end{cases}$

12. $\begin{cases} u + v + w = 9 \\ 27u + 9v + 3w = 93 \\ 8u + 4v + 2w = 36 \end{cases}$

13. $\begin{cases} 2r - s + t = 3 \\ -r + 2s - t = 1 \\ 3r + s + 2t = -1 \end{cases}$

14. $\begin{cases} x + y + 2z = 4 \\ x + y - 2z = 0 \\ x - y = 0 \end{cases}$

15. $\begin{cases} 2x - 3y = 4 \\ x + y - 2z = 1 \\ x - y - z = 5 \end{cases}$

16. $\begin{cases} a + b + c = 4 \\ a - b + 2c = 8 \\ 2a + b - c = 3 \end{cases}$

17. $\begin{cases} 2u + 3v + w = 6 \\ u - 2v + 3w = -3 \\ 3u + v - w = 8 \end{cases}$

18. $\begin{cases} 3r + 2s + 2t = 6 \\ r - 5s + 6t = 2 \\ 6r - 8s = 12 \end{cases}$

19. $\begin{cases} x + y = 1 \\ y + z = 9 \\ x + z = 0 \end{cases}$

20. $\begin{cases} x - y + z = 3 \\ 2x + 3y - 2z = 5 \\ 3x + y - 4z = 12 \end{cases}$

8.5 Applications Involving Linear Systems

In Section 4.5, we worked applied problems that gave rise to linear equations in one variable. Problems in applied mathematics often contain two or more variables, rather than one. We can solve word problems involving linear systems with two or more variables by using a slight variation of the procedure in Chapter 4 (page 172).

EXAMPLE 1   The difference between two numbers is 12. Also, the sum of the larger number and twice the smaller number is 75. Find the numbers.

SOLUTION   Let

$$x = \text{the larger number,}$$

and let

$$y = \text{the smaller number.}$$

We have the system of linear equations

$$\begin{cases} x + 2y = 75 \\ x - y = 12. \end{cases}$$

To solve this system, we subtract the second equation from the first equation:

$$\begin{cases} x + 2y = 75 \\ \underline{x - y = 12} \\ 3y = 63 \\ y = 21 \end{cases}$$

and

$$x - 21 = 12 \quad \text{or} \quad x = 33.$$

Therefore, the two numbers are 33 and 21.

Check

$$33 + 2(21) = 75 \quad \text{and} \quad 33 - 21 = 12.$$

EXAMPLE 2 A cash register contains 58 bills with a total value of $178. If the money is all in $1 and $5 bills, how many bills of each denomination are there?

SOLUTION Let

$$x = \text{the number of \$1 bills}$$

$$y = \text{the number of \$5 bills.}$$

We have the following system of linear equations:

$$\begin{cases} x + 5y = 178 \\ x + y = 58. \end{cases}$$

We solve this system by subtracting:

$$\begin{cases} x + 5y = 178 \\ x + \ y = \ 58 \end{cases}$$
$$\overline{ 4y = 120}$$
$$y = 30.$$

Substituting $y = 30$ in the equation $x + y = 58$, we obtain

$$x + 30 = 58 \quad \text{or} \quad x = 28.$$

Therefore, the cash register contains 28 $1 bills and 30 $5 bills.

Check
$$28 + 30 = 58 \text{ bills}$$
and
$$28(1) + 30(5) = \$28 + \$150$$
$$= \$178.$$

EXAMPLE 3 The proprietor of a television repair shop charges a fixed amount plus an hourly rate to repair a set. If he charged $50 to repair a set that was worked on for 1 hour, and $90 to repair a set that was worked on for 3 hours, find the fixed charge and hourly rate.

SOLUTION Let

$$x = \text{the fixed charge}$$

$$y = \text{the hourly rate.}$$

The fixed charge plus the hourly rate for 1 hour equals $50, so

$$x + y = 50.$$

Also, the fixed charge plus the hourly rate for 3 hours equals \$90, so

$$x + 3y = 90.$$

Thus, the system of equations is

$$\begin{cases} x + y = 50 \\ x + 3y = 90. \end{cases}$$

We solve this system by subtracting:

$$\begin{cases} x + y = 50 \\ x + 3y = 90 \end{cases}$$
$$\overline{ -2y = -40}$$
$$y = 20.$$

We substitute $y = 20$ in the equation $x + y = 50$:

$$x + 20 = 50$$
$$x = 30.$$

Therefore, the fixed charge is \$30 and the hourly rate is \$20.

Check

$$30 + 20 = 50 \qquad \text{and} \qquad 30 + 3(20) = 90.$$

EXAMPLE 4   A bank customer bought two commercial papers for \$40,000. One commercial paper paid 17% simple annual interest and the other paid 18% simple annual interest.

    If the total interest from both papers, at the end of 1 year, is \$7,050, how much did the customer pay for each paper?

SOLUTION   Let

$$x = \text{the amount (in dollars) paid for the 17\% paper}$$
$$y = \text{the amount (in dollars) paid for the 18\% paper.}$$

Then

$$x + y = 40,000,$$

and the total interest is

$$0.17x + 0.18y = 7,050.$$

The system is

$$\begin{cases} x + y = 40{,}000 \\ 0.17x + 0.18y = 7{,}050 \end{cases}$$

or

$$\begin{cases} x + y = 40{,}000 \\ 17x + 18y = 705{,}000. \end{cases}$$

We solve this system as follows:

$$\begin{cases} x + y = 40{,}000 \\ 17x + 18y = 705{,}000 \end{cases} \xrightarrow[\text{side by } -17.]{\text{We multiply each}} \begin{cases} -17x - 17y = -680{,}000 \\ \underline{17x + 18y = 705{,}000} \quad \text{add} \\ y = 25{,}000. \end{cases}$$

We substitute $y = 25{,}000$ in the equation $x + y = 40{,}000$:

$$x + 25{,}000 = 40{,}000$$

$$x = 15{,}000.$$

Therefore, the bank customer paid \$15,000 for the commercial paper carrying 17% interest and \$25,000 for the commercial paper carrying 18% interest.

Check

$$0.17(15{,}000) + 0.18(25{,}000) = 2{,}550 + 4{,}500 = 7{,}050$$

$$15{,}000 + 25{,}000 = 40{,}000.$$

EXAMPLE 5   A concert was held in a sports arena that can seat 20,000 people. Tickets were available for \$20, \$15, and \$10. There were twice as many \$15 seats as \$20 seats. If the concert was sold out and grossed \$260,000, how many tickets of each kind were sold?

SOLUTION   Let

$$x = \text{the number of \$10 tickets}$$

$$y = \text{the number of \$15 tickets}$$

$$z = \text{the number of \$20 tickets.}$$

Then we have the following system of equations:

$$\begin{cases} x + y + z = 20{,}000 \\ y = 2z \\ 10x + 15y + 20z = 260{,}000. \end{cases}$$

We solve this system as follows:

We multiply each side by -10.

$$\begin{cases} x + y + z = 20{,}000 \\ 10x + 15y + 20z = 260{,}000 \end{cases} \longrightarrow \begin{cases} -10x - 10y - 10z = -200{,}000 \\ \underline{10x + 15y + 20z = 260{,}000} \quad \text{add} \\ 5y + 10z = 60{,}000. \end{cases}$$

so that we now have the system

$$\begin{cases} 5y + 10z = 60{,}000 \\ y - 2z = 0 \end{cases} \xrightarrow{\text{We multiply each side by 5.}} \begin{cases} 5y + 10z = 60{,}000 \\ \underline{5y - 10z = 0} \quad \text{add} \\ 10y = 60{,}000 \\ y = 6{,}000. \end{cases}$$

We substitute $y = 6{,}000$ in the equation $y = 2z$:

$$6{,}000 = 2z$$

$$3{,}000 = z.$$

Next we substitute $y = 6{,}000$ and $z = 3{,}000$ in $x + y + z = 20{,}000$:

$$x + 6{,}000 + 3{,}000 = 20{,}000$$

$$x = 11{,}000.$$

Thus, there were 11,000 tickets sold at $10, 6,000 tickets sold at $15, and 3,000 tickets sold at $20.

Check

$$11{,}000 + 6{,}000 + 3{,}000 = 20{,}000$$

$$6{,}000 = 2(3{,}000)$$

$$10(11{,}000) + 15(6{,}000) + 20(3{,}000) = 260{,}000.$$

PROBLEM SET 8.5

© In many of the following problems, a calculator may be useful to speed up the arithmetic.

1. The sum of two numbers is 12. If one of the numbers is multiplied by 5, and the other is multiplied by 8, the sum of the products is 75. Find the numbers.

2. Find two numbers such that twice the first plus five times the second is 20, and four times the first less three times the second is 14.

3. The sum of the ages of Pete and Sal is 15. If Pete is four times as old as Sal, how old is each boy?

4. A bus has 31 passengers. If the number of adults exceeds the number of children by 5, how many adults are in the bus?

5. In a certain habitat, the number of prey exceeds the number of predators by 4,200, while together they total 5,650. How many prey and how many predators are in the habitat?

6. The sum of the digits of a two-place number is 13, and the number is increased by 9 when the order of the digits is reversed. Find the number.

7. Joe and Dawn are arguing over the difference in their ages. Joe says that he is 3 years older than Dawn. Dawn claims that 4 years ago the sum of their ages was 23. If they are both right, how old is each now?

8. Joan is 3 years older than Steve; 8 years ago she was four times as old as Steve. How old is each now?

9. Raul has 18 coins worth $3.60 in his pocket, and the coins are all quarters and dimes. How many quarters and how many dimes does he have?

10. Nadia has 11 coins, consisting of nickels and dimes. How many of each coin does she have if the total amount is 90 cents?

11. A cashier has 45 coins in dimes and quarters. If she has $8.70, how many coins of each type does she have?

12. After counting his cash, a bookstore cashier finds that he has three times as many dimes as nickels. If the value of the dimes and nickels is $4.20, how many coins of each type does he have?

13. A bank teller has 78 bills in $5 and $10 denominations. The total value is $465. How many bills of each denomination does the teller have?

14. A theater sells tickets at $2.50 for children and $4 for adults. If 375 tickets were sold to one show and $1,398 was collected, how many tickets of each kind were sold?

15. A plumber charges a fixed charge plus an hourly rate for service on a house call. The plumber charged $70 to repair a water tank that required 2 hours of labor and $100 to repair a water tank that took 3.5 hours. Find the plumber's fixed charge and hourly rate.

16. An agency providing temporary help charges $16 per hour for a bookkeeper and $10 per hour for a typist. A school called the agency to obtain a bookkeeper and a typist. The bill from the agency for their services came to $176. If the typist worked 2 more hours than the bookkeeper, how many hours did each employee work, and how much did each of them earn?

17. If Tom and John work together, they can complete a job in 4 hours. When Tom works by himself, it takes him twice the time that it takes John, working alone, to finish the same job. How long does it take each man, working alone, to complete the job?

18. A 100-gallon vat in a chemical plant has two intake pipes, a large one carrying water and a small one carrying acid. If both pipes are turned on together, the vat can be filled in 40 minutes. Ordinarily, however, the water pipe is allowed to run for 45 minutes and is shut off before the acid pipe is turned on. If it then takes the acid pipe 20 minutes to fill the tank, what are the delivery rates of the water pipe and the acid pipe?

19. An electrician works for 8 hours and his assistant works for 6 hours on a job that pays them a total of $440. If the electrician is paid twice as much per hour as his assistant, what is the hourly rate for each?

20. A grocer bought a number of cases of vegetables for $221. He obtained some of the vegetables at the rate of three cases for $17 and the rest at the rate of four cases for $17. When he sold all the vegetables at $10 per case, he cleared $209. How many cases of each kind did the grocer buy?

21. A businesswoman has invested a total of $40,000 in two certificates. The first certificate pays 10.5% simple annual interest, and the second pays 13.5%. At the end of 1 year, her combined interest on the two certificates is $4,650. How much did she originally invest in each certificate?

22. A person invests part of $35,000 in a certificate that yields 11.5% simple annual interest, and the rest in a certificate that yields 13.2%. At the end of 1 year, the combined interest on the two certificates is $4,382. How much was invested in each certificate?

23. A man invested a total of $4,000 in securities. Part was invested at 8.5% simple annual interest and the rest was invested at 7.9%. His annual income from both investments was $329.50. What amount did he invest at each rate?

24. A chemist has in her laboratory the same acid in two strengths. Six parts of the first mixed with four parts of the second gives a mixture that is 86% pure, and four parts of the first mixed with six parts of the second gives a mixture that is 84% pure. What is the percentage of purity of each of the original solutions of acid?

25. Victor has loaned part of $30,000 at 18% simple annual interest and the rest at 19.5%. If after 1 year the income from the money loaned at 18% is $1,650 more than the income from the money loaned at 19.5%, what was the amount of each loan?

26. An 80% acid solution is mixed with an 18% acid solution. The result is 3 gallons of a solution that is one-third acid. How much of each solution was used?

27. A cashier has $48 consisting of 294 coins in half-dollars, quarters, and dimes. There are $3\frac{1}{2}$ times as many dimes as quarters. How many coins of each kind are there?

28. Judy bought three different bonds for $20,000, one paying a 6% annual dividend, one paying a 7% dividend, and the other paying an 8% dividend. If the sum of the dividends from the 6% and the 7% bonds amounts to $940 in a year, and the sum of the dividends from the 6% and the 8% bonds is $720, how much did she invest in each bond?

29. A department store has sold 80 men's suits of three different types at a discount. If the suits had been sold at their original prices—type I suits for $80, type II suits for $90, and type III suits for $95—the total receipts would have been $6,825. However, the suits were sold for $75, $80, and $85, respectively, and the total receipts amounted to $6,250. Determine the number of suits of each type sold during the sale.

30. A watch, a chain, and a ring together cost $225. The watch cost $50 more than the chain, and the ring cost $25 more than the watch and the chain together. What was the cost of each item?

8.6 Systems of Linear Inequalities

In Section 7.6, we graphed linear inequalities. Now we turn our attention to systems of inequalities in two variables. We define the **graph** of a system of linear inequalities as the set of all points (x, y) in the xy plane whose coordinates satisfy every inequality in the system. Such a graph is obtained by sketching the graphs of all inequalities on the same coordinate axes. The region where the half-planes intersect is the **solution** of the system.

In Examples 1–4, sketch the graph of each system of inequalities.

EXAMPLE 1
$$\begin{cases} x + y > 3 \\ 3x - y \geq 6 \end{cases}$$

SOLUTION   First we sketch the graphs of $x + y = 3$ and $3x - y = 6$ on the same coordinate axes (Figure 1). The graph of $x + y > 3$ is the half-plane above the line $x + y = 3$, and the graph of $3x - y \geq 6$ is the half-plane below and including the line $3x - y = 6$. The two half-planes overlap in the region shaded in Figure 1. This shaded region is the graph of the system of inequalities.

Figure 1

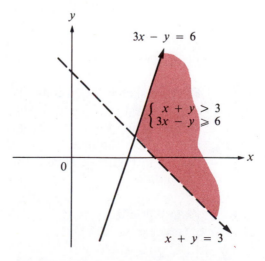

EXAMPLE 2 $\begin{cases} 2x - y \geq 4 \\ 2x + 3y \leq 6 \end{cases}$

SOLUTION   First we sketch the graphs of $2x - y = 4$ and $2x + 3y = 6$ on the same coordinate system (Figure 2). The graph of $2x - y \geq 4$ is the half-plane below and including the line $2x - y = 4$, and the graph of $2x + 3y \leq 6$ is the half-plane below and including the line $2x + 3y = 6$. The two half-planes overlap in the region shaded in Figure 2. Hence, this shaded region is the graph of the system of inequalities.

Figure 2

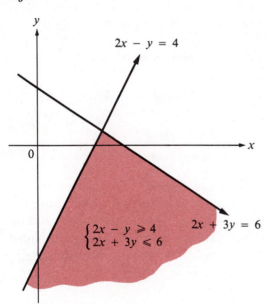

Figure 3

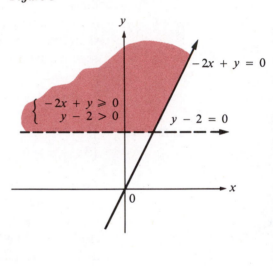

EXAMPLE 3 $\begin{cases} -2x + y \geq 0 \\ y - 2 > 0 \end{cases}$

SOLUTION   First we sketch the graphs of $-2x + y = 0$ and $y - 2 = 0$ on the same coordinate system (Figure 3). The graph of $-2x + y \geq 0$ is the half-plane above and including the line $-2x + y = 0$ and the graph of $y - 2 > 0$ is the half-plane above the line $y - 2 = 0$. The two half-planes overlap in the region shaded in Figure 3. Hence, this shaded region is the graph of the system of inequalities.

EXAMPLE 4 $\begin{cases} x \geq 0, y \geq 0 \\ 2x + y \leq 5 \\ x + 2y \leq 4. \end{cases}$

SOLUTION The restrictions $x \geq 0$ and $y \geq 0$ limit us to the first quadrant, including the x and y axes. First we sketch the graphs of $2x + y = 5$ and $x + 2y = 4$. The graph of $2x + y \leq 5$ is the half-plane below and including the line whose equation is $2x + y = 5$. The graph of $x + 2y \leq 4$ is the half-plane below and including the line $x + 2y = 4$. The two half-planes overlap in the region in the first quadrant in Figure 4. Hence the shaded region is the graph of the system of inequalities.

Figure 4

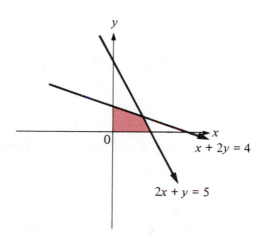

$x + 2y = 4$

$2x + y = 5$

PROBLEM SET 8.6

In problems 1–16, sketch the graph of each system of linear inequalities.

1. $\begin{cases} x + y < 2 \\ 2x - y < -1 \end{cases}$
2. $\begin{cases} -x + y < 1 \\ -x + 3y \leq 5 \end{cases}$
3. $\begin{cases} x + y \leq 3 \\ x - y \geq 3 \end{cases}$
4. $\begin{cases} x + y > 5 \\ x - y < 9 \end{cases}$

5. $\begin{cases} x + 2y \leq 12 \\ x - y > 6 \end{cases}$
6. $\begin{cases} x + 2y > 12 \\ 3x - y < 1 \end{cases}$
7. $\begin{cases} 3x + y > 6 \\ x - y \geq 1 \end{cases}$
8. $\begin{cases} 2x - y \leq -2 \\ x - 2y \geq -2 \end{cases}$

9. $\begin{cases} 2x - 3y \leq -3 \\ 5x - 2y > 9 \end{cases}$
10. $\begin{cases} y \leq 2x + 4 \\ y \geq 3 - x \end{cases}$
11. $\begin{cases} x + y \leq 2 \\ -x + 3y \geq 4 \end{cases}$
12. $\begin{cases} 2x + y \geq 2 \\ x - 2y \leq 3 \end{cases}$

13. $\begin{cases} x \geq 0, \, y \geq 0 \\ 2x + y \leq 2 \\ x + 2y \leq 2 \end{cases}$
14. $\begin{cases} x \geq 0, \, y \geq 0 \\ x + y \geq 12 \\ 3x - y \geq 16 \end{cases}$
15. $\begin{cases} x \geq 0, \, y \geq 0 \\ x + 2y \geq 6 \\ 2x - 3y \geq 10 \end{cases}$
16. $\begin{cases} x \geq 0, \, y \geq 0 \\ x + y \leq 4 \\ 3x - y \leq 7 \end{cases}$

17. An investment firm has at most $200,000 to invest in two stocks, A and B. Stock A sells at $56 a share and stock B at $86 a share; and the total number of shares to be purchased of both stocks cannot exceed 3,000. If x is the number of shares to be purchased of stock A and y is the number of shares to be purchased of stock B, write a system of inequalities that indicates the restrictions on x and y. Graph the system showing the region of permissible values for x and y.

18. A typesetter estimates that it costs $500 per day to set type for a textbook using computer A, and $400 per day using computer B. The total number of days required to typeset the book cannot exceed 45, only one computer per day can be used, and no more than \$20,000 can be spent. If x is the number of days required to typeset the book using computer A, and y is the number of days required to typeset using computer B, write a system of inequalities that indicates the restrictions on x and y. Graph the system showing the region of permissible values for x and y.

REVIEW PROBLEM SET

In problems 1–4, use graphs to determine whether the system is dependent, inconsistent, or independent.

1. $\begin{cases} y = -2x + 2 \\ y = x - 4 \end{cases}$
2. $\begin{cases} y = 5x + 2 \\ 10x - 2y + 4 = 0 \end{cases}$
3. $\begin{cases} 3u + 2v = 1 \\ 3u + 2v = 3 \end{cases}$
4. $\begin{cases} r + s = 4 \\ 2r - s = 8 \end{cases}$

In problems 5–8, use the substitution method to solve each system.

5. $\begin{cases} 2x - y = 5 \\ x + 3y = 6 \end{cases}$
6. $\begin{cases} 3p + q = -1 \\ 2p - 2q = -14 \end{cases}$
7. $\begin{cases} 2x + y + z = 2 \\ x + 3y - z = 9 \\ 3x - y - 2z = 5 \end{cases}$
8. $\begin{cases} r + s + 2t = 0 \\ 2r - 3s - t = 5 \\ 3r + 7s - 2t = 8 \end{cases}$

In problems 9–14, use the elimination method to solve each system.

9. $\begin{cases} x - y = 3 \\ 2x + y = 3 \end{cases}$
10. $\begin{cases} 5m + 2p = 3 \\ 2m - 3p = 5 \end{cases}$
11. $\begin{cases} u - v + 2w = 0 \\ 3u + v + w = 2 \\ 2u - v + 5w = 5 \end{cases}$

12. $\begin{cases} 3x + 2y - z = -4 \\ x - y + 2z = 13 \\ 5x + 3y - 4z = -15 \end{cases}$
13. $\begin{cases} \dfrac{2}{x} + \dfrac{1}{y} = 8 \\ \dfrac{3}{x} - \dfrac{1}{y} = 7 \end{cases}$
14. $\begin{cases} \dfrac{1}{r} + \dfrac{2}{s} + \dfrac{3}{t} = 0 \\ \dfrac{2}{r} - \dfrac{3}{s} + \dfrac{1}{t} = 2 \\ \dfrac{5}{r} - \dfrac{4}{s} - \dfrac{2}{t} = -3 \end{cases}$

In problems 15–18, evaluate each determinant.

15. $\begin{vmatrix} 3 & 4 \\ 1 & 5 \end{vmatrix}$
16. $\begin{vmatrix} 4 & 2 \\ 1 & -1 \end{vmatrix}$
17. $\begin{vmatrix} 4 & 2 & 1 \\ 5 & 7 & 1 \\ 6 & 2 & 3 \end{vmatrix}$
18. $\begin{vmatrix} -1 & 4 & 3 \\ 7 & 1 & 4 \\ 1 & 3 & 5 \end{vmatrix}$

In problems 19 and 20, solve for x.

19. $\begin{vmatrix} x - 1 & -3 \\ 2 & x + 3 \end{vmatrix} = 6$
20. $\begin{vmatrix} x + 1 & 3 \\ 4 & x \end{vmatrix} < 0$

In problems 21–24, use Cramer's rule to solve each system.

21. $\begin{cases} x - y = 3 \\ 2x + y = 3 \end{cases}$
22. $\begin{cases} 5p + 2q = 3 \\ 2p + 3q = -1 \end{cases}$

23. $\begin{cases} x - y + 2z = 0 \\ 3x + y + z = 2 \\ 2x - y + 5z = 5 \end{cases}$

24. $\begin{cases} 3r + 2s - t = -4 \\ r - s + 2t = 13 \\ 5r + 3s - 4t = -15 \end{cases}$

In problems 25–28, sketch the graph of each system of linear inequalities.

25. $\begin{cases} 2x + 4y < 3 \\ -2x + y > 4 \end{cases}$

26. $\begin{cases} 3x - 5y \geq 15 \\ -2x + 6y \geq 12 \end{cases}$

27. $\begin{cases} x \geq 0, \ y \geq 0 \\ 2x + 3y \leq 6 \\ 5x + y \leq 5 \end{cases}$

28. $\begin{cases} x \geq 0, \ y \geq 0 \\ x + y \geq 4 \\ 2x - y \geq 3 \end{cases}$

In problems 29–36, use a system of linear equations to solve each problem.

29. The sum of two numbers is 2. If one of the numbers is multiplied by 3 and the other is multiplied by 2, the sum of the products is 11. Find the numbers.

30. The sum of three numbers is 11. One number is twice the smallest number, and the remaining number is 3 more than the smallest number. Find the numbers.

31. A college mailed 100 letters, some requiring 22 cents postage and the rest requiring 30 cents postage. If the total bill was $24.96, find the number of letters sent at each rate.

32. The specific gravity of an object is defined to be its weight in air divided by its loss of weight when it is submerged in water. An object made partly of gold (which has specific gravity 16) and partly of silver (which has specific gravity 10.8) weighs 8 grams in air and 7.3 grams when it is submerged in water. How many grams of gold and how many grams of silver does the object contain?

33. A coin collection containing nickels and dimes consisted of 32 coins. If the total value of the coins was $2.65, how many coins of each kind were in the collection?

34. A piggy bank contained 35 coins, all nickels, dimes, and quarters. If there were twice as many nickels as quarters, and one-fourth as many dimes as nickels, how many coins of each kind were in the bank?

35. Sylvia invested a total of $30,000 in three different certificates, one paying 14% simple annual interest, one paying 12%, and the other paying 15%. If the total annual interest from the three certificates was $3,950, and if she invested twice as much in the 14% certificate as she did in the 15% certificate, how much did she invest in each kind of certificate?

36. Max and Joshua started a job that had to be finished in two days. The first day, Max worked 9 hours and Joshua worked 8 hours, and they finished half of the job. The next day, Max worked 6 hours and Joshua worked 12 hours to complete the job. How long would it take each, working alone, to do the job?

CHAPTER 8 TEST

1. Use graphs to determine whether the system is dependent, inconsistent, or independent.

(a) $\begin{cases} 3x - 2y = 6 \\ 6x + 4y = 12 \end{cases}$ (b) $\begin{cases} 2x + y = 1 \\ 4x + 2y = 3 \end{cases}$ (c) $\begin{cases} -x + 2y = 1 \\ 3x - 6y = -3 \end{cases}$

2. Use the substitution method to solve each system:

(a) $\begin{cases} 2x - y = 3 \\ -3x + 2y = -7 \end{cases}$ (b) $\begin{cases} 2x - y + z = 6 \\ x + 2y + 3z = 3 \\ -4x - y + 2z = 2 \end{cases}$

(c) Solve the systems in parts (a) and (b) by the elimination method.

3. Evaluate each determinant:

(a) $\begin{vmatrix} -5 & 7 \\ -1 & 2 \end{vmatrix}$ (b) $\begin{vmatrix} 2 & -3 & 1 \\ 1 & 5 & -2 \\ 3 & -1 & 1 \end{vmatrix}$

4. Solve for x:

$$\begin{vmatrix} x + 1 & 3 \\ 1 & x + 5 \end{vmatrix} = 9.$$

5. Use Cramer's rule to solve each system.

(a) $\begin{cases} 3x + 2y = 4 \\ 2x - 4y = 0 \end{cases}$ (b) $\begin{cases} x + y + z = 6 \\ 3x - y + 2z = 7 \\ -2x + 3y - z = 1 \end{cases}$

6. Sketch the graph of each system of linear inequalities.

(a) $\begin{cases} 3x - 5y \le 15 \\ -2x + y < 6 \end{cases}$ (b) $\begin{cases} x \ge 0, \ y \ge 0 \\ 3x + 2y \le 6 \\ x + 5y \le 5 \end{cases}$

7. A coin collection containing nickels and dimes consists of 20 coins. The total value of the coins is $1.35. Use a system of equations to determine the number of nickels and dimes.

9 Logarithms

In this chapter we use the properties of exponents as the basis for establishing the properties of logarithms. Together, exponents and logarithms provide us with a tool to work applications in business, economics, engineering, and biology. They are also used to simplify many types of calculations.

9.1 Exponential Equations and Logarithms

Equations that contain a variable in an exponent are called **exponential equations.** Examples of exponential equations are

$$2^x = 8, \qquad 2^{t/5} = 32, \qquad \text{and} \qquad 5^{3y-7} = 125.$$

Some exponential equations may be solved by using the following property:

$$b^x = b^y \text{ if and only if } x = y \text{ for } b > 0 \text{ and } b \neq 1.$$

In Examples 1–4, solve each equation.

EXAMPLE 1 $2^x = 16$

SOLUTION   First we express 16 as a power of 2, so that both sides of the equation will have the same base. Because $16 = 2^4$, we can rewrite

$$2^x = 16 \qquad \text{as} \qquad 2^x = 2^4.$$

Now we set the exponents equal to one another:

$$x = 4.$$

Therefore, 4 is the solution.

EXAMPLE 2 $9^t = 27^{4t-1}$

SOLUTION   Because $9 = 3^2$ and $27 = 3^3$, we can rewrite the given equation as

$$(3^2)^t = (3^3)^{4t-1} \qquad \text{or} \qquad 3^{2t} = 3^{3(4t-1)}.$$

We set the exponents equal to one another:

$$2t = 3(4t - 1)$$
$$2t = 12t - 3$$
$$-10t = -3$$
$$t = \frac{3}{10}.$$

Therefore, $\frac{3}{10}$ is the solution.

EXAMPLE 3 $(\frac{1}{5})^y = 25^{3y-1}$

SOLUTION   Because $\frac{1}{5} = 5^{-1}$ and $25 = 5^2$, we can rewrite the given equation as

$$(5^{-1})^y = (5^2)^{3y-1} \qquad \text{or} \qquad 5^{-y} = 5^{2(3y-1)}.$$

It follows that

$$-y = 2(3y - 1)$$
$$-y = 6y - 2$$
$$-7y = -2$$
$$y = \frac{2}{7}.$$

Therefore, $\frac{2}{7}$ is the solution.

EXAMPLE 4 $7^{w^2+w} = 49$

SOLUTION   Because $49 = 7^2$, we can rewrite the given equation as

$$7^{w^2+w} = 7^2,$$

so that

$$w^2 + w = 2.$$

We solve this equation:

$$w^2 + w - 2 = 0 \qquad \text{or} \qquad (w + 2)(w - 1) = 0$$

so that

$$w + 2 = 0 \qquad\qquad w - 1 = 0$$
$$w = -2 \qquad\qquad w = 1.$$

Therefore, -2 and 1 are the solutions.

In the previous examples, we were able to express each side of the given equation as an exponential expression with the same base. Now suppose we are given an equation for which we are not able to do this. For example, consider the equation

$$10^x = 5.$$

We cannot simplify this equation so that both sides will be expressed in terms of the same base, as we have done previously. So, how do we solve this equation for x?

We can show that the equation $10^x = 5$ has a real solution, although the proof is beyond the scope of this text. We shall denote this solution by

$$x = \log_{10} 5,$$

which is read "x equals the logarithm of 5 to the base 10." That is, $\log_{10} 5$ is an exponent. It is the power to which we raise 10 in order to get 5. The actual value of $\log_{10} 5$ is an irrational number. (We explain how to obtain this irrational number later in the chapter.)

In general, we have the following definition:

DEFINITION 1   **Logarithm**

If $b > 0$, $b \neq 1$, and $c > 0$, then

$$x = \log_b c \text{ is equivalent to } b^x = c.$$

You can use the preceding definition to convert equations from exponential form to logarithmic form and vice versa, as the following table shows:

Exponential Form	Logarithmic Form
$3^2 = 9$	$2 = \log_3 9$
$5^4 = 625$	$4 = \log_5 625$
$64^{1/3} = 4$	$\frac{1}{3} = \log_{64} 4$
$1{,}000 = 10^3$	$\log_{10} 1{,}000 = 3$
$4 = \left(\frac{1}{2}\right)^{-2}$	$\log_{1/2} 4 = -2$
$\frac{1}{49} = 7^{-2}$	$\log_7 \frac{1}{49} = -2$
$5^0 = 1$	$0 = \log_5 1$
$x^k = d$	$k = \log_x d \qquad (x > 0,\, x \neq 1,\, d > 0)$

EXAMPLE 5   Write each exponential equation in logarithmic form.

(a) $4^{3/2} = 8$ (b) $6^3 = 216$

SOLUTION   Using Definition 1, we have:

(a) $4^{3/2} = 8$ is equivalent to $\frac{3}{2} = \log_4 8$.

(b) $6^3 = 216$ is equivalent to $3 = \log_6 216$.

EXAMPLE 6   Write each equation in exponential form.

(a) $\log_8 4 = \frac{2}{3}$ (b) $\log_{10} 0.01 = -2$

SOLUTION   Using Definition 1, we have:

(a) $\log_8 4 = \frac{2}{3}$ is equivalent to $4 = 8^{2/3}$.

(b) $\log_{10} 0.01 = -2$ is equivalent to $0.01 = 10^{-2}$.

EXAMPLE 7   Find each value.

(a) $\log_3 81$ (b) $\log_2 32$ (c) $\log_8(\frac{1}{64})$

(d) $\log_{32} 8$ (e) $\log_7 7$ (f) $\log_5 1$

SOLUTION   (a) Let $x = \log_3 81$, so that $3^x = 81$. Then,

$$3^x = 3^4 \quad \text{or} \quad x = 4.$$

Therefore, $\log_3 81 = 4$.

(b) Let $x = \log_2 32$, so that $2^x = 32$. Then

$$2^x = 2^5 \quad \text{or} \quad x = 5.$$

Therefore, $\log_2 32 = 5$.

(c) Let $x = \log_8(\frac{1}{64})$, so that $8^x = \frac{1}{64}$. Then

$$8^x = \frac{1}{8^2} = 8^{-2} \quad \text{or} \quad x = -2.$$

Therefore, $\log_8(\frac{1}{64}) = -2$.

(d) Let $x = \log_{32} 8$, so that $32^x = 8$, or $(2^5)^x = 2^{5x} = 8$. Then

$$2^{5x} = 2^3 \quad \text{or} \quad 5x = 3 \quad \text{or} \quad x = \frac{3}{5}.$$

Therefore, $\log_{32} 8 = \frac{3}{5}$.

(e) Let $x = \log_7 7$, so that $7^x = 7$. Then

$$7^x = 7^1 \qquad \text{or} \qquad x = 1.$$

Therefore, $\log_7 7 = 1$.

(f) Let $x = \log_5 1$, so that $5^x = 1$. Then

$$5^x = 1 = 5^0 \qquad \text{or} \qquad x = 0,$$

Therefore, $\log_5 1 = 0$.

Examples 7(e) and (f) can be generalized as follows:

If $b > 0$ and $b \neq 1$, then

$$\text{(i) } \log_b b = 1 \qquad \text{(ii) } \log_b 1 = 0$$

Definition 1, on page 421, can be used to solve some equations involving logarithms. Keep in mind that if $x = \log_b c$, the base b and the quantity c (whose logarithm we are taking) *must* be positive.

In Examples 8–10, solve each equation.

EXAMPLE 8 $\log_2(5x - 3) = 5$

SOLUTION   The equation $\log_2(5x - 3) = 5$ is equivalent to

$$2^5 = 5x - 3.$$

Because $2^5 = 32$, we have

$$5x - 3 = 32$$
$$5x = 35$$
$$x = 7.$$

Therefore, 7 is the solution.

EXAMPLE 9 $\log_4 x^2 = 2$

SOLUTION   The equation $\log_4 x^2 = 2$ is equivalent to $4^2 = x^2$. Therefore,

$$x^2 = 16$$
$$x = \pm\sqrt{16} = \pm 4.$$

The solutions are $x = -4$ and $x = 4$. Notice that both numbers -4 and 4 satisfy the equation because the expression x^2 is positive when either value is substituted for x.

EXAMPLE **10** $\log_x 9 = 2$

SOLUTION   The equation $\log_x 9 = 2$ is equivalent to $x^2 = 9$. Definition 1 requires that the base of a logarithm be positive and not equal to 1. Therefore, we have the restriction that $x > 0$ and $x \neq 1$. Hence, the only solution is

$$x = \sqrt{9}$$
$$= 3.$$

PROBLEM SET 9.1

In problems 1–14, solve each exponential equation.

1. $5^x = 25$ **2.** $3^{2x} = 81$ **3.** $2^{2t} = 16$ **4.** $2^{-4p} = 64$ **5.** $3^{u-5} = 27$

6. $5^{2w-1} = 25$ **7.** $4^{3x} = 8^{x-1}$ **8.** $25^{3t} = 125^{2t-3}$ **9.** $3^{4c-5} = 81$ **10.** $5^{2x-1} = 125$

11. $3^{w^2+2w} = 27$ **12.** $2^{3u^2-2u} = 32$ ✱**13.** $2^{x^2-6x} = (\frac{1}{2})^3$ **14.** $(\frac{1}{3})^{2x+7} = 9^2$

In problems 15–30, write each equation in logarithmic form.

15. $5^3 = 125$ **16.** $4^4 = 256$ **17.** $10^5 = 100,000$ **18.** $49^{0.5} = 7$

19. $4^{-2} = \frac{1}{16}$ **20.** $(\frac{1}{3})^{-2} = 9$ **21.** $6^{-2} = \frac{1}{36}$ **22.** $2^{-3} = 0.125$

23. $\sqrt{9} = 3$ **24.** $\sqrt[5]{32} = 2$ **25.** $(100)^{-3/2} = 0.001$ **26.** $(\frac{1}{8})^{-2/3} = 4$

27. $7^0 = 1$ **28.** $15^0 = 1$ **29.** $x^3 = a$ **30.** $\pi^t = z$

In problems 31–44, write each equation in exponential form.

31. $\log_9 81 = 2$ **32.** $\log_6 36 = 2$ **33.** $\log_{27} 9 = \frac{2}{3}$ **34.** $\log_{27} \frac{1}{9} = -\frac{2}{3}$

35. $\log_{10} 0.001 = -3$ **36.** $\log_{10} \frac{1}{10} = -1$ **37.** $\log_{1/3} 9 = -2$ **38.** $\log_{36} 216 = \frac{3}{2}$

39. $\log_{10} 4.35 = 0.64$ **40.** $\log_{10} 9.14 = 0.96$ **41.** $\log_{\sqrt{16}} 2 = \frac{1}{2}$ **42.** $\log_{4/9} \frac{27}{8} = -\frac{3}{2}$

43. $\log_x 1 = 0$ **44.** $\log_x 2 = 4$

In problems 45–64, find the value of each logarithm.

45. $\log_2 64$ **46.** $\log_4 \frac{1}{16}$ ✔ **47.** $\log_9 3$ **48.** $\log_4 8$ ✗ **49.** $\log_9 \frac{1}{3}$

50. $\log_{1/2} \frac{1}{8}$ ✗ **51.** $\log_{1/9} \frac{1}{81}$ **52.** $\log_2 \frac{1}{32}$ **53.** $\log_5 \frac{1}{125}$ **54.** $\log_{10} 0.00001$

55. $\log_3 9\sqrt{3}$ **56.** $\log_2 4\sqrt{2}$ ➤ **57.** $\log_{10} \frac{1}{10,000}$ **58.** $\log_7 343$ **59.** $\log_3 729$

60. $\log_5 5$ **61.** $\log_7 1$ **62.** $\log_b 1$ **63.** $\log_6 \frac{1}{216}$ **64.** $\log_b b^3$

In problems 65–76, solve each equation.

65. $\log_{10}(x+1) = 1$ **66.** $\log_5(2y-7) = 0$ **67.** $\log_4(3w+1) = 2$ **68.** $\log_7(2u-3) = 2$

69. $\log_5 N = 2$ **70.** $\log_b 36 = 2$ **71.** $\log_x 81 = 4$ **72.** $\log_c 16 = -\frac{4}{3}$

✗ **73.** $\log_b 3 = \frac{1}{5}$ **74.** $\log_3 y = 4$ **75.** $\log_2(\log_5 5) = x$ **76.** $\log_4(\log_8 8) = y$

In chemistry, the **pH** of a substance is defined by

$$pH = -\log_{10}[H^+],$$

where $[H^+]$ is the concentration of hydrogen ions in the substance, measured in moles per liter. The pH of distilled water is 7. A substance with a pH of less than 7 is known as an *acid*, whereas a substance with a pH of greater than 7 is known as a *base*. In problems 77–80, find the pH of each substance for the given concentration of hydrogen ions.

77. Acid rain whose $[H^+]$ is $10^{-5.6}$.

78. A vinegar whose $[H^+]$ is 10^{-3}.

79. An orange juice whose $[H^+]$ is 10^{-4}.

80. A tomato whose $[H^+]$ is 10^{-4}.

The seismologist Charles Richter established the **Richter scale** for measuring the magnitude of an earthquake M in terms of its intensity I by means of the equation

$$M = \log_{10}\left(\frac{I}{I_0}\right),$$

where I_0 is a certain minimum intensity. In problems 81–83, find the magnitude M of an earthquake of the given intensity I.

81. 100 times that of I_0.

82. 10,000 times that of I_0.

83. 100,000 times that of I_0.

84. The magnitude of the San Francisco earthquake of 1906 was approximately 8 on the Richter scale. What is the corresponding intensity in terms of I_0?

9.2 Basic Properties of Logarithms

In this section, we state the basic properties of logarithms and illustrate how these properties can be applied. We shall see that the properties of logarithms are based upon the properties of exponents. First we should note that it is possible to extend the definition of b^x so that all real numbers can be used as exponents. It is this that makes logarithms such a valuable computational tool. We conclude with proofs of some of the properties of logarithms.

Properties of Logarithms

Let M, N, and b be positive numbers, $b \neq 1$, and let r be any real number. Then

(i) $\log_b(MN) = \log_b M + \log_b N$ (ii) $\log_b\left(\dfrac{M}{N}\right) = \log_b M - \log_b N$

(iii) $\log_b N^r = r \log_b N$ (iv) $\log_b b^r = r$

Use the properties of logarithms to work Examples 1–4.

EXAMPLE 1 Write each expression as a sum or difference of multiples of logarithms.

(a) $\log_3 5y$ (b) $\log_8 \frac{17}{5}$ (c) $\log_5 \left(\frac{x^2 y}{2} \right)$ (d) $\log_2 \sqrt{\frac{t}{t+1}}$

SOLUTION We assume that all quantities whose logarithms are taken are positive. Then:

(a) $\log_3 5y = \log_3 5 + \log_3 y$ [Property (i)]

(b) $\log_8 \frac{17}{5} = \log_8 17 - \log_8 5$ [Property (ii)]

(c) $\log_5 \left(\dfrac{x^2 y}{2} \right) = \log_5 (x^2 y) - \log_5 2$ [Property (ii)]

$$= \log_5 x^2 + \log_5 y - \log_5 2 \qquad \text{[Property (i)]}$$
$$= 2 \log_5 x + \log_5 y - \log_5 2 \qquad \text{[Property (iii)]}$$

(d) $\log_2 \sqrt{\dfrac{t}{t+1}} = \log_2 \left(\dfrac{t}{t+1} \right)^{1/2} = \dfrac{1}{2} \log_2 \left(\dfrac{t}{t+1} \right)$ [Property (iii)]

$$= \frac{1}{2} \left[\log_2 t - \log_2 (t+1) \right] \qquad \text{[Property (ii)]}$$

$$= \frac{1}{2} \log_2 t - \frac{1}{2} \log_2 (t+1)$$

EXAMPLE 2 Write each expression as a single logarithm.

(a) $\log_3 15 + \log_3 13$ (b) $\log_5 216 - \log_5 54$

(c) $3 \log_2 x - \log_2 5x$ (d) $\log_4(t-2) + \log_4(t+2)$

SOLUTION We assume that all quantities whose logarithms are taken are positive.

(a) $\log_3 15 + \log_3 13 = \log_3 15(13)$ [Property (i)]
$$= \log_3 195$$

(b) $\log_5 216 - \log_5 54 = \log_5 \frac{216}{54}$ [Property (ii)]
$$= \log_5 4$$

(c) $3 \log_2 x - \log_2 5x = \log_2 x^3 - \log_2 5x$ [Property (iii)]
$$= \log_2 \frac{x^3}{5x} = \log_2 \frac{x^2}{5} \qquad \text{[Property (ii)]}$$

(d) $\log_4(t-2) + \log_4(t+2) = \log_4(t-2)(t+2)$ [Property (i)]
$$= \log_4(t^2 - 4)$$

EXAMPLE 3 Evaluate the following expressions.

(a) $\log_3 3^7$ (b) $\log_{11} \sqrt[3]{11}$

SOLUTION We use Property (iv):

(a) $\log_3 3^7 = 7$ (b) $\log_{11} \sqrt[3]{11} = \log_{11} 11^{1/3} = \frac{1}{3}$

EXAMPLE 4 Use $\log_b 2 = 0.35$ and $\log_b 3 = 0.55$ to find the value of each expression.

(a) $\log_b 6$ (b) $\log_b \frac{2}{3}$ (c) $\log_b 8$

(d) $\log_b \sqrt{\frac{2}{3}}$ (e) $\log_b 24$ (f) $\dfrac{\log_b 2}{\log_b 3}$

SOLUTION

(a) $\log_b 6 = \log_b 2(3) = \log_b 2 + \log_b 3 = 0.35 + 0.55 = 0.90$

(b) $\log_b \frac{2}{3} = \log_b 2 - \log_b 3 = 0.35 - 0.55 = -0.20$

(c) $\log_b 8 = \log_b 2^3 = 3 \log_b 2 = 3(0.35) = 1.05$

(d) $\log_b \sqrt{\frac{2}{3}} = \log_b(\frac{2}{3})^{1/2} = \frac{1}{2} \log_b \frac{2}{3} = \frac{1}{2}(\log_b 2 - \log_b 3) = \frac{1}{2}(0.35 - 0.55)$
$$= -0.10$$

(e) $\log_b 24 = \log_b 8(3) = \log_b(2^3)(3) = \log_b 2^3 + \log_b 3$
$$= 3 \log_b 2 + \log_b 3 = 3(0.35) + 0.55 = 1.05 + 0.55 = 1.60$$

(f) $\dfrac{\log_b 2}{\log_b 3} = \dfrac{0.35}{0.55} = \dfrac{7}{11} = 0.64$

We can combine the basic properties of logarithms and Definition 1 of Section 9.1 to solve equations containing logarithms.

In Examples 5 and 6, solve each equation.

EXAMPLE 5 $\log_3(x + 1) + \log_3(x + 3) = 1$

SOLUTION Because

$$\log_3(x + 1) + \log_3(x + 3) = \log_3[(x + 1)(x + 3)] \qquad \text{[Property (i)]}$$

we have

$$\log_3[(x + 1)(x + 3)] = 1$$

or

$$(x + 1)(x + 3) = 3^1 \qquad \text{[Definition 1, Section 9.1]}$$

$$x^2 + 4x + 3 = 3$$

$$x^2 + 4x = 0$$

$$x(x + 4) = 0$$

$$x = 0 \quad \bigg| \quad x + 4 = 0$$

$$x = -4.$$

We cannot take the logarithm of a negative number. Therefore, -4 cannot be a solution. The only solution is 0.

EXAMPLE 6 $\log_4(x + 3) - \log_4 x = 1$

SOLUTION   Because

$$\log_4(x + 3) - \log_4 x = \log_4\left(\frac{x + 3}{x}\right) \qquad \text{[Property (ii)]}$$

we have

$$\log_4\left(\frac{x + 3}{x}\right) = 1$$

$$\frac{x + 3}{x} = 4^1 = 4$$

$$x + 3 = 4x$$

$$3 = 3x$$

$$1 = x.$$

Therefore, the solution is 1.

We now prove the first three properties of logarithms.

Proofs of the Properties of Logarithms

(i) To prove that

$$\log_b(MN) = \log_b M + \log_b N,$$

let

$$x = \log_b M \qquad \text{and} \qquad y = \log_b N,$$

so that

$$b^x = M \quad \text{and} \quad b^y = N.$$

It follows that

$$MN = b^x \cdot b^y = b^{x+y},$$

or

$$\log_b(MN) = x + y.$$

However, $x = \log_b M$ and $y = \log_b N$. Therefore,

$$\log_b(MN) = \log_b M + \log_b N.$$

(ii) To prove that

$$\log_b\left(\frac{M}{N}\right) = \log_b M - \log_b N,$$

we write

$$\frac{M}{N} \cdot N = M,$$

so that

$$\log_b\left(\frac{M}{N} \cdot N\right) = \log_b M.$$

We use Property (i):

$$\log_b\left(\frac{M}{N}\right) + \log_b N = \log_b M.$$

Thus,

$$\log_b\left(\frac{M}{N}\right) = \log_b M - \log_b N.$$

(iii) To prove that

$$\log_b N^r = r \log_b N,$$

we let

$$y = \log_b N.$$

Thus,

$$N = b^y.$$

It follows that

$$N^r = (b^y)^r = b^{yr},$$

or

$$\log_b N^r = yr.$$

Because $y = \log_b N$,

$$\log_b N^r = r \log_b N.$$

The proof of Property (iv) is straightforward and is left as an exercise (Problem 73).

PROBLEM SET 9.2

In problems 1–28, use the properties of logarithms to write each expression as a sum or difference of multiples of logarithms. Assume that all variables represent positive real numbers.

1. $\log_4 5y$ 　　　　**2.** $\log_5 7x$ 　　　　**3.** $\log_3 uv$ 　　　　**4.** $\log_7 cd$

5. $\log_5 \dfrac{x}{3}$ 　　**6.** $\log_5 \dfrac{z}{11}$ 　　**7.** $\log_2 \tfrac{7}{15}$ 　　**8.** $\log_3 \dfrac{x}{y}$

9. $\log_7 3^5$ 　　**10.** $\log_8 7^{1.4}$ 　　**11.** $\log_3 c^5$ 　　**12.** $\log_7 y^4$

13. $\log_4 \sqrt{w}$ 　　**14.** $\log_5 \sqrt[3]{y}$ 　　**15.** $\log_7 3^4 \cdot 5^2$ 　　**16.** $\log_3 x^4 y^5$

17. $\log_{11} x^4 y^2$ 　　**18.** $\log_7 \sqrt[3]{x y^3}$ 　　**19.** $\log_2 \sqrt[5]{xy}$ 　　**20.** $\log_4 \sqrt[7]{x^2 y^5}$

21. $\log_5 \dfrac{a^2}{b^4}$ 　　**22.** $\log_5 \dfrac{x^7}{y^8}$ 　　**23.** $\log_7 \dfrac{x^3 \sqrt[4]{y}}{z^3}$ 　　**24.** $\log_5 \dfrac{\sqrt[5]{xy^4}}{w^4}$

25. $\log_4 \dfrac{u^4 v^5}{\sqrt[4]{z^3}}$ 　　**26.** $\log_{10} \dfrac{c^7 \sqrt[9]{d^2}}{5\sqrt{f}}$ 　　**27.** $\log_3 \sqrt[7]{\dfrac{y}{y+7}}$ 　　**28.** $\log_{10} x(x+2)$

In problems 29–50, use the properties of logarithms to write each expression as a single logarithm. Assume that all variables represent positive real numbers.

29. $\log_5 4 - \log_5 3$ 　　　　**30.** $\log_2 \tfrac{5}{7} + \log_2 \tfrac{14}{70}$ 　　　　**31.** $\log_7 \tfrac{3}{8} - \log_7 \tfrac{9}{4}$

32. $\log_3 \tfrac{3}{4} - \log_3 \tfrac{5}{8}$ 　　　**33.** $\log_3 5 + \log_3 z - \log_3 y$ 　　**34.** $\log_6 x + \log_6 y - \log_6 z$

35. $2\log_3 \tfrac{4}{5} + 3\log_3 \tfrac{1}{2}$ 　　**36.** $3\log_5 \tfrac{3}{4} + 2\log_5 \tfrac{1}{5}$ 　　**37.** $3\log_2 x + 7\log_2 y$

38. $3\log_3 z - 2\log_3 y$ 　　**39.** $\log_a \dfrac{x}{y} + \log_a \dfrac{y^2}{3x}$ 　　**40.** $\log_b \dfrac{x^2}{y} - \log_b \dfrac{x^4}{y^2}$

41. $\tfrac{1}{2}\log_4 a - 3\log_4 b - 4\log_4 z$ 　　**42.** $5\log_{10} x - \tfrac{5}{3}\log_{10} y - 7\log_{10} z$ 　　**43.** $\log_7(x-1) + \log_7(x+1)$

44. $\log_3 \dfrac{a}{a-1} + \log_3 \dfrac{a^2-1}{a}$ 　　**45.** $\log_e(y^2 - 25) - \log_e(y-5)$ 　　**46.** $\log_e \dfrac{x+y}{z} - \log_e \dfrac{1}{x+y}$

47. $\log_7 7^4$ 　　　　**48.** $\log_y y^5$ 　　　　**49.** $\log_p \sqrt[5]{p}$

50. $\log_q \sqrt[7]{q}$

In problems 51–62, use $\log_{10} 2 = 0.3010$ and $\log_{10} 3 = 0.4771$ to find the value of each expression.

51. $\log_{10} 6$ **52.** $\log_{10} 12$ **53.** $\log_{10} 18$ **54.** $\log_{10} 24$

55. $\log_{10} \frac{3}{2}$ **56.** $\log_{10} \frac{1}{3}$ **57.** $\log_{10} 5$ **58.** $\log_{10} 32$

59. $\log_{10} 81$ **60.** $\log_{10} \sqrt[5]{2}$ **61.** $\log_{10} 0.5$ **62.** $\log_{10} 60$

In problems 63–72, solve each equation.

63. $\log_7 x + \log_7 14 = 1$ **64.** $\log_4 y + \log_4 3 = 2$ **65.** $\log_3 w - \log_3 2 = 2$

66. $\log_7 z - \log_7 2 = 2$ **67.** $\log_5 x + \log_5(x - 4) = 1$ **68.** $\log_2 x + \log_2(x - 1) = 1$

69. $\log_4(y + 2) - \log_4(y - 1) = 2$ **70.** $\log_3(u + 3) - \log_3(u - 1) = 2$

71. $\log_{10}(z^2 - 9) - \log_{10}(z + 3) = 2$ **72.** $\log_4 x + \log_4(6x + 11) = 1$

73. Prove that $\log_b b^r = r$, where $b > 0$, $b \neq 1$, and r is any real number.

74. Use the properties of logarithms to show that the formula

$$S = P\left(1 + \frac{r}{n}\right)^{nt}$$

can be written as

$$\log_{10} S = \log_{10} P + nt \cdot \log_{10}\left(1 + \frac{r}{n}\right).$$

75. Use the properties of logarithms to show that the formula

$$P = S\left(1 + \frac{r}{n}\right)^{-nt}$$

can be written as

$$\log_{10} P = \log_{10} S - nt \cdot \log\left(1 + \frac{r}{n}\right).$$

76. The formula

$$S = 10 \cdot \log_{10} \frac{P}{P_0}$$

is used to measure the loudness (in decibels) produced by a sound wave of intensity P (watts per square meter at the eardrum), where

$$P_0 = 10^{-12} \text{ watts per square meter.}$$

Show that the formula can be written as

$$S = 10[\log_{10} P + 12].$$

In problems 77–80, use the properties of logarithms to transform the left side of each equation into the right side. Assume that all variables represent positive real numbers.

77. $\log_3\left[\dfrac{x^3 + 1}{x^3 - x^2 + x}\right] = \log_3(x + 1) - \log_3 x$

78. $\log_5 \left[\dfrac{\sqrt{y^3} - \sqrt{y}}{y^2 - 2y + 1} \right] = \dfrac{1}{2} \log_5 y - \log_5(y - 1)$

79. $\log_7 \left[\dfrac{\sqrt{x} + \sqrt{y}}{\sqrt{x} - \sqrt{y}} \right] = 2 \log_7[\sqrt{x} + \sqrt{y}] - \log_7(x - y)$

80. $\log_6 \left[\dfrac{\sqrt{x^2 + y^2} - x}{\sqrt{x^2 + y^2} + x} \right] = 2 \log_6[\sqrt{x^2 + y^2} - x] - 2 \log_6 y$

9.3 Common Logarithms

The bases of all the logarithms we have used so far are positive numbers such as 2, 3, 5, 6, etc. Since any positive number except 1 can serve as a base for logarithms, people have decided to simplify working with logarithms by choosing the same base. Because the decimal system is based on 10, logarithms to base 10 are the most useful for computational purposes, and are called **common logarithms.** The symbol "log x" (with no subscript) is often used as an abbreviation for $\log_{10} x$. That is,

$$\log_{10} x \qquad \text{is written as} \qquad \log x.$$

Scientific problems often involve very large or very small numbers. We have seen how to express these numbers in a special form called scientific notation.

Using a Logarithmic Table

We use base 10 in computational work because every positive real number x can be written in scientific notation as

$$x = s \times 10^n$$

where $1 \leq s < 10$ and n is an integer. If we apply Property (i) of Section 9.2 to the preceding equation, we have

$$\log x = \log[s \times 10^n]$$
$$= \log s + \log 10^n.$$

Since $\log 10^n = n$, we have

$$\boxed{\log x = \log s + n.}$$

This last equation is called the **standard form** of log x. The number log s, where $1 \leq s < 10$, is called the **mantissa** of log x, and the integer n is called the **characteristic** of log x.

Notice that for $1 \leq s < 10$, we have

$$\log 1 \leq \log s < \log 10,$$

or, equivalently,

$$0 \leq \log s < 1.$$

That is, the mantissa is either 0 or a positive number between 0 and 1. Therefore, to determine the value of log x, we simply determine the value of log s, where s is always between 1 and 10. We can obtain the approximate value of log s from a **table of common logarithms** (Table I in Appendix I).

The value of the characteristic of the logarithm of a number x is n. (In scientific notation we substitute $s \times 10^n$ for x.) Therefore, we know that $n \geq 0$ if $x \geq 1$, and that $n < 0$ if $0 < x < 1$. To illustrate, we list the values of n for several different values of x in the table below:

x	$s \times 10^n$	n
480	4.8×10^2	2
25	2.5×10^1	1
1.3	1.3×10^0	0
0.28	2.8×10^{-1}	-1
0.031	3.1×10^{-2}	-2

In Examples 1 and 2, use Table I in Appendix I to find the values of the given common logarithms. In each case, indicate the characteristic.

EXAMPLE 1   (a) log 53,900   (b) log 385   (c) log 28.4

SOLUTION   To find the characteristics n of the logarithms, we write the numbers x in scientific notation:

x	$s \times 10^n$	n
(a) 53,900	5.39×10^4	4
(b) 385	3.85×10^2	2
(c) 28.4	2.84×10^1	1

Using Table I, we have

(a) $\log 53{,}900 = \log 5.39 + 4 = 0.7316 + 4 = 4.7316$

(b) $\log 385 = \log 3.85 + 2 = 0.5855 + 2 = 2.5855$

(c) $\log 28.4 = \log 2.84 + 1 = 0.4533 + 1 = 1.4533$

EXAMPLE 2   (a) $\log 4.06$ (b) $\log 0.628$ (c) $\log 0.0035$

SOLUTION   The numbers x are written in scientific notation and their characteristics n are indicated below:

x	$s \times 10^n$	n
(a) 4.06	4.06×10^0	0
(b) 0.628	6.28×10^{-1}	-1
(c) 0.0035	3.5×10^{-3}	-3

Therefore, using Table I, we have

(a) $\log 4.06 = \log 4.06 + 0 = 0.6085 + 0 = 0.6085$

(b) $\log 0.628 = \log 6.28 + (-1) = 0.7980 + (-1) = -0.2020$

(c) $\log 0.0035 = \log 3.50 + (-3) = 0.5441 + (-3) = -2.4559$

The only logarithms that we can find directly from Table I, Appendix I, are logarithms of numbers that contain, at most, three significant digits. If a number has four significant digits, it is possible to obtain an approximation of its logarithm by using a method known as **interpolation** or **linear interpolation.**

Interpolation depends on the following property:

> If x, y, and z are positive numbers such that $z < x < y$, then
>
> $$\log z < \log x < \log y.$$

Interpolation also depends on the following assumption: For small differences in numbers, the change in the mantissas is proportional to the change in the numbers. (This assumption does not always hold, but the results are sufficiently accurate for our purposes, especially since the differences between consecutive mantissas in Table I are small.)

To illustrate the method of interpolation, let us find log 1.234. We see from Table I that

$$\log 1.23 = 0.0899 \quad \text{and} \quad \log 1.24 = 0.0934.$$

We selected these numbers because

$$1.23 < 1.234 < 1.24.$$

It follows that

$$\log 1.23 < \log 1.234 < \log 1.24.$$

The difference between the two consecutive logarithms in the table is

$$0.0934 - 0.0899 = 0.0035.$$

Also,

$$1.234 - 1.23 = 0.004 \quad \text{and} \quad 1.24 - 1.23 = 0.01.$$

Because log 1.234 > log 1.23, there is a "correction" d such that

$$\log 1.234 = \log 1.23 + d.$$

The value of d can be determined by using the following ratio:

$$\frac{d}{0.0035} = \frac{0.004}{0.01}.$$

Hence,

$$d = \frac{0.0035(0.004)}{0.01} = 0.0014.$$

Therefore,

$$\log 1.234 = \log 1.23 + d$$
$$= 0.0899 + 0.0014 = 0.0913.$$

EXAMPLE 3   Use interpolation to find log 52.33.

SOLUTION   We obtain the values for

$$\log 52.3 = \log 5.23 + 1 \quad \text{and} \quad \log 52.4 = \log 5.24 + 1,$$

from Table I:

$$\log 52.3 = \log 5.23 + 1$$
$$= 1.7185$$

and

$$\log 52.4 = \log 5.24 + 1$$
$$= 1.7193.$$

We now find the difference between the two logarithms, so that

$$1.7193 - 1.7185 = 0.0008.$$

We also find the difference between the two numbers whose logarithms these are:

$$52.4 - 52.3 = 0.1$$

and between the smaller number and the number whose logarithm we are seeking:

$$52.33 - 52.3 = 0.03.$$

Because $\log 52.3 < \log 52.33 < \log 52.4$, we arrange our work as follows:

$$0.1 \left[\begin{array}{c} 0.03 \left[\begin{array}{l} \log 52.3 \;\; = 1.7185 \\ \log 52.33 = ? \end{array} \right] d \\ \log 52.4 \;\; = 1.7193 \end{array} \right] 0.0008$$

Thus,

$$\frac{0.03}{0.1} = \frac{d}{0.0008} \quad \text{or}$$

$$d = \frac{0.03(0.0008)}{0.1}$$
$$= 0.0002.$$

Therefore,

$$\log 52.33 = \log 52.3 + d$$
$$= 1.7185 + 0.0002$$
$$= 1.7187.$$

Antilogarithms

The process of finding a number whose logarithm is given is the reverse of finding the logarithm of a number. The number found is called the **antilogarithm** of the given logarithm. For instance, if we are given a number r, we can determine the value of x such that $\log x = r$. The number x is called the **antilogarithm of r** and is abbreviated **antilog r**. Thus:

$$\text{If } \log x = r, \text{ then } x = \text{antilog } r.$$

To find the antilog of 4.4969 (or to determine the solution of the equation $\log x = 4.4969$), we write $\log x = 4.4969$ in standard form—that is, as the sum of a number between 0 and 1 and an integer. The given logarithm

$$4.4969 = 0.4969 + 4$$

has the mantissa 0.4969 and the characteristic 4. We use Table I in Appendix I to find a value s such that $\log s = 0.4969$. We find that $s = 3.14$. The characteristic of $\log x$ is 4. Therefore,

$$\begin{aligned}
\log x &= 0.4969 + 4 \\
&= \log 3.14 + 4 = \log 3.14 + \log 10^4 \\
&= \log(3.14 \times 10^4) = \log 31{,}400.
\end{aligned}$$

Therefore, $x = \text{antilog } 4.4969 = 31{,}400$.

In Examples 4–6, find the values of the given antilogarithms.

EXAMPLE 4   antilog 2.7210

SOLUTION   Let $x = \text{antilog } 2.7210$, so that $\log x = 2.7210$. Then

$$\log x = 0.7210 + 2.$$

Using Table I, we find that $\log 5.26 = 0.7210$. Thus,

$$\begin{aligned}
\log x &= \log 5.26 + 2 = \log 5.26 + \log 10^2 \\
&= \log(5.26 \times 10^2) = \log 526.
\end{aligned}$$

Therefore, $x = \text{antilog } 2.7210 = 526$.

EXAMPLE 5   antilog$[0.5105 + (-3)]$

SOLUTION   Let $x = \text{antilog}[0.5105 + (-3)]$, so that

$$\log x = 0.5105 + (-3).$$

Using Table I, we find that $\log 3.24 = 0.5105$. Thus,

$$\begin{aligned}
\log x &= \log 3.24 + (-3) = \log 3.24 + \log 10^{-3} \\
&= \log(3.24 \times 10^{-3}) = \log 0.00324.
\end{aligned}$$

Therefore, $x = \text{antilog}[0.5105 + (-3)] = 0.00324$.

EXAMPLE 6   antilog(-2.0804)

SOLUTION   Let $x = $ antilog(-2.0804), so that $\log x = -2.0804$. The mantissa must always be positive. Therefore, we write

$$-2.0804 = (-2.0804 + 3) - 3 = 0.9196 + (-3).$$

Thus,

$$\log x = 0.9196 + (-3).$$

Using Table I, we find that $\log 8.31 = 0.9196$, and

$$\log x = \log 8.31 + (-3) = \log 8.31 + \log 10^{-3}$$
$$= \log(8.31 \times 10^{-3}) = \log 0.00831.$$

Therefore, $x = $ antilog(-2.0804) $= 0.00831$.

The interpolation method can also be used to find antilogarithms.

EXAMPLE 7   Use interpolation to find antilog(-1.7186).

SOLUTION   Let $x = $ antilog(-1.7186), so that $\log x = -1.7186$. The mantissa must be positive. Therefore, we write

$$-1.7186 = (-1.7186 + 2) - 2$$
$$= 0.2814 + (-2).$$

Thus,

$$\log x = 0.2814 + (-2).$$

From Table I, we find that the values closest to 0.2814 are 0.2810 ($\log 1.91$) and 0.2833 ($\log 1.92$),

$$[0.2810 + (-2)] < [0.2814 + (-2)] < [0.2833 + (-2)].$$

Therefore,

$$\log 1.91 + (-2) < \log x < \log 1.92 + (-2)$$

or

$$\log 0.0191 < \log x < \log 0.0192.$$

We arrange the work as follows:

$$
0.0001 \left[\begin{array}{l} \rule{0pt}{2.2ex}\!\!\!\! \text{---} \log 0.0192 = 0.2833 + (-2) \text{---} \\ d \left[\begin{array}{l} \text{---} \log x \quad\;\; = 0.2814 + (-2) \text{---} \\ \text{---} \log 0.0191 = 0.2810 + (-2) \text{---} \end{array} \right] 0.0004 \end{array} \right] 0.0023
$$

The value of d can be determined by using the following ratio:

$$\frac{d}{0.0001} = \frac{0.0004}{0.0023} \quad \text{or} \quad d = \frac{0.0001(0.0004)}{0.0023} = 0.00002.$$

Therefore,

$$x = \text{antilog}[0.2814 + (-2)]$$
$$= \text{antilog}[0.2810 + (-2)] + d$$
$$= 0.0191 + 0.00002 = 0.01912.$$

PROBLEM SET 9.3

In problems 1–20, rewrite each number in scientific notation. Use the results to determine the characteristic value of the logarithm of each number.

1. 3,782	**2.** 0.000132	**3.** 0.00381	**4.** 38,173
5. 375,000	**6.** 137,100,000	**7.** 0.0001321	**8.** 681,000,000
9. 0.000271312	**10.** 0.00127142281	**11.** 210	**12.** 8,600
13. 0.0000314	**14.** 0.0075	**15.** 11,300	**16.** 720,000
17. 0.00000541	**18.** 0.0001871	**19.** 0.0003127	**20.** 0.0000194

In problems 21–38, use Table I in Appendix I to find the value of each common logarithm.

21. log 317	**22.** log 3,910	**23.** log 53,400	**24.** log 348,000
25. log 17.1	**26.** log 5	**27.** log 6.81	**28.** log 7.59
29. log 1.18	**30.** log 9.81	**31.** log 0.315	**32.** log 0.712
33. log 0.0713	**34.** log 0.00512	**35.** log 0.000178	**36.** log 0.00081
37. log 0.000007	**38.** log 0.00000137		

In problems 39–48, use Table I in Appendix I and interpolation to find the value of each common logarithm.

39. log 1,545	**40.** log 333.3	**41.** log 79.56	**42.** log 62.95	**43.** log 5.312
44. log 1.785	**45.** log 0.5725	**46.** log 0.7125	**47.** log 0.05342	**48.** log 0.006487

In problems 49–68, use Table I in Appendix I to find the value of each antilogarithm.

49. antilog 0.4133	**50.** antilog 0.4871	**51.** antilog 1.2945
52. antilog 1.7825	**53.** antilog 2.7427	**54.** antilog 2.9795
55. antilog 3.5514	**56.** antilog 3.8993	**57.** antilog[0.7348 + (−1)]
58. antilog[0.8082 + (−2)]	**59.** antilog[0.8993 + (−3)]	**60.** antilog[0.5922 + (−4)]
61. antilog(−1.6289)	**62.** antilog(−2.4157)	**63.** antilog(−3.4881)

64. antilog(-4.8153) **65.** antilog(-0.1574) **66.** antilog(-0.3251)
67. antilog(-2.1475) **68.** antilog(-3.1884)

In problems 69–76, use Table I in Appendix I and interpolation to find the value of each antilogarithm.

69. antilog 0.1452 **70.** antilog 1.5375 **71.** antilog 1.5425
72. antilog(-1.1275) **73.** antilog[$0.2259 + (-2)$] **74.** antilog[$0.4950 + (-2)$]
75. antilog(-4.4625) **76.** antilog(-4.565)

9.4 Using a Calculator to Evaluate Logarithmic and Exponential Expressions

In the days before calculators and computers, tables of logarithms were used extensively to speed up numerical computations. Today, scientific calculators with both a y^x key and a log key are usually employed.

For example, using a 10-digit calculator, we find log 364 by entering 364 and pressing the log key, to obtain

$$\log 364 = 2.561101384.$$

Similarly,

$$\log 478 = 2.679427897$$

$$\log 86.2 = 1.935507266$$

$$\log 0.568 = -0.2456516643$$

$$\log 0.0841 = -1.075204004$$

$$\log 0.34827 = -0.4580839341.$$

To find antilogarithms we use the y^x key. For example, using a 10-digit calculator and rounding off to four significant digits, we find antilog 1.7959 by entering 10, pressing the y^x key, entering 1.7959, and pressing the equals key, to obtain

$$\text{antilog } 1.7959 = 62.50.$$

We are commanding the calculator to evaluate $10^{1.7959}$. This is another way of expressing antilog 1.7959 (the number whose log is 1.7959). Other examples are:

$$\text{antilog } 2.6372 = 433.7$$
$$\text{antilog } 0.1381 = 1.374$$
$$\text{antilog } 0.0254 = 1.060.$$

c *In Examples 1 and 2, use a calculator with* log *and* y^x *keys to evaluate each of the following.*

EXAMPLE 1   (a) log 6.23   (b) log 8,921   (c) log 0.007316

SOLUTION   On a 10-digit calculator, we obtain:

(a) log 6.23 = 0.7944880467   (b) log 8,921 = 3.950413539

(c) log 0.007316 = −2.135726303

EXAMPLE 2   (a) antilog 1.7782   (b) antilog 0.9741   (c) antilog(−1.8742)

SOLUTION   We use a 10-digit calculator and round off to four significant digits:

(a) antilog 1.7782 = 60.01   (b) antilog 0.9741 = 9.421

(c) antilog(−1.8742) = 0.0134

In many applications of mathematics there are formulas that become much simpler if we use logarithms and exponential expressions with base

$$e \approx 2.718281828.$$

Logarithms with base e are called **natural logarithms.** The symbol ln x, which is read "log of x to the base e" or "natural log of x," is often used as an abbreviation for $\log_e x$. Thus:

$$\ln x = \log_e x \text{ for } x > 0.$$

In other words, for $x > 0$,

$$\ln x = c \text{ if and only if } e^c = x.$$

EXAMPLE 3   ⓒ Use a calculator ln key to evaluate each of the following:

(a) ln 8,132   (b) ln 0.041326

SOLUTION   We enter the number and press the ln key. On a 10-digit calculator we obtain:

(a) ln 8,132 = 9.003562175   (b) ln 0.041326 = −3.186263437

EXAMPLE 4   ⓒ Use a calculator with an e^x key to evaluate each of the following:

(a) $e^{4.3}$ (b) $e^{\sqrt{5}}$ (c) $e^{-0.67}$

SOLUTION   We enter the exponent and press the e^x key. On a 10-digit calculator we obtain:

(a) $e^{4.3} = 73.6997937$ (b) $e^{\sqrt{5}} = 9.356469012$
(c) $e^{-0.67} = 0.5117085778$

EXAMPLE 5   ⓒ Use a calculator with a y^x key to evaluate $(0.6423)^{-0.271}$.

SOLUTION   We enter 0.6423, press the y^x key, enter −0.271, and press the equals key. On a 10-digit calculator we obtain

$$(0.6423)^{-0.271} = 1.127464881.$$

EXAMPLE 6   ⓒ Use a calculator with either an $\sqrt[x]{y}$ key or a y^x key to evaluate $\sqrt[5]{17}$.

SOLUTION   On a 10-digit calculator we obtain

$$\sqrt[5]{17} = (17)^{1/5} = (17)^{0.2} = 1.762340348.$$

EXAMPLE 7   ⓒ Evaluate each expression. Round off the answer to three decimal places.

(a) $\dfrac{\log 3}{\log 4.12}$ (b) $\dfrac{\ln 7}{3 \ln 2.35}$

SOLUTION   Using a calculator, we have

(a) $\dfrac{\log 3}{\log 4.12} = \dfrac{0.4771213}{0.6148972} = 0.776$ (b) $\dfrac{\ln 7}{3 \ln 2.35} = \dfrac{1.9459101}{2.5632460} = 0.759$

Solving Exponential Equations by Using Logarithms

We can often solve an exponential equation by taking the logarithm of both sides of the equation and applying the properties of logarithms to simplify the results. For this purpose, we can use either the common or the natural logarithm.

C *In Examples 8–10, solve each equation. Round off your answers to four significant digits.*

EXAMPLE 8 $5^x = 7$

SOLUTION   We take the common logarithm of both sides of the equation:

$$5^x = 7$$
$$\log 5^x = \log 7$$
$$x \log 5 = \log 7$$
$$x = \frac{\log 7}{\log 5}.$$

Therefore, using a 10-digit calculator we obtain

$$x = \frac{\log 7}{\log 5} = \frac{0.8450980400}{0.6989700041}$$
$$= 1.209.$$

EXAMPLE 9 $e^{3x+7} = 8$

SOLUTION   Taking the natural logarithm of both sides of the equation, we have

$$e^{3x+7} = 8$$
$$\ln e^{3x+7} = \ln 8$$
$$(3x + 7) \ln e = \ln 8$$
$$3x + 7 = \ln 8 \qquad (\ln e = 1)$$
$$3x = \ln 8 - 7$$
$$x = \frac{\ln 8 - 7}{3}$$
$$= \frac{2.079441542 - 7}{3}$$
$$= -1.640.$$

EXAMPLE 10 $7^{2x} = 4^{x+1}$

SOLUTION   We take the common logarithms of both sides:

$$7^{2x} = 4^{x+1}$$

$$\log 7^{2x} = \log 4^{x+1}$$

$$2x \log 7 = (x + 1) \log 4$$

$$2x \log 7 = x \log 4 + \log 4$$

$$2x \log 7 - x \log 4 = \log 4$$

$$x(2 \log 7 - \log 4) = \log 4$$

$$x = \frac{\log 4}{2 \log 7 - \log 4}$$

$$= \frac{0.6020599913}{2(0.84509804) - 0.6020599913}$$

$$= \frac{0.6020599913}{1.088136089}$$

$$= 0.5533.$$

PROBLEM SET 9.4

C In problems 1–54, use a calculator to find the value of each of the following expressions. Round off your answers to four significant digits.

1. log 32.94
2. log 281.5
3. log 6.183
4. log 792.83
5. log 603.75
6. log 834.72
7. log 0.001342
8. log 0.005217
9. log 0.003561
10. log 0.0002794
11. log 0.00004175
12. log 0.000023517
13. antilog 1.9281
14. antilog 2.9741
15. antilog 0.09481
16. antilog 5.0546
17. antilog 1.47372
18. antilog 3.10375
19. antilog 4.40661
20. antilog 3.60437
21. antilog(-1.4837)
22. antilog(-1.0254)
23. antilog(-2.8459)
24. antilog(-3.20951)
25. ln 7,324
26. ln 543.1
27. ln 9.942
28. ln 0.6984
29. ln 0.5342
30. ln 0.90471
31. $e^{2.1}$
32. $e^{3.7}$
33. $e^{\sqrt{2}}$
34. $e^{\sqrt{3}}$
35. $e^{-0.15}$
36. $e^{-0.73}$
37. $e^{-3.1}$
38. $e^{-1.712}$
39. $e^{3.714}$
40. $e^{-\sqrt{7}}$
41. $(0.4014)^{3.2}$
42. $(4.81)^{1.5}$
43. $(5.977)^{-1.8}$
44. $(2.477)^{-3.7}$
45. $(3.912)^2$
46. $(21.85)^{-3}$
47. $(1.716)^{4.3}$
48. $(0.0763)^{0.34}$
49. $\sqrt[4]{7.18}$
50. $\sqrt[3]{5.32}$
51. $\sqrt[5]{91.81}$
52. $\sqrt[10]{2.63}$
53. $\sqrt[4]{0.293}$
54. $\sqrt[5]{(1.53)^3}$

C In problems 55–60, evaluate each expression. Round off the answer to three decimal places.

55. $\dfrac{\log 11}{\log 4.85}$ **56.** $\dfrac{\log 8}{\log 2.17}$ **57.** $\dfrac{\ln 5}{2 \ln 1.15}$ **58.** $\dfrac{\ln 4}{3 \ln 2.47}$ **59.** $\dfrac{3 \log 6}{5 \log 2.38}$ **60.** $\dfrac{5 \ln 3}{2 \ln 1.45}$

C In problems 61–72, solve each equation. Round off your answers to four significant digits.

61. $2^x = 7$ **62.** $3^{2x} = 5$ **63.** $4^{-x} = 13$ **64.** $5^{-x} = 10$

65. $e^{2x} = 3$ **66.** $e^{-4x} = 7$ **67.** $e^{2x-1} = 6$ **68.** $7^{3x-1} = 9$

69. $3^{5-2t} = 8^{t-4}$ **70.** $4^{y+1} = 6^{8-3y}$ **71.** $3^{8p-1} = 5^{1-3p}$ **72.** $e^{3-7w} = 4^{1-5w}$

Recall the formula

$$\text{pH} = -\log[\text{H}^+]$$

where $[\text{H}^+]$ is the concentration of hydrogen ions in the substance measured in moles per liter. In problems 73 and 74, find the pH of each substance (rounded off to two decimal places).

C **73.** Eggs: $[\text{H}^+] = 1.6 \times 10^{-8}$ moles per liter.   C **74.** Milk: $[\text{H}^+] = 4 \times 10^{-7}$ moles per liter.

In problems 75 and 76, find the concentration of hydrogen ions $[\text{H}^+]$ for the given pH.

75. Rain: pH = 5.6   **76.** Beer: pH = 4.3

The **base-changing formula**

$$\log_a c = \frac{\log_b c}{\log_b a}$$

is often used to rewrite a logarithm in terms of logarithms to other bases. In problems C 77–80, find the value of each expression.

77. $\log_5 7$ **78.** $\log_4 9$ **79.** $\log_6 11$ **80.** $\log_7 8$

9.5 Applications

In this section, we examine some useful formulas that are expressed as exponential equations. For example, we show how exponential equations are used to calculate the growth of money accumulation due to interest payments. You will find the use of a calculator indispensable in this section.

Compound Interest

Bankers use the **compound interest** formula

$$S = P\left(1 + \frac{r}{t}\right)^{nt}$$

to determine the amount of dollars S that will accrue from a principal of P dollars, invested for a term of n years at a nominal annual interest rate r compounded t times per year.

EXAMPLE 1   © If you invest \$5,000 at a nominal annual interest rate of 14%, how much money do you accumulate after 4 years if the interest is compounded (a) annually and (b) quarterly?

SOLUTION   Here $P = 5,000$, $r = 0.14$, and $n = 4$.

(a) For interest compounded annually, $t = 1$. Therefore,

$$S = P\left(1 + \frac{r}{t}\right)^{nt}$$

$$= 5,000\left(1 + \frac{0.14}{1}\right)^{4(1)} = 5,000(1.14)^4.$$

Using a calculator, we find that

$$S = 5,000(1.14)^4 = 8,444.80.$$

The amount accumulated is \$8,444.80.

(b) For interest compounded quarterly, $t = 4$. Therefore,

$$S = P\left(1 + \frac{r}{t}\right)^{nt}$$

$$= 5,000\left(1 + \frac{0.14}{4}\right)^{4(4)} = 5,000(1.035)^{16} = 8,669.93.$$

The amount accumulated is \$8,669.93.

Effective Simple Annual Interest Rate

When a bank offers compound interest, it usually specifies not only the nominal annual interest rate r but also the **effective** simple annual interest rate R, that is, the rate of simple annual interest that would yield the same amount, accumulated over a 1-year term, as the compound interest yields. The formula

$$R = \left(1 + \frac{r}{t}\right)^t - 1$$

is used to calculate R in terms of r.

EXAMPLE 2 © Find the effective simple annual interest rate R corresponding to a nominal annual interest rate of 14% compounded quarterly.

SOLUTION   Here $r = 0.14$ and $t = 4$. Therefore,

$$R = \left(1 + \frac{r}{t}\right)^t - 1 = \left(1 + \frac{0.14}{4}\right)^4 - 1$$
$$= (1.035)^4 - 1 = 0.1475.$$

In other words, the effective simple annual interest rate is 14.75%.

Present Value

Money that you will receive in the future is worth *less* to you than the same amount of money received now. This is because you will miss out on the interest you could accumulate if you invested the money now. For this reason, the **present value** of money to be received in the future is the actual amount of money discounted by the prevailing interest rate during the period.

Suppose, for example, that you have an opportunity to invest P dollars at a nominal annual interest rate r, compounded t times a year. This principal plus the interest it earns will amount to S dollars after n years. This is stated in the compound interest formula

$$S = P\left(1 + \frac{r}{t}\right)^{nt}.$$

Now, consider what this means. It means that P dollars today is worth S dollars to be received n years in the future. We can solve the compound interest equation for P to determine the **present value** of an offer of S dollars to be received n years in the future:

$$P = S\left(1 + \frac{r}{t}\right)^{-nt}.$$

EXAMPLE 3   c To dissolve a partnership that owns a small parking lot, one partner buys the shares of the other partners for a total of $40,000, to be paid 2 years in the future. If, during this period, investments earn a nominal annual interest rate of 16% compounded quarterly, find the present value of the $40,000.

SOLUTION   Here $S = 40,000$, $r = 0.16$, $t = 4$, and $n = 2$, so that

$$P = S\left(1 + \frac{r}{t}\right)^{-nt}$$

$$= 40,000\left(1 + \frac{0.16}{4}\right)^{-2(4)} = 40,000(1.04)^{-8} = 29,227.61.$$

Therefore, the present value of the $40,000 is $29,227.61.

Population Growth

We can also apply exponential equations to solve problems dealing with population growth. We use the formula

$$P = P_0 e^{kt}$$

where P is the population after time t, P_0 is the original population, and k is the rate of growth per unit of time.

EXAMPLE 4   c The population of a small country was 10 million in 1980, and has been growing at 3% per year. Predict the population in the year 2000.

SOLUTION   Here $P_0 = 10$ million, $k = 0.03$, and $t = 20$ years. Thus,

$$P = P_0 e^{kt}$$
$$= 10e^{0.03(20)}$$
$$= 10e^{0.6} = 18.22.$$

Therefore, the population will be approximately 18.22 million in the year 2000.

Science

An example of the use of exponentials is found in an application of Boyle's law, in chemistry.

EXAMPLE 5   Ⓒ **Boyle's law** for adiabatic expansion of air is given by the equation $PV^{1.4} = C$, where P is the pressure of the air, V is its volume, and C is a constant. At a certain instant, the volume of the air is 75.2 cubic inches and $C = 12,600$ pounds per inch. Find the pressure P.

SOLUTION   We have

$$PV^{1.4} = C \qquad \text{or} \qquad P = \frac{C}{V^{1.4}} = CV^{-1.4}$$

Here $V = 75.2$ and $C = 12,600$ so that

$$P = 12,600(75.2)^{-1.4}$$
$$= 29.76.$$

Therefore, the pressure is 29.76 pounds per square inch.

PROBLEM SET 9.5

Ⓒ **1.** If you invest $7,000 in a savings certificate at a nominal annual interest rate of 13% compounded annually, how much money is accumulated after 3 years?

Ⓒ **2.** If $8,000 is invested at a nominal annual interest rate of 14% compounded quarterly, how much money is accumulated after 4 years?

Ⓒ **3.** If $10,000 is invested at a nominal annual interest rate of 11.8% compounded monthly, how much money is accumulated after 6 years?

Ⓒ **4.** Find the amount of interest earned on $4,000 that was placed in a bank at a nominal annual interest rate of $7\frac{1}{4}\%$ compounded semiannually for 5 years.

Ⓒ **5.** Find the effective simple annual interest rate R corresponding to a nominal annual interest rate of 11% compounded annually.

Ⓒ **6.** Find the effective simple annual interest rate R corresponding to a nominal annual interest rate of 12% compounded semiannually.

Ⓒ **7.** Find the effective simple annual interest rate R corresponding to a nominal annual interest rate of 8% compounded quarterly.

Ⓒ **8.** Find the effective simple annual interest rate R corresponding to a nominal annual interest rate of 10% compounded monthly.

Ⓒ **9.** Find the present value of $10,000, to be paid 3 years in the future, if money could be invested at 14.5% compounded annually during that time.

Ⓒ **10.** Find the present value of $5,000 to be paid to you 7 years in the future, if investments during this period will earn a nominal annual interest rate of 10% compounded quarterly.

Ⓒ **11.** On Gus's 18th birthday, his father promises to give him $25,000 when he turns 21, to help set him up in business. Savings certificates are available at a nominal annual interest rate of 12.8% compounded quarterly. How much does his father need to invest on Gus's 18th birthday in order to fulfill his promise?

©️ **12.** Suppose that someone owes you \$130,000 to be paid to you 3 years in the future. What is the present value of this money to you if investments are earning a nominal interest rate of 10.5% compounded annually?

©️ **13.** The population of a small country was 2 million in 1983 and has been growing according to the equation $P = P_0 e^{kt}$ at 4% per year. Predict the population in the year 1993.

©️ **14.** A biologist finds that there are 2,000 bacteria in a culture and that the culture has been growing according to the equation $P = P_0 e^{kt}$ at 7% per hour. How many bacteria will be present after 12 hours?

©️ **15.** A bank offers a savings account with continuously compounded interest at a nominal annual rate of 7%. Such an account is worth an amount $P = P_0 e^{kt}$ after t years. If a principal $P = \$1,000$ is deposited when $t = 0$, what is the final value of this investment after 10 years?

©️ **16.** A manufacturing plant estimates that the value V, in dollars, of a machine is decreasing according to the equation $V = 45,000 e^{-0.13t}$, where t is the number of years since the machine was placed in service. Find the value of the machine after 8 years.

©️ **17.** According to **Newton's law of cooling,** under certain conditions the temperature T (in degrees Celsius) of an object is given by the equation $T = 75 e^{-2t}$, where t is the time in hours. Find the temperature of the object after 2.5 hours.

©️ **18.** The required area A of the cross section of the interior of a chimney is given by $A = 0.06 p h^{-1.2}$, where p is the number of pounds of coal burned each hour and h is the height of the chimney in feet. Find the required cross-sectional area (in square feet) of a chimney 72 feet high if 750 pounds of coal are burned each hour.

©️ **19.** If a fully charged electrical condenser is allowed to discharge, after t seconds the remaining charge Q (in coulombs) is given by the equation $Q = 750(2.7)^{-0.4t}$. What is the charge Q after 23 seconds?

©️ **20.** Under certain conditions, the atmospheric pressure P, in inches of mercury, at altitude h, in feet, is given by the equation $P = 29(2.6)^{-0.000034h}$. What is the pressure at an altitude of 35,000 feet?

In determining the price of a used car, car dealers often use the formula

$$\log(1 - r) = \frac{1}{t} \log \frac{W}{p},$$

where P(dollars) is the purchase price of the car when it was new, W(dollars) is its value now, t years later, and r is the annual rate of depreciation.

©️ In problems 21–24, find the annual rate of depreciation.

21. New car purchased for \$11,400 and sold 5 years later for \$5,600.

22. New car purchased for \$16,500 and sold 3 years later for \$11,200.

23. New car purchased for \$8,600 and sold 3 years later for \$3,800.

24. New car purchased for \$7,200 and sold 4 years later for \$2,700.

⏣ **25.** In a psychological experiment, it is estimated that a worker with t weeks of experience can produce N items per day, where

$$N = 50 - 25(0.8)^t.$$

(a) How many items can a worker with 4 weeks experience produce?

(b) How much experience does a worker need before he or she can produce 45 items per day?

⏣ **26.** Sociologists estimate that if a person starts a rumor in a small town whose population is 2,000, the rumor will spread so that after t hours approximately N people will have heard the rumor, where

$$N = 2,000(1 + 1,999e^{-2t})^{-1}.$$

How long will it be before 1,500 people have heard the rumor?

REVIEW PROBLEM SET

In problems 1–8, solve each exponential equation.

1. $3^{2x-1} = 9$ **2.** $4^{1-t} = 2^{t+2}$ **3.** $8^{y+1} = 4^y$ **4.** $5^{u+2} = 625$

5. $27^{v-1} = 9$ **6.** $6^{3x+7} = 216^{3-x}$ **7.** $3^{2x+1} = 27^{x-1}$ **8.** $(1.2)^{2c+1} = 1.44$

In problems 9–16, write each equation in logarithmic form.

9. $2^5 = 32$ **10.** $9^{3/2} = 27$ **11.** $8^{2/3} = 4$ **12.** $32^{-4/5} = \frac{1}{16}$

13. $27^{-2/3} = \frac{1}{9}$ **14.** $13^0 = 1$ **15.** $z^n = w$ **16.** $c^t = d$

In problems 17–24, write each equation in exponential form.

17. $\log_2 8 = 3$ **18.** $\log_2 64 = 6$ **19.** $\log_{10} 100 = 2$ **20.** $\log_9 27 = \frac{3}{2}$

21. $\log_{17} 1 = 0$ **22.** $\log_{125} 625 = \frac{4}{3}$ **23.** $\log_{10} \frac{1}{10} = -1$ **24.** $\log_a z = w$

In problems 25–36, find the value of each logarithm.

25. $\log_3 9$ **26.** $\log_4 8$ **27.** $\log_6 1$ **28.** $\log_5 0.04$

29. $\log_{100} 0.001$ **30.** $\log_9 \frac{1}{3}$ **31.** $\log_4 \frac{1}{128}$ **32.** $\log_2 16^{-2}$

33. $\log_5 625^{-1}$ **34.** $\log_2 1024^{-1}$ **35.** $\log_4 8\sqrt{2}$ **36.** $\log_{6/5} \frac{25}{36}$

In problems 37–42, solve each equation.

37. $\log_4 16 = t$ **38.** $\log_3 9\sqrt{3} = y$ **39.** $\log_2(7x - 1) = 4$

40. $\log_5(3u - 11) = 2$ **41.** $\log_9(17z - 33) = 0$ **42.** $\log_5\left(\frac{x}{2} - \frac{3}{2}\right) = 1$

In problems 43–50, use the properties of logarithms to write each expression as a sum or difference of multiples of logarithms. Assume that all variables represent positive real numbers.

43. $\log_6 7u$

44. $\log_2 4x^3$

45. $\log_2 3^6 \cdot 4^7$

46. $\log_8 \dfrac{5^7}{9^3}$

47. $\log_4 xy^5$

48. $\log_3 \sqrt[7]{5^3 \cdot 8^6}$

49. $\log_a \dfrac{w^2}{z^4}$

50. $\log_b x^3y^2z^4$

In problems 51–58, use the properties of logarithms to write each expression as a single logarithm or number. Assume that all variables represent positive real numbers.

51. $\log_2 \frac{3}{7} + \log_2 \frac{14}{27}$

52. $\log_3 \frac{5}{12} + \log_3 \frac{4}{15}$

53. $\log_5 \frac{6}{7} - \log_5 \frac{27}{4} + \log_5 \frac{21}{16}$

54. $\log_9 \frac{11}{5} + \log_9 \frac{14}{3} - \log_9 \frac{22}{15}$

55. $5 \log_a x - 3 \log_a y$

56. $2 \log_c x^3 + \log_c \dfrac{2}{x} - \log_c \dfrac{2}{x^4}$

57. $\log_9 9^5 + \log_9 3^{-7}$

58. $\log_t \sqrt[3]{t} + \log_t \sqrt[3]{t^2}$

In problems 59–64, use the properties of logarithms to solve each equation.

59. $\log_3 z + \log_3 4 = 2$

60. $\log_2(t + 1) + \log_2 3 = 4$

61. $\log_5 x - \log_5 3 = 1$

62. $\log_3 4y - \log_3 2 = 5$

63. $\log_5(2x - 1) + \log_5(2x + 1) = 2$

64. $\log_{1/2}(4t^2 - 1) - \log_{1/2}(2t + 1) = 1$

In problems 65–72, express each number in scientific notation. Use the result to determine the characteristic value of the logarithm of each number.

65. 46,800,000

66. 432,000,000

67. 0.0000012

68. 0.0000000326

69. 5,600

70. 32,100

71. 0.000192

72. 0.00000837

In problems 73–82, use Table I in Appendix I to find the value of each common logarithm. Use interpolation when necessary.

73. log 846

74. log 55.6

75. log 75.2

76. log 7.38

77. log 3.184

78. log 0.5315

79. log 0.473

80. log 0.0392

81. log 0.005867

82. log 0.0009254

In problems 83–92, use Table I in Appendix I to find the value of each antilogarithm. Interpolate when necessary.

83. antilog 0.7466

84. antilog 0.5514

85. antilog 1.9533

86. antilog 2.9243

87. antilog 3.2375

88. antilog 4.1152

89. antilog[0.4518 + (−2)]

90. antilog[0.9289 + (−3)]

91. antilog(−3.4076)

92. antilog(−2.3478)

© In problems 93–108, use a calculator to find the value of each of the following expressions. Round off your answers to four significant digits.

93. log 14.73

94. log 289.7

95. log 0.000315

96. log 0.000008792

97. antilog 3.4917

98. antilog 4.2163

99. antilog(−2.3168)

100. antilog(−1.9721)

101. $\ln 45.31$ **102.** $\ln 178.9$ **103.** $e^{\sqrt{5}}$ **104.** $e^{-0.51}$

105. $(2.91)^{-1.7}$ **106.** $(0.87)^{-2.3}$ **107.** $\sqrt[4]{23.89}$ **108.** $\sqrt[5]{173.8}$

© In problems 109–114, solve each equation.

109. $2^x = 5$ **110.** $5^{2x} = 3$ **111.** $4^{3-t} = 5^{t+1}$ **112.** $2^{2y+1} = 4^{y-2}$

113. $e^{3u-2} = 7$ **114.** $e^{4x+2} = 6^{3-2x}$

© **115.** A sum of \$5,000 is invested in a stock whose average growth rate is 15% compounded annually. Assuming that the rate of growth continues, find the investment's value after 6 years.

© **116.** If \$10,000 is invested at a nominal annual interest rate of 12% compounded quarterly, how much money is accumulated after 8 years?

© **117.** Find the effective simple annual interest rate R corresponding to a nominal annual interest rate of 14% compounded semiannually.

© **118.** Find the effective simple annual interest rate R corresponding to a nominal annual interest rate of 12.6% compounded quarterly.

© **119.** Find the present value of \$7,000 to be paid to you 5 years in the future if investments during this period will be earning a nominal annual interest rate of 11.6% compounded every 2 months.

© **120.** Find the present value of the \$20,000 you will receive in 10 years if investments during this period will be earning a nominal annual interest rate of 14% compounded quarterly.

© **121.** The population of a city was 520,000 in 1983, and it has been growing according to the equation $P = P_0 e^{kt}$ at 3.5% per year. Predict the city's population in the year 1990.

© **122.** At the start of an experiment, the number of bacteria present in a culture is 3,000. If the culture is growing according to the equation $P = P_0 e^{kt}$ at 6% per hour, how many bacteria will be present after 8 hours?

CHAPTER 9 TEST

1. Solve each exponential equation.

(a) $4^{2x+1} = 16$ (b) $9^{1-x} = 27^{x-3}$

2. Write each equation in logarithmic form.

(a) $3^4 = 81$ (b) $z^c = b$

3. Write each equation in exponential form.

(a) $\log_6 216 = 3$ (b) $\log_a B = t$

4. Find the value of each logarithm.

(a) $\log_2 64$ (b) $\log_9 27$ (c) $\log_{4/3} \dfrac{27}{64}$

5. Use the properties of logarithms to write each expression as a sum or difference of multiples of logarithms.

(a) $\log_5 8x$ (b) $\log_3 \dfrac{z}{7}$ (c) $\log_4 x^3 y^4$

6. Use the properties of logarithms to write each expression as a single logarithm.

(a) $\log_7 \dfrac{3}{8} + \log_7 \dfrac{16}{15}$ (b) $3 \log_2 x - 2 \log_2 y$

7. Solve each equation:

(a) $\log_2 8\sqrt{2} = x$ (b) $\log_4(x + 1) + \log_4(2x - 2) = 2$

Use a common log table for problems 8 and 9. (Use interpolation as necessary.)

8. (a) $\log 735$ (b) $\log 0.00892$ (c) $\log 17.63$

9. (a) antilog 2.5478   (b) antilog 1.7866   (c) antilog(-2.6737)

C Use a calculator to find the value of each expression. Round off answers to four significant digits.

10. (a) $\log 27.89$ (b) antilog(-3.8192) (c) $\ln 13.94$ (d) $e^{-0.72}$

C **11.** Solve each equation. Round off answers to four significant digits.

(a) $2^{x-1} = 7^{2x}$ (b) $e^{3t} = 2^{t+1}$

C **12.** If \$20,000 is invested at a nominal annual interest rate of 8% compounded quarterly, how much money is accumulated after 5 years?

10 Functions and Related Curves

Scientific advances often result from the discovery that things depend upon one another in definite ways. An example is Einstein's famous equation, $E = mc^2$, which relates mass m to energy E, using the speed of light, c, squared. Such relationships among quantities are referred to as *functions*. In this chapter, we discuss briefly some functions and their graphs. We also learn how to graph special curves, known as conic sections, which may be used to solve nonlinear systems of equations.

10.1 Functions

You have encountered the concept of a function many times in everyday life. For example, the amount of sales tax charged on a purchase is a function of the amount of the purchase; the number of textbooks to be ordered by the bookstore for a course is a function of the number of students enrolled in the course, and the number of congressional representatives for a particular state is a function of the population of the state. A function suggests some kind of *correspondence*. In each of the preceding examples, there is an established correspondence between numbers: between the amount of the sales tax and the amount of the purchase; between the number of books and the number of students; and between the number of congressional representatives and the number of people in each state.

In mathematics, the general idea of a function is simple. Suppose that one variable quantity—say, y—depends, in a definite way, on another variable quantity—say, x. Then for each particular value of x, there is *one* corresponding value of y. Such a correspondence defines a **function,** and we say that (the variable) y is a function of (the variable) x.

For example, if x is used to represent the length of the side of a square and y is used to represent the area of this square, then y depends on x in a definite way, namely, $y = x^2$. Therefore, we say that the area y of the square is a function of the length of its side x.

In a sense, the value of y *depends* on the value assigned to x. For this reason, we sometimes refer to x as the **independent variable** and to y as the **dependent variable.** Thus, if $y = x^2$, then when $x = 2$, $y = 4$; when $x = 5$, $y = 25$; and when $x = -11$, $y = 121$. More formally, we have the following definition:

DEFINITION 1   ### A Function as a Correspondence

> A **function** is a correspondence that assigns to each member in a certain set, called the *domain* of the function, one and only one member in a second set, called the *range* of the function.

The **independent variable** of the function can take on any value in the domain of the function. The range of the function is the set of all possible corresponding values that the **dependent variable** can assume.

Suppose that a function is denoted by the letter f and is determined by the equation $3r + 5t = 3$. If r represents the independent variable for the function f determined by $3r + 5t = 3$, we say that t is a function of r. We indicate this fact by the function notation $t = f(r)$. This equation, which is read "t equals f of r," means that t is a function of r, in other words that r represents the independent variable and t represents the dependent variable. If we solve for t in terms of r, we obtain $t = f(r) = -\frac{3}{5}r + \frac{3}{5}$.

On the other hand, if we write $r = f(t)$, we are saying that t represents the independent variable and r the dependent variable. If we then solve for r in terms of t, we have $r = f(t) = -\frac{5}{3}t + 1$.

We often use letters other than f to represent functions. For example, g and h as well as F, G, and H are favorites for this purpose. *If f is a function and x represents a member of the domain, then $f(x)$ represents the corresponding member of the range.* Note that $f(x)$ is *not* the function f. However, to save time, we frequently use the phrase "the function $y = f(x)$." There is no great harm in this practice. Indeed, we use this phrase whenever it is convenient. However, it is important to remember that "the function $y = f(x)$" actually means "the function f determined by the equation $y = f(x)$."

In the functions

$$g(t) = t^2, \qquad h(x) = x + 7, \qquad V(r) = \frac{4}{3}\pi r^3, \qquad \text{and} \qquad F(s) = \sqrt{s},$$

the independent variables are represented by t, x, r, and s, respectively. If $g(t) = t^2$, then we can determine $g(3)$ by substituting 3 for t:

$$g(3) = 3^2 = 9.$$

Similarly, we can find $g(x + h)$ by substituting $x + h$ for t:

$$g(x + h) = (x + h)^2$$
$$= x^2 + 2xh + h^2.$$

EXAMPLE 1   Let $f(x) = 5x + 6$. Find the following values.

(a) $f(1)$ 　　　　 (b) $f(-3)$ 　　　　 (c) $f(\sqrt{2})$ 　　　　 (d) $\sqrt{f(2)}$

(e) $[f(-4)]^2$ 　　 (f) $f(a)$ 　　　　　 (g) $f(a + 4)$ 　　　 (h) $f(3b)$

(i) $f(-c)$ 　　　　 (j) $f(3t + 7)$

SOLUTION   (a) $f(1) = 5(1) + 6 = 11$

(b) $f(-3) = 5(-3) + 6 = -9$

(c) $f(\sqrt{2}) = 5\sqrt{2} + 6$

(d) $\sqrt{f(2)} = \sqrt{5(2) + 6} = \sqrt{16} = 4$

(e) $[f(-4)]^2 = [5(-4) + 6]^2 = (-14)^2 = 196$

(f) $f(a) = 5a + 6$

(g) $f(a + 4) = 5(a + 4) + 6 = 5a + 20 + 6 = 5a + 26$

(h) $f(3b) = 5(3b) + 6 = 15b + 6$

(i) $f(-c) = 5(-c) + 6 = -5c + 6$

(j) $f(3t + 7) = 5(3t + 7) + 6 = 15t + 35 + 6 = 15t + 41$

EXAMPLE 2   Let $f(x) = \sqrt{25 - x^2}$. Find the following values.

(a) $f(0)$ 　　 (b) $f(3)$ 　　 (c) $f(4)$ 　　 (d) $f(5)$

SOLUTION   (a) $f(0) = \sqrt{25 - 0^2} = \sqrt{25} = 5$

(b) $f(3) = \sqrt{25 - 3^2} = \sqrt{25 - 9} = \sqrt{16} = 4$

(c) $f(4) = \sqrt{25 - 4^2} = \sqrt{25 - 16} = \sqrt{9} = 3$

(d) $f(5) = \sqrt{25 - 5^2} = \sqrt{25 - 25} = \sqrt{0} = 0$

EXAMPLE 3   Let $f(x) = |x|$. Find the following values.

(a) $f(-2)$ 　　 (b) $f(0)$ 　　 (c) $f(2)$

SOLUTION   (a) $f(-2) = |-2| = 2$ 　　　 (b) $f(0) = |0| = 0$ 　　　 (c) $f(2) = |2| = 2$

EXAMPLE 4   The expression

$$\frac{f(x + h) - f(x)}{h}, \qquad h \neq 0$$

is called the **difference quotient** for a function f. Find the difference quotient for $f(x) = -5x + 3$.

SOLUTION   We have

$$\frac{f(x + h) - f(x)}{h} = \frac{[-5(x + h) + 3] - (-5x + 3)}{h}$$

$$= \frac{-5x - 5h + 3 + 5x - 3}{h}$$

$$= \frac{-5h}{h} = -5.$$

Whenever a function f is defined by an equation $y = f(x)$, you may assume (unless you are told otherwise) that its domain consists of all values of x for which the equation makes sense and determines a corresponding real number y. The range of the function is then automatically determined, since it consists of the set of all values of y that correspond, by the equation that defines the function, to values of x in the domain.

In Examples 5–7, find the domain of the function determined by each equation.

EXAMPLE 5 $f(x) = 2x + 1$

SOLUTION   The expression $2x + 1$ is defined for all real values of x. Therefore, the domain of f is the set of all real numbers.

EXAMPLE 6 $g(x) = \dfrac{1}{x + 2}$

SOLUTION   The domain of g is the set of all real numbers except -2 because $x + 2 = 0$ for $x = -2$, and division by 0 is undefined.

EXAMPLE 7 $h(x) = \sqrt{1 - x}$

SOLUTION   The expression $\sqrt{1 - x}$ represents a real number if and only if $1 - x \geq 0$, that is, if and only if $x \leq 1$. Therefore, the domain of h is the set of all real numbers x such that $x \leq 1$.

The **graph** of a function f is the graph of the corresponding equation $y = f(x)$. The graph of f is the set of all points (x, y) in the Cartesian plane such that x is in the domain of f and $y = f(x)$.

In Examples 8 and 9, sketch the graph of each function.

EXAMPLE 8 $f(x) = x$ (the **identity** function)

SOLUTION   The domain of f consists of all real numbers. The graph of the function f is the graph of the equation $y = x$. The graph of $y = x$ is the line with slope 1 and y intercept 0 (Figure 1).

Figure 1

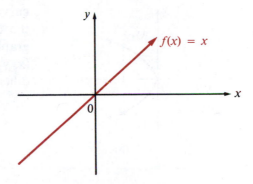

EXAMPLE 9 $f(x) = |x|$ (the **absolute value** function)

SOLUTION   The domain of f consists of all real numbers. The graph of the function f is the graph of the equation $y = |x|$. If we apply the definition of absolute value,

$$y = \begin{cases} x & \text{for } x \geq 0. \\ -x & \text{for } x < 0. \end{cases}$$

The graphs are portions of straight lines with slopes 1 and -1, respectively (Figure 2).

Figure 2

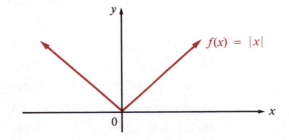

It is important to realize that *not every curve* in the Cartesian plane is the *graph* of a function. Indeed, the definition of a function requires that there be one and only one value of y corresponding to each value of x in the domain. Thus, we cannot have two points (x_1, y_1) and (x_1, y_2) on the graph of a function with the same abscissa x_1 and different ordinates y_1 and y_2.

EXAMPLE 10   Which of the curves in Figure 3 is the graph of a function?

SOLUTION   Notice that the dashed vertical line in Figure 3a intersects the curve in exactly one point. This is true of every vertical line that would intersect this curve. However, the dashed vertical line in Figure 3b intersects the curve in two points. Thus, the curve in Figure 3a is the graph of a function, but the curve in Figure 3b is not, because on the graph of a function we cannot have two points (x, y_1) and (x, y_2) with the same abscissa x and different ordinates y_1 and y_2.

Figure 3

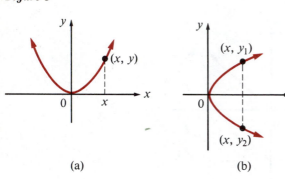

(a)   (b)

Example 10 can be generalized in the following test.

Vertical-Line Test

> A curve in the Cartesian plane is the graph of a function if and only if no vertical straight line intersects the curve more than once.

PROBLEM SET 10.1

In problems 1–10, let $f(x) = 3x + 1$ and find the following values.

1. $f(1)$ **2.** $f(\frac{2}{3})$ **3.** $f(-2)$ **4.** $f(u + v)$

5. $f(0)$ **6.** $f(2z)$ **7.** $f(a + z)$ **8.** $f(a + b) - f(a)$

9. $[f(4)]^2$ **10.** $\sqrt{f(3)}$

In problems 11–18, let $g(x) = \sqrt{16 - x^2}$ and find the following values.

11. $g(0)$ **12.** $g(2)$ **13.** $g(4)$ **14.** $g(-3)$

15. $g(-4)$ **16.** $g(x + 4)$ **17.** $g(\sqrt{7})$ **18.** $[g(2)]^2$

In problems 19–26, let $h(x) = |x - 2|$ and find the following values.

19. $h(7)$ **20.** $h(w + 2)$ **21.** $h(-3)$ **22.** $h(b + 3)$

23. $h(0)$ **24.** $h(2\frac{1}{2})$ **25.** $h(-a)$ **26.** $\sqrt{h(6)}$

In problems 27–36, find the domain of the function determined by each equation.

27. $f(x) = -3x + 2$ **28.** $f(x) = \frac{2}{3}x + \frac{1}{3}$ **29.** $f(x) = \frac{1}{x}$ **30.** $f(x) = \sqrt[3]{-2x + 1}$

31. $f(x) = \dfrac{1}{\sqrt{x - 1}}$ **32.** $f(x) = 2x^2 - 5$ **33.** $f(x) = \sqrt{2 - x}$ **34.** $f(x) = \dfrac{4}{\sqrt{x^2 - 1}}$

35. $f(x) = x^2$ **36.** $f(x) = x^3 - 1$

In problems 37–40, sketch the graph of each function.

37. $f(x) = x + 1$ **38.** $f(x) = x - 1$ **39.** $f(x) = |x - 1|$ **40.** $f(x) = |x + 3|$

In problems 41–44, evaluate the difference quotient $\dfrac{f(x + h) - f(x)}{h}$ for each function.

41. $f(x) = 3x + 2$ **42.** $f(x) = 2 - 3x$ **43.** $f(x) = -4x + 7$ **44.** $f(x) = 7x + 5$

In problems 45–54, let f, g, and h be defined by $f(x) = 2x - 1$, $g(x) = 3x + 5$, and $h(x) = -4x + 7$. Write an expression for each of the given functions.

45. $f(x) + g(x)$ **46.** $f(x) + h(x)$ **47.** $g(x) - h(x)$ **48.** $f(x) - h(x)$ **49.** $f(x) \cdot g(x)$

50. $f(x) \cdot h(x)$ **51.** $\dfrac{f(x)}{g(x)}$ **52.** $\dfrac{g(x)}{h(x)}$ **53.** $f[g(x)]$ **54.** $g[h(x)]$

55. A computer company sells software for home computers. They find they can sell y diskettes per day at x dollars per diskette, according to the equation $y = 4{,}000 - 20x$. If $y = f(x)$, find $f(50)$ and $f(70)$.

56. The cost C (in dollars) to the Style Clothing Company for manufacturing x items of a particular type is given by the function

$$C(x) = x^2 - 100x + 2{,}750,$$

where x is the number of items produced. What is the cost of manufacturing 100 items?

In problems 57–62, determine which curve is the graph of a function.

57.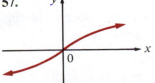

58.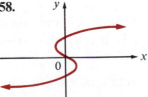

59.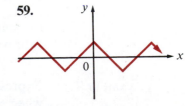

60.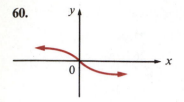

61.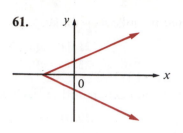

62.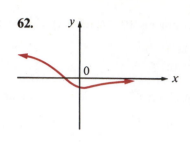

10.2 Variation

We often encounter a function described by an equation of the form $y = kx$, where k is some constant number. When this occurs, we have what is called a **direct variation.**

DEFINITION 1   **Direct Variation**

> Let x and y be two variable quantities. We say that **y varies directly with x** or that **y is directly proportional to x,** if there is a constant k such that
>
> $$y = kx.$$
>
> The number k is called the **constant of variation** or the **constant of proportionality.**

For example, if an airplane is flying at a speed of 650 miles per hour, then the distance d that it travels in t hours is $d = 650t$. The distance d is directly proportional to the time t, and the constant of proportionality is 650.

At times, we deal with two variables x and y, related by an equation of the form

$$y = kx^n$$

for some constant number k and some positive rational number n. In this case we say that **y is directly proportional to the nth power of x.**

EXAMPLE 1   Express y as a function of x if y is directly proportional to x, and if $y = 4$ when $x = 1$.

SOLUTION   Since y is directly proportional to x, there is some constant k such that

$$y = kx.$$

Because $y = 4$ when $x = 1$, we have $4 = k(1)$ or $k = 4$, so that the equation becomes

$$y = 4x.$$

EXAMPLE 2   The surface area of a sphere is directly proportional to the square of its radius. If a sphere with a radius of 4 inches has a surface area of 64π square inches, express the surface area of a sphere as a function of its radius.

SOLUTION   Let S represent the surface area of a sphere, and let r represent the radius. Because S is directly proportional to r^2, there is a constant k such that

$$S = kr^2.$$

Because $S = 64\pi$ when $r = 4$, then

$$64\pi = 16k,$$

and

$$k = \frac{64}{16}\pi = 4\pi.$$

Hence,

$$S = 4\pi r^2.$$

EXAMPLE 3   It has been estimated that the amount of pollution A entering the atmosphere is directly proportional to the number of people N living in a certain area. If a city with a population of 140,000 people produces 100,000 tons of atmospheric pollutants in a year, how many tons of pollutants are likely to enter the atmosphere annually in a city with a population of 1,350,000 people?

SOLUTION   Because the pollution A is directly proportional to the number of people N, there is some constant k such that

$$A = kN.$$

Because $A = 100,000$ when $N = 140,000$, we have

$$100,000 = 140,000k.$$

We solve for k: $k = \frac{5}{7}$, so that the function becomes

$$A = \frac{5}{7}N.$$

Hence, if $N = 1,350,000$, then

$$A = \frac{5}{7}(1,350,000) = 964,285.71.$$

This means that about 964,286 tons of pollutants will probably enter the atmosphere.

Consider the function determined by the equation

$$y = \frac{k}{x},$$

where k is a positive constant. Notice that as the value of x increases, the value of y decreases. This is an example of an **inverse variation.**

DEFINITION 2   **Inverse Variation**

Let x and y be two variable quantities. We say that **y varies inversely with x** or that **y is inversely proportional to x** if there is a constant k such that

$$y = \frac{k}{x} \qquad \text{for} \qquad x \neq 0.$$

For example, if y is inversely proportional to x, and $y = 0.4$ when $x = 0.8$, then

$$y = \frac{k}{x}$$

and

$$0.4 = \frac{k}{0.8} \qquad \text{or} \qquad k = 0.32.$$

Therefore, the function that relates y to x is given by the equation

$$y = \frac{0.32}{x}.$$

If

$$y = \frac{k}{x^n}$$

for some constant number k and some positive rational number n, we say that **y is inversely proportional to the nth power of x.**

EXAMPLE 4 Express y as a function of x if y is inversely proportional to x^2, and if $y = 12$ when $x = 2$.

SOLUTION Because y is inversely proportional to x^2, there is a number k such that

$$y = \frac{k}{x^2}.$$

Given that $y = 12$ when $x = 2$, then $12 = k/2^2$ and $k = 48$. Thus,

$$y = \frac{48}{x^2}.$$

EXAMPLE 5 Boyle's law states that the pressure P of an ideal gas at a constant temperature is inversely proportional to its volume V. Find the constant of variation if the pressure P of a gas is 30 pounds per square inch when the volume V is 100 cubic inches.

SOLUTION Since the pressure P is inversely proportional to V, there is a number k such that

$$P = \frac{k}{V} \quad \text{or} \quad k = PV.$$

At $P = 30$ pounds per square inch, $V = 100$ cubic inches, so that we have $k = 30(100) = 3,000$ pounds inches.

Consider the area of a rectangle with length ℓ units and width w units. The area A is given by the equation $A = \ell w$. In this situation, we say that A varies jointly with ℓ and w. We have the following formal definition:

DEFINITION 3 **Joint Variation**

Let x, y, and z be variable quantities. We say that z **varies jointly with x and y** if there is a constant k such that

$$z = kxy.$$

We call the relationship between z and the product xy **joint variation.**

Several kinds of variations can occur together. For example, if

$$w = \frac{kx^2y^3}{z^4},$$

then w varies jointly with the square of x and with the cube of y, and inversely with the fourth power of z.

EXAMPLE 6   Suppose that z varies directly with the cube of x and inversely with y. If $z = 8$ when $x = 2$ and $y = 4$, find z when $x = 5$ and $y = 10$.

SOLUTION   Because z varies directly with the cube of x and inversely with y, there is some constant k such that

$$z = \frac{kx^3}{y}.$$

We substitute for z, x, and y:

$$8 = \frac{k(2)^3}{4} \quad \text{or} \quad k = 4$$

so that $z = 4x^3/y$. Thus, for $x = 5$ and $y = 10$,

$$z = \frac{4(5)^3}{10} = 50.$$

EXAMPLE 7   The volume V of a right circular cone varies jointly with its height h and the square of its base radius r. If $V = 12\pi$ cubic inches when $r = 3$ inches and $h = 4$ inches, find V in terms of r and h.

SOLUTION   Because V varies jointly with r^2 and h, there is a real number k such that

$$V = kr^2h.$$

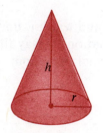

We substitute for V, r, and h:

$$12\pi = k(9)(4) \quad \text{or} \quad k = \frac{12\pi}{36} = \frac{\pi}{3}.$$

Hence, the required formula is

$$V = \frac{1}{3}\pi r^2 h.$$

PROBLEM SET 10.2

1. Let y be directly proportional to x. Express y as a function of x. If $y = 8$ when $x = 4$, find $f(x + 2)$, $f(2) + f(3)$, and $[f(x + h) - f(x)]/h$, where $h \neq 0$.

2. If y is directly proportional to x^2 and $y = f(x)$, does $f(ax) = af(x)$?

In problems 3–8, y is directly proportional to x^3. Express y as a function of x in each case.

3. $y = 4$ when $x = 2$ **4.** $y = 12$ when $x = -2$ **5.** $y = 3$ when $x = 1$

6. $y = -2$ when $x = 3$ **7.** $y = 14$ when $x = 11$ **8.** $y = 10$ when $x = -3$

9. If y is inversely proportional to x^2, and if $y = 9$ when $x = 2$, find y when $x = 3$.

10. If y is inversely proportional to $\sqrt[3]{x}$ and $y = 9$ when $x = 8$, find y when $x = 216$.

11. If T is directly proportional to x and inversely proportional to y, and if $T = 0.01$ when $x = 20$ and $y = 20$, express T as a function of x and y.

12. If y is inversely proportional to x^2, and if $y = 8$ when $x = 10$, find y when $x = 2$.

13. If y is inversely proportional to x^3, and if $y = 3$ when $x = 4$, express y as a function of x.

14. If V varies directly with T and inversely with P, and if $V = 40$ when $T = 300$ and $P = 30$, find V when $T = 324$ and $P = 24$.

15. The surface areas of two spheres have the ratio of 9 to 4. (The *ratio* 9 to 4 may be written as the fraction $\frac{9}{4}$ or as 9:4.) What is the ratio of their radii? Their volumes? (*Hint:* $S = 4\pi r^2$ and $V = \frac{4}{3}\pi r^3$.)

16. Coulomb's law states that the magnitude of the force F (in newtons) that acts on two charges q_1 and q_2 varies directly with the product of q_1 and q_2 (in coulombs) and inversely with the square of the distance r (in meters) between them. If two charges, each having a magnitude of 1 coulomb, are separated in air by a distance of 0.1 meter and if the force on the two charges is 9×10^{11} newtons, find the force when the charges are separated by 0.2 meter.

17. The total surface area S of a cube is directly proportional to the square of an edge x. If the cube with an edge of 3 inches has a surface area of 54 square inches, express the surface area S as a function of x. Then find the surface area of a cube with an edge of 12 inches.

18. Newton's law of gravitational attraction states that the force F with which two particles of mass m_1 and m_2 attract each other varies directly with the product of the masses and inversely with the square of the distance r between them. If one of the masses is tripled, and the distance between the masses is also tripled, what happens to the force?

19. In the life sciences, it is found that the number N of gene mutations as a result of x-ray exposure is directly proportional to the size of the x-ray dose m. What is the effect on N if m is tripled?

20. Psychologists define intelligence quotient (IQ) as a quantity that varies directly with people's mental age (MA) and inversely with their chronological age (CA). If a 12-year-old girl with a mental age of 15 has an IQ of 125, what will be the IQ of a 10-year-old boy with an MA of 15.7?

10.3 Linear and Quadratic Functions

In this section, we explore a variety of *linear* and *quadratic* functions by examining their graphs. These functions have many important applications.

Linear Functions

If m and b are constants, then the graph of the function f defined by

$$f(x) = mx + b$$

is the same as the graph of the equation

$$y = mx + b.$$

As we saw in Section 7.5, $y = mx + b$ is a straight line with slope m and y intercept b. Thus, we have the following definition:

DEFINITION 1   **Linear Function**

> A function f of the form
>
> $$f(x) = mx + b$$
>
> is called a **linear function.**

For example,

$$f(x) = 2x - 1, \qquad g(x) = -3x + 7, \qquad \text{and} \qquad h(x) = 1 - 7x$$

are linear functions.

If we substitute $m = 0$ in the equation $f(x) = mx + b$, we have a specific type of linear function:

DEFINITION 2   **Constant Function**

> A function of the form
>
> $$f(x) = b$$
>
> is called a **constant function.**

The graph of a constant function $f(x) = b$ is a horizontal straight line (Figure 1).

Figure 1

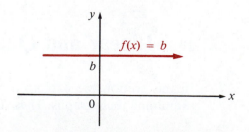

In Examples 1 and 2, sketch the graph of each linear function. Find the slope of the graph, and the domain and range of each function.

EXAMPLE 1 $f(x) = 2x + 5$

SOLUTION   The graph of f is the same as the graph of the equation

$$y = 2x + 5.$$

This graph is a line with slope $m = 2$ and y intercept 5. The x intercept is $-\frac{5}{2}$ (Figure 2). We see from the graph that the domain of f is the set $\mathbb{R}$ of all real numbers, and that the range of f is also $\mathbb{R}$.

Figure 2

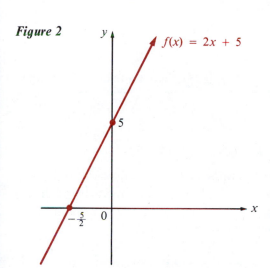

EXAMPLE 2 $g(x) = -\frac{9}{5}x + \frac{7}{5}$

SOLUTION   The graph of g is the same as the graph of the equation

$$y = -\frac{9}{5}x + \frac{7}{5}.$$

This graph is a line with slope $m = -\frac{9}{5}$ and y intercept $\frac{7}{5}$. The x intercept is $\frac{7}{9}$ (Figure 3). We see from the graph that the domain and the range of g are both $\mathbb{R}$.

Figure 3

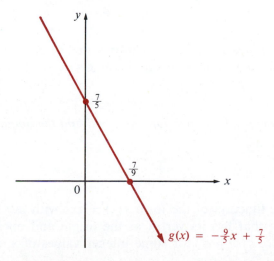

Quadratic Functions

Quadratic functions are used often in applied mathematics. This type of function is defined as follows:

DEFINITION 3   **Quadratic Function**

A function f of the form

$$f(x) = ax^2 + bx + c,$$

where a, b, and c are real numbers and $a \neq 0$, is called a **quadratic function.**

The simplest quadratic function is

$$f(x) = ax^2.$$

The graph of this function is called a **parabola.** It opens upward and has a lowest point at $(0, 0)$ if $a > 0$ (Figure 4a). It opens downward and has a highest point at $(0, 0)$ if $a < 0$ (Figure 4b). The highest or lowest point of the graph of $f(x) = ax^2$ is called the **vertex** of the parabola.

Figure 4

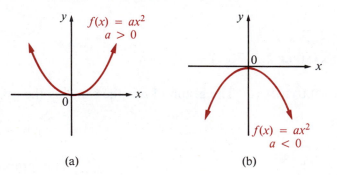

(a)   (b)

Graphs of quadratic functions resemble the graphs in Figure 4. We can obtain these graphs by plotting some points and connecting them with a smooth curve.

In Examples 3 and 4, sketch the graph of each function, and find the domain and range of the function.

EXAMPLE 3 $f(x) = x^2$

SOLUTION   The function f is a quadratic function of the form $f(x) = ax^2$ with $a > 0$. The graph of f is a parabola that has its vertex at the origin and opens upward. We calculate values of $f(x) = x^2$ for some integer values of x, as shown in the table in Figure 5. Then we plot these points and connect them with a smooth curve to obtain the graph of f (Figure 5). The graph of f reveals that the domain is $\mathbb{R}$, and that the range consists of all nonnegative real numbers.

Figure 5

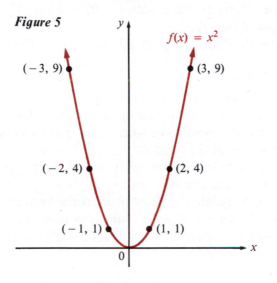

x	$y = f(x) = x^2$
-3	9
-2	4
-1	1
0	0
1	1
2	4
3	9

EXAMPLE 4 $g(x) = -x^2$

SOLUTION   The function g is of the form $f(x) = ax^2$ with $a < 0$. The graph of g is a parabola that has its vertex at the origin and opens downward. We calculate values of $g(x) = -x^2$ for some integer values of x, as shown in the table in Figure 6. Then we plot the corresponding points and connect them by a smooth curve to obtain the graph of g (Figure 6). The graph of g reveals

Figure 6

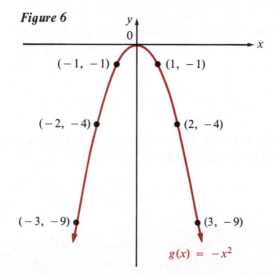

x	$y = g(x) = -x^2$
-3	-9
-2	-4
-1	-1
0	0
1	-1
2	-4
3	-9

that the domain is $\mathbb{R}$ and that the range consists of all nonpositive real numbers.

The graph of $f(x) = ax^2 + bx + c$ intersects the y axis at the point $(0, c)$ where $x = 0$. We call c the **y intercept** of the graph. If the graph intersects the x axis at the points $(x_1, 0)$ and $(x_2, 0)$, we call x_1 and x_2 the **x intercepts** of the graph.

The graph of any quadratic function of the form

$$f(x) = ax^2 + bx + c$$

will have the same general shape as the curves in Figure 4. If $a > 0$, the graph of f is a parabola that opens upward. If $a < 0$, the graph of f is a parabola that opens downward. The location of the graph will vary, depending upon specific values of a, b, and c.

The vertex of the parabola is one of the key points that we locate when we graph a quadratic function.

If we designate the point (h, k) as the vertex, we can find the values for the coordinates h and k by using the procedure given in Section 6.2 for completing the squares. We will find (see problem 34) that the vertex has the coordinates

$$(h, k) = \left(-\frac{b}{2a}, f\left(-\frac{b}{2a} \right) \right) = \left(-\frac{b}{2a}, \frac{4ac - b^2}{4a} \right).$$

We can find the maximum and minimum values of a quadratic function by using the preceding vertex formula. Indeed, if $a > 0$, the number $f(-b/2a)$ is the **minimum** value of f, and if $a < 0$, the number $f(-b/2a)$ is the **maximum** value of f.

In Examples 5 and 6, find the vertex and the x and y intercepts of the graph of each function and sketch the graph. What are the domain and the range of the function?

EXAMPLE 5 $f(x) = x^2 - 3x + 2$

SOLUTION   Here we have $a = 1$, $b = -3$, and $c = 2$. The graph is a parabola that opens upward. The vertex (h, k) has coordinates

$$h = -\frac{b}{2a} = -\frac{(-3)}{2(1)} = \frac{3}{2}$$

and

$$k = f\left(-\frac{b}{2a} \right) = \frac{4ac - b^2}{4a} = \frac{4(1)(2) - (-3)^2}{4(1)} = -\frac{1}{4}$$

so that

$$(h, k) = \left(\frac{3}{2}, -\frac{1}{4}\right).$$

The y intercept is obtained when $x = 0$. Therefore, the y intercept is 2. The x intercepts are obtained by solving the equation

$$x^2 - 3x + 2 = 0$$

or

$$(x - 1)(x - 2) = 0.$$

Thus, the x intercepts are 1 and 2. We can plot these four points and connect them with a smooth curve to obtain the graph of f (Figure 7). The graph of f reveals that the domain of f is $\mathbb{R}$, and that the range consists of all real numbers $y \geq -\frac{1}{4}$.

Figure 7

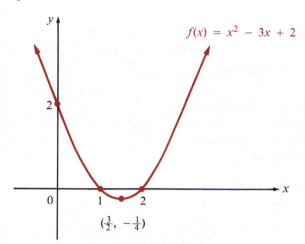

$f(x) = x^2 - 3x + 2$

$(\frac{3}{2}, -\frac{1}{4})$

EXAMPLE 6 $g(x) = -x^2 + 2x$

SOLUTION Here $a = -1$, $b = 2$, and $c = 0$. The graph is a parabola that opens downward. Its vertex (h, k) has the coordinates

$$h = -\frac{b}{2a} = -\frac{2}{2(-1)} = 1$$

and

$$k = f\left(-\frac{b}{2a}\right) = \frac{4ac - b^2}{4a} = \frac{0 - 2^2}{4(-1)} = 1$$

so that

$$(h, k) = (1, 1).$$

The y intercept is 0, and the x intercepts are 0 and 2. (Why?) The graph of g shows that the domain of g is $\mathbb{R}$ and that the range consists of all real numbers $y \leq 1$ (Figure 8).

Figure 8

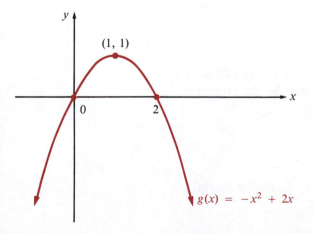

(1, 1)

$g(x) = -x^2 + 2x$

PROBLEM SET 10.3

In problems 1–10, sketch the graph of each linear function. Find the slope of the graph and the domain and the range of the function.

1. $f(x) = -3x + 5$
2. $f(x) = 5x + 1$
3. $f(x) = -\frac{3}{4}x + 1$
4. $f(x) = \frac{1}{4}x + 3$
5. $f(x) = -1$
6. $f(x) = -3x$
7. $f(x) = 2(x - 2) + 1$
8. $f(x) = 7$
9. $f(x) = 4x$
10. $f(x) = 3 - 2(1 - x)$

In problems 11–16, find a linear function f such that each condition is satisfied.

11. $f(1) = 3$ and $f(2) = 5$
12. $f(-2) = 7$ and $f(3) = -5$
13. $f(0) = 4$ and $f(3) = 0$
14. $f(1) = -8$ and $f(\frac{1}{2}) = -6$
15. $f(-7) = 3$ and $f(5) = 3$
16. $f(2) = -4$ and $f(-10) = -4$
17. Find a linear function f such that $2f(x) = f(2x)$ for every real number x.
18. Find a linear function f such that $f(3x + 2) = f(3x) + 2$.
19. Find a linear function f such that $f(1) = 3$ and the graph of f is parallel to the graph of the line determined by the points $(-2, 1)$ and $(3, 2)$.
20. Find a linear function f such that $f(1) = 3$ and the graph of f is perpendicular to the graph of the line determined by the points $(-2, 1)$ and $(3, 2)$.

In problems 21–32, find the vertex and the x and y intercepts of the graph of each quadratic function, and sketch the graph. What are the domain and the range of each function?

21. $f(x) = 2x^2$
22. $f(x) = -\frac{1}{2}x^2$
23. $f(x) = 2x^2 - 3$
24. $f(x) = x^2 - 3$
25. $f(x) = -x^2 - 2x - 1$
26. $f(x) = (x - 5)^2$
27. $f(x) = x^2 + 5x + 6$
28. $f(x) = -x^2 - 1$
29. $f(x) = 2x^2 - 3x$
30. $f(x) = -(x + 1)^2$
31. $f(x) = x^2 + 4x + 3$
32. $f(x) = -x^2 + x - 5$
33. Sketch the graph of each of the following functions on the same coordinate system: $f(x) = x^2 - 1$, $f(x) = x^2 + 1$, $f(x) = x^2 - 2$, and $f(x) = x^2 + 2$.
34. Show that the coordinates of the vertex of $y = ax^2 + bx + c$, $a \neq 0$, are

$$(h, k) = \left(-\frac{b}{2a}, \frac{4ac - b^2}{4a}\right).$$

(*Hint:* Use the method of completing the square, page 284.)

35. A ball is thrown vertically upward from the ground with an initial speed of 64 feet per second. The distance s from the starting point on the ground at t seconds is given by the function

$$s = f(t) = -16t^2 + 64t.$$

(a) How high will the ball be after 3 seconds?
(b) How many seconds does it take the ball to reach its maximum height?

36. The annual gross earnings S (in millions of dollars) of a particular corporation t years from now are given by the function

$$S = f(t) = -\frac{2}{5}t^2 + 2t + 10.$$

(a) Find the earnings S after 4 years.

(b) How many years will it take the corporation to maximize its earnings?

10.4 Exponential and Logarithmic Functions

In this section, we introduce two additional functions that are of great practical importance in many applications of mathematics: exponential and logarithmic functions. Bankers use exponential functions, for instance, to compute compound interest, and scientists use them to determine the rate of radioactive decay of material in order to determine its age.

Exponential Functions

In Section 10.3, we studied quadratic functions which take the form

$$p(x) = ax^2,$$

where the base is the variable x and the exponent 2 is constant. We now consider functions of the form

$$f(x) = b^x,$$

where the exponent is the variable x and the base b is constant. Such a function is called an **exponential function.**

DEFINITION 1 **Exponential Function**

If b is a positive number, the function f defined by

$$f(x) = b^x,$$

where $b \neq 1$, is called an **exponential function** with base b.

Examples of exponential functions are

$$f(x) = 2^x,$$

$$g(x) = \left(\frac{1}{3}\right)^x,$$

and

$$h(x) = (5^{-1})^x.$$

If b is a positive constant, and if you plot some points (x, b^x) for rational values of x, you will notice that these points lie along a smooth curve. The curve is the graph of the function $f(x) = b^x$ with base b. For example, if $b = 2$, and if you plot a few points $(x, 2^x)$ for rational values of x, you obtain a collection of points similar to those in Figure 1a. In Figure 1b we have connected these points with a smooth curve to obtain the graph of $f(x) = 2^x$. This graph reveals that the domain of the function $f(x) = 2^x$ is the set of real numbers $\mathbb{R}$ and that the range is the set of all positive real numbers.

Figure 1

x	$f(x)$
-3	$\frac{1}{8}$
-2	$\frac{1}{4}$
-1	$\frac{1}{2}$
0	1
1	2
2	4
3	8

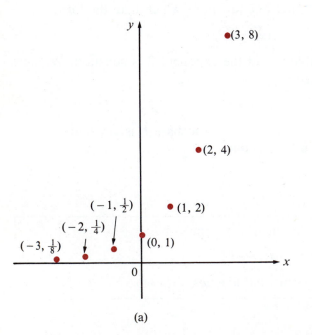

(a)

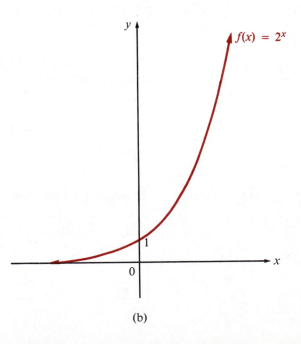

(b)

In Examples 1–3, sketch the graph of each function. Indicate the domain and the range of the function.

EXAMPLE 1 $f(x) = 3^x$

SOLUTION   We begin by calculating values of $f(x) = 3^x$ for some integer values of x, as shown in the table in Figure 2. Then we plot the corresponding points and connect them with a smooth curve to obtain the graph of f (Figure 2). From the graph we see that the domain of f is the set of real numbers $\mathbb{R}$, and that the range consists of all positive real numbers.

Figure 2

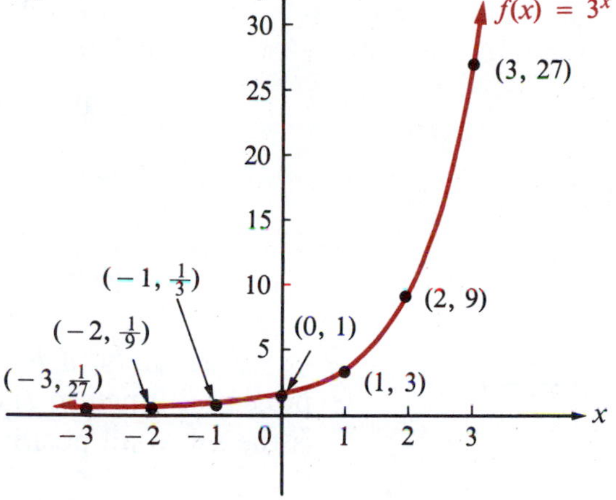

x	$f(x)$
-3	$\frac{1}{27}$
-2	$\frac{1}{9}$
-1	$\frac{1}{3}$
0	1
1	3
2	9
3	27

EXAMPLE 2 $g(x) = \left(\frac{1}{4}\right)^x$

SOLUTION   We calculate values of $g(x) = \left(\frac{1}{4}\right)^x$ for some integer values of x and obtain the table in Figure 3. We plot the corresponding points and connect them by a smooth curve. We obtain the graph of g (Figure 3). The graph reveals that the domain of g is $\mathbb{R}$, and that the range consists of all positive real numbers.

Figure 3

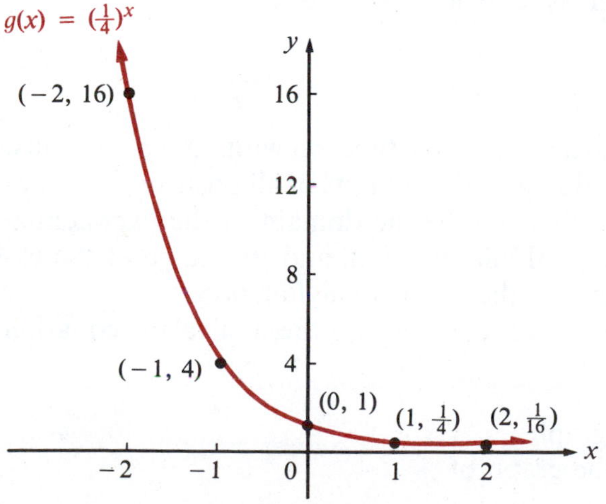

x	$g(x)$
-2	16
-1	4
0	1
1	$\frac{1}{4}$
2	$\frac{1}{16}$

EXAMPLE 3   Ⓒ $h(x) = \pi^x$

SOLUTION   Using a calculator with a y^x key, we evaluate $h(x) = \pi^x$ for some integer values of x, as shown in the table in Figure 4. We plot the corresponding points and connect them with a smooth curve to obtain the graph of h (Figure 4). From the graph we see that the domain of h is $\mathbb{R}$, and that the range consists of all positive real numbers.

Figure 4

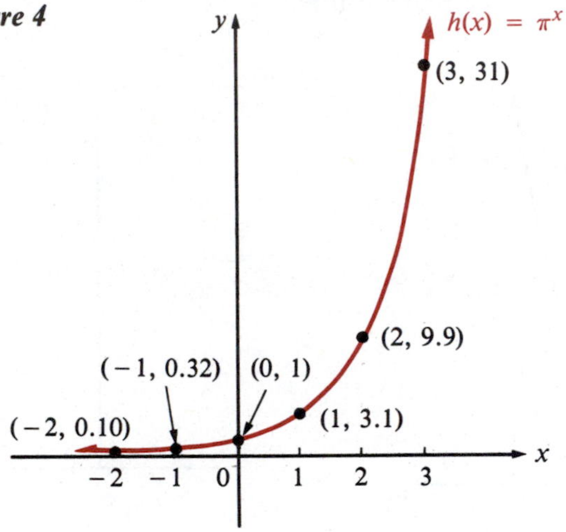

x	$h(x)$
-2	0.10
-1	0.32
0	1
1	3.1
2	9.9
3	31

Generally, if $b > 0$, the domain of the exponential function $f(x) = b^x$ is $\mathbb{R}$, and the range is the set of all positive real numbers.

Logarithmic Functions

In Section 9.1, we showed the connection between the equations $x = b^y$ and $y = \log_b x$, that is,

$$x = b^y \qquad \text{if and only if} \qquad y = \log_b x.$$

A function f of the form

$$f(x) = \log_b x,$$

where $b > 0$ and $b \neq 1$, is called a **logarithmic function** with base b. We must also have $x > 0$. Therefore, the domain of f consists of all positive real numbers. The range of f is $\mathbb{R}$. In other words, the domain of the exponential function is the range of the logarithmic function, and the range of the exponential function is the domain of the logarithmic function.

To sketch the graph of $f(x) = \log_b x$, we use the equivalent equation

$$x = b^{f(x)}$$

to locate some points on the graph.

For example, to sketch the graph of

$$f(x) = \log_3 x,$$

we use the equivalent equation

$$x = 3^{f(x)} = 3^y, \qquad \text{where } y = f(x).$$

Thus,

$$\text{if } f(x) = 0, \qquad \text{then } x = 3^0 = 1$$

$$\text{if } f(x) = 1, \qquad \text{then } x = 3^1 = 3$$

$$\text{if } f(x) = 2, \qquad \text{then } x = 3^2 = 9.$$

Note that we have reversed the usual technique for finding points on the graph: we have selected values of y first and then determined the corresponding values of x, as shown in the table in Figure 5. The graph shows that the domain of $f(x) = \log_3 x$ consists of all positive real numbers and that the range of f is the set of real numbers $\mathbb{R}$.

Figure 5

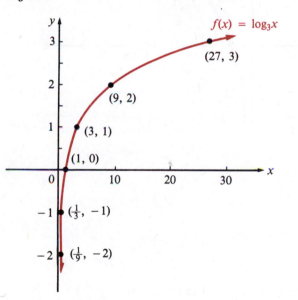

$x = 3^y$	y
$\frac{1}{9}$	-2
$\frac{1}{3}$	-1
1	0
3	1
9	2
27	3

In Examples 4 and 5, sketch the graph of each function. Indicate the domain and the range of the function.

EXAMPLE 4 $f(x) = \log_2 x$

SOLUTION   The equation

$$f(x) = \log_2 x$$

is equivalent to the equation

$$x = 2^{f(x)} = 2^y, \qquad \text{where} \qquad y = f(x).$$

We calculate values of $x = 2^y$ for some integer values of y, as shown in the table in Figure 6. Then we plot the corresponding points and connect them by a smooth curve to obtain the graph of $f(x) = \log_2 x$ (Figure 6). The graph of f reveals that the domain consists of all positive real numbers and that the range is $\mathbb{R}$.

Figure 6

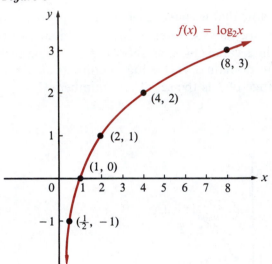

$x = 2^y$	y
$\frac{1}{2}$	-1
1	0
2	1
4	2
8	3

EXAMPLE 5 $f(x) = \log_{1/4} x$

SOLUTION   The equation

$$f(x) = \log_{1/4} x$$

is equivalent to the equation

$$x = \left(\frac{1}{4}\right)^{f(x)}$$

$$= \left(\frac{1}{4}\right)^{y}, \qquad \text{where } y = f(x).$$

We calculate values of $x = (\frac{1}{4})^y$ for some values of y as shown in the table in Figure 7. Then we plot the corresponding points and connect them by a smooth curve to obtain the graph of $f(x) = \log_{1/4} x$ (Figure 7). From the graph of f we see that the domain consists of all positive real numbers, and that the range is $\mathbb{R}$.

Figure 7

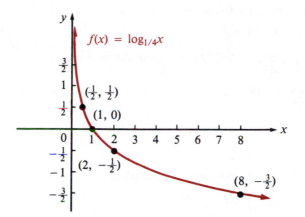

$x = \left(\frac{1}{4}\right)^y$	y
$\frac{1}{2}$	$\frac{1}{2}$
1	0
2	$-\frac{1}{2}$
8	$-\frac{3}{2}$

PROBLEM SET 10.4

In problems 1–10, sketch the graph of each exponential function. Indicate the domain and the range.

1. $f(x) = 4^x$ **2.** $f(x) = 4^{x-1}$ **3.** $f(x) = 3^{x+1}$ **4.** $f(x) = -2^x$ **5.** $f(x) = -\left(\frac{1}{3}\right)^x$

6. $f(x) = 2^{-x}$ **7.** $f(x) = \left(\frac{1}{5}\right)^{-x}$ **8.** $f(x) = (0.1)^x$ **9.** $f(x) = 5(3^x)$ **10.** $f(x) = -(4)^x$

In problems 11–16, determine the base of the exponential function $f(x) = b^x$ if the graph of $f(x)$ contains the given points.

11. $(2, 9)$ **12.** $(3, 27)$ **13.** $(2, 16)$ **14.** $(3, 125)$ **15.** $(0, 1)$ **16.** $\left(\frac{1}{2}, \sqrt{10}\right)$

17. Let f be an exponential function with base b. Show that $f(u - v) = f(u) \div f(v)$ for any real numbers u and v.

18. Use the graph in Figure 2 (page 477) to approximate the value of $3^{3/2}$.

In problems 19–24, sketch the graph of each function. Indicate the domain and the range.

19. $f(x) = -\log_2 x$ **20.** $f(x) = \log_5 x$ **21.** $f(x) = \log_{1/2} x$

22. $f(x) = \log_4 x$ **23.** $f(x) = \log_6 x$ **24.** $f(x) = \log_{1/3} x$

In problems 25–30, determine the base of the logarithmic function $f(x) = \log_b x$ if the graph of $f(x)$ contains the given points.

25. $(8, 3)$ **26.** $(125, 3)$ **27.** $\left(\frac{1}{16}, -2\right)$ **28.** $\left(8, \frac{3}{2}\right)$ **29.** $\left(3, \frac{1}{2}\right)$ **30.** $(c^{3/2}, 3)$

31. Let f be a logarithmic function with base b. Show that $f(u) + f(v) = f(uv)$ for any positive real numbers u and v.

32. Let f be a logarithmic function with base b. Show that $f(u) - f(v) = f(u/v)$ for any positive real numbers u and v.

c 33. Ecologists have determined that the approximate population N of bears in a certain protected forest area is given by the formula

$$N = 225e^{0.02t},$$

where t is the elapsed time in years, since 1980. Estimate the number of bears that will inhabit the region in 1995.

c 34. A biologist finds that the number N of bacteria in a culture after t hours is given by

$$N = 2000e^{0.7t}.$$

How many bacteria will be present after $t = 14$ hours?

c 35. A particle is moving on a straight line according to the function

$$s = (t + 1)^2 \ln (t + 1),$$

where s is the directed distance of the particle from the starting point at t seconds. Find the distance traveled by the particle after 5 seconds.

10.5 Graphs of Special Curves— Conic Sections

The **graph** of an equation in two unknowns x and y is defined to be the set of all points $P = (x, y)$ in the Cartesian plane whose coordinates x and y satisfy the equation. Many (but not all) equations in x and y have graphs that are smooth curves in the plane. If we have an equation whose graph is a given curve in the Cartesian plane, the equation is called an **equation of the curve.** In this section, we discuss certain types of equations that play a special role in geometry and in many other applications. We also examine certain types of curves. We introduce these curves by establishing the standard forms for the equations of their graphs. The curves that we consider are **circles, ellipses, hyperbolas,** and **parabolas.** Each of these curves is obtained by sectioning or cutting a circular cone with a plane. Therefore, the curves are called **conic sections.**

Ellipses

We saw in Section 7.2 that the equation of a circle of radius r and center (h, k) is

$$(x - h)^2 + (y - k)^2 = r^2.$$

By setting $(h, k) = (0, 0)$ in this equation, we can see that the graph of

$$x^2 + y^2 = r^2$$

is a circle of radius r units with its center at the origin 0 (Figure 1).

Figure 1

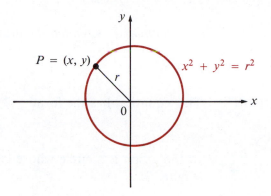

The equation of a circle $x^2 + y^2 = r^2$ can be rewritten in the form

$$\frac{x^2}{r^2} + \frac{y^2}{r^2} = 1.$$

We can modify this last equation to read:

$$\frac{x^2}{a^2} + \frac{y^2}{b^2} = 1,$$

where $a > 0$, $b > 0$, and $a \neq b$. This produces an equation of an ellipse (Figure 2). The ellipse intersects the x axis at the points $(-a, 0)$ and $(a, 0)$. It

Figure 2

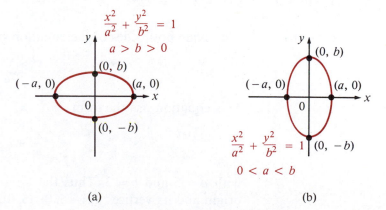

(a)   (b)

intersects the y axis at the points $(0, -b)$ and $(0, b)$. These four points are called the **vertices** of the ellipse. The two line segments joining opposite pairs of vertices are called the **axes** of the ellipse. The longer axis is called the **major axis,** the shorter axis is called the **minor axis,** and the two axes intersect at the center of the ellipse. If $a > b > 0$, the major axis is horizontal (Figure 2a). If $0 < a < b$, the major axis is vertical (Figure 2b).

In Examples 1–3, sketch the graph of each equation.

EXAMPLE 1 $x^2 + y^2 = 4$

SOLUTION   This equation has the form

$$x^2 + y^2 = r^2,$$

so the graph is a circle whose center is at the origin and whose radius is 2 (Figure 3).

Figure 3

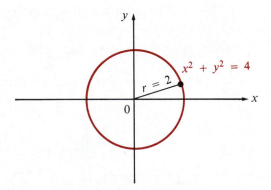

EXAMPLE 2 $9x^2 + 25y^2 = 225$

SOLUTION   We divide both sides of the equation by 225 to obtain

$$\frac{x^2}{25} + \frac{y^2}{9} = 1.$$

This equation has the form

$$\frac{x^2}{a^2} + \frac{y^2}{b^2} = 1$$

with $a = 5$ and $b = 3$. Thus the graph is an ellipse with its center at the origin and its vertices at $(-5, 0)$, $(5, 0)$, $(0, -3)$, and $(0, 3)$ (Figure 4).

Figure 4

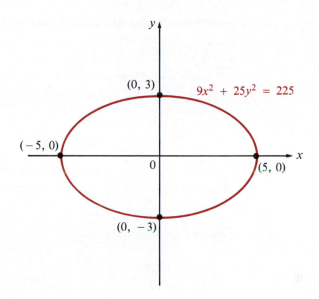

EXAMPLE 3 $9x^2 + 4y^2 = 36$

SOLUTION   We divide both sides of the equation by 36:

Figure 5

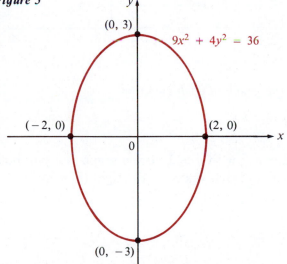

$$\frac{x^2}{4} + \frac{y^2}{9} = 1.$$

This equation has the form

$$\frac{x^2}{a^2} + \frac{y^2}{b^2} = 1$$

with $a = 2$ and $b = 3$. Therefore, the graph is an ellipse with its center at the origin and its vertices at $(-2, 0)$, $(2, 0)$, $(0, -3)$, and $(0, 3)$ (Figure 5).

Parabolas

In Section 10.3, we saw that the graph of the equation $y = ax^2 + bx + c$, $a \neq 0$, is a parabola that *opens upward if $a > 0$ and downward if $a < 0$*. If we interchange the variables x and y, we obtain the equation

$$x = ay^2 + by + c.$$

If $b = c = 0$, then

$$x = ay^2, \qquad a \neq 0.$$

The graph of this equation is a **parabola** that has its vertex at the origin, and that opens to the right if $a > 0$ (Figure 6a) and to the left if $a < 0$ (Figure 6b).

Figure 6

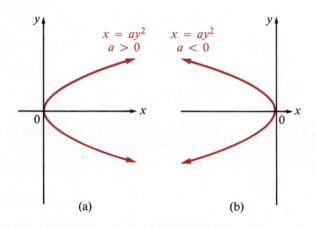

(a)   (b)

In Examples 4 and 5, sketch the graph of each equation.

EXAMPLE 4 $x = 4y^2$

SOLUTION   The equation has the form $x = ay^2$ with $a > 0$, so the graph is a parabola that has its vertex at the origin and that opens to the right (Figure 7).

Figure 7

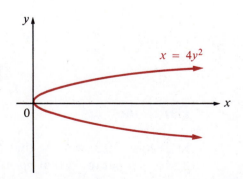

EXAMPLE 5 $x = -3y^2$

SOLUTION   The equation has the form $x = ay^2$ with $a < 0$, so the graph is a parabola that has its vertex at the origin and that opens to the left (Figure 8).

Figure 8

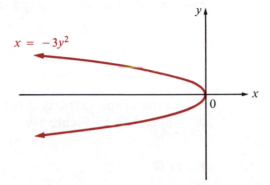

Hyperbolas

The graph of the equation

$$\frac{x^2}{a^2} - \frac{y^2}{b^2} = 1, \qquad a > 0, b > 0$$

is a **hyperbola** with its center at the origin (Figure 9). The hyperbola has two branches—one opening to the right and one opening to the left. The

Figure 9

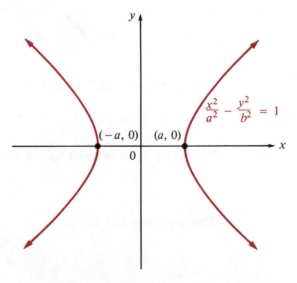

two points $(-a, 0)$ and $(a, 0)$, where these branches intersect the x axis, are called the **vertices** of the hyperbola. The line segment between the two vertices is called the **transverse** axis. The midpoint of the transverse axis is called the **center** of the hyperbola.

If we position the hyperbola so that the y axis contains the transverse axis, the equation of the graph is:

$$\frac{y^2}{b^2} - \frac{x^2}{a^2} = 1, \qquad a > 0, b > 0.$$

The center of this graph is at the origin, and the vertices are now the points $(0, -b)$ and $(0, b)$ (Figure 10).

Figure 10

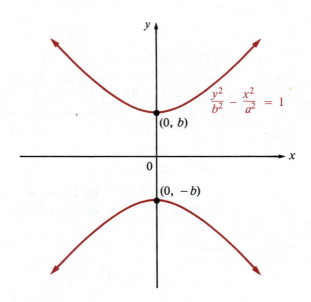

In Examples 6 and 7, sketch the graph of each equation.

EXAMPLE 6 $4x^2 - 9y^2 = 36$

SOLUTION   We divide both sides of the equation by 36 to obtain

$$\frac{x^2}{9} - \frac{y^2}{4} = 1.$$

This equation has the form

$$\frac{x^2}{a^2} - \frac{y^2}{b^2} = 1$$

Figure 11

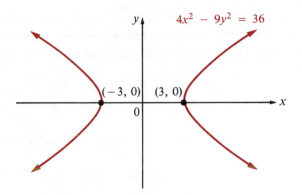

with $a = 3$ and $b = 2$. Thus the graph is a hyperbola with a horizontal transverse axis and vertices $(-3, 0)$ and $(3, 0)$ (Figure 11).

EXAMPLE 7 $4y^2 - 25x^2 = 100$

SOLUTION   We divide both sides of the equation by 100 to obtain

$$\frac{y^2}{25} - \frac{x^2}{4} = 1.$$

Figure 12

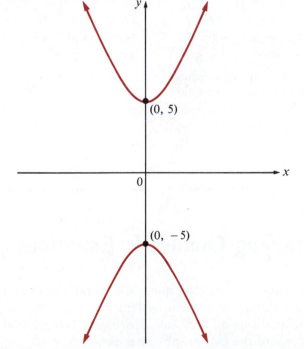

The equation has the form

$$\frac{y^2}{b^2} - \frac{x^2}{a^2} = 1$$

with $a = 2$ and $b = 5$. Therefore, the graph is a hyperbola with vertical transverse axis and vertices $(0, -5)$ and $(0, 5)$ (Figure 12).

PROBLEM SET 10.5

In problems 1–6, sketch the graph of each circle and find its radius.

1. $x^2 + y^2 = 9$ **2.** $x^2 + y^2 = 16$ **3.** $4x^2 + 4y^2 = 25$ **4.** $9x^2 + 9y^2 = 36$

5. $\dfrac{x^2}{3} + \dfrac{y^2}{3} = 2$ **6.** $x^2 + y^2 = k,\, k > 0$

In problems 7–12, sketch the graph of each ellipse and find its vertices.

7. $\dfrac{x^2}{9} + \dfrac{y^2}{4} = 1$ **8.** $\dfrac{x^2}{49} + \dfrac{y^2}{81} = 1$ **9.** $25x^2 + 4y^2 = 100$ **10.** $16x^2 + 25y^2 = 400$

11. $7x^2 + 8y^2 = 56$ **12.** $11x^2 + 5y^2 = 55$

In problems 13–18, sketch the graph of each parabola.

13. $x = 2y^2$ **14.** $x = -3y^2$ **15.** $x = -\frac{1}{2}y^2$ **16.** $x = \frac{4}{5}y^2$

17. $4x = 3y^2$ **18.** $2x = 7y^2$

In problems 19–24, sketch the graph of each hyperbola and find its vertices.

19. $\dfrac{x^2}{16} - \dfrac{y^2}{7} = 1$ **20.** $\dfrac{y^2}{49} - \dfrac{x^2}{81} = 1$ **21.** $\dfrac{y^2}{4} - \dfrac{x^2}{16} = 1$ **22.** $\dfrac{x^2}{64} - \dfrac{y^2}{25} = 1$

23. $5x^2 - 9y^2 = 45$ **24.** $7y^2 - 4x^2 = 28$

In problems 25–40, sketch the graph of the given conic.

25. $y^2 - x^2 = 1$ **26.** $y^2 + 2x = 0$ **27.** $4x^2 + 9y^2 = 1$ **28.** $x^2 + y^2 = 81$

29. $x^2 - 4y = 0$ **30.** $16y^2 - 4x^2 = 48$ **31.** $x^2 + y^2 = 49$ **32.** $y^2 + 4x^2 = 16$

33. $36x^2 - 9y^2 = 1$ **34.** $3x^2 + 3y^2 = 24$ **35.** $3x^2 + 2y = 0$ **36.** $9x^2 - y^2 = 9$

37. $4x^2 + 4y^2 = 9$ **38.** $-2x^2 + 5y = 0$ **39.** $16x^2 + 4y^2 = 64$ **40.** $9x^2 + 25y^2 = 1$

41. Olympic Stadium in Montreal, Canada, is constructed in the shape of an ellipse with major and minor axes of 480 and 280 meters, respectively. Find an equation of this ellipse.

42. A radar bowl in the form of a rotated parabola is 25 meters in diameter and 5 meters deep. Find an equation of the parabola, assuming that it contains the origin of a coordinate system and that it opens upward.

10.6 Systems Containing Quadratic Equations

In this section, we consider systems containing quadratic equations in two unknowns. An approximate solution of such a system can be found by sketching graphs of the two equations on the same coordinate system, and then determining the points where the two graphs intersect.

The substitution and elimination methods, introduced in Section 8.1, can often be used to solve systems containing quadratic equations. Even then, graphs can be used to determine the number of solutions to the system and to provide a rough check on the calculations.

In Examples 1–3, sketch graphs to determine the number of solutions to each system, and then solve the system algebraically.

EXAMPLE 1

$$\begin{cases} 2x + 3y = 8 \\ 2x^2 - 3y^2 = -10 \end{cases}$$

SOLUTION

The graph of $2x + 3y = 8$ is a line. The graph of $2x^2 - 3y^2 = -10$ is a hyperbola that intersects the line at two points (Figure 1). Thus, there are two solutions to the system. To find these solutions algebraically, we use the method of substitution. We solve the first equation for x:

$$x = \frac{8 - 3y}{2}.$$

Substituting $(8 - 3y)/2$ for x in the second equation, we obtain

$$2\left(\frac{8 - 3y}{2}\right)^2 - 3y^2 = -10.$$

We simplify this equation:

$$3y^2 - 48y + 84 = 0 \qquad \text{or} \qquad y^2 - 16y + 28 = 0.$$

Factoring, we have

$$(y - 2)(y - 14) = 0.$$

Figure 1

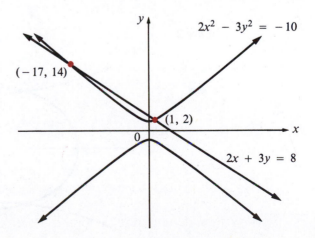

$2x^2 - 3y^2 = -10$

$(-17, 14)$

$(1, 2)$

$2x + 3y = 8$

We set each factor equal to zero and solve for y:

$$y = 2 \quad \text{or} \quad y = 14.$$

For each of these values of y there is a corresponding value for x given by

$$x = \frac{8 - 3y}{2}.$$

That is, for $y = 2$, $x = 1$ and for $y = 14$, $x = -17$. Therefore, the solutions are $(1, 2)$ and $(-17, 14)$.

EXAMPLE 2 $\begin{cases} 4x^2 + 7y^2 = 32 \\ -3x^2 + 11y^2 = 41 \end{cases}$

SOLUTION   The graph of $4x^2 + 7y^2 = 32$ is an ellipse and the graph of $-3x^2 + 11y^2 = 41$ is a hyperbola that intersects the ellipse at four points (Figure 2). Thus, there are four solutions to the system. We use the method of elimination to find these solutions algebraically.

$\begin{cases} 4x^2 + 7y^2 = 32 \\ -3x^2 + 11y^2 = 41 \end{cases}$ $\xrightarrow{\text{We multiply each side by 3.}}$ $\begin{cases} 12x^2 + 21y^2 = 96 \\ -12x^2 + 44y^2 = 164 \quad \text{add} \end{cases}$

$$ 65y^2 = 260$$

so that $y^2 = 4$. Hence,

$$y = -2 \quad \text{or} \quad y = 2.$$

We substitute $y = -2$ into the first equation $4x^2 + 7y^2 = 32$:

$$4x^2 + 7(-2)^2 = 32 \quad \text{or} \quad x^2 = 1$$

Figure 2

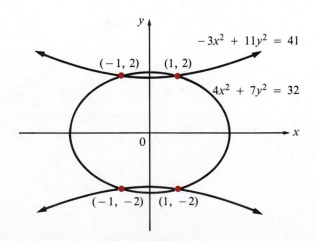

so that

$$x = -1 \quad \text{or} \quad x = 1.$$

We substitute $y = 2$ into $4x^2 + 7y^2 = 32$:

$$4x^2 + 7(2)^2 = 32 \quad \text{or} \quad x^2 = 1$$

so that

$$x = -1 \quad \text{or} \quad x = 1.$$

Therefore, the solutions are $(1, 2)$, $(-1, 2)$, $(1, -2)$, and $(-1, -2)$.

EXAMPLE 3
$$\begin{cases} x^2 + 2y^2 = 22 \\ 2x^2 + y^2 = 17 \end{cases}$$

SOLUTION   The graph of $x^2 + 2y^2 = 22$ is an ellipse and the graph of $2x^2 + y^2 = 17$ is an ellipse that intersects the first ellipse at four points (Figure 3). Thus, there are four solutions, which can be found algebraically by the method of elimination.

$$\begin{cases} x^2 + 2y^2 = 22 \\ 2x^2 + y^2 = 17 \end{cases} \xrightarrow{\text{We multiply each side by 2.}} \begin{array}{l} 2x^2 + 4y^2 = 44 \\ \underline{2x^2 + y^2 = 17} \quad \text{subtract} \\ \phantom{2x^2 + {}} 3y^2 = 27 \end{array}$$

so that $y^2 = 9$. Hence,

$$y = -3 \quad \text{or} \quad y = 3.$$

We substitute $y = 3$ into the equation $x^2 + 2y^2 = 22$:

$$x^2 + 2(3)^2 = 22 \quad \text{or} \quad x^2 = 4$$

so that

$$x = -2 \quad \text{or} \quad x = 2.$$

Figure 3

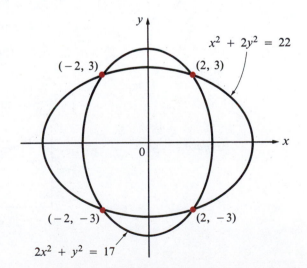

We substitute $y = -3$ into $x^2 + 2y^2 = 22$:

$$x^2 + 2(-3)^2 = 22 \qquad \text{or} \qquad x^2 = 4$$

so that

$$x = -2 \qquad \text{or} \qquad x = 2.$$

Therefore, the solutions are $(2, 3)$, $(2, -3)$, $(-2, 3)$, and $(-2, -3)$.

EXAMPLE 4   A manufacturer of art supplies makes templates by cutting out right triangles with perimeters of 60 centimeters from plastic sheets. If the hypotenuse of each triangle is 25 centimeters, find the lengths of the two sides of the triangle.

SOLUTION   Let x and y represent the lengths of the other two sides of the right triangle (Figure 4). Then

$$x + y + 25 = 60 \qquad \text{(perimeter of triangle)}$$

and

$$x^2 + y^2 = 25^2 \qquad \text{(Pythagorean theorem)}.$$

This gives us the system of equations

$$\begin{cases} x^2 + y^2 = 625 \\ x + y = 35. \end{cases}$$

We solve the second equation for y and obtain $y = 35 - x$. We substitute this result into the first equation:

$$x^2 + (35 - x)^2 = 625$$
$$2x^2 - 70x + 1{,}225 = 625$$
$$2x^2 - 70x + 600 = 0$$
$$x^2 - 35x + 300 = 0$$
$$(x - 15)(x - 20) = 0$$

so that

$$x = 15 \qquad \text{or} \qquad x = 20.$$

Figure 4

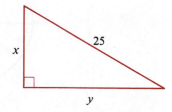

When $x = 15$, then $y = 35 - 15 = 20$; when $x = 20$, then $y = 35 - 20 = 15$. Hence, the lengths of the other two sides of each triangle are 15 and 20 centimeters.

PROBLEM SET 10.6

In problems 1–6, use the graphs of the equations to determine the number of solutions of each system of equations. Then solve the system.

1. $\begin{cases} x - y = 1 \\ x^2 + y^2 = 5 \end{cases}$ **2.** $\begin{cases} x - 2y = 3 \\ x^2 - y^2 = 24 \end{cases}$ **3.** $\begin{cases} 3x - y = 2 \\ x^2 + y^2 = 20 \end{cases}$

4. $\begin{cases} x + y = 3 \\ 3x^2 - y^2 = \frac{9}{2} \end{cases}$ **5.** $\begin{cases} 3x + 2y = 1 \\ 3x^2 - y^2 = -4 \end{cases}$ **6.** $\begin{cases} x + y = 6 \\ x^2 + y^2 = 20 \end{cases}$

In problems 7–30, solve each system of equations by the elimination or substitution method.

7. $\begin{cases} 5x - 3y = 10 \\ x^2 - y^2 = 6 \end{cases}$ **8.** $\begin{cases} 2x + y = 10 \\ xy = 12 \end{cases}$ **9.** $\begin{cases} 2x + 3y = 7 \\ x^2 + y^2 + 4y + 4 = 0 \end{cases}$

10. $\begin{cases} x - y + 4 = 0 \\ x^2 + 3y^2 = 12 \end{cases}$ **11.** $\begin{cases} 5x - y = 21 \\ y = x^2 - 5x + 4 \end{cases}$ **12.** $\begin{cases} x^2 - 25y^2 = 20 \\ 2x^2 + 25y^2 = 88 \end{cases}$

13. $\begin{cases} x - y^2 = 0 \\ x^2 + 2y^2 = 24 \end{cases}$ **14.** $\begin{cases} 3x^2 - 8y^2 = 40 \\ 5x^2 + y^2 = 81 \end{cases}$ **15.** $\begin{cases} 2x^2 - 3y^2 = 6 \\ 3x^2 + 2y^2 = 35 \end{cases}$

16. $\begin{cases} x^2 - y^2 = 7 \\ x^2 + y^2 = 25 \end{cases}$ **17.** $\begin{cases} x^2 + 9y^2 = 33 \\ x^2 + y^2 = 25 \end{cases}$ **18.** $\begin{cases} x^2 + 5y^2 = 70 \\ 3x^2 - 5y^2 = 30 \end{cases}$

19. $\begin{cases} 4x^2 - y^2 = 4 \\ 4x^2 + \frac{5}{3}y^2 = 36 \end{cases}$ **20.** $\begin{cases} x^2 - 2y^2 = 17 \\ 2x^2 + y^2 = 54 \end{cases}$ **21.** $\begin{cases} 2x^2 - 3y^2 = 20 \\ x^2 + 2y = 20 \end{cases}$

22. $\begin{cases} 4x^2 + 3y^2 = 43 \\ 3x^2 - y^2 = 3 \end{cases}$ **23.** $\begin{cases} x^2 - 2y^2 = 1 \\ x^2 + 4y^2 = 25 \end{cases}$ **24.** $\begin{cases} 2x^2 - 5y^2 + 8 = 0 \\ x^2 - 7y^2 + 4 = 0 \end{cases}$

25. $\begin{cases} x^2 + 4y = 8 \\ x^2 + y^2 = 5 \end{cases}$ **26.** $\begin{cases} 3x - 2y = 9 \\ 9x = y^2 \end{cases}$ **27.** $\begin{cases} x^2 + y^2 = 16 \\ x^2 - y^2 = -34 \end{cases}$

28. $\begin{cases} x^2 - 4y^2 = -15 \\ -x^2 + 3y^2 = 11 \end{cases}$ **29.** $\begin{cases} x^2 + y^2 = 25 \\ (x - 5)^2 + y^2 = 9 \end{cases}$ **30.** $\begin{cases} x^2 - y = 0 \\ x^2 + (y - 6)^2 = 36 \end{cases}$

31. Find the dimensions of a rectangle whose area is 96 square centimeters and whose perimeter is 40 centimeters.

32. Suppose that the demand and supply curves of a certain product are given by the equations $p + 2q^2 = 8$ and $p - q = 5$, respectively, where p is the price and q is the quantity. Find the (point of) equilibrium of price and quantity (the point of intersection of the two curves where both p and q are nonnegative).

33. A woman receives $170 interest on a sum of money she lent for a year at simple annual interest. If the interest rate had been 1 percent higher, she would have received $238. What was the amount lent and what was the interest rate?

34. The sum of the squares of two positive numbers is 73 and the difference of their squares is 5. What are the numbers?

35. A rectangular parking lot has an area of 27,000 square feet. If a strip 10 feet wide is eliminated on each of the ends and on each of the sides, the available parking space is reduced to 20,800 square feet. What are the dimensions of the original parking lot?

REVIEW PROBLEM SET

In problems 1–12, let $f(x) = 2x + 3$, $g(x) = \sqrt{4 - x^2}$, and $h(x) = |x + 5|$. Find the following values.

1. $f(3)$

2. $g(0)$

3. $g(-2)$

4. $\sqrt{h(-14)}$

5. $h(-8)$

6. $f(u + v) - f(u)$

7. $[f(1)]^2$

8. $\dfrac{1}{g(0)}$

9. $f(u + 1)$

10. $4f(4)$

11. $h(2) - h(-2)$

12. $h(-5) - h(0)$

In problems 13–18, find the domain of the function determined by each equation.

13. $f(x) = 7x - 2$

14. $f(x) = 2x^2$

15. $f(x) = \dfrac{2}{(x - 1)(x + 2)}$

16. $f(x) = \sqrt[3]{3x + 1}$

17. $f(x) = \sqrt{3 - 2x}$

18. $f(x) = \dfrac{x}{\sqrt{2 - x}}$

In problems 19–22, determine whether the curve is the graph of a function.

19.

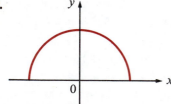

20.

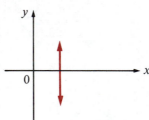

21.

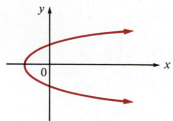

22.

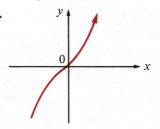

In problems 23–28, express y as a function of x, that is, $y = f(x)$, and graph the function.

23. If y is directly proportional to x and if $y = 8$ when $x = 12$.

24. If y is directly proportional to x^2 and if $y = 18$ when $x = 3$.

25. If y is directly proportional to $\sqrt{x}$ and if $y = 16$ when $x = 16$.

26. If y is directly proportional to $\sqrt{x}$ and if $y = 9$ when $x = 16$.

27. If y is inversely proportional to x and if $y = 4$ when $x = 5$.

28. If y is inversely proportional to x and if $y = 12$ when $x = \frac{3}{4}$.

29. Hooke's law states that the extension of an elastic spring beyond its natural length is directly proportional to the force applied. If a weight of 8 pounds causes a spring to stretch from a length of 9 inches to a length of 9.5 inches, what weight will cause it to stretch to a length of 1 foot?

30. The power required to operate a fan is directly proportional to the speed of the fan. If 1 horsepower will drive the fan at a speed of 480 revolutions per minute, how fast will 8 horsepower drive it? What power will be required to give the fan a speed of 600 revolutions per minute?

In problems 31–34, sketch the graph of each linear function. Find the slope of the graph and the domain and the range of the function.

31. $f(x) = 2x - 2$ **32.** $f(x) = \frac{3}{4}x + 7$ **33.** $f(x) = \frac{3}{2}x$ **34.** $f(x) = -3(x + 1) + 4$

In problems 35–40, find the vertex and the x and y intercepts of the graph of each quadratic function. Sketch the graph of the function and determine the domain and the range of the function.

35. $f(x) = 6x^2 - 5x - 4$ **36.** $f(x) = 2x^2 - x - 6$ **37.** $f(x) = x^2 + 6x + 9$

38. $f(x) = x^2 - 8x + 16$ **39.** $f(x) = -3 - 10x - 8x^2$ **40.** $f(x) = 10 + 3x - x^2$

In problems 41–46, sketch the graph of each exponential function. Indicate the domain and the range.

41. $f(x) = 5^x$ **42.** $f(x) = 7^x$ **43.** $f(x) = 3(2^x)$ **44.** $f(x) = -2(3^x)$

45. $f(x) = (\frac{1}{2})^x$ **46.** $f(x) = (\frac{1}{3})^x$

In problems 47–50, sketch the graph of each logarithmic function. Indicate the domain and the range.

47. $g(x) = \log_{1/2} x$ **48.** $g(x) = \log_2(x + 1)$ **49.** $f(x) = \log_3(x + 2)$ **50.** $f(x) = \log_3|x|$

In problems 51–54, let $f(x) = 4^x$. Find the values.

51. $f(0)$ **52.** $f(2)$ **53.** $f(-\frac{1}{2})$ **54.** $f(\frac{5}{2})$

In problems 55–58, let $f(x) = \log_{16} x$. Find the values.

55. $f(32)$ **56.** $f(64)$ **57.** $f(\sqrt[5]{2})$ **58.** $f(\sqrt[3]{4})$

In problems 59–66, sketch the graph of each equation. Find the x and y intercepts and identify the graph.

59. $2x^2 + 2y^2 = 50$ 60. $2x^2 + 3y^2 = 18$ 61. $9x^2 + 16y^2 = 36$ 62. $y^2 - x^2 = 4$

63. $y^2 = 4x$ 64. $x^2 = 2y$ 65. $16x^2 - 11y^2 = 64$ 66. $\dfrac{x^2}{10} + \dfrac{y^2}{10} = 1$

In problems 67–72, solve each system of equations algebraically. Check your solutions by sketching the graph of each equation and approximating the points of intersection.

67. $\begin{cases} 3x - 4y = 25 \\ x^2 + y^2 = 25 \end{cases}$ 68. $\begin{cases} 2x - y = 2 \\ x^2 + 2y^2 = 12 \end{cases}$ 69. $\begin{cases} x + y^2 = 6 \\ x^2 + y^2 = 36 \end{cases}$ 70. $\begin{cases} 3x^2 - 2y^2 = 27 \\ 7x^2 + 5y^2 = 63 \end{cases}$

71. $\begin{cases} x^2 + y^2 = 29 \\ x^2 - y^2 = 21 \end{cases}$ 72. $\begin{cases} 3x^2 - 2y^2 = 190 \\ 2x^2 + 5y^2 = 133 \end{cases}$

73. The diagonal of a rectangle is 34 feet and the perimeter is 92 feet. Find the dimensions of the rectangle.

74. Suppose that the demand and supply curves of a certain product are given by the equations $p - q^2 + 16q = 35$ and $p - q^2 - 4q = -5$, respectively, where p is the price and q is the quantity. Find the (point of) equilibrium of price and quantity (the point of intersection of the two curves where both p and q are nonnegative).

75. A rectangle has an area of 20 square centimeters and a perimeter of 18 centimeters. Find the length and width of the rectangle.

76. The sum of the squares of two numbers is 117, and the difference of the two numbers is 3. What are the numbers?

CHAPTER 10 TEST

1. Let $f(x) = 3x + 2$ and $g(x) = \sqrt{12 - x^2}$. Find the following values.
   (a) $f(-4)$ (b) $g(2)$ (c) $|f(-4)|$ (d) $f(x + y) - f(x)$

2. Find the domain of the function determined by each equation.
   (a) $f(x) = \dfrac{3}{x - 2}$ (b) $g(x) = \sqrt{5 - 3x}$

3. Express y as a function of x, that is, $y = f(x)$.
   (a) If y is directly proportional to x and $y = 15$ when $x = 3$.
   (b) If y is inversely proportional to x and $y = 4$ when $x = 2$.

4. Sketch the graph of each linear function. Find the slope of the graph and the domain and range of the function.
   (a) $f(x) = -3x + 4$ (b) $f(x) = \tfrac{1}{3}x - 2$

5. Find the vertex and the x and y intercepts of each quadratic function. Sketch the graph of the function and determine the domain and range of the function.

(a) $f(x) = 2x^2 + x - 6$ (b) $f(x) = -2 - 7x - 3x^2$

6. Sketch the graph of each function. Indicate the domain and range.

(a) $f(x) = 4^x$ (b) $f(x) = 2^{-x}$

(c) $f(x) = \log_2 x$ (d) $f(x) = \log_3(x + 1)$

7. Let $f(x) = 5^x$ and $g(x) = \log_8 x$. Find the values.

(a) $f(-2)$ (b) $f(0)$ (c) $g(64)$ (d) $g(16)$

8. Solve each system algebraically.

(a) $\begin{cases} x^2 + y^2 = 13 \\ x^2 - y^2 = 5 \end{cases}$ (b) $\begin{cases} x + y = 3 \\ x^2 + 2y^2 = 6 \end{cases}$

9. A rectangle has an area of 50 square feet and a perimeter of 30 feet. Find the length and width of the rectangle by using a system of equations.

11   Topics in Algebra

This chapter provides a brief introduction to some topics that supplement ideas presented in previous chapters. These topics include sequences, series, and the binomial theorem.

11.1 Sequences

A **sequence** is a function whose domain is the set of positive integers. We often represent the values of the function with the symbol a_n. We call these values the **terms** of the sequence. Thus, a_1 is the **first term,** a_2 is the **second term,** and a_n is the **general term,** or **nth term,** of the sequence.

The symbols

$$a_1, a_2, a_3, \ldots, a_n, \ldots$$

are typically used to denote the sequence. This notation suggests a never-ending list, in which the terms appear in order, with the nth term in the nth position for each positive integer n. The set of three dots is read "and so on." We also use the notation $\{a_n\}$ to denote the sequence whose nth term is a_n.

We can specify a particular sequence by giving a rule for determining the nth term a_n. This is often done by using a formula.

Examples of sequences are $\{a_n\}$ and $\{b_n\}$, where

$$a_n = n^2 - n \qquad \text{and} \qquad b_n = \frac{2}{n}.$$

EXAMPLE 1   Find the first five terms of each sequence.

(a) $a_n = (-1)^n$ (b) $a_n = 3 - \dfrac{1}{n}$

SOLUTION   To find the first five terms of each sequence we substitute the positive integers 1, 2, 3, 4, and 5 in turn for n in the formula for the general term.

(a) We have

$$a_1 = (-1)^1 = -1, \qquad a_2 = (-1)^2 = 1, \qquad a_3 = (-1)^3 = -1,$$
$$a_4 = (-1)^4 = 1, \qquad \text{and} \qquad a_5 = (-1)^5 = -1.$$

Thus, the first five terms of the sequence $\{a_n\}$ are $-1, 1, -1, 1,$ and -1.

(b) We have

$$a_1 = 3 - \frac{1}{1} = 2, \qquad a_2 = 3 - \frac{1}{2} = \frac{5}{2}, \qquad a_3 = 3 - \frac{1}{3} = \frac{8}{3},$$
$$a_4 = 3 - \frac{1}{4} = \frac{11}{4}, \qquad \text{and} \qquad a_5 = 3 - \frac{1}{5} = \frac{14}{5}.$$

Therefore, the first five terms of the sequence $\{a_n\}$ are $2, \frac{5}{2}, \frac{8}{3}, \frac{11}{4},$ and $\frac{14}{5}$.

Arithmetic and Geometric Sequences

We now consider special types of sequences. The first type we consider, the arithmetic sequence, is defined as follows:

DEFINITION 1   **Arithmetic Sequence**

A sequence $a_1, a_2, a_3, \ldots, a_n, \ldots$ is called an **arithmetic sequence** (or an **arithmetic progression**) if each term (after the first term) differs from the preceding term by a fixed amount.

For example, the sequence $\{a_n\}$, whose general term a_n is given by

$$a_n = 1 + 2n,$$

is an arithmetic sequence. The terms of $\{a_n\}$ are

$$a_1 = 3, \qquad a_2 = 5, \qquad a_3 = 7, \qquad a_4 = 9, \ldots, a_n = 1 + 2n, \ldots$$

Note that after the first term a_1, each term in the sequence is always 2 more than the preceding term:

$$a_1 = 3 \qquad\qquad\qquad\qquad = 3$$
$$a_2 = 3 + 2 \qquad\qquad\qquad = 3 + 1 \cdot 2$$
$$a_3 = (3 + 2) + 2 \qquad\qquad = 3 + 2 \cdot 2$$
$$a_4 = [(3 + 2) + 2] + 2 \qquad = 3 + 3 \cdot 2$$
$$\vdots$$
$$a_n = [(3 + 2) + 2] + 2 + \cdots + 2 = 3 + (n - 1) \cdot 2.$$

In general, a sequence

$$a_1, a_2, a_3, \ldots, a_n, \ldots$$

is an **arithmetic sequence** (or an **arithmetic progression**) if it can be expressed in the form

$$a_1, \quad a_1 + d, \quad a_1 + 2d, \quad a_1 + 3d, \ldots, a_1 + (n-1)d, \ldots$$

for every positive integer n. The number d is called the **common difference** associated with the arithmetic sequence. The nth term a_n of such a sequence is given by the following equation:

$$a_n = a_1 + (n-1)d.$$

EXAMPLE 2   Find the tenth term of the arithmetic progression whose first four terms are 2, -1, -4, and -7.

SOLUTION   The common difference is $d = -3$. We substitute $a_1 = 2, d = -3$, and $n = 10$ in the formula $a_n = a_1 + (n-1)d$:

$$
\begin{aligned}
a_{10} &= 2 + (10 - 1)(-3) \\
&= 2 - 27 \\
&= -25.
\end{aligned}
$$

EXAMPLE 3   If the third term of an arithmetic progression is 7 and the seventh term is 15, find the fifth term.

SOLUTION   We substitute $n = 3$ and $n = 7$ in the formula $a_n = a_1 + (n-1)d$. We use the fact that $a_3 = 7$ and $a_7 = 15$ to obtain the following system of linear equations in the unknowns a_1 and d:

$$
\begin{aligned}
7 &= a_1 + (3 - 1)d &\quad \text{or} \quad& a_1 + 2d = 7 \\
15 &= a_1 + (7 - 1)d &\quad \text{or} \quad& a_1 + 6d = 15.
\end{aligned}
$$

We solve for d and a_1:

$$4d = 8 \quad \text{or} \quad d = 2 \quad \text{and}$$
$$a_1 = 7 - 2(2) = 3.$$

Therefore,

$$
\begin{aligned}
a_5 &= 3 + (5 - 1)(2) \\
&= 3 + 8 \\
&= 11.
\end{aligned}
$$

Another important type of sequence is defined as follows:

DEFINITION 2   **Geometric Sequence**

> A sequence $a_1, a_2, a_3, \ldots, a_n, \ldots$ is called a **geometric sequence** (or a **geometric progression**) if each term (after the first) is obtained by multiplying the preceding term by a fixed amount.

For example, the sequence $\{a_n\}$ whose general term a_n is given by

$$a_n = 3(2)^{n-1}$$

is a geometric sequence. The terms of $\{a_n\}$ are:

$$3, 3(2), 3(2^2), 3(2^3), \ldots$$

Note that each term in this geometric sequence (after the first term) is obtained by multiplying the preceding term by 2.

In general, a sequence of the form

$$a_1, a_1 r, a_1 r^2, \ldots, a_1 r^{n-1}, \ldots$$

for every positive integer n is a **geometric sequence** or a **geometric progression.** The number r is called the **common ratio** associated with the geometric progression, and the nth term a_n is given by

$$a_n = a_1 r^{n-1}.$$

EXAMPLE 4   Find the tenth term of the geometric progression having the first term $a_1 = \frac{1}{2}$ and common ratio $r = \frac{1}{2}$.

SOLUTION   Substituting $a_1 = \frac{1}{2}$, $r = \frac{1}{2}$, and $n = 10$ in the formula

$$a_n = a_1 r^{n-1},$$

we have

$$a_{10} = \frac{1}{2}\left(\frac{1}{2}\right)^{10-1}$$
$$= \frac{1}{2}\left(\frac{1}{2}\right)^9 = \frac{1}{1,024}.$$

EXAMPLE 5   If the first two terms of a geometric sequence are 2 and 4, respectively, which term of the sequence is equal to 512?

SOLUTION   Since $a_1 = 2$ and $a_2 = 4$, we conclude that $r = 2$. If $a_n = 512$, then

$$512 = 2(2^{n-1})$$

or

$$2^9 = 2^{1+n-1} = 2^n,$$

so that

$$n = 9.$$

Therefore, the ninth term of the sequence is 512.

PROBLEM SET 11.1

In problems 1–6, find the first five terms in each sequence.

1. $a_n = \dfrac{n(n+2)}{2}$ **2.** $b_n = \dfrac{n+4}{n}$ **3.** $c_n = \dfrac{n(n-3)}{2}$ **4.** $a_n = \dfrac{3}{n(n+1)}$

5. $a_n = (-1)^n + 3$ **6.** $c_n = \dfrac{n^2 - 2}{2}$

In problems 7–14, determine which sequences are arithmetic progressions, and find the common difference d for each arithmetic progression.

7. 2, 5, 8, 11, . . .   **8.** 3, 5, 7, 9, . . .   **9.** 7, 12, 17, 22, . . .

10. $11a + 7b, 7a + 2b, 3a - 3b, . . .$ **11.** 67, 54, 41, 28, . . .   **12.** $9a^2, 16a^2, 23a^2, 30a^2, . . .$

13. 5.7, 6.9, 8.1, 9.3, . . .   **14.** 1.4, 4.5, 7.6, 10.7, . . .

15. Find the tenth and fifteenth terms of the arithmetic progression $-13, -6, 1, 8, ...$

16. Find the twelfth and thirty-fifth terms of the arithmetic progression 19, 17, 15, 13, ...

17. Find the sixth and ninth terms of the arithmetic progression $a + 24b, 4a + 20b, 7a + 16b, ...$

18. Find the third and sixteenth terms of the arithmetic progression $7a^2 - 4b, 2a^2 + 7b, -3a^2 + 18b, ...$

In problems 19–26, determine which sequences are geometric progressions and give the value of the common ratio r for each geometric progression.

19. 2, 6, 18, . . .   **20.** $1, \frac{1}{5}, \frac{1}{25}, . . .$ **21.** $1, -2, 4, . . .$ **22.** $\frac{4}{9}, \frac{1}{6}, \frac{1}{16}, . . .$

23. 81, 54, 36, . . .   **24.** $147, -21, 3, . . .$ **25.** $9, -6, 4, . . .$ **26.** $64, -32, 16, . . .$

In problems 27–32, find the indicated term of each geometric progression.

27. The tenth term of $-4, 2, -1, \frac{1}{2}, \ldots$

28. The eighth term of $\frac{1}{8}, \frac{1}{4}, \frac{1}{2}, \ldots$

29. The fifth term of $32, 16, 8, \ldots$

30. The eleventh term of $1, 1.03, (1.03)^2, \ldots$

31. The nth term of $1, 1 + a, (1 + a)^2, \ldots$

32. The twelfth term of $10^{-5}, 10^{-7}, 10^{-9}, \ldots$

33. Find the sixth and tenth terms of the geometric progression $6, 12, 24, 48, \ldots$.

34. Find the sixth and eighth terms of the geometric progression $2, 6, 18, \ldots$.

35. Find the fifth term of the geometric progression $3, 6, 12, \ldots$.

36. Find the eleventh term of the geometric progression $10, 10^2, 10^3, \ldots$.

37. If the first two terms of a geometric sequence are 2 and 1, respectively, which term of the sequence is equal to $\frac{1}{16}$?

38. If the first two terms of a geometric sequence are $3\sqrt{3}$ and 9, respectively, which term of the sequence is equal to $243\sqrt{3}$?

39. Suppose that you are offered a job with a starting salary of $1,000 per month and you are told that the average yearly raise is 6%. On this basis, what will your monthly salary be during the fifth year, if you accept the job offer?

40. Suppose that the population of a certain city increases at the rate of 5% per year and that the present population is 300,000. Find the population of the city at the end of the fifth year.

11.2 Series

Some applications of mathematics involve finding the sum of the terms of a sequence. This sum is called a **series.**

Summation Notation

Although the sum of the first n terms of a sequence

$$a_1, a_2, a_3, \ldots, a_n, \ldots$$

can be written as

$$a_1 + a_2 + a_3 + \cdots + a_n,$$

a more compact notation is useful. The Greek capital letter $\sum$ (sigma) is used for this purpose. We write the sum in **sigma notation** as

$$\sum_{k=1}^{n} a_k = a_1 + a_2 + a_3 + \cdots + a_n.$$

Here $\sum$ indicates a sum, and the symbols above and below the $\sum$ indicate that k is an integer from 1 to n inclusive; k is called the **index of summation.** There is no particular reason to use k for the index of summation—any letter will do; however, i, j, k, and n are the most commonly used indexes. For instance,

$$\sum_{k=1}^{n} 5^k = \sum_{i=1}^{n} 5^i$$

$$= \sum_{j=1}^{n} 5^j = 5^1 + 5^2 + 5^3 + \cdots + 5^n.$$

In Examples 1 and 2, evaluate each sum.

EXAMPLE 1 $\displaystyle\sum_{k=1}^{3} (4k^2 - 3k)$

SOLUTION   Here we have $a_k = 4k^2 - 3k$. To find the indicated sum, we substitute the integers 1, 2, and 3 for k in succession, and then add the resulting numbers. Thus,

$$\sum_{k=1}^{3} (4k^2 - 3k) = [4(1^2) - 3(1)] + [4(2^2) - 3(2)] + [4(3^2) - 3(3)]$$

$$= 1 + 10 + 27 = 38.$$

EXAMPLE 2 $\displaystyle\sum_{k=2}^{5} \frac{k-1}{k+1}$

SOLUTION   Here we have $a_k = (k - 1)/(k + 1)$, and

$$\sum_{k=2}^{5} \frac{k-1}{k+1} = \left(\frac{2-1}{2+1}\right) + \left(\frac{3-1}{3+1}\right) + \left(\frac{4-1}{4+1}\right) + \left(\frac{5-1}{5+1}\right)$$

$$= \frac{1}{3} + \frac{2}{4} + \frac{3}{5} + \frac{4}{6} = \frac{21}{10}.$$

The Sum of the First n Terms of Arithmetic and Geometric Sequences

We now derive formulas for the sum of the first n terms of an arithmetic sequence or a geometric sequence. We begin with an arithmetic progression. Let

$$a_1, a_2, a_3, \ldots, a_n, \ldots$$

be an arithmetic sequence with a first term a_1 and a common difference d. Let S_n represent the sum of the first n terms, that is,

$$S_n = \sum_{k=1}^{n} a_k.$$

The sum can be written out as

$$S_n = a_1 + (a_1 + d) + (a_1 + 2d) + (a_1 + 3d) + \cdots + [a_1 + (n-3)d]$$
$$+ [a_1 + (n-2)d] + [a_1 + (n-1)d].$$

If we reverse the order of these terms, we have

$$S_n = [a_1 + (n-1)d] + [a_1 + (n-2)d] + [a_1 + (n-3)d] + \cdots + (a_1 + 3d)$$
$$+ (a_1 + 2d) + (a_1 + d) + a_1.$$

Now observe what happens when we add the two representations of S_n term by term:

$$
\begin{array}{llll}
S_n = & a_1 & + & (a_1 + d) & + \cdots + [a_1 + (n-2)d] & + & [a_1 + (n-1)d] \\
+ S_n = & [a_1 + (n-1)d] & + & [a_1 + (n-2)d] & + \cdots + (a_1 + d) & + & a_1 \\
\hline
2S_n = & [2a_1 + (n-1)d] & + & [2a_1 + d + (n-2)d] & + \cdots + [2a_1 + d + (n-2)d] & + & [2a_1 + (n-1)d]
\end{array}
$$

$$\underbrace{}_{n \text{ times}}$$

$$= [2a_1 + (n-1)d] + [2a_1 + (n-1)d] + \cdots + [2a_1 + (n-1)d] + [2a_1 + (n-1)d]$$

$$2S_n = n[2a_1 + (n-1)d].$$

We divide both sides of the equation by 2:

$$S_n = \frac{n}{2}[2a_1 + (n-1)d].$$

Therefore, the formula for the sum of the first n terms of an arithmetic sequence is:

$$S_n = \sum_{k=1}^{n} a_k$$
$$= \frac{n}{2}[2a_1 + (n-1)d].$$

Using the fact that $a_n = a_1 + (n-1)d$, we can rewrite this formula as follows:

$$S_n = \frac{n}{2}[2a_1 + (n-1)d]$$

$$= \frac{n}{2}[a_1 + a_1 + (n-1)d]$$

$$= \frac{n}{2}(a_1 + a_n).$$

Therefore, an alternate form of the formula for the sum of the first n terms of an arithmetic sequence is:

$$S_n = \sum_{k=1}^{n} a_k$$
$$= \frac{n}{2}(a_1 + a_n).$$

EXAMPLE 3   Find the sum of the first 20 terms of an arithmetic sequence whose first term is 2 and whose common difference is 4.

SOLUTION   We substitute $a_1 = 2$, $d = 4$, and $n = 20$ in $S_n = (n/2)[2a_1 + (n-1)d]$:

$$S_{20} = \frac{20}{2}[2(2) + (20-1)4]$$
$$= 10(4 + 76)$$
$$= 800.$$

EXAMPLE 4   The sum of the first 10 terms of an arithmetic sequence is 351 and the tenth term is 51. Find the first term and the common difference.

SOLUTION   Using the formula $S_n = (n/2)(a_1 + a_n)$, we have

$$S_{10} = \frac{10}{2}(a_1 + a_{10}).$$

We know the values of S_{10} and a_{10}, so the equation is

$$351 = \frac{10}{2}(a_1 + 51) = 5a_1 + 255.$$

We solve for a_1:

$$5a_1 = 96 \quad \text{or} \quad a_1 = 19.2.$$

To solve for d, we use the formula $a_n = a_1 + (n-1)d$:

$$a_{10} = a_1 + 9d.$$

Thus,

$$51 = 19.2 + 9d, \quad 9d = 31.8, \quad \text{and} \quad d = \frac{31.8}{9} = \frac{53}{15}.$$

Therefore, the first term is 19.2 and the common difference is $\frac{53}{15}$.

EXAMPLE 5   How many terms are there in the arithmetic sequence for which $a_1 = 3$, $d = 5$, and $S_n = 255$?

SOLUTION   Using the formula

$$S_n = \frac{n}{2}(a_1 + a_n),$$

we have

$$255 = \frac{n}{2}(3 + a_n) \qquad \text{or} \qquad 510 = n(3 + a_n).$$

We can obtain another equation in the variables n and a_n by using the formula

$$a_n = a_1 + (n - 1)d.$$

We substitute the given values for a_1 and d in this formula:

$$a_n = 3 + (n - 1)5 = 5n - 2.$$

We can now solve the system

$$\begin{cases} 510 = n(3 + a_n) \\ a_n = 5n - 2 \end{cases}$$

by using the substitution method:

$$510 = n[3 + (5n - 2)] = n(5n + 1) = 5n^2 + n$$

or

$$5n^2 + n - 510 = 0.$$

This equation can be factored as

$$(5n + 51)(n - 10) = 0.$$

This gives us $n = 10$ or $n = -\frac{51}{5}$. Hence, the sequence has 10 terms (n must be a positive integer).

We now consider a geometric sequence with first term a_1 and a common ratio r,

$$a_1, a_1r, a_1r^2, \ldots, a_1r^{n-1}, \ldots.$$

To find a formula for the sum

$$S_n = \sum_{k=1}^{n} a_1 r^{k-1},$$

we start with the expression for S_n in expanded form:

$$S_n = a_1 + a_1r + a_1r^2 + \cdots + a_1r^{n-1}.$$

We then multiply both sides of the equation by r:

$$rS_n = a_1 r + a_1 r^2 + a_1 r^3 + \cdots + a_1 r^n.$$

Next we subtract rS_n from S_n:

$$S_n - rS_n = a_1 - a_1 r^n.$$

So the equation becomes

$$(1 - r)S_n = a_1(1 - r^n) \qquad \text{or} \qquad S_n = \frac{a_1(1 - r^n)}{1 - r}.$$

Therefore, the formula for the sum of the first n terms of a geometric sequence is:

$$S_n = \sum_{k=1}^{n} a_1 r^{k-1} = \frac{a_1(1 - r^n)}{1 - r}, \qquad r \neq 1.$$

EXAMPLE 6   Find the sum of the first 10 terms of the geometric sequence whose first term is $\frac{1}{2}$ and whose common ratio is 2.

SOLUTION   We substitute $n = 10$, $a_1 = \frac{1}{2}$, and $r = 2$ in the formula

$$S_n = \frac{a_1(1 - r^n)}{1 - r}$$

and obtain

$$S_{10} = \frac{\frac{1}{2}(1 - 2^{10})}{1 - 2} = \frac{\frac{1}{2}(-1{,}023)}{-1} = 511.5.$$

EXAMPLE 7   The sum of the first 5 terms of a geometric sequence is $2\frac{7}{27}$ and the common ratio is $-\frac{1}{3}$. Find the first four terms of the sequence.

SOLUTION   Using the formula for S_n, we have

$$2\frac{7}{27} = \frac{a_1[1 - (-\frac{1}{3})^5]}{1 - (-\frac{1}{3})}$$

so that

$$2\frac{7}{27} = \left(\frac{\frac{244}{243}}{\frac{4}{3}}\right)a_1, \qquad \frac{61}{27} = \frac{61}{81}a_1, \qquad \text{and} \qquad a_1 = \frac{\frac{61}{27}}{\frac{61}{81}} = 3.$$

Hence, the first four terms of the sequence are 3, $3(-\frac{1}{3})$, $3(-\frac{1}{3})^2$, and $3(-\frac{1}{3})^3$, or 3, -1, $\frac{1}{3}$, and $-\frac{1}{9}$.

PROBLEM SET 11.2

In problems 1–14, evaluate each sum.

1. $\displaystyle\sum_{k=1}^{5} k$

2. $\displaystyle\sum_{k=0}^{4} \frac{2^4}{k+1}$

3. $\displaystyle\sum_{i=1}^{10} 2i(i-1)$

4. $\displaystyle\sum_{k=0}^{4} 3^{2k}$

5. $\displaystyle\sum_{k=2}^{5} 2^{k-2}$

6. $\displaystyle\sum_{i=2}^{6} \frac{1}{i}$

7. $\displaystyle\sum_{k=1}^{3} (2k+1)$

8. $\displaystyle\sum_{k=1}^{5} (3k^2 - 5k + 1)$

9. $\displaystyle\sum_{i=1}^{4} \frac{i}{i+1}$

10. $\displaystyle\sum_{k=1}^{4} k^k$

11. $\displaystyle\sum_{k=1}^{100} 5$

12. $\displaystyle\sum_{i=3}^{7} (i+2)$

13. $\displaystyle\sum_{k=1}^{5} \frac{1}{k(k+1)}$

14. $\displaystyle\sum_{k=1}^{4} \frac{3}{k}$

15. Find the sum of the first 10 terms of an arithmetic sequence whose first term is 1 and whose common difference is 3.

16. Find the sum of the first 15 terms of an arithmetic sequence whose first term is $\frac{1}{2}$ and whose common difference is $\frac{1}{2}$.

17. Find the sum of the first 8 terms of an arithmetic sequence whose first term is -5 and whose common difference is $\frac{3}{7}$.

18. Find the sum of the first 12 terms of an arithmetic sequence whose first term is 11 and whose common difference is -2.

19. Find S_7 for the arithmetic sequence $6, 3b + 1, 6b - 4, \ldots$.

20. Find S_{10} for the arithmetic sequence $x + 2y, 3y, -x + 4y, \ldots$.

In problems 21–26, certain elements of an arithmetic sequence are given. Find the indicated unknown elements.

21. $a_1 = 6$; $d = 3$; a_{10} and S_{10}

22. $a_1 = 38$; $d = -2$; $n = 25$; S_n

23. $a_1 = 17$; $S_{18} = 2{,}310$; d and a_{18}

24. $d = 3$; $S_{25} = 400$; a_1 and a_{25}

25. $a_1 = 27$; $a_n = 48$; $S_n = 1{,}200$; n and d

26. $a_1 = -3$; $d = 2$; $S_n = 140$; n

27. Find the sum of the first 6 terms of the geometric sequence whose first term is $\frac{3}{2}$ and whose common ratio is 2.

28. Find the sum of the first 10 terms of the geometric sequence whose first term is 6 and whose common ratio is $\frac{1}{2}$.

29. Find the sum of the first 12 terms of the geometric sequence whose first term is -4 and whose common ratio is -2.

30. Find the sum of the first 8 terms of the geometric sequence whose first term is 5 and whose common ratio is $-\frac{1}{2}$.

31. Find S_6 for the geometric sequence $10, 10a, 10a^2, 10a^3, \ldots$.

32. Find S_8 for the geometric sequence $k, \dfrac{k}{b}, \dfrac{k}{b^2}, \dfrac{k}{b^3}, \ldots$.

In problems 33–40, find the indicated element in each geometric progression with the given elements.

33. $a_1 = 2$; $n = 3$; $S_n = 26$; r

34. $r = 2$; $n = 5$; $a_n = -48$; a_1 and S_n

35. $a_1 = 3$; $a_n = 192$; $n = 7$; r

36. $a_6 = 3$; $a_9 = -81$; r and a_1

37. $a_5 = \frac{1}{8}$; $r = -\frac{1}{2}$; a_9 and S_8

38. $a_1 = 1$; $r = (1.03)^{-1}$; $a_9 = (1.03)^{-8}$; S_8

39. $a_1 = \frac{1}{16}$; $r = 2$; $a_n = 32$; n and S_n

40. $a_1 = 250$; $r = \frac{3}{5}$; $a_n = 32\frac{2}{5}$; n and S_n

11.3 The Binomial Theorem

In Chapter 2, Section 2.4, we considered the special products $(a + b)^2$ and $(a + b)^3$. We often work with expressions of the form $(a + b)^n$, where n is a positive integer. Since the expression $a + b$ is a binomial, the formula for expanding $(a + b)^n$ is called the **binomial theorem.**

We can expand $(a + b)^n$, for small values of n, by using direct calculation. For instance,

$$(a + b)^1 = a + b$$
$$(a + b)^2 = a^2 + 2ab + b^2$$
$$(a + b)^3 = a^3 + 3a^2b + 3ab^2 + b^3$$
$$(a + b)^4 = a^4 + 4a^3b + 6a^2b^2 + 4ab^3 + b^4$$
$$(a + b)^5 = a^5 + 5a^4b + 10a^3b^2 + 10a^2b^3 + 5ab^4 + b^5$$

This pattern holds for the expansion of $(a + b)^n$, where n is any positive integer. The following rules are used for this expansion:

1. There are $n + 1$ terms. The first term is a^n and the last term is b^n.

2. The powers of a decrease by 1 and the powers of b increase by 1 for each term. The sum of the exponents of a and b is n for each term.

One way to display the coefficients in the expansion of $(a + b)^n$ for $n = 1, 2, 3, \ldots$ is the following array of numbers, known as **Pascal's triangle:**

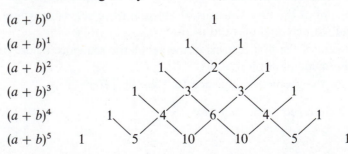

$(a + b)^0$	1
$(a + b)^1$	1 1
$(a + b)^2$	1 2 1
$(a + b)^3$	1 3 3 1
$(a + b)^4$	1 4 6 4 1
$(a + b)^5$	1 5 10 10 5 1

The first and last numbers in each line are always 1. The other numbers can be found by adding the pair of numbers from the preceding line, as indicated by the V's. For example,

indicates that 10 was obtained by adding 4 and 6.

It is easier to detect the pattern for determining the coefficients in the expansion of $(a + b)^n$ if we use the following notation for the product of all positive integers from 1 to n inclusive. The symbol $n!$ (read "n factorial" or "factorial n") is defined by:

$$n! = 1 \cdot 2 \cdot 3 \cdots (n - 1)n$$

or

$$n! = n(n - 1)(n - 2) \cdots 2 \cdot 1.$$

Thus, $4! = 4 \cdot 3 \cdot 2 \cdot 1 = 24$ and $6! = 6 \cdot 5 \cdot 4 \cdot 3 \cdot 2 \cdot 1 = 720$.

We have defined $n!$ for positive integers n as

$$n! = n(n - 1)(n - 2) \cdots 4 \cdot 3 \cdot 2 \cdot 1.$$

Therefore,

$$(n - 1)! = (n - 1)(n - 2)(n - 3) \cdots 4 \cdot 3 \cdot 2 \cdot 1.$$

If we multiply both sides of this equation by n, we find that

$$n! = n(n - 1)! \qquad \text{if} \qquad n \neq 1.$$

We want this relationship to hold for $n = 1$:

$$1! = 1(1 - 1)! \qquad \text{or} \qquad 1! = 1 \cdot 0!;$$

therefore, we define

$$0! = 1.$$

Table 1

$0! = 1$	
$1! = 1$	
$2! = 2$	
$3! = 6$	
$4! = 24$	
$5! = 120$	
$6! = 720$	
$7! = 5040$	

We can use this definition together with the preceding formula, $n! = n(n - 1)!$, to make the table of values of $n!$ shown in Table 1.

Expressions involving factorial notation may be simplified as follows:

EXAMPLE 1   Simplify:

(a) $\dfrac{7!}{5!}$ (b) $\dfrac{8!}{3! \cdot 5!}$ (c) $\dfrac{(n + 1)!}{(n - 1)!}$

SOLUTION   (a) $\dfrac{7!}{5} = \dfrac{7 \cdot 6 \cdot 5 \cdot 4 \cdot 3 \cdot 2 \cdot 1}{5 \cdot 4 \cdot 3 \cdot 2 \cdot 1} = 7 \cdot 6 = 42$

(b) $\dfrac{8!}{3! \cdot 5!} = \dfrac{8 \cdot 7 \cdot 6 \cdot 5 \cdot 4 \cdot 3 \cdot 2 \cdot 1}{(3 \cdot 2 \cdot 1)(5 \cdot 4 \cdot 3 \cdot 2 \cdot 1)} = 8 \cdot 7 = 56$

(c) $\dfrac{(n + 1)!}{(n - 1)!} = \dfrac{(n + 1)(n)[(n - 1)!]}{(n - 1)!} = (n + 1)n = n^2 + n$

It is helpful to introduce factorial notation to provide a general description of the binomial expansion $(a + b)^n$ without relying on Pascal's triangle. The variables in the expansion of $(a + b)^n$ have the following pattern:

$$a^n, a^{n-1}b, a^{n-2}b^2, a^{n-3}b^3, \ldots, ab^{n-1}, b^n.$$

Notice that the sum of the exponents of a and b is n for each term. In addition:

1. The first term is a^n, and the coefficient is 1.

2. The second term contains $a^{n-1}b$, and the coefficient is

$$\frac{n}{1!}.$$

3. The third term contains $a^{n-2}b^2$, and the coefficient is

$$\frac{n(n - 1)}{2!}.$$

4. The fourth term contains $a^{n-3}b^3$, and the coefficient is

$$\frac{n(n - 1)(n - 2)}{3!}.$$

The following formula provides the general expansion of $(a + b)^n$:

THEOREM 1   **The Binomial Theorem**

$$(a + b)^n = a^n + \frac{n}{1!} a^{n-1}b + \frac{n(n - 1)}{2!} a^{n-2}b^2$$

$$+ \frac{n(n - 1)(n - 2)}{3!} a^{n-3}b^3 + \cdots$$

$$+ \frac{n(n - 1)(n - 2) \cdots (n - k + 2)}{(k - 1)!} a^{n-k+1}b^{k-1} + \cdots + b^n,$$

where k is an integer such that $1 \leq k \leq n + 1$.

For example, if we substitute x for a, $2y^2$ for b, and 5 for n in the preceding theorem, we have

$$(x + 2y^2)^5 = [x + (2y^2)]^5$$

$$= x^5 + \frac{5}{1!} x^4(2y^2) + \frac{5 \cdot 4}{2!} x^3(2y^2)^2 + \frac{5 \cdot 4 \cdot 3}{3!} x^2(2y^2)^3$$

$$+ \frac{5 \cdot 4 \cdot 3 \cdot 2}{4!} x(2y^2)^4 + \frac{5 \cdot 4 \cdot 3 \cdot 2 \cdot 1}{5!} (2y^2)^5$$

$$= x^5 + 10x^4y^2 + 40x^3y^4 + 80x^2y^6 + 80xy^8 + 32y^{10}.$$

Note that the kth term of the binomial expansion $(a + b)^n$, denoted by u_k, is given by the equation

$$u_k = \frac{n(n - 1)(n - 2) \cdots (n - k + 2)}{(k - 1)!} a^{n-k+1}b^{k-1}.$$

This information will help us to determine a particular term of the binomial expansion or to find the term where b has a particular exponent. For example, the sixth term u_6 of $(x^2 + 2y)^{12}$ is

$$\frac{12 \cdot 11 \cdot 10 \cdot 9 \cdot 8}{5 \cdot 4 \cdot 3 \cdot 2 \cdot 1} (x^2)^7(2y)^5 = 25{,}344x^{14}y^5.$$

If we replace $k - 1$ by k in the expression for u_k, we obtain an expression for the $(k + 1)$th term u_{k+1} that contains the factor b^k:

$$u_{k+1} = \frac{n(n - 1)(n - 2) \cdots (n - k + 1)}{k!} a^{n-k}b^k$$

$$= \frac{n(n - 1)(n - 2) \cdots (n - k + 1)(n - k)!}{k!(n - k)!} a^{n-k}b^k \quad \text{[We multiplied the numerator and denominator of the coefficient by } (n - k)!.]}$$

$$= \frac{n(n - 1)(n - 2) \cdots (n - k + 1)(n - k)(n - k - 1)(n - k - 2) \cdots 3 \cdot 2 \cdot 1}{k!(n - k)!} a^{n-k}b^k$$

$$= \frac{n!}{k!(n - k)!} a^{n-k}b^k.$$

Thus:

$$u_{k+1} = \frac{n!}{k!(n - k)!} a^{n-k}b^k.$$

For example, to find the term involving y^4 in the expansion of $(x^2 + 2y)^{12}$, we have $b^k = (2y)^k$. We choose a value of k that will give us the variable factor y^4: $k = 4$. We substitute 4 for k in the preceding formula:

$$u_5 = \frac{12!}{4!8!}(x^2)^8(2y)^4 = \frac{12 \cdot 11 \cdot 10 \cdot 9 \cdot 8!}{4!8!}x^{16}(16y^4)$$

$$= \frac{12 \cdot 11 \cdot 10 \cdot 9}{4 \cdot 3 \cdot 2 \cdot 1}(16)x^{16}y^4$$

$$= 7{,}920x^{16}y^4.$$

EXAMPLE 2   Expand $(x + y)^7$.

SOLUTION   We substitute $a = x$, $b = y$, and $n = 7$ in the binomial theorem:

$$(x + y)^7 = x^7 + \frac{7}{1!}x^6y + \frac{7 \cdot 6}{2!}x^5y^2 + \frac{7 \cdot 6 \cdot 5}{3!}x^4y^3 + \frac{7 \cdot 6 \cdot 5 \cdot 4}{4!}x^3y^4$$

$$+ \frac{7 \cdot 6 \cdot 5 \cdot 4 \cdot 3}{5!}x^2y^5 + \frac{7 \cdot 6 \cdot 5 \cdot 4 \cdot 3 \cdot 2}{6!}xy^6 + y^7$$

$$= x^7 + 7x^6y + 21x^5y^2 + 35x^4y^3 + 35x^3y^4 + 21x^2y^5 + 7xy^6 + y^7.$$

EXAMPLE 3   Find the eighth term in the expansion of $(x - y)^{12}$.

SOLUTION   The kth term u_k of $(a + b)^n$ is given by

$$u_k = \frac{n(n - 1)(n - 2) \cdots (n - k + 2)}{(k - 1)!}a^{n-k+1}b^{k-1}.$$

Substituting $a = x$, $b = -y$, $k = 8$, and $n = 12$ in the above formula, we have

$$n - k + 2 = 12 - 8 + 2 = 6$$

$$n - k + 1 = 12 - 8 + 1 = 5$$

$$k - 1 = 8 - 1 = 7.$$

Therefore,

$$u_8 = \frac{12 \cdot 11 \cdot 10 \cdot 9 \cdot 8 \cdot 7 \cdot 6}{7!}x^5(-y)^7$$

$$= -792x^5y^7.$$

EXAMPLE 4   Find and simplify the term involving x^7 in the expansion of $(2 - x)^{12}$.

SOLUTION   We use the formula for the $(k + 1)$th term,

$$u_{k+1} = \frac{n!}{k!(n - k)!}a^{n-k}b^k,$$

with $k = 7$, $a = 2$, $b = -x$, and $n = 12$. We have

$$\frac{12!}{7!(12-7)!} (2)^5(-x)^7 = \frac{\cancel{12} \cdot 11 \cdot \cancel{10} \cdot 9 \cdot 8 \cdot \cancel{7!}}{\cancel{7!} \cdot \cancel{5} \cdot \cancel{4} \cdot \cancel{3} \cdot 2 \cdot 1} (-32x^7)$$
$$= -25{,}344x^7.$$

PROBLEM SET 11.3

In problems 1–10, write each expression in expanded form and simplify the results.

1. $\dfrac{4!}{6!}$ **2.** $\dfrac{10!}{5! \cdot 7!}$ **3.** $\dfrac{2!}{4! - 3!}$ **4.** $\dfrac{1}{4!} + \dfrac{1}{3!}$ **5.** $\dfrac{3! \cdot 8!}{4! \cdot 7!}$

6. $\dfrac{4! \cdot 6!}{8! - 5!}$ **7.** $\dfrac{0}{0!}$ **8.** $\dfrac{(n-2)!}{(n+1)!}$ **9.** $\dfrac{(n+1)!}{(n-3)!}$ **10.** $\dfrac{(n+k)!}{(n+k-2)!}$

In problems 11–18, expand each expression by using the binomial theorem and simplify each term.

11. $(x + 2)^5$ **12.** $(a - 2b)^4$ **13.** $(x^2 + 4y^2)^3$ **14.** $(1 - a^{-1})^5$

15. $(a^3 - a^{-1})^6$ **16.** $\left(1 - \dfrac{x}{y^2}\right)^5$ **17.** $\left(2 + \dfrac{x}{y}\right)^5$ **18.** $(x + y + z)^3$

In problems 19–22, use the binomial theorem to expand each expression and check the results using Pascal's triangle.

19. $(2z + x)^8$ **20.** $(x - 3)^8$ **21.** $(y^2 - 2x)^5$ **22.** $\left(\dfrac{1}{a} + \dfrac{x}{2}\right)^5$

In problems 23–26, find the first four terms of each expansion.

23. $(x^2 - 2a)^{10}$ **24.** $\left(2a - \dfrac{1}{b}\right)^6$ **25.** $\left(\sqrt{\dfrac{x}{2}} + 2y\right)^7$ **26.** $\left(\dfrac{1}{a} + \dfrac{x}{2}\right)^{11}$

In problems 27–34, find the first five terms in each expansion and simplify.

27. $(x + y)^{16}$ **28.** $(a^2 + b^2)^{12}$ **29.** $(a - 2b^2)^{11}$ **30.** $(a + 2y^2)^8$

31. $(x - 2y)^7$ **32.** $\left(1 - \dfrac{x}{y^2}\right)^8$ **33.** $(a^3 - a^2)^9$ **34.** $\left(x + \dfrac{1}{2y}\right)^{15}$

In problems 35–40, find the indicated term for each expression.

35. $\left(\dfrac{x^2}{2} + a\right)^{15}$, fourth term   **36.** $(y^2 - 2z)^{10}$, sixth term

37. $\left(2x^2 - \dfrac{a^2}{3}\right)^9$, seventh term   **38.** $(x + \sqrt{a})^{12}$, middle term

39. $\left(a + \dfrac{x^3}{3}\right)^9$, term containing x^{12} **40.** $\left(2\sqrt{y} - \dfrac{x}{2}\right)^{10}$, term containing y^4

REVIEW PROBLEM SET

In problems 1–6, write the first four terms of each sequence.

1. $a_n = 3 + (-1)^{n+1}$ **2.** $b_n = 2^n$ **3.** $c_n = 5 - \dfrac{3}{n}$ **4.** $a_n = \dfrac{2}{n+1}$

5. $a_n = \dfrac{n(4n+1)}{5}$ **6.** $c_n = (-1)^n 2^{n-1}$

In problems 7–12, for each arithmetic progression, find the indicated term and the indicated sum.

7. $4, 9, 14, \ldots$; ninth term and S_9
9. $42, 39, 36, \ldots$; eleventh term and S_{11}
11. $\frac{1}{6}, \frac{1}{3}, \frac{1}{2}, \ldots$; twenty-fourth term and S_{24}

8. $21, 19, 17, \ldots$; tenth term and S_{10}
10. $0.3, 1.2, 2.1, \ldots$; fifteenth term and S_{15}
12. $\frac{1}{6}, \frac{1}{4}, \frac{1}{3}, \ldots$; thirtieth term and S_{30}

In problems 13–16, find the value of x so that the terms given will be the first three terms of an arithmetic progression.

13. $2, 1 + 2x, 21 - 3x, \ldots$
15. $3x, 2x + 1, x^2 - 4, \ldots$

14. $2x, \frac{1}{2}x + 3, 3x - 10. \ldots$
16. $1, x + 1, 3x - 5, \ldots$

In problems 17–22, for each geometric progression, find the indicated term and the indicated sum.

17. $3, 12, 48, \ldots$; eighth term and S_8
19. $81, -27, 9, \ldots$; sixth term and S_6
21. $3, -3\sqrt{2}, 6, \ldots$; eighteenth term and S_{18}

18. $16, 8, 4, \ldots$; ninth term and S_9
20. $\sqrt{2}, 2, 2\sqrt{2}, \ldots$; tenth term and S_{10}
22. $2, -2\sqrt{2}, 4, \ldots$; twentieth term and S_{20}

In problems 23–26, determine the value of x so that the three terms given will be the first three terms of a geometric progression.

23. $x - 6, x + 6, 2x + 2, \ldots$
25. $x - 7, x + 5, 8x - 5, \ldots$

24. $\frac{1}{2}x, x + 2, 3x + 1, \ldots$
26. $x + 1, x + 2, x - 3, \ldots$

In problems 27–32, evaluate each sum.

27. $\displaystyle\sum_{k=1}^{5} k(2k - 1)$ **28.** $\displaystyle\sum_{k=5}^{10} (2k - 1)^2$ **29.** $\displaystyle\sum_{k=1}^{4} 2k^2(k - 3)$ **30.** $\displaystyle\sum_{k=1}^{6} 3^{k+1}$

31. $\displaystyle\sum_{k=2}^{6} (k + 1)(k + 2)$ **32.** $\displaystyle\sum_{k=4}^{7} \dfrac{1}{k(k - 3)}$

In problems 33–44, use the binomial theorem to expand each expression.

33. $(x + 2y)^4$ **34.** $(x - 3y)^4$ **35.** $(1 + x)^5$ **36.** $(2x + 1)^5$

37. $(1 - 2x)^6$ **38.** $(a - b)^6$ **39.** $(3x + y)^4$ **40.** $\left(x - \dfrac{1}{x}\right)^8$

41. $(3x + \sqrt{x})^5$ **42.** $\left(3y + \dfrac{1}{3\sqrt{y}}\right)^6$ **43.** $\left(2x + \dfrac{1}{y}\right)^3$ **44.** $\left(x^3 - \dfrac{1}{\sqrt{x}}\right)^9$

In problems 45–50, find the indicated term in each binomial expansion.

45. fifth term of $(x + y)^{10}$ **46.** sixth term of $(x - y)^{11}$ **47.** fifth term of $(2x + y)^{10}$

48. sixth term of $(x - 3y)^9$ **49.** fourth term of $(3x + y)^{11}$ **50.** third term of $(2x + y)^{20}$

CHAPTER 11 TEST

1. Write the first five terms of the sequence

$$a_n = \frac{n}{n + 1}.$$

2. Find the twelfth term and S_{12} for the arithmetic progression $3, 1, -1, \ldots$.

3. Find the value of x so that $3, 2 + 3x, 15 - 2x, \ldots$ will be the first three terms of an arithmetic progression.

4. Find the seventh term and S_7 for the geometric progression $16, -8, 4, \ldots$.

5. Determine the value of x so that $x + 3, x + 2, x - 1, \ldots$ will be the first three terms of a geometric progression.

6. Evaluate the sum

$$\sum_{k=1}^{4} (k + 1)(k + 3).$$

7. Use the binomial theorem to expand $(2x + y)^5$.

8. Find the fourth term of $(x - 3y)^9$.

12 Geometry

The word *geometry* comes from the Greek words *ge* and *metria*, which mean "earth measuring." The ancient Egyptians, Greeks, and Romans used geometry for surveying, navigation, astronomy, and other practical applications. The Greeks systematized the facts of geometry, and the Greek mathematician Euclid codified the system about 325 B.C. in his textbook, the *Elements*.

In this chapter, we present a concise review of geometry. Not intended to be mathematically rigorous, this material is designed to provide you with some intuitive understanding of the concepts of geometry and to acquaint you with its terminology and methods.

12.1 Basic Elements of Geometry

The basic elements of geometry are *points*, *lines*, and *planes*. We accept these elements as concepts with which we are familiar.

Intuitively, a **point** in geometry is suggested by a pinpoint made on a piece of paper by a sharp pencil. Points are designated by capital letters, such as A, B, and C. The entire set of points is called a **plane,** and certain subsets of the plane are called **lines.** The following properties characterize lines.

1. A **line** may be determined by any two distinct points. For example, the line in Figure 1 is labeled as $\overleftrightarrow{AB}$ or AB.

Figure 1

2. If two pairs of points determine the same line, the lines are said to be **equal.** For example, in Figure 2, $\overleftrightarrow{AB} = \overleftrightarrow{CD}$; also $\overleftrightarrow{BC} = \overleftrightarrow{CB}$.

Figure 2

3. If A and B are different points on a line, the portion of the line that starts at A (and includes A) and continues indefinitely through B is called a **ray** with **endpoint** A (Figure 3). It is labeled $\overrightarrow{AB}$.

Figure 3

A B

4. A **line segment** is part of a line that contains two points on a line and all the points between them. Figure 4 shows the line segment that contains A and B. It is labeled $\overline{AB}$. Recall from Section 7.1 that when referring to the length of a line segment $\overline{AB}$, we use the designation $|\overline{AB}|$.

Figure 4 $A \bullet$————————————$\bullet B$

EXAMPLE 1

(a) Is line AB in Figure 2 the same as line BA?
(b) Do the rays $\overrightarrow{BD}$ and $\overrightarrow{CA}$ in Figure 2 intersect? If so, name the intersection.

SOLUTION

(a) Yes, because a line is determined by two points.
(b) Yes; the intersection of the rays $\overrightarrow{BD}$ and $\overrightarrow{CA}$ is the line segment $\overline{BC}$.

Figures 5 and 6 illustrate some facts about lines:

1. Two different lines p and ℓ are said to be **intersecting** lines if and only if they have exactly one point in common (Figure 5).

2. Two different lines p and ℓ that have no points in common are called **parallel** lines (Figure 6). The notation we use for this fact is the symbol $p \parallel \ell$.

Figure 5   **Figure 6**

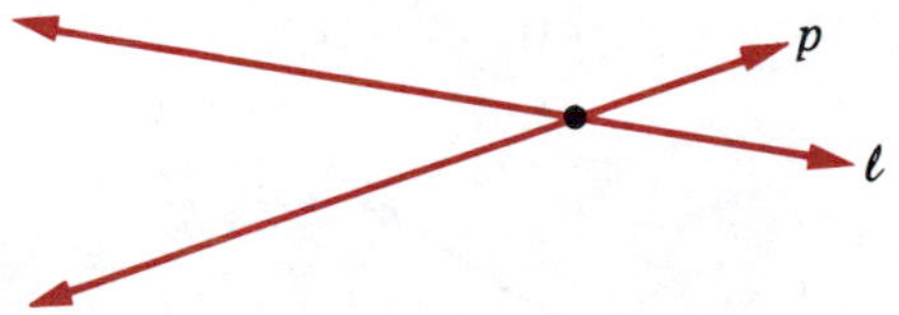

Angles

An **angle** is determined by rotating a ray about its endpoint (called the **vertex** of the angle) from some initial position (called the **initial side** of the angle) to a terminal position (called a **terminal side** of the angle).

In Figure 7, the angle determined by rotating $\overrightarrow{AC}$ to the position of $\overrightarrow{AB}$ is indicated by $\angle CAB$ or $\angle BAC$ or simply by $\angle A$.

Figure 7

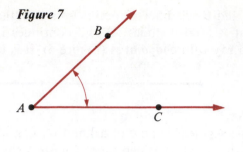

Figure 8

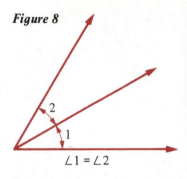

$\angle 1 = \angle 2$

These expressions are read, respectively, "angle *CAB*" "angle *BAC*" and "angle *A*." Note that the point which is the vertex occupies the central position in the *name* of the angle.

Angles are classified according to their *measure*—that is, according to the amount of "opening" between two rays or the amount of "turning" that occurs as one of the rays moves from a position coinciding with the first ray to the position of the second ray. A basic unit commonly used for measuring angles is the **degree**. An angle of **one degree** is denoted by 1°. A complete rotation about a point is an opening of 360°. *One degree* is then $\frac{1}{360}$ of a complete rotation.

A line that **bisects** an angle divides it into equal angles (Figure 8). We use the following terms to define different types of angles.

DEFINITION

> (i) A **right angle** is a 90° angle (Figure 9a).
>
> (ii) An **acute angle** is an angle that is less than 90° (Figure 9b).
>
> (iii) An **obtuse angle** is an angle that is more than 90° and less than 180° (Figure 9c).
>
> (iv) A **straight angle** is an angle that is equal to 180° (Figure 9d).
>
> *Figure 9*
>
> (a)   (b)
>
> (c)   (d)

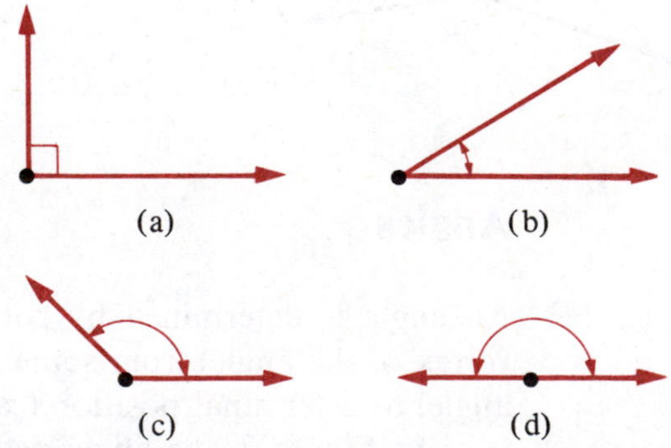

When two lines ℓ and p intersect so that the angles formed are right angles, we say that the lines are **perpendicular lines** (Figure 10a). A common notation for this fact is the symbol $p \perp \ell$. The line ℓ that is perpendicular to $\overline{AB}$ and intersects $\overline{AB}$ at its midpoint M is called the **perpendicular bisector** of $\overline{AB}$ (Figure 10b).

Figure 10

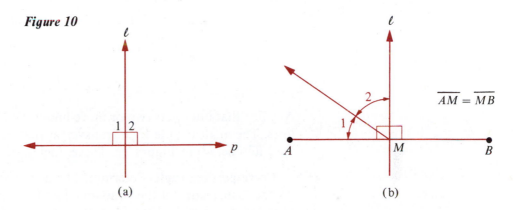

(a)   (b)

Two intersecting lines form four nonstraight angles in the plane. In Figure 11, the four angles are $\angle 1$, $\angle 2$, $\angle 3$, and $\angle 4$. Other angles are classified as follows:

1. Angles 1 and 2 (Figure 11) are called **adjacent angles.** Adjacent angles have a common vertex and a common side between them.

Figure 11

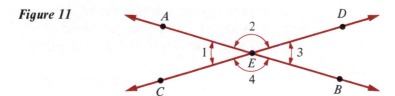

2. Nonadjacent angles formed by two intersecting lines such as angles 1 and 3 or angles 2 and 4 are called **vertical angles** or **opposite angles** (Figure 11). Vertical angles formed by intersecting lines are *equal.*

3. Two angles are called **supplementary angles** if the sum of their measure is 180°. In Figure 11, angles 1 and 2, 2 and 3, 3 and 4, and 4 and 1 are supplementary angles.

4. Two angles are called **complementary angles** if the sum of their measure is 90°. Angles 1 and 2 in Figure 10b are complementary angles.

Figure 12

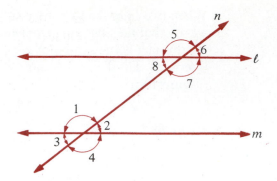

Any line that intersects two or more lines is called a **transversal** of these lines. The angles made by a transversal that cuts two lines are shown (and numbered) in Figure 12. These angles are named as follows:

5. **Corresponding angles** determined by a transversal are pairs of angles on the "same side" of the transversal and on the "same relative sides" of the given lines. In Figure 12, the pairs of corresponding angles are $\angle 6$ and $\angle 2$, $\angle 7$ and $\angle 4$, $\angle 5$ and $\angle 1$, and $\angle 8$ and $\angle 3$.

6. **Interior angles** are those angles between the lines cut by the transversal. Angles 1, 2, 7, and 8 are interior angles.

7. **Alternate interior angles** are certain pairs of angles on opposite sides of the transversal. In Figure 12, the pairs of alternate interior angles are 1 and 7, and 2 and 8.

In Figure 13, line *a* is parallel to line *b*, and *c* is a transversal; angles 2 and 7 and angles 1 and 8 are alternate interior angles; and angles 5 and 7, angles 6 and 8, angles 2 and 3, and angles 1 and 4 are corresponding angles.

Figure 13

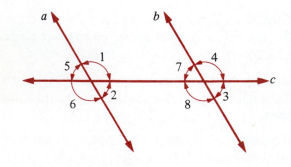

If we assume that the

1. **corresponding angles** (determined by a transversal) of parallel lines are *equal* (Figure 13), and conversely, if two lines are cut by a transversal so that the corresponding angles are equal, then the lines are parallel.

Then it can be shown that the

2. **alternate interior angles** (determined by a transversal) of parallel lines are *equal* (Figure 13), and conversely, if two lines are cut by a transversal so that alternate interior angles are equal, then the lines are parallel.

EXAMPLE 2   In Figure 14, C, E, and F lie on a straight line, and so do D, E, and A. The angle BEF is a right angle.

(a) Name a pair of vertical angles.

(b) Name an obtuse angle.

(c) Name an acute angle.

(d) Which angle is the complement of $\angle CED$?

(e) Which angle is the supplement of $\angle FEA$?

SOLUTION   (a) Angles AEF and CED are a pair of vertical angles.

(b) Angle BED is an obtuse angle.

(c) Angle AEF is an acute angle.

(d) The complement of $\angle CED$ is $\angle BEA$.

(e) The supplement of $\angle FEA$ is $\angle AEC$.

EXAMPLE 3   In Figure 14, if $\angle BEA = 45°$, find

(a) $\angle DEC$ (b) $\angle CEA$.

SOLUTION   (a) Because $\angle BEA + \angle AEF = 90°$, and because $\angle BEA = 45°$, then $\angle AEF = 45°$. Also, $\angle DEC = \angle AEF$, so that $\angle DEC = 45°$.

(b) Because $\angle BEA = \angle AEF = 45°$, and $\angle CEA + \angle AEF = 180°$, then $\angle CEA + 45° = 180°$ or $\angle CEA = 135°$.

Figure 14

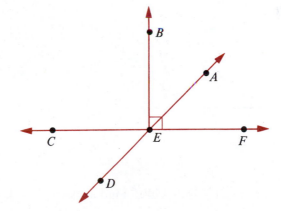

PROBLEM SET 12.1

1. Draw figures to show that the intersection of two given rays may be any of the following:

(a) one of the given rays   (b) a point   (c) a line segment

2. Explain the difference, if any, between AB, $\overline{AB}$, $\overrightarrow{BA}$, $\overline{BA}$, $\overrightarrow{AB}$, and BA.

3. Referring to Figure 15, identify each angle whose sides are the given rays.

(a) $\overrightarrow{CA}$ and $\overrightarrow{CB}$ (b) $\overrightarrow{CA}$ and $\overrightarrow{CD}$ (c) $\overrightarrow{CD}$ and $\overrightarrow{CB}$

Figure 15

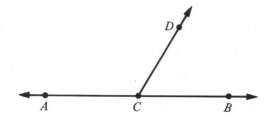

4. Answer true or false:

(a) The supplement of an acute angle is another acute angle.

(b) The complement of an acute angle is another acute angle.

(c) The supplement of a right angle is an obtuse angle.

(d) The supplement of an obtuse angle is an acute angle.

5. Look at Figure 16. If ∠ 1 is the complement of ∠ 3, how many degrees are there in ∠ 2?

6. Indicate which pairs of angles in Figure 17 are adjacent and which are vertical.

Figure 16

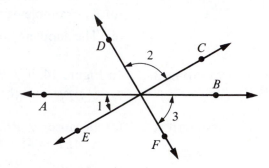

Figure 17

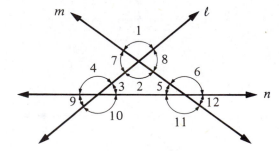

7. In Figure 18, lines *m* and *n* are parallel. Find angles 1, 2, 3, 4, and 5.

Figure 18

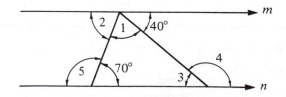

8. In Figure 19, find each of the following if $\overline{AB} \perp \overline{BD}$.

   (a) $\angle ADC$ if $\angle 1 = 45°$ and $\angle 2 = 85°$ (b) $\angle AEB$ if $\angle 5 = 60°$

   (c) $\angle EBD$ if $\angle 4 = 15°$ (d) $\angle ABC$ if $\angle 3 = 42°$

Figure 19

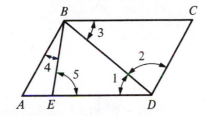

9. In Figure 20, classify each pair of angles if *AB* and *CD* are straight lines and $EO \perp AB$.

   (a) $\angle 1$ and $\angle 4$ (b) $\angle 3$ and $\angle 4$ (c) $\angle 1$ and $\angle 2$

   (d) $\angle 4$ and $\angle 5$ (e) $\angle 1$ and $\angle 3$

Figure 20

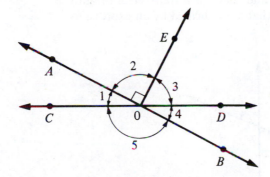

10. In Figure 21:

   (a) Name two pairs of supplementary angles.

   (b) Name two pairs of adjacent angles.

   (c) Find the degree measure of $\angle CDB$, $\angle ABC$, $\angle EDF$, and $\angle BDE$.

Figure 21

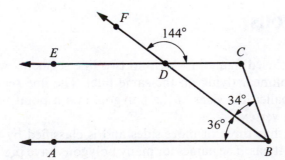

11. In Figure 22, $\overline{DE}$ is parallel to $\overline{CB}$, $\overline{FE}$ is parallel to $\overline{AC}$, and $\overline{DF}$ is parallel to AB. If $\angle 1 = 40°$ and $\angle 2 = 70°$, find:

(a) $\angle 3$ (b) $\angle A$ (c) $\angle B$ (d) $\angle C$

Figure 22

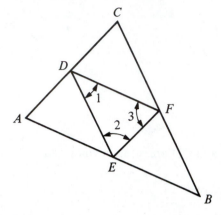

12. Suppose that the lines ℓ and m are parallel and that n is a transversal (Figure 23). Find the degree measures of the angles that are indicated by an expression involving x.

Figure 23

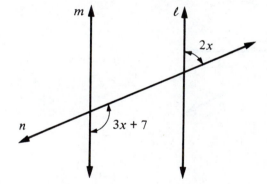

12.2 Polygons

A **polygon** is a closed figure in which no two line segments with a common endpoint are collinear (lying on the same line). The line segments forming a polygon are called the **sides** of the polygon, and a point where two sides meet is called a **vertex.**

    A polygon has three or more sides and is classified by the number of sides (or angles) it has. The names for many polygons have prefixes indicating the number of sides (or angles). For example, a polygon with four sides is called a **quadrilateral** (Figure 1a), and a polygon with five sides is called a **pentagon** (Figure 1b).

Figure 1

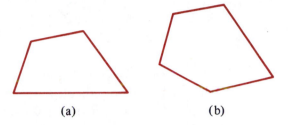

          (a)   (b)

The names of the most common quadrilaterals are:

1. A **trapezoid** is a quadrilateral with one pair of opposite sides parallel (Figure 2a).

2. A **parallelogram** is a quadrilateral with each pair of opposite sides parallel and equal (Figure 2b).

3. A **rhombus** is a parallelogram with all its sides equal (Figure 2c).

4. A **rectangle** is a parallelogram in which each of the angles is a right angle (Figure 2d).

5. A **square** is a rectangle with all four sides equal (Figure 2e).

Figure 2

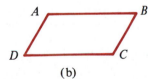

(a)

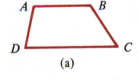

(b)

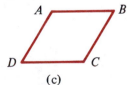

(c)

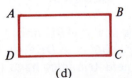

(d)

(e)

Figure 3

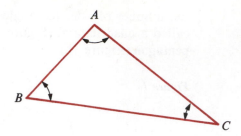

A **triangle** is a polygon with three sides. The symbol △ and three letters are often used to indicate a triangle, for example, △ABC in Figure 3, whose angles are ∠A, ∠B, and ∠C. In any triangle the sum of the angle measures is 180°.

Triangles are classified as follows:

1. A triangle is a **right triangle** if one of its angles is a right angle (Figure 4a).

2. A triangle is an **isosceles triangle** if two of its sides are equal in length (Figure 4b). It can be shown that the angles opposite the equal sides of an isosceles triangle are equal. In Figure 4b ∠B = ∠C.

3. A triangle is an **equilateral triangle** if all three of its sides are equal in length (Figure 4c). All three angles are equal.

Figure 4

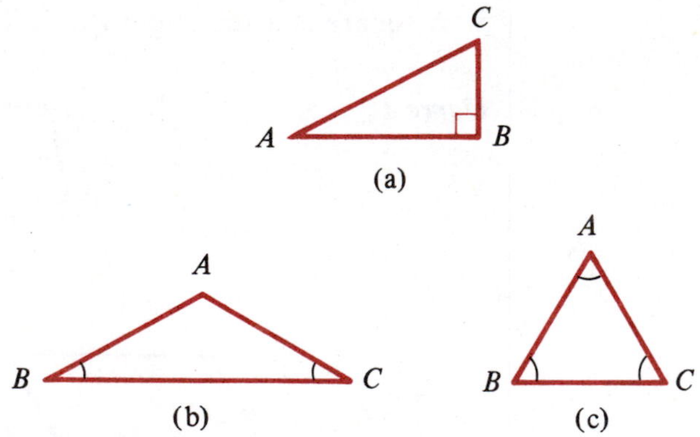

(a)

(b)

(c)

Congruent Triangles

Two triangles, say △ABC and △DEF, are **congruent,** which is written in symbols as △ABC ≅ △DEF, if they can be made to coincide. In other words, congruent triangles have the same size and shape (Figure 5). That is, △ABC ≅ △DEF means that each of the following holds:

Figure 5

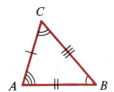

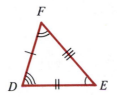

$$\angle A = \angle D \qquad |\overline{CB}| = |\overline{FE}|$$
$$\angle B = \angle E \qquad |\overline{AC}| = |\overline{DF}|$$
$$\angle C = \angle F \qquad |\overline{AB}| = |\overline{DE}|.$$

Note that *equal sides are opposite equal angles.* (We use a shorthand in Figure 5 to indicate equal sides and equal angles. The single lines drawn through $\overline{AC}$ and $\overline{DF}$, for example, indicate that these sides are equal, as do the double lines through $\overline{AB}$ and $\overline{DE}$, etc.)

The concept of congruence does not give us any practical way of deciding when two triangles are congruent. The following properties are useful:

1. If three sides of one triangle equal three sides of another triangle (SSS), then the triangles are *congruent.*

2. If two sides and the included angle of one triangle equal two sides and the included angle of another triangle (SAS), then the triangles are *congruent.*

3. If two angles and the included side of one triangle equal two angles and the included side of another triangle (ASA), then the triangles are *congruent.*

EXAMPLE 1   In Figure 6, let $\angle 1 = \angle 2$, and $\angle EAB = \angle ABC$. Show that $\triangle AEB \cong \triangle ABC$. What can be concluded about the remaining corresponding parts?

Figure 6

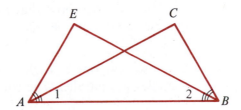

SOLUTION   In the figure, we have

$$|\overline{AB}| = |\overline{AB}|$$
$$\angle 1 = \angle 2$$
$$\angle EAB = \angle ABC.$$

Therefore

$$\triangle AEB \cong \triangle ABC. \qquad (\text{ASA})$$

Thus,

$$|\overline{AE}| = |\overline{CB}|, \qquad |\overline{AC}| = |\overline{BE}|, \qquad \text{and} \qquad \angle E = \angle C. \qquad (\text{why?})$$

Similar Triangles

Two triangles are **similar** if they have the same shape. Hence, $\triangle ABC$ is similar to $\triangle DEF$, which is denoted by $\triangle ABC \sim \triangle DEF$, if the triangles have equal angles and the corresponding sides are proportional (Figure 7).

Figure 7

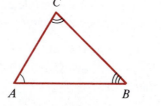

Since $\triangle ABC \sim \triangle DEF$, it follows that:

1. $\angle A = \angle D \qquad \angle B = \angle E \qquad \angle C = \angle F.$

2. $\dfrac{|\overline{AB}|}{|\overline{DE}|} = \dfrac{|\overline{AC}|}{|\overline{DF}|} = \dfrac{|\overline{BC}|}{|\overline{EF}|}.$

Note that corresponding sides are opposite equal angles. $\overline{AB}$ corresponds to $\overline{DE}$, since $\overline{AB}$ lies opposite $\angle C$, $\overline{DE}$ lies opposite $\angle F$, and $\angle C = \angle F$.

In order to show that *two triangles are similar, it is enough to show that two angles of one triangle equal two angles of the other triangle.*

EXAMPLE 2   Suppose that one triangle is inside another triangle and the sides of the triangles are parallel. The smaller triangle has sides 5, 8, and 10 inches in

Figure 8

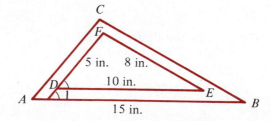

length, and the larger triangle has a side of 15 inches parallel to the 10-inch side (Figure 8). Show that the triangles are similar and find the lengths of the other two sides of the larger triangle.

SOLUTION   In Figure 8, $\angle A = \angle 1$ and $\angle D = \angle 1$ because corresponding angles of parallel lines cut by a transversal are equal. Hence, $\angle D = \angle A$. Similarly, $\angle E = \angle B$ (why?), so that $\triangle DEF \sim \triangle ABC$. Therefore,

$$\frac{|\overline{DE}|}{|\overline{AB}|} = \frac{|\overline{FE}|}{|CB|} = \frac{|\overline{DF}|}{|AC|} \qquad \text{or} \qquad \frac{10}{15} = \frac{8}{|CB|} = \frac{5}{|AC|}$$

so that

$$|\overline{AC}| = 7.5 \text{ inches} \qquad \text{and} \qquad |CB| = 12 \text{ inches.}$$

EXAMPLE 3   A vertical flagpole casts a shadow 35 feet long. A 4-foot stick, held vertically with its end on the ground, casts a shadow of 7 feet (Figure 9). How tall is the flagpole?

Figure 9

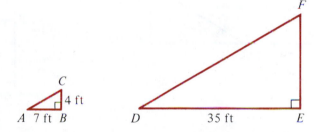

SOLUTION   In Figure 9, $\angle C = \angle F$ and $\angle B = \angle E$, so that $\triangle ABC \sim \triangle DEF$. Therefore,

$$\frac{|\overline{FE}|}{|\overline{BC}|} = \frac{|\overline{DE}|}{|AB|} \qquad \text{(why?)}$$

or

$$\frac{|\overline{FE}|}{4} = \frac{35}{7}.$$

Hence

$$|\overline{FE}| = \frac{4(35)}{7} \qquad \text{or} \qquad |\overline{FE}| = 20 \text{ feet.}$$

PROBLEM SET 12.2

In problems 1–4, determine from the information given in Figures 10–13 if it is possible to conclude that $\triangle I$ and $\triangle II$ are congruent.

1. In Figure 10, $\angle 1 = \angle 2$ and $\angle 3 = \angle 4$.

2. In Figure 11, $|\overline{AC}| = |\overline{CB}|$ and $|\overline{AD}| = |\overline{DB}|$.

Figure 10

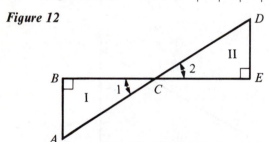

Figure 11

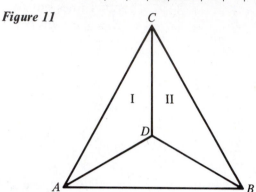

3. In Figure 12, $\angle 1 = \angle 2$, and $|\overline{BC}| = |\overline{CE}|$.

4. In Figure 13, $|\overline{BC}| = |\overline{CD}|$ and $|\overline{AB}| = |\overline{DE}|$.

Figure 12

Figure 13

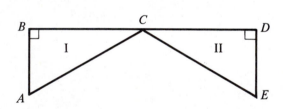

5. In Figure 14, $\overline{DE} \parallel \overline{AB}$, $\overline{DB}$ and $\overline{AE}$ are transversals intersecting at C, and $|DE| = |\overline{AB}|$. Show that $\triangle ABC \cong \triangle CDE$.

Figure 14

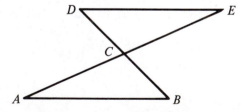

6. In Figure 15, $ABCD$ is a parallelogram. Show that $\triangle ABC \cong \triangle BDC$.

Figure 15

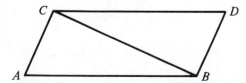

7. In Figure 16, suppose that $\triangle ABC \sim \triangle DEF$, $|\overline{BC}| = 13.1$, $|\overline{AB}| = 16$, $|\overline{DE}| = 5$, and $\angle BAC = \angle FDE = 55°$. If $\angle C = 90°$ and $\angle F = 90°$, find:

(a) $|\overline{AC}|$ (b) $|\overline{EF}|$ (c) $\angle B$ (d) $|\overline{DF}|$

Figure 16

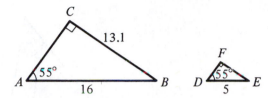

8. Show that congruence is transitive. That is, given $\triangle ABC \cong \triangle DEF$ and $\triangle DEF \cong \triangle RST$, show that $\triangle ABC \cong \triangle RST$.

9. Given $\triangle ABC$ (Figure 17), with $\overline{DE} \parallel \overline{BC}$, find the length of the indicated line segment in each of the following cases:

(a) $|\overline{CE}|$ if $|\overline{AD}| = 3$, $|\overline{BD}| = 5$, and $|\overline{AE}| = 4$

(b) $|\overline{AE}|$ if $|\overline{AD}| = 2$, $|\overline{BD}| = 3$, and $|\overline{AC}| = 10$

(c) $|\overline{AB}|$ if $|\overline{AD}| = 4$, $|\overline{BC}| = 30$, and $|\overline{DE}| = 10$

(d) $|\overline{AD}|$ if $|\overline{AD}| = |\overline{EC}|$, $|\overline{BD}| = 4$, and $|\overline{AE}| = 9$

Figure 17

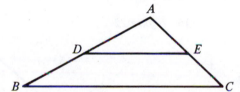

10. Given a triangle $\triangle ABC$, with $\overline{DE} \parallel \overline{BC}$ (Figure 18), find the following:

(a) $\dfrac{|\overline{AD}|}{|\overline{AB}|}$ if $\dfrac{|\overline{AD}|}{|\overline{DB}|} = \dfrac{2}{3}$

(b) $\dfrac{|\overline{AE}|}{|\overline{EC}|}$ if $\dfrac{|\overline{AE}|}{|\overline{AC}|} = \dfrac{2}{7}$

(c) $\dfrac{|\overline{AE}|}{|\overline{EC}|}$ if $\dfrac{|\overline{AD}|}{|\overline{DB}|} = \dfrac{3}{4}$

Figure 18

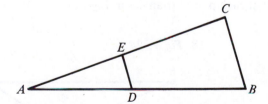

11. Suppose that $\triangle ABC$ is similar to $\triangle DEF$. If $\angle A = 48°$, $\angle B = 97°$, and $\angle F = 35°$, find the measures of the other angles.

12. A man 6 feet tall casts a shadow 13 feet long. He is standing beside a building that has a shadow 45 feet long. How high is the building?

In problems 13–16, indicate which pairs of triangles are similar. Assume all lines are straight lines.

13. *Figure 19*

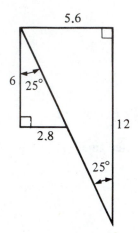

14. *Figure 20*

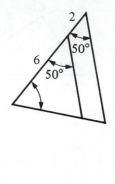

15. *Figure 21*

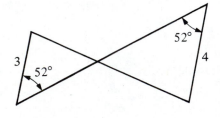

16. *Figure 22*

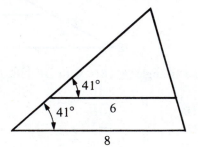

In problems 17 and 18, find the value of *x* such that each pair of triangles in Figures 23 and 24 is similar.

17. *Figure 23*

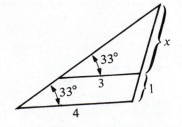

18. *Figure 24*

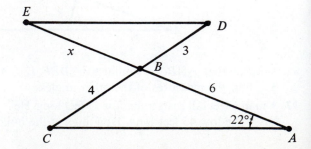

19. Use similar triangles to show that the line segment joining the midpoints of two sides of a triangle is parallel to the third side and is equal to one-half of it.

20. Let $ABCD$ be a quadrilateral with $|\overline{AD}| = |\overline{CD}|$ and $\angle A = \angle C$ (Figure 25). Show that $|\overline{AB}| = |\overline{CB}|$.

Figure 25

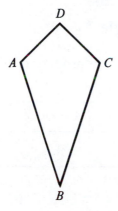

21. Show that the four midpoints of the sides of a quadrilateral are the vertices of a parallelogram. (*Hint*: Use the results of problem 19.)

22. Prove that the diagonals of a square bisect each other.

23. Let $ABCD$ be a quadrilateral with $|\overline{BC}| = |\overline{AD}|$, where E and F are the midpoints of $\overline{AB}$ and $\overline{DC}$, respectively. If the extensions of $\overline{BC}$ and $\overline{AD}$ meet EF at points G and H (Figure 26), show that $\angle G = \angle H$. [*Hint*: Bisect the diagonal $\overline{DB}$ at point K, $\overline{EK} \parallel \overline{AD}$ and $\overline{FK} \parallel \overline{BC}$; hence, $|\overline{EK}| = |\overline{FK}|$. (why?)]

Figure 26

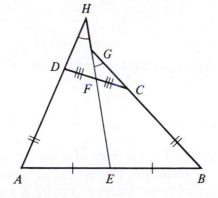

12.3 **Areas of Plane Figures**

So far we have considered and classified polygons. In this section, we study the **area** of particular polygons such as rectangles, squares, parallelograms, and triangles. The most commonly used unit for measuring area is the *square*.

For example, a square measuring one inch on each side has an area of one square inch, and a square measuring one meter on each side has an area of one square meter.

The Area of a Rectangle

The **area A of a rectangle** is the product of the length ℓ and the width w (Figure 1), that is:

$$A = \ell w.$$

Figure 1

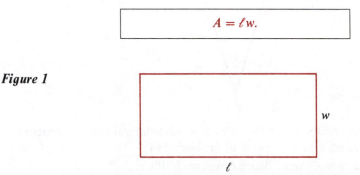

EXAMPLE 1   Find the area of a rectangle with a length of 7 centimeters and a width of 5 centimeters.

SOLUTION
$$
\begin{aligned}
A &= \ell w \\
&= 7 \cdot 5 \\
&= 35.
\end{aligned}
$$

The area of the rectangle is 35 square centimeters.

Figure 2

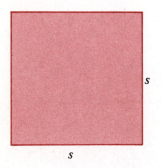

The Area of a Square

As you know, the formula for the **area of a square** is very similar to the formula for the area of a rectangle. This is because a square is a special kind of rectangle. If the length of each side of a square is s units (Figure 2), then the area A is given by the formula

$$A = s^2.$$

EXAMPLE 2   Find the area of a square with sides of 4.3 meters.

SOLUTION

$$A = s^2$$
$$= (4.3)^2 = 18.49.$$

The area of the square is 18.49 square meters.

The Area of a Parallelogram

The **area A of a parallelogram** is given by the formula

$$A = bh,$$

where b is the base and h is the height of the parallelogram (Figure 3).

Figure 3

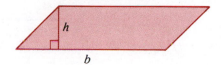

EXAMPLE 3   Find the area of a parallelogram with a base of 5 feet and a height of 3 feet.

SOLUTION   The area of a parallelogram is

$$A = bh$$
$$= 5 \cdot 3 = 15.$$

Therefore, $A = 15$ square feet.

The Area of a Triangle

The **area A of a triangle** is equal to one-half the product of the base b and the height h (Figure 4):

$$A = \tfrac{1}{2}bh.$$

Figure 4

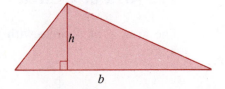

EXAMPLE 4   Find the area of a triangle with a base of 10 centimeters and a height of 3.7 centimeters.

SOLUTION   The area of the triangle is

$$A = \tfrac{1}{2} \cdot 10 \cdot 3.7 = 18.5$$

The area is 18.5 square centimeters.

The Area of a Trapezoid

The **area A of a trapezoid** is equal to one-half the product of the height h and the sum of the bases b_1 and b_2 (Figure 5):

$$A = \tfrac{1}{2}h(b_1 + b_2).$$

Figure 5

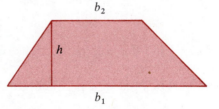

EXAMPLE 5   Find the area of a trapezoid with bases of 12 inches and 8 inches, and with a height of 5 inches.

SOLUTION   We substitute $h = 5$, $b_1 = 12$, and $b_2 = 8$ in the formula

$$A = \tfrac{1}{2}h(b_1 + b_2),$$

and we have

$$A = \tfrac{1}{2}(5)(12 + 8) = 50.$$

The area is 50 square inches.

Figure 6

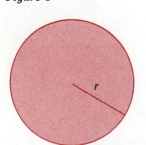

The Area of a Circle

The **area A of a circle** with radius r is π times r^2, in which $\pi \approx 3.1416\ldots$ (Figure 6):

$$A = \pi r^2.$$

EXAMPLE 6 Find the area of a circle with a radius of 10 inches.

SOLUTION
$$A = \pi r^2$$
$$= \pi(10)^2 = 100\pi \approx 314.16.$$

Therefore, the area of the circle is approximately 314.16 square inches.

PROBLEM SET 12.3

In problems 1–6, find the area of the rectangle with the given length ℓ and width w.

1. $\ell = 5$ inches and $w = 3$ inches

2. $\ell = 7$ yards and $w = 2$ yards

3. $\ell = 6.2$ centimeters and $w = 4.1$ centimeters

4. $\ell = 9.3$ feet and $w = 8.5$ feet

5. $\ell = 7.3$ feet and $w = 5.6$ feet

6. $\ell = 14.3$ meters and $w = 11.6$ meters

In problems 7–12, find the area of the square with the given side s.

7. $s = 8$ centimeters

8. $s = 13$ inches

9. $s = 6.3$ feet

10. $s = 11.3$ meters

11. $s = 9.7$ yards

12. $s = 6\frac{2}{3}$ feet

In problems 13–18, find the area of the parallelogram with the given base b and height h.

13. $b = 5$ feet and $h = 3.4$ feet

14. $b = 12$ centimeters and $h = 8$ centimeters

15. $b = 11.7$ meters and $h = 5.3$ meters

16. $b = 6.3$ yards and $h = 5.8$ yards

17. $b = 4\frac{1}{3}$ inches and $h = 6$ inches

18. $b = 3\frac{1}{4}$ inches and $h = 5\frac{2}{3}$ inches

In problems 19–24, find the area of the triangle with the given base b and height h.

19. $b = 6$ feet and $h = 3$ feet

20. $b = 8$ inches and $h = 10$ inches

21. $b = 5.1$ meters and $h = 7.3$ meters

22. $b = 11.7$ yards and $h = 39.3$ yards

23. $b = 8.9$ inches and $h = 4.2$ inches

24. $b = 6\frac{2}{5}$ feet and $h = 7\frac{3}{4}$ feet

In problems 25–30, find the area of the trapezoid with the given height h and bases b_1 and b_2.

25. $h = 5$ feet, $b_1 = 7$ feet, and $b_2 = 9$ feet

26. $h = 3.2$ yards, $b_1 = 5.9$ yards, and $b_2 = 7.2$ yards

27. $h = 3.8$ inches, $b_1 = 4.7$ inches, and $b_2 = 8.2$ inches

28. $h = 11.3$ meters, $b_1 = 9.7$ meters, and $b_2 = 6.5$ meters

29. $h = 6\frac{1}{3}$ feet, $b_1 = 7\frac{1}{2}$ feet, and $b_2 = 5\frac{1}{4}$ feet

30. $h = 5\frac{5}{6}$ yards, $b_1 = 3\frac{1}{2}$ yards, and $b_2 = 4\frac{2}{3}$ yards

In problems 31–36, find the area of the circle with the given radius r.

31. $r = 5$ feet

32. $r = 7$ yards

33. $r = 3.1$ meters

34. $r = 6.3$ feet

35. $r = 4.4$ inches

36. $r = 8.2$ centimeters

12.4 Perimeters of Plane Figures

The **perimeter** of a polygon is the sum of the lengths of its sides. The basic units for measuring perimeters are inches, meters, centimeters, and so on.

The Perimeter of a Rectangle

The **perimeter P of a rectangle** is the sum of twice the length ℓ and twice the width w (Figure 1):

$$P = 2\ell + 2w.$$

Figure 1

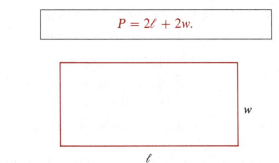

EXAMPLE 1   Find the perimeter of a rectangle with a length of 8 meters and a width of 5 meters.

SOLUTION
$$P = 2\ell + 2w$$
$$= 2(8) + 2(5) = 16 + 10 = 26.$$

The perimeter is 26 meters.

Figure 2

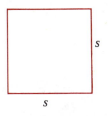

The Perimeter of a Square

The **perimeter P of a square** is four times the length of its side (Figure 2):

$$P = 4s.$$

EXAMPLE 2   Find the perimeter of a square with sides of 8 centimeters.

SOLUTION
$$P = 4s$$
$$= 4(8) = 32.$$

The perimeter of the square is 32 centimeters.

The Perimeter of a Parallelogram

The **perimeter P of a parallelogram** is the sum of twice the length ℓ and twice the width w (Figure 3):

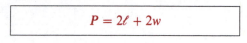

$$P = 2\ell + 2w$$

Figure 3

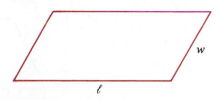

EXAMPLE 3   Find the perimeter of a parallelogram with a length of 10 inches and a width of 6 inches.

SOLUTION
$$P = 2\ell + 2w$$
$$= 2(10) + 2(6) = 20 + 12 = 32.$$

The perimeter of the parallelogram is 32 inches.

The Perimeter of a Triangle

The **perimeter P of a triangle** is the sum of the lengths of its sides a, b, and c (Figure 4):

$$P = a + b + c.$$

Figure 4

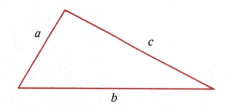

EXAMPLE 4   Find the perimeter of a triangle with sides 5 inches, 6 inches, and 8 inches.

SOLUTION
$$P = a + b + c$$
$$= 5 + 6 + 8 = 19.$$

The perimeter of the triangle is 19 inches.

The Perimeter of a Trapezoid

The **perimeter P of a trapezoid** is the sum of the lengths of the sides a, b, c, and d (Figure 5):

$$P = a + b + c + d.$$

Figure 5

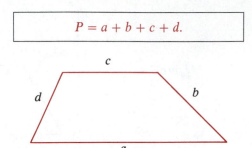

EXAMPLE 5   Find the perimeter of a trapezoid with sides 8 inches, 5 inches, 6 inches, and 5 inches.

SOLUTION
$$P = a + b + c + d$$
$$= 8 + 5 + 6 + 5 = 24.$$

The perimeter of the trapezoid is 24 inches.

The Circumference of a Circle

The **circumference C of a circle** of radius r is 2π times r (Figure 6):

$$C = 2\pi r.$$

Figure 6

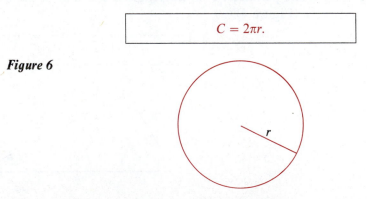

EXAMPLE 6   Find the circumference of a circle with a radius of 5 inches.

SOLUTION
$$C = 2\pi r = 2\pi(5) = 10\pi \approx 31.42.$$

The circumference of the circle is approximately 31.42 inches.

PROBLEM SET 12.4

In problems 1–6, find the perimeter of the rectangle with the given length ℓ and width w.

1. $\ell = 5$ inches and $w = 3$ inches   **2.** $\ell = 12$ yards and $w = 7$ yards

3. $\ell = 6.2$ centimeters and $w = 4.1$ centimeters   **4.** $\ell = 9.3$ feet and $w = 8.5$ feet

5. $\ell = 7.3$ feet and $w = 5.6$ feet   **6.** $\ell = 14.3$ meters and $w = 11.6$ meters

In problems 7–12, find the perimeter of the square with the given side s.

7. $s = 8$ centimeters   **8.** $s = 13$ inches   **9.** $s = 6.3$ feet

10. $s = 11.3$ meters   **11.** $s = 9.7$ yards   **12.** $s = 6\frac{2}{3}$ feet

In problems 13–18, find the perimeter of the parallelogram with the given length ℓ and width w.

13. $\ell = 5$ feet and $w = 3.4$ feet   **14.** $\ell = 12$ centimeters and $w = 8$ centimeters

15. $\ell = 11.7$ meters and $w = 5.3$ meters   **16.** $\ell = 6.3$ yards and $w = 5.8$ yards

17. $\ell = 6$ inches and $w = 4\frac{1}{3}$ inches   **18.** $\ell = 5\frac{2}{3}$ inches and $w = 3\frac{1}{4}$ inches

In problems 19–24, find the perimeter of the triangle with the given sides a, b, and c.

19. $a = 3$ inches, $b = 4$ inches, and $c = 5$ inches

20. $a = 5$ feet, $b = 12$ feet, and $c = 13$ feet

21. $a = 5.6$ meters, $b = 4.9$ meters, and $c = 7.2$ meters

22. $a = 10.7$ yards, $b = 13.2$ yards, and $c = 6.1$ yards

23. $a = 3\frac{1}{4}$ inches, $b = 4\frac{2}{3}$ inches, and $c = 2\frac{5}{6}$ inches

24. $a = 12\frac{1}{5}$ feet, $b = 5\frac{11}{15}$ feet, and $c = 10\frac{2}{3}$ feet

In problems 25–30, find the perimeter of the trapezoid with the given sides a, b, c, and d.

25. $a = 10$ inches, $b = 4$ inches, $c = 6$ inches, and $d = 4$ inches

26. $a = 7\frac{2}{3}$ feet, $b = 4\frac{1}{3}$ feet, $c = 3$ feet, and $d = 5$ feet

27. $a = 23.9$ meters, $b = 9.1$ meters, $c = 9.3$ meters, and $d = 8.1$ meters

28. $a = 8.2$ yards, $b = 2.8$ yards, $c = 4.7$ yards, and $d = 3.2$ yards

29. $a = \frac{1}{2}$ foot, $b = \frac{1}{4}$ foot, $c = \frac{1}{3}$ foot, and $d = \frac{1}{4}$ foot

30. $a = 3\frac{3}{7}$ inches, $b = 2\frac{1}{7}$ inches, $c = 1\frac{3}{7}$ inches, and $d = 1\frac{6}{7}$ inches

In problems 31–36, find the circumference of the circle with given radius r.

31. $r = 7$ feet   **32.** $r = 15$ feet   **33.** $r = 5.2$ inches

34. $r = 7.9$ meters   **35.** $r = 16.4$ centimeters   **36.** $r = 3\frac{1}{2}$ inches

12.5 Volumes and Surface Areas

The geometric figures we have studied so far have been plane figures. In this section, we consider geometric closed surfaces and solids.

The Volume and Surface Area of a Rectangular Solid (Box)

The **volume V of a rectangular solid** is the product of the length ℓ, the width w, and the height h (Figure 1):

Figure 1

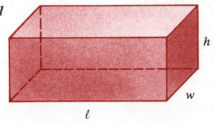

$$V = \ell wh.$$

The total surface area S of such a rectangular **solid** is the sum of the areas of its faces and its bases:

$$S = 2(\ell w + \ell h + wh).$$

EXAMPLE 1   Find the volume and the total surface area of a rectangular box with dimensions 7, 5, and 3 centimeters.

SOLUTION   The volume is given by the formula $V = \ell wh = 7(5)(3) = 105$.
Therefore, the volume of the box is 105 cubic centimeters.
　　The total surface area is given by the formula

$$\begin{aligned}
S &= 2[\ell w + \ell h + wh]\\
&= 2[7(5) + 7(3) + 5(3)]\\
&= 2(35 + 21 + 15) = 142.
\end{aligned}$$

Therefore, the total surface area is 142 square centimeters.

The Volume and the Surface Area of a Right Circular Cylinder

The **volume V of a right circular cylinder** is the product of the height h and the area πr^2 of a base of the cylinder (Figure 2):

$$V = \pi r^2 h.$$

Figure 2

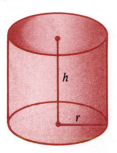

The **lateral surface area** *LS* of a right circular cylinder is the product of the height *h* of the cylinder and the circumference $2\pi r$ of a base:

$$LS = 2\pi rh.$$

The **total surface area** *S* of a right circular cylinder is the sum of the area $2\pi r^2$ of the two bases and the lateral surface area $2\pi rh$ of the cylinder:

$$S = 2\pi r^2 + 2\pi rh.$$

EXAMPLE 2   Find the lateral surface area, the total surface area, and the volume of a right circular cylinder with a radius of 3 centimeters and a height of 6 centimeters.

SOLUTION $$LS = 2\pi rh = 2\pi(3)(6) = 36\pi.$$

Thus, the lateral surface area of the cylinder is 36π square centimeters.

$$S = 2\pi r^2 + 2\pi rh$$
$$= 2\pi(3)^2 + 36\pi = 18\pi + 36\pi = 54\pi.$$

Therefore, the total surface area of the cylinder is 54π square centimeters.

$$V = \pi r^2 h = \pi(3)^2(6) = 54\pi.$$

The volume of the cylinder is 54π cubic centimeters.

The Volume and Surface Area of a Pyramid

A pyramid is a three-dimensional figure whose base is a polygon and whose lateral faces are triangles (Figure 3). If the base of the pyramid is a regular polygon and the pyramid has equal lateral edges, it is called a regular

Figure 3

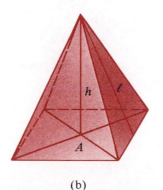

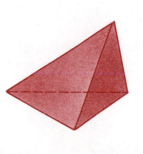

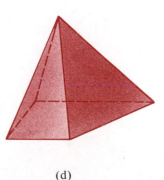

(a)   (b)   (c)   (d)

pyramid. As you can see, the pyramids shown in Figure 3a and 3b are regular. Those shown in 3c and 3d are not.

The **volume V of a regular pyramid** is one-third the product of its height h and the area A of its base (Figure 3b):

$$V = \tfrac{1}{3}Ah.$$

The **lateral surface area LS** of a right pyramid is one-half the product of the perimeter P of the pyramid's base and the slant height ℓ (the height of a lateral face) (Figure 3b):

$$LS = \tfrac{1}{2}P\ell.$$

EXAMPLE 3   Find the lateral surface area and the volume of a regular square pyramid with a base edge of 12 inches, a height of $2\sqrt{7}$ inches, and a slant height of 8 inches (Figure 4).

Figure 4

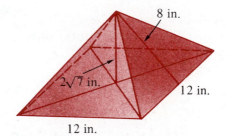

SOLUTION   The perimeter of the pyramid's base is

$$P = 4(12) = 48.$$

Thus, the lateral surface area is given by

$$LS = \tfrac{1}{2}P\ell$$
$$= \tfrac{1}{2}(48)(8) = 192.$$

Therefore, the lateral surface area of the pyramid is 192 square inches.
The volume of the pyramid is

$$V = \tfrac{1}{3}Ah$$
$$= \tfrac{1}{3}(144)(2\sqrt{7})$$
$$= 96\sqrt{7}.$$

Therefore, the volume is $96\sqrt{7}$ cubic inches.

Figure 5

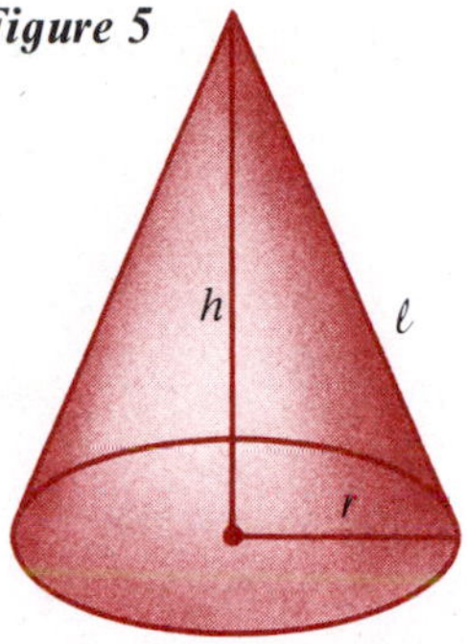

The Volume and Surface Area of a Right Circular Cone

The right circular cone resembles the pyramid, except that the base of the cone is a circle (Figure 5). The **volume** V and the **lateral surface area** LS of a right circular cone are given by the following formulas:

$$V = \tfrac{1}{3}\pi r^2 h \quad \text{and} \quad LS = \pi r \ell,$$

in which r is the radius of the base, h is the height, and ℓ is the slant height.

EXAMPLE 4   Find the lateral surface area and the volume of a right circular cone with radius 6 inches, height 8 inches, and a slant height of 10 inches.

SOLUTION

$$LS = \pi r \ell = \pi(6)(10) = 60\pi.$$

Therefore, the lateral surface area of the cone is 60π square inches.

$$\begin{aligned} V &= \tfrac{1}{3}\pi r^2 h \\ &= \tfrac{1}{3}\pi(6)^2(8) \\ &= 96\pi. \end{aligned}$$

The volume of the cone is 96π cubic inches.

The Volume and Surface Area of a Sphere

A sphere is the set of all points in space at a given distance, r, from a point (Figure 6). If r is the radius of the sphere, then the **surface area** S and the **volume** V are given by the following formulas:

$$S = 4\pi r^2 \quad \text{and} \quad V = \tfrac{4}{3}\pi r^3.$$

Figure 6

EXAMPLE 5   Find the surface area and the volume of a sphere with a radius of 3 meters.

SOLUTION

$$S = 4\pi r^2$$
$$= 4\pi(3)^2$$
$$= 36\pi.$$

Therefore, the surface area of the sphere is 36π square meters.

$$V = \tfrac{4}{3}\pi r^3$$
$$= \tfrac{4}{3}\pi(3)^3$$
$$= 36\pi.$$

The volume of the sphere is 36π cubic meters.

PROBLEM SET 12.5

In problems 1–6, find the volume and the surface area of the rectangular solid (box) with the given length ℓ, width w, and height h.

1. $\ell = 5$ inches, $w = 4$ inches, and $h = 3$ inches

2. $\ell = 7$ meters, $w = 6$ meters, and $h = 8$ meters

3. $\ell = 5\tfrac{1}{3}$ yards, $w = 3\tfrac{1}{4}$ yards, and $h = 4\tfrac{2}{3}$ yards

4. $\ell = 8.9$ centimeters, $w = 6.7$ centimeters, and $h = 5.3$ centimeters

5. $\ell = 11.9$ feet, $w = 10.3$ feet, and $h = 7.5$ feet

6. $\ell = 11.21$ meters, $w = 9.03$ meters, and $h = 4.17$ meters

In problems 7–12, find the volume, the lateral surface area, and the total surface area of the right circular cylinder with the given radius r and height h.

7. $r = 5$ inches and $h = 10$ inches

8. $r = 7$ yards and $h = 9$ yards

9. $r = 4.1$ meters and $h = 11.3$ meters

10. $r = 8.2$ centimeters and $h = 6.5$ centimeters

11. $r = 3.4$ feet and $h = 7.6$ feet

12. $r = 4\tfrac{1}{3}$ feet and $h = 5\tfrac{2}{3}$ feet

In problems 13–18, find the volume and the lateral surface area of the given pyramid.

13. A regular square pyramid with a base edge of 10 inches, a height of 12 inches, and a slant height of 13 inches.

14. A regular square pyramid with a base edge of 1.8 meters, a height of 1.2 meters, and a slant height of 1.5 meters.

15. A regular triangular pyramid with a base edge of 8 feet, a base area of $16\sqrt{3}$ square feet, a height of 10 feet, and a slant height of 10.3 feet.

16. A regular triangular pyramid with a base edge of 14.3 centimeters, a base area of 88.6 square centimeters, a height of 5.6 centimeters, and a slant height of 6.9 centimeters.

17. A regular pentagonal pyramid with a base edge of 12 inches, a height of 9 inches, a base area of 247.8 square inches, and a slant height of 12.2 inches.

18. A regular pentagonal pyramid with a base edge of 8 feet, a height of 15 feet, a base area of 110.1 square feet, and a slant height of 16 feet.

In problems 19–24, find the volume and the lateral surface area of the right circular cone with the given radius r, height h, and slant height ℓ.

19. $r = 4$ inches, $h = 3$ inches, and $\ell = 5$ inches
20. $r = 5$ meters, $h = 12$ meters, and $\ell = 13$ meters
21. $r = 9.2$ feet, $h = 6.9$ feet, and $\ell = 11.5$ feet
22. $r = 8.5$ yards, $h = 20.4$ yards, and $\ell = 22.1$ yards
23. $r = 3.2$ centimeters, $h = 2.4$ centimeters, and $\ell = 4$ centimeters
24. $r = 3\sqrt{3}$ feet, $h = 3$ feet, and $\ell = 6$ feet

In problems 25–30, find the volume and the surface area of a sphere with the given radius r.

25. $r = 4$ feet   **26.** $r = 3.1$ centimeters   **27.** $r = 5.6$ meters
28. $r = 2.5$ yards   **29.** $r = 10$ inches   **30.** $r = 8.7$ meters

REVIEW PROBLEM SET

1. In Figure 1:
   (a) What are the line segments?
   (b) What are the line segments that intersect at C?
   (c) What is the point of intersection of $\overline{AB}$ and $\overline{BC}$?

Figure 1

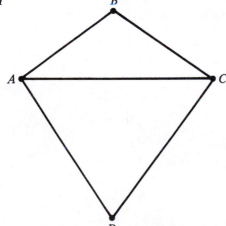

2. Suppose ℓ is a line and A is a point not on ℓ. How many lines intersecting ℓ may be drawn from A?

3. How many rays are determined by three collinear points, if each ray contains the three points?

4. How many lines are determined by three non-collinear points?

5. In Figure 2, if $\angle 1 = 47°$, find:

   (a) $\angle 3$ (b) $\angle 2$ (c) $\angle 4$

6. In Figure 3, find $\angle x$ and $\angle y$.

Figure 2

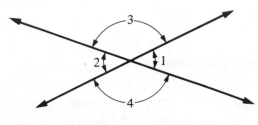

Figure 3

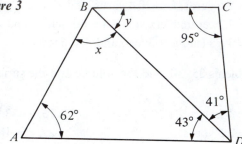

7. Find the measure of the supplement and the complement of each of the following angles:

   (a) $\angle A = 60°$ (b) $\angle B = 35°$ (c) $\angle C = 2°$ (d) $\angle D = 0°$

8. Find the degree measures of the missing angles in each of the following triangles:

   (a) $\triangle ABC$, $\angle A = 48°$, and $\angle C = 95°$

   (b) $\triangle ABC$, $\angle A = 35°$, and $\angle B = 35°$

   (c) $\triangle ABC$, $\angle A = \angle B = \angle C$

9. Given $\triangle ABC$, in which $\angle B = 90°$, assume that $\overline{BD}$ is perpendicular to $\overline{AC}$, where D is on $\overline{AC}$. If $\angle A = 37°$, find:

   (a) $\angle C$ (b) $\angle DBA$ (c) $\angle DBC$

10. Let ABC be a triangle. Draw a ray DAF parallel to BC, as in Figure 4. If $\angle DAB = 85°$ and $\angle FAC = 110°$, find:

   (a) $\angle BAC$ (b) $\angle B$ (c) $\angle C$

Figure 4

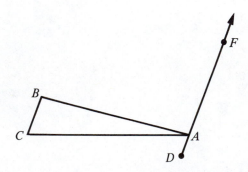

11. In Figure 5, find x.

Figure 5

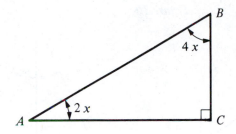

12. In Figure 6, find x and y.

Figure 6

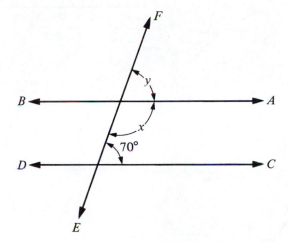

In problems 13–16, determine from the information given in Figures 7, 8, 9, and 10, if it is possible to conclude that triangles I and II are congruent. Justify each answer.

13. See Figure 7.   **14.** See Figure 8.

Figure 7   *Figure 8*

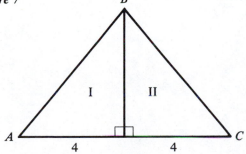   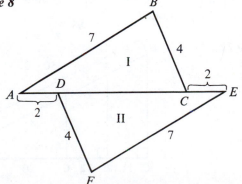

15. See Figure 9.

Figure 9

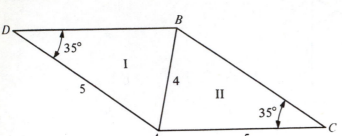

16. See Figure 10.

Figure 10

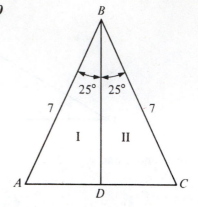

17. In Figure 11, if $\overline{DE} \parallel \overline{AB}$, find x.

Figure 11

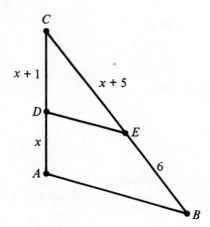

18. In Figure 12, if $\overline{DC} \parallel \overline{EF} \parallel \overline{GH} \parallel \overline{AB}$, find x, y and z. (*Hint:* Extend $\overline{DA}$ and $\overline{CB}$ to meet at a point I.)

Figure 12

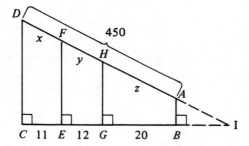

19. Find the area of Figure 13, if *ABCD* is a rectangle and the curve is a semicircle of radius 7 feet.

Figure 13

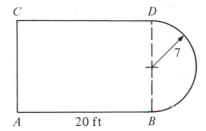

20. Find the area of △*DEF* where *D*, *E*, and *F* are the midpoints of the sides of △*ABC* whose height is 10 inches and whose corresponding base is 15 inches (Figure 14).

Figure 14

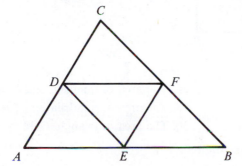

21. Find the area of the shaded region determined by circles *A*, *B*, and *0* if $|\overline{AB}| = 2$ (Figure 15).

Figure 15

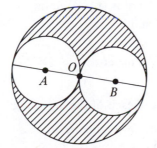

22. Let *ABCD* be a square whose area is 100 square units. What is the length of its side? What is the area of the square whose vertices are the midpoints of the sides of *ABCD*?

23. Let *ABC* be an isosceles triangle with $|\overline{AB}| = |\overline{AC}| = 13$. If $|\overline{BC}| = 10$, find the length of the perpendicular line *AD* from *A* to *BC*. Find the area of △ABC.

24. If a floor is 80 feet long and 20 feet wide, how many tiles are needed to cover it if each tile is a square 4 inches on each side?

In problems 25–30, given that $ABCD$ is a square (Figure 16), $|\overline{DE}| = |\overline{EC}|$, $\overline{EF} \parallel \overline{CG}$, $|\overline{AB}| = 9$ inches, and $\overline{HB}/\overline{HC} = \frac{1}{2}$, find:

Figure 16

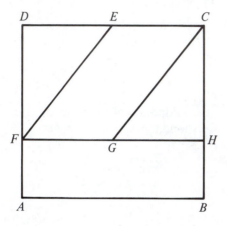

25. The area of square $ABCD$

26. The area of quadrilateral $FHCD$

27. The area of quadrilateral $FGCE$

28. The area of quadrilateral $FHCE$

29. The area of quadrilateral $ABHF$

30. The area of triangle GHC

In problems 31–38, find the area of the given plane figure.

31. A rectangle, if the base is 25 centimeters and the perimeter is 90 centimeters.

32. A rectangle, if the perimeter is 50 feet and the ratio of the sides is $\frac{2}{3}$.

33. A square, if the perimeter is 81 meters.

34. A parallelogram, if the base is $(x - 5)$ units and the height is $(x + 5)$ units

35. A triangle, if the base is $5x$ units and the height is $4x$ units.

36. An equilateral triangle, if the perimeter is 36 inches.

37. A trapezoid, if $a = 20$ meters, $b = 40$ meters, and $h = 16$ meters.

38. A circle, if the circumference is 36π centimeters.

In problems 39–44, find the volume and the surface area of each of the following:

39. A rectangular box, if $\ell = 6$, $w = 2$, and $h = 7$ (units are in centimeters).

40. A rectangular box whose dimensions are 6, 8, and 10 meters.

41. A right circular cylinder, with radius of 3 inches and height of 5 inches.

42. A right circular cone, with radius of 4 inches and height of 5 inches.

43. A sphere, with radius of 5 inches.

44. A cube, whose side is 10 centimeters.

CHAPTER 12 TEST

1. Complete each statement.
   (a) A line segment has _____ endpoint(s).
   (b) A ray has _____ endpoint(s).

2. In Figure 1, let AOB be a straight line. If $\angle 4 = 45°$, find $\angle AOE$.

Figure 1

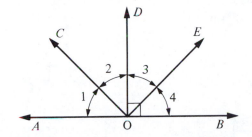

3. If $\angle A$ is complementary to $\angle C$, where $\angle C = 50°$ and $\angle B$ is complementary to $\angle D$, where $\angle D = 50°$, what do you know about $\angle A$ and $\angle B$?

4. If $\overline{AC} \| \overline{BD}$, name the pairs of equal angles in Figure 2.

Figure 2

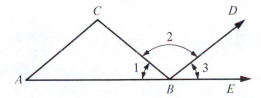

5. Given quadrilateral $ABCD$, with $\angle BAC = \angle DCA$, and $\angle BCA = \angle CAD$ (Figure 3), is $\triangle ABC \cong \triangle CAD$?

Figure 3

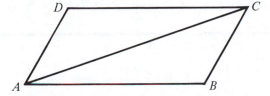

6. In Figure 4, $\overline{AB}$ and $\overline{CD}$ intersect at E. Find x if $\overline{AC} \| \overline{DB}$.

Figure 4

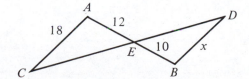

7. Find the area of a square if the perimeter is 20 centimeters.

8. Find the height of a parallelogram if its area is 22 square inches and its base is 1.1 inches.

9. Find the height of a triangle if its base is 10 inches and its area is equal to the area of a parallelogram whose base is 10 inches and whose height is 8 inches.

10. Find the height of a trapezoid if its area is 40 square centimeters and the lengths of its bases are 13 centimeters and 7 centimeters.

11. Find the area of a circle if its circumference is 49π meters.

12. Find the area of Figure 5.

Figure 5

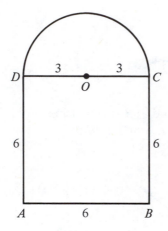

In problems 13–16, find (a) the volume and (b) the surface area.

13. A sphere of radius 6 centimeters.

14. A right circular cone of radius 2 inches, height 5 inches, and slant height $\sqrt{29}$ inches.

15. A rectangular box if $\ell = 8$ meters, $w = 3$ meters, and $h = 4$ meters.

16. A right circular cylinder of radius 6 centimeters and height 10 centimeters.

Appendices

TABLE I COMMON LOGARITHMS

x	0.00	0.01	0.02	0.03	0.04	0.05	0.06	0.07	0.08	0.09
1.0	0.0000	0.0043	0.0086	0.0128	0.0170	0.0212	0.0253	0.0294	0.0334	0.0374
1.1	0.0414	0.0453	0.0492	0.0531	0.0569	0.0607	0.0645	0.0682	0.0719	0.0755
1.2	0.0792	0.0828	0.0864	0.0899	0.0934	0.0969	0.1004	0.1038	0.1072	0.1106
1.3	0.1139	0.1173	0.1206	0.1239	0.1271	0.1303	0.1335	0.1367	0.1399	0.1430
1.4	0.1461	0.1492	0.1523	0.1553	0.1584	0.1614	0.1644	0.1673	0.1703	0.1732
1.5	0.1761	0.1790	0.1818	0.1847	0.1875	0.1903	0.1931	0.1959	0.1987	0.2014
1.6	0.2041	0.2068	0.2095	0.2122	0.2148	0.2175	0.2201	0.2227	0.2253	0.2279
1.7	0.2304	0.2330	0.2355	0.2380	0.2405	0.2430	0.2455	0.2480	0.2504	0.2529
1.8	0.2553	0.2577	0.2601	0.2625	0.2648	0.2672	0.2695	0.2718	0.2742	0.2765
1.9	0.2788	0.2810	0.2833	0.2856	0.2878	0.2900	0.2923	0.2945	0.2967	0.2989
2.0	0.3010	0.3032	0.3054	0.3075	0.3096	0.3118	0.3139	0.3160	0.3181	0.3201
2.1	0.3222	0.3243	0.3263	0.3284	0.3304	0.3324	0.3345	0.3365	0.3385	0.3404
2.2	0.3424	0.3444	0.3464	0.3483	0.3502	0.3522	0.3541	0.3560	0.3579	0.3598
2.3	0.3617	0.3636	0.3655	0.3674	0.3692	0.3711	0.3729	0.3747	0.3766	0.3784
2.4	0.3802	0.3820	0.3838	0.3856	0.3874	0.3892	0.3909	0.3927	0.3945	0.3962
2.5	0.3979	0.3997	0.4014	0.4031	0.4048	0.4065	0.4082	0.4099	0.4116	0.4133
2.6	0.4150	0.4166	0.4183	0.4200	0.4216	0.4232	0.4249	0.4265	0.4281	0.4298
2.7	0.4314	0.4330	0.4346	0.4362	0.4378	0.4393	0.4409	0.4425	0.4440	0.4456
2.8	0.4472	0.4487	0.4502	0.4518	0.4533	0.4548	0.4564	0.4579	0.4594	0.4609
2.9	0.4624	0.4639	0.4654	0.4669	0.4683	0.4698	0.4713	0.4728	0.4742	0.4757
3.0	0.4771	0.4786	0.4800	0.4814	0.4829	0.4843	0.4857	0.4871	0.4886	0.4900
3.1	0.4914	0.4928	0.4942	0.4955	0.4969	0.4983	0.4997	0.5011	0.5024	0.5038
3.2	0.5051	0.5065	0.5079	0.5092	0.5105	0.5119	0.5132	0.5145	0.5159	0.5172
3.3	0.5185	0.5198	0.5211	0.5224	0.5237	0.5250	0.5263	0.5276	0.5289	0.5302
3.4	0.5315	0.5328	0.5340	0.5353	0.5366	0.5378	0.5391	0.5403	0.5416	0.5428
3.5	0.5441	0.5453	0.5465	0.5478	0.5490	0.5502	0.5514	0.5527	0.5539	0.5551
3.6	0.5563	0.5575	0.5587	0.5599	0.5611	0.5623	0.5635	0.5647	0.5658	0.5670
3.7	0.5682	0.5694	0.5705	0.5717	0.5729	0.5740	0.5752	0.5763	0.5775	0.5786
3.8	0.5798	0.5809	0.5821	0.5832	0.5843	0.5855	0.5866	0.5877	0.5888	0.5899
3.9	0.5911	0.5922	0.5933	0.5944	0.5955	0.5966	0.5977	0.5988	0.5999	0.6010
4.0	0.6021	0.6031	0.6042	0.6053	0.6064	0.6075	0.6085	0.6096	0.6107	0.6117
4.1	0.6128	0.6138	0.6149	0.6160	0.6170	0.6180	0.6191	0.6201	0.6212	0.6222
4.2	0.6232	0.6243	0.6253	0.6263	0.6274	0.6284	0.6294	0.6304	0.6314	0.6325
4.3	0.6335	0.6345	0.6355	0.6365	0.6375	0.6385	0.6395	0.6405	0.6415	0.6425
4.4	0.6435	0.6444	0.6454	0.6464	0.6474	0.6484	0.6493	0.6503	0.6513	0.6522
4.5	0.6532	0.6542	0.6551	0.6561	0.6571	0.6580	0.6590	0.6599	0.6609	0.6618
4.6	0.6628	0.6637	0.6646	0.6656	0.6665	0.6675	0.6684	0.6693	0.6702	0.6712
4.7	0.6721	0.6730	0.6739	0.6749	0.6758	0.6767	0.6776	0.6785	0.6794	0.6803
4.8	0.6812	0.6821	0.6830	0.6839	0.6848	0.6857	0.6866	0.6875	0.6884	0.6893
4.9	0.6902	0.6911	0.6920	0.6928	0.6937	0.6946	0.6955	0.6964	0.6972	0.6981
5.0	0.6990	0.6998	0.7007	0.7016	0.7024	0.7033	0.7042	0.7050	0.7059	0.7067
5.1	0.7076	0.7084	0.7093	0.7101	0.7110	0.7118	0.7126	0.7135	0.7143	0.7152
5.2	0.7160	0.7168	0.7177	0.7185	0.7193	0.7202	0.7210	0.7218	0.7226	0.7235
5.3	0.7243	0.7251	0.7259	0.7267	0.7275	0.7284	0.7292	0.7300	0.7308	0.7316
5.4	0.7324	0.7332	0.7340	0.7348	0.7356	0.7364	0.7372	0.7380	0.7388	0.7396
5.5	0.7404	0.7412	0.7419	0.7427	0.7435	0.7443	0.7451	0.7459	0.7466	0.7474
5.6	0.7482	0.7490	0.7497	0.7505	0.7513	0.7520	0.7528	0.7536	0.7543	0.7551
5.7	0.7559	0.7566	0.7574	0.7582	0.7589	0.7597	0.7604	0.7612	0.7619	0.7627
5.8	0.7634	0.7642	0.7649	0.7657	0.7664	0.7672	0.7679	0.7686	0.7694	0.7701
5.9	0.7709	0.7716	0.7723	0.7731	0.7738	0.7745	0.7752	0.7760	0.7767	0.7774

x	0.00	0.01	0.02	0.03	0.04	0.05	0.06	0.07	0.08	0.09
6.0	0.7782	0.7789	0.7796	0.7803	0.7810	0.7818	0.7825	0.7832	0.7839	0.7846
6.1	0.7853	0.7860	0.7868	0.7875	0.7882	0.7889	0.7896	0.7903	0.7910	0.7917
6.2	0.7924	0.7931	0.7938	0.7945	0.7952	0.7959	0.7966	0.7973	0.7980	0.7987
6.3	0.7993	0.8000	0.8007	0.8014	0.8021	0.8028	0.8035	0.8041	0.8048	0.8055
6.4	0.8062	0.8069	0.8075	0.8082	0.8089	0.8096	0.8102	0.8109	0.8116	0.8122
6.5	0.8129	0.8136	0.8142	0.8149	0.8156	0.8162	0.8169	0.8176	0.8182	0.8189
6.6	0.8195	0.8202	0.8209	0.8215	0.8222	0.8228	0.8235	0.8241	0.8248	0.8254
6.7	0.8261	0.8267	0.8274	0.8280	0.8287	0.8293	0.8299	0.8306	0.8312	0.8319
6.8	0.8325	0.8331	0.8338	0.8344	0.8351	0.8357	0.8363	0.8370	0.8376	0.8382
6.9	0.8388	0.8395	0.8401	0.8407	0.8414	0.8420	0.8426	0.8432	0.8439	0.8445
7.0	0.8451	0.8457	0.8463	0.8470	0.8476	0.8482	0.8488	0.8494	0.8500	0.8506
7.1	0.8513	0.8519	0.8525	0.8531	0.8537	0.8543	0.8549	0.8555	0.8561	0.8567
7.2	0.8573	0.8579	0.8585	0.8591	0.8597	0.8603	0.8609	0.8615	0.8621	0.8627
7.3	0.8633	0.8639	0.8645	0.8651	0.8657	0.8663	0.8669	0.8675	0.8681	0.8686
7.4	0.8692	0.8698	0.8704	0.8710	0.8716	0.8722	0.8727	0.8733	0.8739	0.8745
7.5	0.8751	0.8756	0.8762	0.8768	0.8774	0.8779	0.8785	0.8791	0.8797	0.8802
7.6	0.8808	0.8814	0.8820	0.8825	0.8831	0.8837	0.8842	0.8848	0.8854	0.8859
7.7	0.8865	0.8871	0.8876	0.8882	0.8887	0.8893	0.8899	0.8904	0.8910	0.8915
7.8	0.8921	0.8927	0.8932	0.8938	0.8943	0.8949	0.8954	0.8960	0.8965	0.8971
7.9	0.8976	0.8982	0.8987	0.8993	0.8998	0.9004	0.9009	0.9015	0.9020	0.9025
8.0	0.9031	0.9036	0.9042	0.9047	0.9053	0.9058	0.9063	0.9069	0.9074	0.9079
8.1	0.9085	0.9090	0.9096	0.9101	0.9106	0.9112	0.9117	0.9122	0.9128	0.9133
8.2	0.9138	0.9143	0.9149	0.9154	0.9159	0.9165	0.9170	0.9175	0.9180	0.9186
8.3	0.9191	0.9196	0.9201	0.9206	0.9212	0.9217	0.9222	0.9227	0.9232	0.9238
8.4	0.9243	0.9248	0.9253	0.9258	0.9263	0.9269	0.9274	0.9279	0.9284	0.9289
8.5	0.9294	0.9299	0.9304	0.9309	0.9315	0.9320	0.9325	0.9330	0.9335	0.9340
8.6	0.9345	0.9350	0.9355	0.9360	0.9365	0.9370	0.9375	0.9380	0.9385	0.9390
8.7	0.9395	0.9400	0.9405	0.9410	0.9415	0.9420	0.9425	0.9430	0.9435	0.9440
8.8	0.9445	0.9450	0.9455	0.9460	0.9465	0.9469	0.9474	0.9479	0.9484	0.9489
8.9	0.9494	0.9499	0.9504	0.9509	0.9513	0.9518	0.9523	0.9528	0.9533	0.9538
9.0	0.9542	0.9547	0.9552	0.9557	0.9562	0.9566	0.9571	0.9576	0.9581	0.9586
9.1	0.9590	0.9595	0.9600	0.9605	0.9609	0.9614	0.9619	0.9624	0.9628	0.9633
9.2	0.9638	0.9643	0.9647	0.9652	0.9657	0.9661	0.9666	0.9671	0.9675	0.9680
9.3	0.9685	0.9689	0.9694	0.9699	0.9703	0.9708	0.9713	0.9717	0.9722	0.9727
9.4	0.9731	0.9736	0.9741	0.9745	0.9750	0.9754	0.9759	0.9763	0.9768	0.9773
9.5	0.9777	0.9782	0.9786	0.9791	0.9795	0.9800	0.9805	0.9809	0.9814	0.9818
9.6	0.9823	0.9827	0.9832	0.9836	0.9841	0.9845	0.9850	0.9854	0.9859	0.9863
9.7	0.9868	0.9872	0.9877	0.9881	0.9886	0.9890	0.9894	0.9899	0.9903	0.9908
9.8	0.9912	0.9917	0.9921	0.9926	0.9930	0.9934	0.9939	0.9943	0.9948	0.9952
9.9	0.9956	0.9961	0.9965	0.9969	0.9974	0.9978	0.9983	0.9987	0.9991	0.9996

TABLE II POWERS AND ROOTS

Number	Square	Square Root	Cube	Cube Root	Number	Square	Square Root	Cube	Cube Root
1	1	1.000	1	1.000	51	2,601	7.141	132,651	3.708
2	4	1.414	8	1.260	52	2,704	7.211	140,608	3.733
3	9	1.732	27	1.442	53	2,809	7.280	148,877	3.756
4	16	2.000	64	1.587	54	2,916	7.348	157,464	3.780
5	25	2.236	125	1.710	55	3,025	7.416	166,375	3.803
6	36	2.449	216	1.817	56	3,136	7.483	175,616	3.826
7	49	2.646	343	1.913	57	3,249	7.550	185,193	3.849
8	64	2.828	512	2.000	58	3,364	7.616	195,112	3.871
9	81	3.000	729	2.080	59	3,481	7.681	205,379	3.893
10	100	3.162	1,000	2.154	60	3,600	7.746	216,000	3.915
11	121	3.317	1,331	2.224	61	3,721	7.810	226,981	3.936
12	144	3.464	1,728	2.289	62	3,844	7.874	238,328	3.958
13	169	3.606	2,197	2.351	63	3,969	7.937	250,047	3.979
14	196	3.742	2,744	2.410	64	4,096	8.000	262,144	4.000
15	225	3.873	3,375	2.466	65	4,225	8.062	274,625	4.021
16	256	4.000	4,096	2.520	66	4,356	8.124	287,496	4.041
17	289	4.123	4,913	2.571	67	4,489	8.185	300,763	4.062
18	324	4.243	5,832	2.621	68	4,624	8.246	314,432	4.082
19	361	4.359	6,859	2.668	69	4,761	8.307	328,509	4.102
20	400	4.472	8,000	2.714	70	4,900	8.367	343,000	4.121
21	441	4.583	9,261	2.759	71	5,041	8.426	357,911	4.141
22	484	4.690	10,648	2.802	72	5,184	8.485	373,248	4.160
23	529	4.796	12,167	2.844	73	5,329	8.544	389,017	4.179
24	576	4.899	13,824	2.884	74	5,476	8.602	405,224	4.198
25	625	5.000	15,625	2.924	75	5,625	8.660	421,875	4.217
26	676	5.099	17,576	2.962	76	5,776	8.718	438,976	4.236
27	729	5.196	19,683	3.000	77	5,929	8.775	456,533	4.254
28	784	5.292	21,952	3.037	78	6,084	8.832	474,552	4.273
29	841	5.385	24,389	3.072	79	6,241	8.888	493,039	4.291
30	900	5.477	27,000	3.107	80	6,400	8.944	512,000	4.309
31	961	5.568	29,791	3.141	81	6,561	9.000	531,441	4.327
32	1,024	5.657	32,768	3.175	82	6,724	9.055	551,368	4.344
33	1,089	5.745	35,937	3.208	83	6,889	9.110	571,787	4.362
34	1,156	5.831	39,304	3.240	84	7,056	9.165	592,704	4.380
35	1,225	5.916	42,875	3.271	85	7,225	9.220	614,125	4.397
36	1,296	6.000	46,656	3.302	86	7,396	9.274	636,056	4.414
37	1,369	6.083	50,653	3.332	87	7,569	9.327	658,503	4.431
38	1,444	6.164	54,872	3.362	88	7,744	9.381	681,472	4.448
39	1,521	6.245	59,319	3.391	89	7,921	9.434	704,969	4.465
40	1,600	6.325	64,000	3.420	90	8,100	9.498	729,000	4.481
41	1,681	6.403	68,921	3.448	91	8,281	9.539	753,571	4.498
42	1,764	6.481	74,088	3.476	92	8,464	9.592	778,688	4.514
43	1,849	6.557	79,507	3.503	93	8,649	9.644	804,357	4.531
44	1,936	6.633	85,184	3.530	94	8,836	9.695	830,584	4.547
45	2,025	6.708	91,125	3.557	95	9,025	9.747	857,375	4.563
46	2,116	6.782	97,336	3.583	96	9,216	9.798	884,736	4.579
47	2,209	6.856	103,823	3.609	97	9,409	9.849	912,673	4.595
48	2,304	6.928	110,592	3.634	98	9,604	9.899	941,192	4.610
49	2,401	7.000	117,649	3.659	99	9,801	9.950	970,299	4.626
50	2,500	7.071	125,000	3.684	100	10,000	10.000	1,000,000	4.642

Answers to Selected Problems

Chapter 1

PROBLEM SET 1.1   page 6

1. $y + 3 = 15$ **3.** $t - 3 = 7$ **5.** $x - 2 > 5$ **7.** $7x - 4 \leq 35$ **9.** $y \div 3 = y - 5$ **11.** $2x + 3(x + 2) < 5$
13. $3(8y) \leq 7(y + 8)$ **15.** The sum of x and 2 is less than 10.
17. The product of 3 and x is less than or equal to 15.   **19.** The sum of 3 times t and 2 is greater than 8.
21. Two less than the product of 10 and x does not equal 15.
23. One more than the product of 5 and y is greater than or equal to 6.
25. Three more than the quotient of u and 4 equals 7.
27. Five-eighths less than the quotient of x and 3 equals $\frac{3}{8}$.   **29.** Base = 5; exponent = 2; $5^2 = 5 \cdot 5 = 25$
31. Base = 10; exponent = 3; $10^3 = 10 \cdot 10 \cdot 10 = 1,000$
33. Base = 2; exponent = 7; $2^7 = 2 \cdot 2 \cdot 2 \cdot 2 \cdot 2 \cdot 2 \cdot 2 = 128$
35. Base = 3; exponent = 5; $3^5 = 3 \cdot 3 \cdot 3 \cdot 3 \cdot 3 = 243$
37. Base = 11; exponent = 2; $11^2 = 11 \cdot 11 = 121$ **39.** Base = 9; exponent = 3; $9^3 = 9 \cdot 9 \cdot 9 = 729$
41. 31   **43.** 63   **45.** 19   **47.** 13   **49.** 7   **51.** 86   **53.** 6   **55.** 57   **57.** 41   **59.** 13   **61.** 45
63. 28   **65.** 25   **67.** 16   **69.** 29   **71.** 65   **73.** 102   **75.** 1848   **77.** 150   **79.** 53   **81.** $\frac{36}{25}$

83. 173   **85.** $\left[\dfrac{(5 + 3)6}{2} - 9\right] \div 3 = 5$; $\left[\dfrac{(7 + 3)6}{2} - 9\right] \div 3 = 7$

PROBLEM SET 1.2   page 13

1. $5 \in A$; $5 \notin B$; $10 \in A$; $4 \notin B$; $10 \in B$; $12 \notin A$; $8 \in B$; $9 \in A$ **3.** $A = \{\text{March, May}\}$
5. $C = \{x \mid x$ is a counting number and $4 < x < 17\}$, finite   **7.** $\emptyset$; $\{2\}$; $\{3\}$; $\{2, 3\}$
9. $\emptyset$; $\{5\}$; $\{6\}$; $\{7\}$; $\{8\}$; $\{5, 6\}$; $\{5, 7\}$; $\{5, 8\}$; $\{6, 7\}$; $\{6, 8\}$; $\{7, 8\}$; $\{5, 6, 7\}$; $\{5, 6, 8\}$; $\{5, 7, 8\}$; $\{6, 7, 8\}$; $\{5, 6, 7, 8\}$
11.

13.

15.

17.

19.

21. 0.6   **23.** 1.5   **25.** 0.8   **27.** -1.25

29. $-2.\overline{3}$ **31.** $\frac{27}{100}$ **33.** $\frac{66}{25}$ **35.** $-\frac{1}{8}$ **37.** $\frac{527}{10,000}$ **39.** $-\frac{329}{10,000}$ **41.** Rational   **43.** Irrational
45. Rational   **47.** Irrational   **49.** Rational   **51.** Rational   **53.** Rational

PROBLEM SET 1.3   page 19

1. 32   **3.** 72   **5.** 1.1   **7.** 0.328   **9.** 15   **11.** 120   **13.** 14.2   **15.** 0.134   **17.** 39   **19.** 48
21. 9.45   **23.** Commutative property for addition   **25.** Associative property for multiplication
27. Distributive property   **29.** Multiplicative inverse   **31.** Negative property (iii)
33. Zero-factor property (ii)   **35.** Cancellation property for multiplication   **37.** Zero-factor property (i)
39. Commutative property for addition   **41.** Identity property for addition
43. (i) Identity property for multiplication and the negative property; (ii) distributive property;
(iii) additive inverse; (iv) zero-factor property (i)   **45.** $x = -3$ **47.** $3t + 2t = 5t$

PROBLEM SET 1.4   page 29

1. 3   **3.** 11   **5.** -16 **7.** -21 **9.** $-\frac{5}{7}$ **11.** 4   **13.** -14 **15.** -12 **17.** 0   **19.** -7
21. -12 **23.** -53 **25.** 1.5   **27.** -18.2 **29.** -9 **31.** -7 **33.** -4 **35.** 37   **37.** 20
39. -15 **41.** -16 **43.** 12   **45.** 0.059   **47.** 8   **49.** -94 **51.** 154   **53.** -15 **55.** 16
57. -2.7 **59.** 0   **61.** -24 **63.** 252   **65.** 30   **67.** 210   **69.** -2 **71.** -2 **73.** 3   **75.** -13
77. 15   **79.** $-1,500$ **81.** 5   **83.** 4   **85.** 4   **87.** -18 **89.** 25   **91.** $\frac{21}{2}$ **93.** -18 **95.** -15
97. 11   **99.** -72 **101.** 45   **103.** -6 **105.** 29   **107.** -10 **109.** -71 **111.** 9   **113.** 121
115. $15°C$ **117.** A loss of 2 yards

PROBLEM SET 1.5   page 38

1. $\frac{1}{3}$ **3.** $\frac{7}{11}$ **5.** $-\frac{5}{7}$ **7.** $-\frac{5}{7}$ **9.** $\frac{5}{8}$ **11.** $\frac{3}{4}$ **13.** $\frac{3}{5}$ **15.** $\frac{2}{11}$ **17.** $\frac{5}{6}$ **19.** $\frac{1}{5}$ **21.** $\frac{7}{8}$ **23.** $\frac{1}{6}$
25. $\frac{65}{48}$ **27.** $\frac{8}{15}$ **29.** $\frac{17}{20}$ **31.** $-\frac{1}{36}$ **33.** $\frac{13}{12}$ **35.** $\frac{23}{42}$ **37.** $-\frac{17}{36}$ **39.** $\frac{19}{24}$ **41.** $\frac{2}{3}$ **43.** $-\frac{5}{16}$ **45.** $-\frac{9}{2}$
47. $-\frac{1}{3}$ **49.** -1 **51.** $\frac{2}{5}$ **53.** $\frac{4}{15}$ **55.** 1   **57.** $-\frac{10}{9}$ **59.** $-\frac{12}{7}$ **61.** $-\frac{8}{9}$ **63.** $\frac{2}{15}$ **65.** $-\frac{6}{5}$
67. $\frac{16}{9}$ **69.** -1 **71.** $\frac{1}{2}$ **73.** $-\frac{63}{40}$ **75.** $\frac{5}{9}$ **77.** $-\frac{8}{35}$ **79.** $-\frac{37}{180}$ **81.** $\frac{107}{140}$ **83.** 220   **85.** $\frac{3}{20}$ hour

PROBLEM SET 1.6   page 44

1. 0.13793103   **3.** 0.88235294   **5.** 5.0990195   **7.** 8.4439327   **9.** 0.3   **11.** 5.3   **13.** 8.0   **15.** 15.0
17. 24.1   **19.** 3.19   **21.** 14.36   **23.** 21.00   **25.** 16.51   **27.** 23.70   **29.** 1.73   **31.** 14.3   **33.** 368
35. 5140   **37.** 28.0   **39.** 1.33   **41.** 0.143   **43.** 2.38   **45.** 1.83   **47.** 5.48   **49.** 1.732   **51.** 2.646
53. 4.796   **55.** 6.856   **57.** 10.54   **59.** 3.782×10^3 **61.** 3.84×10^5 **63.** 7.8×10^8 **65.** 210
67. 750,000   **69.** 31,200,000   **71.** 8.7×10^5 **73.** 5.98×10^{24}

PROBLEM SET 1.7   page 50

1. 4   **3.** 5   **5.** $-\frac{319}{20}$ **7.** 0.00128503   **9.** 5^3 **11.** x^4 **13.** $(-t)^5$ **15.** u^3v^5 **17.** $3^2x^4y^2$
19. $8 \cdot 8 \cdot 8 \cdot 8$ **21.** $y \cdot y \cdot y \cdot y \cdot y$ **23.** $5 \cdot 5 \cdot 5 \cdot t \cdot t \cdot t \cdot t$ **25.** $(-x) \cdot (-x) \cdot (-x) \cdot (-x) \cdot y \cdot y \cdot y$
27. $4 \cdot 4 \cdot 4 \cdot u \cdot u \cdot u \cdot u \cdot v$ **29.** $32°$ **31.** $50°$ **33.** $68°$ **35.** (a) 81 square inches; (b) 36 inches
37. (a) 16 square feet; (b) 16 feet   **39.** (a) 2.89 square inches; (b) 6.8 inches
41. (a) 314 square inches; (b) 62.8 inches   **43.** (a) 78.5 square feet; (b) 31.4 feet

45. (a) 120.7 square meters; (b) 38.94 meters   **47.** (a) 35 square inches; (b) 24 inches
49. (a) 24 square meters; (b) 22 meters   **51.** (a) 150 square feet; (b) 50 feet
53. (a) 27 cubic inches; (b) 54 square inches   **55.** (a) 216 cubic centimeters; (b) 216 square centimeters
57. (a) 1,000 cubic centimeters; (b) 600 square centimeters
59. (a) 523.33 cubic centimeters; (b) 314 square centimeters   **61.** (a) 288.55 cubic inches; (b) 211.13 square inches
63. (a) 2,308.39 cubic meters; (b) 844.53 square meters   **65.** 150 miles   **67.** 990 feet   **69.** 7,800 centimeters
71. $11,564.80   **73.** $41,393.44   **75.** 175 beats per minute

REVIEW PROBLEM SET   page 53

1. $x - 11 = 21$ **3.** $2(y + 3) = 27$ **5.** The difference of y and 2 is less than or equal to 11
7. The sum of 4 times x and 3 equals 18   **9.** Base = 3; exponent = 4; $3^4 = 3 \cdot 3 \cdot 3 \cdot 3 = 81$
11. Base = 4; exponent = 5; $4^5 = 4 \cdot 4 \cdot 4 \cdot 4 \cdot 4 = 1,024$ **13.** 33   **15.** 22   **17.** $A = \{4, 5, 6, 7, 8, 9, 10, 11\}$
19.

21.

23. 1.4   **25.** 3.125   **27.** $\frac{17}{50}$ **29.** $\frac{23}{125}$ **31.** $-\frac{3,407}{500}$ **33.** Rational   **35.** Irrational   **37.** Rational
39. Irrational   **41.** Rational   **43.** Rational   **45.** Commutative property for addition
47. Commutative property for multiplication   **49.** Associative property for multiplication
51. Identity property for addition   **53.** Zero-factor property (i)   **55.** Cancellation property for addition
57. Distributive property   **59.** Identity property for addition   **61.** 13   **63.** -21 **65.** -3 **67.** -14
69. 14   **71.** 30   **73.** -12 **75.** 48   **77.** -40 **79.** -32 **81.** 105   **83.** -9 **85.** 8   **87.** -3
89. $\frac{3}{5}$ **91.** $-\frac{3}{4}$ **93.** $\frac{5}{7}$ **95.** $\frac{1}{3}$ **97.** $\frac{11}{12}$ **99.** $\frac{2}{9}$ **101.** $\frac{1}{9}$ **103.** $-\frac{1}{10}$ **105.** 1.08   **107.** 3.82
109. 0.04   **111.** 1.035   **113.** 1.013   **115.** 29.30   **117.** 0.714   **119.** 0.432   **121.** 7.28   **123.** 8.44
125. 0.0442   **127.** 0.0177   **129.** 5.872×10^3 **131.** 1.234×10^5 **133.** 68,720   **135.** 291,000   **137.** 21
139. -6 **141.** $2^2 \cdot x^3$ **143.** $5u^2v^3$ **145.** $(-2)^2(-t)^2(-s)^2$ **147.** $x \cdot x \cdot y \cdot y \cdot y \cdot y$
149. $3 \cdot 3 \cdot 3 \cdot u \cdot u \cdot v \cdot v \cdot v \cdot v$ **151.** $m \cdot m \cdot m \cdot n \cdot n \cdot n \cdot n \cdot p$ **153.** $86°$ **155.** 30 square inches
157. 14 feet   **159.** 153.86 square inches

CHAPTER 1 TEST   page 56

1. (a) $2x + 3 = 15$; (b) $7y < y - 5$ **2.** Base = 4; exponent = 3; $4^3 = 4 \cdot 4 \cdot 4 = 64$ **3.** 0
4. $A = \{3, 4, 5, 6, 7, 8, 9, 10\}$ **5.** 1.6   **6.** $-\frac{173}{50}$ **7.** (a) 7; (b) -15; (c) -9; (d) -12; (e) 2; (f) -24
8. (a) $-\frac{7}{13}$ (b) $\frac{17}{21}$ **9.** (a) $\frac{1}{2}$; (b) $\frac{11}{40}$; (c) $\frac{1}{20}$; (d) $\frac{4}{3}$; (e) $-\frac{3}{4}$ **10.** (a) 3.24; (b) 5.002   **11.** 3.196×10^6
12. 52,700   **13.** 105   **14.** $2^2(-x)^3y^2$ **15.** $4 \cdot u \cdot u \cdot u \cdot v \cdot v \cdot v \cdot v$ **16.** 11.5 centimeters

Chapter 2

PROBLEM SET 2.1   page 62

1. Binomial; degree = 1; 3, -2 **3.** Monomial; degree = 2; 4   **5.** Trinomial; degree = 2; 1, -5, 6
7. Binomial; degree = 7; 2, -13 **9.** Trinomial; degree = 4; -1, -1, 13   **11.** 3   **13.** 2   **15.** 5
17. 6   **19.** 9   **21.** 7   **23.** 19   **25.** -1 **27.** 2   **29.** 20   **31.** 4   **33.** 10   **35.** 0   **37.** No
39. Yes   **41.** (a) 64 feet; (b) 48 feet

PROBLEM SET 2.2 page 68

1. $12x^2$ **3.** $11v^3$ **5.** $4t^2$ **7.** $8x + 7$ **9.** $7z^2 + 5z + 5$ **11.** $4x^2 - 2x - 2$ **13.** $5c^4 + c^3 + 4c^2 + c$
15. $4u$ **17.** $2x^2$ **19.** $6v^3$ **21.** $4t^3 - 2$ **23.** $2x^2 + 2x + 4$ **25.** $4s^4 - s^3 + 4s^2 + 2s - 5$
27. $-4t^3 + 5t^2 + 4t - 18$ **29.** $12xy$ **31.** $3uv$ **33.** $2mn^2 + 2m^2n$ **35.** $-2x^2y$ **37.** $4ts - 6$
39. $-6x^2y + 2xy - 4xy^2$ **41.** $5w^2z + 3wz - 10wz^2$ **43.** $5x^2 + 2x - 2$ **45.** $2t^3 + 2t^2 - t + 2$
47. $w^2 + w + 9$ **49.** $10u^3v^2 - 9u^2v - 5w$ **51.** (a) $P = 50x - 0.15x^2 - 100$; (b) \$3,900
53. $10x + 25(x - 6) = 35x - 150$

PROBLEM SET 2.3 page 76

1. $3^5 = 243$ **3.** $(-2)^5 = -32$ **5.** $-x^{12}$ **7.** t^{12} **9.** v^8 **11.** 64 **13.** 64 **15.** x^{35} **17.** t^{22}
19. w^{12} **21.** $16x^4$ **23.** u^5v^5 **25.** $x^7y^7z^7$ **27.** $-8w^3$ **29.** $27x^3y^3$ **31.** $\frac{9}{16}$ **33.** $-\frac{8}{27}$ **35.** $\frac{x^4}{y^4}$
37. $-\frac{a^5}{b^5}$ **39.** $\frac{t^6}{s^6}$ **41.** 27 **43.** 64 **45.** x^5 **47.** y^5 **49.** w^3 **51.** $81x^8y^{12}$ **53.** u^3v^4 **55.** $\frac{w^6}{8z^3}$
57. $\frac{27a^9}{8b^9}$ **59.** $-\frac{16y^{10}z^9}{x}$ **61.** 9.6×10^{14} **63.** 9×10^{11} **65.** 2×10^4 **67.** \$4,270.36 **69.** 1
71. -1

PROBLEM SET 2.4 page 87

1. $6x^6$ **3.** $-30t^7$ **5.** $-28u^5v^7$ **7.** $12x^3y^3z^4$ **9.** $-24a^3b^3c^4$ **11.** $x^2 + x$ **13.** $t^3 + 2t^2$
15. $6w^3 - 12w$ **17.** $-4x^3y^2 + 6x^2y^3 - 10xy^4$ **19.** $12c^5d^2 - 8c^4d^3 + 4c^3d^4$ **21.** $x^2 + 2xy + y^2 - x - y$
23. $2t^3 + 5t^2s + 4ts^2 + s^3$ **25.** $m^5 + 2m^4 + 10m^2 - 9m + 12$ **27.** $y^4 - y^3 - 10y^2 + 4y + 24$
29. $x^5 - x^4y - 2x^3y^2 + 2x^2y^3 + xy^4 - y^5$ **31.** $x^2 + 3x + 2$ **33.** $u^2 - 9u + 20$ **35.** $t^2 + 3t - 10$
37. $y^2 - 3y - 18$ **39.** $6x^2 + x - 1$ **41.** $5w^2 - w - 4$ **43.** $2x^2 + 7xy + 3y^2$ **45.** $24m^2 - 2mn - 15n^2$
47. $28x^2 - 23xy - 15y^2$ **49.** $50v^2 - 115v + 56$ **51.** $x^2 + 2x + 1$ **53.** $4s^2 + 4st + t^2$
55. $u^2 - 6uv + 9v^2$ **57.** $9x^2 - 30x + 25$ **59.** $16y^2 + 40yz + 25z^2$ **61.** $x^2 - y^2$ **63.** $w^2 - 49$
65. $4m^2 - 81$ **67.** $64x^2 - y^2$ **69.** $25u^2 - 36v^2$ **75.** $x^3 + 3x^2 + 3x + 1$ **77.** $x^3 + 1$
79. $c^3 - 6c^2d + 12cd^2 - 8d^3$ **81.** $u^3 - 27v^3$ **83.** $125t^3 + 150t^2s + 60ts^2 + 8s^3$ **85.** $8x^3 + 27y^3$
87. $x^9 - 6x^6 + 12x^3 - 8$ **89.** $u^6 - v^6$ **91.** $v = 4x^3 - 168x^2 + 1728x$; 5,184 cubic centimeters

PROBLEM SET 2.5 page 92

1. $x(x - 1)$ **3.** $3x(3x + 1)$ **5.** $x(4x + 7y)$ **7.** $ab(a - b)$ **9.** $6pq(p + 4q)$ **11.** $6ab(b + 5a)$
13. $12x^2y(x - 4y)$ **15.** $2ab(a^2 - 4ab - 3b^2)$ **17.** $xy^2(x^2 + xy + 2y^2)$ **19.** $9mn(m + 2n - 3)$
21. $(3x + 5y)(2a + b)$ **23.** $(5x + 9ay + 9by)(a + b)$ **25.** $(m - 1)(x - y)$
27. $[7x + 14(2a + 7b) + (2a + 7b)^2](2a + 7b)$ **29.** $(xy + 2)[y(xy + 2)^2 - 5x(xy + 2) + 7]$ **31.** $(a + b)(x + y)$
33. $(x^4 + 1)(x + 3)$ **35.** $(y - 1)(z + 2)$ **37.** $(b^2 - d)(a - c)$ **39.** $(2a - b)(x - y)$ **41.** $(x - a)(x + b)$
43. $(a + b)(x + y + 1)$ **45.** $(x^2 + y)(2x + y - 1)$

PROBLEM SET 2.6   page 97

1. $(x - 2)(x + 2)$ **3.** $(1 - 3y)(1 + 3y)$ **5.** $(6 - 5t)(6 + 5t)$ **7.** $(4u - 5v)(4u + 5v)$ **9.** $(ab - c)(ab + c)$
11. $(a - b - 10c)(a - b + 10c)$ **13.** $(3x - 1)(3x + 1)(9x^2 + 1)$ **15.** $(u - v)(u + v)(u^2 + v^2)(u^4 + v^4)$
17. $(x + y - a + b)(x + y + a - b)$ **19.** $[t - 3(r + s)][t + 3(r + s)][t^2 + 9(r + s)^2]$ **21.** $(x + 1)(x^2 - x + 1)$
23. $(4 - t)(16 + 4t + t^2)$ **25.** $(3w + z)(9w^2 - 3wz + z^2)$ **27.** $(2x - 3y)(4x^2 + 6xy + 9y^2)$
29. $(w - 2yz)(w^2 + 2wyz + 4y^2z^2)$ **31.** $(x + 2 - y)[(x + 2)^2 + (x + 2)y + y^2]$
33. $(y + w + 3)[(y + 1)^2 - (y + 1)(w + 2) + (w + 2)^2]$ **35.** $(w^2 + 2z^2)(w^4 - 2w^2z^2 + 4z^4)$
37. $(x - 1)(x^2 + x + 1)(x^6 + x^3 + 1)$ **39.** $2x(2x - y)(2x + y)$ **41.** $4y(4 - y)(4 + y)$
43. $3uv(u - 2)(u^2 + 2u + 4)$ **45.** $7xy(x^2 + y^2)(x^4 - x^2y^2 + y^4)$ **47.** $(t - 1)(t + 1)(t^2 + t + 1)(t^2 - t + 1)$
49. $2u(u - 2)(u + 2)(u^2 + 2u + 4)(u^2 - 2u + 4)$ **51.** $(y + 4)^2$ **53.** $(3u - 7v)^2$
55. $(x + y - z - 3)(x + y + z + 3)$ **57.** $(w + 1 - y - z)(w + 1 + y + z)$ **59.** $(x^2 + y^2 - xy)(x^2 + y^2 + xy)$
61. $(2m^2 + n^2 - 2mn)(2m^2 + n^2 + 2mn)$

PROBLEM SET 2.7   page 105

1. $(x + 1)(x + 3)$ **3.** $(t - 1)(t - 2)$ **5.** $(y + 3)(y + 12)$ **7.** $(x + 3)(x - 5)$ **9.** $(u - 7)(u - 9)$
11. $(z + 5w)(z + 6w)$ **13.** $(x + 2)(x - 9)$ **15.** $(m - 10n)(m + 12n)$ **17.** $(2 - x)(6 + x)$ **19.** $(4 - t)(9 + t)$
21. $(2w + 1)(w + 3)$ **23.** $(3x - 1)(x + 2)$ **25.** $(5y - 1)(y - 2)$ **27.** $(3c + d)(c + 2d)$ **29.** $(3x + 2)(2x + 3)$
31. $(3z - 2y)(2z + 3y)$ **33.** $(4v - 1)(3v + 5)$ **35.** $(7x - 6)(8x - 5)$ **37.** $(3 - 2w)(4 + w)$
39. $(5r - 4s)(r + 2s)$ **41.** $5x(x - 4)(x - 7)$ **43.** $2st(8t - s)^2$ **45.** $y^2(x + 3)(x + 7)$ **47.** $4m^2(n - 1)(n + 7)$
49. $wy(x - 2)(x - 7)$

PROBLEM SET 2.8   page 112

1. $3x^3$ **3.** $-\dfrac{3x^2}{y^2}$ **5.** $\dfrac{2w^2}{u^2}$ **7.** $3n^2 - 2m$ **9.** $2xy^2 - 8y^2 + 2$ **11.** $2ab^2 + b - \dfrac{4}{ab^2} + \dfrac{3}{a^2b^3}$ **13.** $x - 2$
15. $v - 2;\ R = -8$ **17.** $w^2 + w - 4;\ R = 3$ **19.** $2t^3 + t^2 - 6t + 8;\ R = -9$ **21.** $x^2 + 3x + 4$
23. $y + 2$ **25.** $x + 2$ **27.** $u^2 - 2uv + v^2;\ R = v^3$ **29.** $2m^2 - mn + 2n^2;\ R = 2n^3$
31. $x^3 + 2x^2y - 2xy^2 + y^3;\ R = -6y^4$ **33.** $x^2 + xy + y^2$

REVIEW PROBLEM SET   page 112

1. Trinomial; degree $= 2$; $4, -3, 2$ **3.** Binomial; degree $= 2$; $-7, 3$ **5.** Monomial; degree $= 5$; 10
7. 17 **9.** 46 **11.** 13 **13.** $9w^2$ **15.** $-3x^2 - 1$ **17.** $v^2 + 4v + 6$ **19.** $4x^2 + 6x - 4$ **21.** x^{11}
23. w^{12} **25.** m^{27} **27.** $-v^3u^6$ **29.** $4y^6z^2$ **31.** $\dfrac{9t^2}{s^6}$ **33.** x^3 **35.** $3z^{10}$ **37.** $6t^5 - 12t^3$
39. $6x^3y^4 - 10x^2y^5 + 2xy^4$ **41.** $a^3 - 3a^2b + 3ab^2 - b^3$ **43.** $w^2 + 10w + 21$ **45.** $2x^2 + 7x - 15$
47. $4u^2 - 17uv + 15v^2$ **49.** $t^2 - 64$ **51.** $9x^2 + 42x + 49$ **53.** $t^3 + 8$ **55.** $y^3 + 9y^2 + 27y + 27$
57. $16 - 24s + 9s^2$ **59.** $27w^3 - 54w^2 + 36w - 8$ **61.** $8x^3 - z^3$ **63.** $7xy(x - 3y^2)$
65. $13a^2b^2(2a + 3a^3b^2 - 4b)$ **67.** $2(y + z)(y - 2x)$ **69.** $(x - y)(3 + z)$ **71.** $(x - 3y)(2u + v)$

73. $(5a - b)(m^2 + n)$ **75.** $(5m - 3n)(5m + 3n)$ **77.** $(x + 4)(x^2 - 4x + 16)$ **79.** $(2t - 3)(2t + 3)(4t^2 + 9)$
81. $(x + y - z + 1)(x + y + z - 1)$ **83.** $9u(u - 3v)(u + 3v)$ **85.** $5s(t + 4)(t^2 - 4t + 16)$
87. $(2x - y)(2x + y)(4x^2 + 2xy + y^2)(4x^2 - 2xy + y^2)$ **89.** $(x + y - w - z)(x + y + w + z)$
91. $(t^2 - t + 1)(t^2 + t + 1)$ **93.** $(x - y)(x + 3y)$ **95.** $(m + 4)(m - 9)$ **97.** $(3u + 2)(u + 5)$
99. $(2y + 3)(y - 2)$ **101.** $(5x - 4y)(4x - 3y)$ **103.** $w(w + 11)(w - 2)$ **105.** $yz(x - 8y)(x + 2y)$
107. $6u^2$ **109.** $8w^2z^2 - 4wz$ **111.** $5x - 1$ **113.** $x - 2$ **115.** $t^2 + 2ts - s^2$ **117.** $w^{10} - w^5 + 1$

CHAPTER 2 TEST   page 114

1. (a) 32; (b) 13   **2.** (a) $5m^2 + 2m - 4$; (b) $-3z^3 - 4z^2 - 4z - 5$; (c) 7   **3.** (a) $-x^5$; (b) $3y^4$; (c) x^8y^{12}; (d) $-\dfrac{2x^4}{y^3}$

4. (a) $6s^5 - 12s^2$; (b) $x^3 - 2x^2y + 4xy^2 - 3y^3$ **5.** (a) $4u^2 - 9v^2$; (b) $9w^2 - 42wz + 49z^2$; (c) $x^2 + 10xy + 25y^2$;
(d) $8x^3 - y^3$; (e) $m^3 + 9m^2n + 27mn^2 + 27n^3$ **6.** (a) $6x^2(2 - 5x^3)$; (b) $5m^3n^2(3n^3 - 5m^2 + 7mn)$
7. (a) $(a + b)(x - y)$; (b) $(z + 3)(2x^2 + y)$ **8.** (a) $(4x - 3y)(4x + 3y)$; (b) $(3w + 2)(9w^2 - 6w + 4)$
9. (a) $(x + 4y)(x - 9y)$; (b) $(3y - 5)(2y + 3)$; (c) $(2x - 3z)^2$; (d) $2xy(x + 7y)(x - 2y)$
10. (a) $-8w^3$; (b) $9x - 6y + 3x^2y^2$; (c) $z^3 + 3z^2 - 2z + 5$; (d) $x^2 - 2x + 4$

Chapter 3

PROBLEM SET 3.1   page 124

1. 3   **3.** -10 **5.** 6; -7 **7.** -2; 8   **9.** -8; -9; 4   **11.** Equivalent   **13.** Not equivalent

15. Equivalent   **17.** Not equivalent   **19.** Equivalent   **21.** $\frac{5}{6}$ **23.** $\dfrac{5y^3}{9x}$ **25.** $\dfrac{m + 1}{m - 1}$ **27.** $\dfrac{t - 1}{t + 2}$

29. $\dfrac{2x + 3}{3x}$ **31.** $\dfrac{v + 4}{v + 7}$ **33.** $\dfrac{x + 1}{x - 3}$ **35.** $\dfrac{2u + 3}{4u^2 + 6u + 9}$ **37.** $-\dfrac{x + y}{2x + y}$ **39.** $\dfrac{z + w}{z + y}$ **41.** 36

43. $20m^7y^4$ **45.** $15u + 15v$ **47.** $6x - 6$ **49.** $4t^2 + 2t$ **51.** $y^2 - 2y - 3$ **53.** $2x^2 + x - 6$

55. $4y^2 - 9x^2$ **57.** $\dfrac{7}{x - y}$ **59.** $\dfrac{-xy}{x - y}$ **61.** $\dfrac{x - 3}{x - y}$ **63.** 3 centimeters

PROBLEM SET 3.2   page 130

1. $\dfrac{3x}{5}$ **3.** $\dfrac{2x}{3y^3}$ **5.** $\dfrac{y}{t}$ **7.** $\dfrac{5}{3t}$ **9.** $\dfrac{6}{x - 8}$ **11.** $-7a - 2$ **13.** $\dfrac{(v + 1)(3v + 1)}{v - 9}$ **15.** $(x - 12)(x - 4)$

17. $\dfrac{(y - 1)^2}{(y + 1)^2}$ **19.** $\dfrac{(a + 2b)(a - b)}{(a + b)(a - 2b)}$ **21.** $x^2 - y^2$ **23.** $\frac{5}{9}$ **25.** $6x^2y$ **27.** $5t$ **29.** $-\dfrac{3y + 9}{7}$

31. $\dfrac{x(x + 1)}{2x - 1}$ **33.** $\dfrac{3(a - 2)}{a + 2}$ **35.** $\dfrac{3(u - 3v)}{(u + 3v)(u - v)}$ **37.** $\dfrac{x - 3}{3x + 1}$ **39.** $\dfrac{4y(2y + 3)}{(3y - 5)(4y + 1)}$ **41.** $(2x - 3)(3x + 1)$

43. $\dfrac{x - 2}{(x + 1)^2}$ **45.** $\dfrac{a - 1}{(3 - a)(b + b^2)}$ **47.** $\dfrac{8}{3(v - 2)}$ **49.** $\dfrac{y + 2}{y + 5}$

PROBLEM SET 3.3 page 139

1. $\dfrac{3}{4x}$ **3.** $\dfrac{4}{x}$ **5.** $\dfrac{1}{3}$ **7.** $\dfrac{t}{2}$ **9.** $\dfrac{2}{x+2}$ **11.** $6+v$ **13.** $\dfrac{3}{m-1}$ **15.** $\dfrac{3}{3x+2}$ **17.** $-\dfrac{3}{2u+1}$

19. $2u-3v$ **21.** $\dfrac{59}{56x}$ **23.** $\dfrac{23}{60t}$ **25.** $\dfrac{47x}{30}$ **27.** $\dfrac{u}{21}$ **29.** $\dfrac{15y-57}{(y-5)(y-3)}$ **31.** $\dfrac{m^2+m+6}{(m-3)(m+3)}$

33. $\dfrac{x^2-x+2}{(3x+2)(x-4)}$ **35.** $\dfrac{10s^2+3t}{2t^2s^3}$ **37.** $\dfrac{36n^4-20m}{15m^3n^5}$ **39.** $\dfrac{5x-3}{x(x+1)(x-1)}$ **41.** $\dfrac{3-4c}{(c-3)(c+3)(c-2)}$

43. $\dfrac{2m^2+4}{m(m+1)(m-6)}$ **45.** $\dfrac{2x^2+14x-6}{(x+8)(x+2)(x+7)}$ **47.** $\dfrac{2y^2-14y}{(y+2)(y+5)(y-4)}$ **49.** $\dfrac{3}{3t-2}$

51. $\dfrac{8x^2+15}{(2x+3)(x-4)(3x-1)}$ **53.** $\dfrac{v^2+4v}{4-v^2}$ **55.** $\dfrac{17t^2+t-9}{27-3t^2}$ **57.** $\dfrac{2x^2+13x-7}{(x-3)^2(x+2)^2}$

59. $\dfrac{4y^2+17y-2}{(3y+2)(y-1)(y+3)}$ **61.** $x+3\left(\dfrac{1}{x}\right)=\dfrac{x^2+3}{x}$ **63.** $\dfrac{1}{x}+\dfrac{1}{x+2}=\dfrac{2x+2}{x(x+2)}$

65. $\dfrac{1}{x+10}+\dfrac{1}{x-2}=\dfrac{2x+8}{(x+10)(x-2)}$

PROBLEM SET 3.4 page 144

1. $\dfrac{5}{6}$ **3.** $\dfrac{15y}{8}$ **5.** $\dfrac{3c+1}{2c-3}$ **7.** $\dfrac{1}{2y-1}$ **9.** $\dfrac{2}{3v+4}$ **11.** $\dfrac{m}{m-2}$ **13.** $\dfrac{x^2-y^2}{x^2+y^2}$ **15.** $-\dfrac{b}{2}$ **17.** $3-x$

19. $\dfrac{3t+2}{2t-1}$ **21.** $\dfrac{-2x}{x^2+1}$ **23.** 1 **25.** $R=\dfrac{R_1R_2}{R_1+R_2}$; 6 ohms

REVIEW PROBLEM SET page 146

1. 5 **3.** $-2;7$ **5.** Equivalent **7.** Equivalent **9.** Not equivalent **11.** $\dfrac{2v^3}{3u^4}$ **13.** $\dfrac{1}{3m-3n}$

15. $\dfrac{x-4}{x+6}$ **17.** $\dfrac{t-5}{2t-5}$ **19.** $\dfrac{x-2}{x}$ **21.** $21x^2y^2$ **23.** $3a(a+b)$ **25.** $(2z+1)(3z+2)$

27. $w(w^2-3w+9)$ **29.** $\dfrac{-4b}{b-a}$ **31.** $\dfrac{-3b}{b-a}$ **33.** $\dfrac{x^2u^2}{10yv^3}$ **35.** $\dfrac{3}{2(b-5)}$ **37.** $\dfrac{3-y}{x(x+1)}$ **39.** $\dfrac{u-v}{2u}$

41. 1 **43.** $\dfrac{2t+1}{t+3}$ **45.** $\dfrac{2}{bx^2}$ **47.** $\dfrac{v-2}{v-7}$ **49.** $\dfrac{m+1}{m-3}$ **51.** $\dfrac{x+2y}{x}$ **53.** $\dfrac{(w-2)(w+2)}{(w-4)(w+1)}$ **55.** $\dfrac{x-3}{x+2}$

57. $\dfrac{u+2}{u-2}$ **59.** $\dfrac{2}{x}$ **61.** $\dfrac{1}{3+t}$ **63.** $\dfrac{3}{y-2}$ **65.** $\dfrac{x^2+y^2}{x^2-y^2}$ **67.** $\dfrac{7}{3-t}$ **69.** $\dfrac{5y+16}{(y+4)(y+6)}$

71. $\dfrac{16n^2+15m}{20m^2n^3}$ **73.** $\dfrac{v^2+3v}{(2v-3)(2v+3)(v+2)}$ **75.** $\dfrac{13x-19}{2(x-3)(x-4)(x-1)}$ **77.** $\dfrac{9m+15}{(2m+3)(m-1)(m+1)}$

79. $\dfrac{y^2-5xy}{(x-3y)(x+3y)(x+y)}$ **81.** $\dfrac{1}{a^2x}$ **83.** $\dfrac{3+x}{4+2y}$ **85.** y **87.** $\dfrac{1}{t-1}$ **89.** $\dfrac{6}{x}$

CHAPTER 3 TEST   page 148

1. 3, $-\frac{5}{3}$ **2.** (a) Equivalent; (b) Not equivalent   **3.** (a) $\dfrac{5x}{6y^2}$; (b) $\dfrac{2(x-y)}{x+y}$

4. (a) $21u^2vw^3$; (b) $2x^3 + 2x^2y + 2xy^2$ **5.** (a) x^2y^2; (b) $\dfrac{(x+4)(x-2)}{x+6}$; (c) $x + 3y$

6. (a) $\dfrac{1}{3xy}$; (b) $\dfrac{3x^2 + 5x - 8}{(x-4)(x+1)}$; (c) $\dfrac{4-x}{(2x+1)(x+2)(x-1)}$; (d) $\dfrac{8z^2 - 2z - 2}{(2z+1)(z+2)(3z-1)}$

7. (a) $\dfrac{5y}{4x}$; (b) $\dfrac{5+2y}{x-3}$; (c) $\dfrac{xy - y^2 + x}{x^2 - xy - y}$

Chapter 4

PROBLEM SET 4.1   page 156

1. 7   **3.** 12   **5.** 6   **7.** 3   **9.** -3 **11.** $-\frac{9}{5}$ **13.** 5   **15.** $\frac{15}{8}$ **17.** $-\frac{2}{3}$ **19.** 7   **21.** 4   **23.** 3
25. $\frac{25}{3}$ **27.** $\frac{5}{4}$ **29.** $\frac{15}{7}$ **31.** -3 **33.** 2   **35.** 3   **37.** 1   **39.** 0   **41.** 5   **43.** 1   **45.** 3   **47.** 3
49. 9   **51.** $\frac{5}{9}$ **53.** $\frac{4}{3}$ **55.** $\frac{1}{2}$ **57.** $-\frac{3,125}{999}$ **59.** -1.20 **61.** -33.572

PROBLEM SET 4.2   page 161

1. -3 **3.** 7   **5.** 1   **7.** 24   **9.** 2   **11.** $\frac{267}{40}$ **13.** 3   **15.** $-\frac{2}{3}$ **17.** 2   **19.** 5   **21.** 7   **23.** $\frac{25}{16}$
25. 2   **27.** $\frac{1}{4}$ **29.** 2   **31.** $-\frac{1}{3}$ **33.** $-\frac{10}{39}$ **35.** 8   **37.** $-\frac{6}{13}$ **39.** $-\frac{5}{2}$ **41.** No solution
43. No solution

PROBLEM SET 4.3   page 166

1. $5c$ **3.** $\dfrac{c-b}{a}$ **5.** $-\dfrac{b}{2}$ **7.** $\dfrac{9c}{2}$ **9.** $-\dfrac{5a}{3}$ **11.** $\dfrac{h}{4}$ **13.** $-\dfrac{c}{2}$ **15.** $-7b$ **17.** $\dfrac{5+2b}{7}$ **19.** $\dfrac{bc+d}{a}$

21. $\dfrac{30a - 35b}{16}$ **23.** $\dfrac{1+2c}{2}$ **25.** $-2a$ **27.** 1   **29.** $\dfrac{9b}{17}$ **31.** $\dfrac{1-3a-a^2}{3-a}$ **33.** $-\dfrac{a}{5}$ **35.** $h = \dfrac{2A}{b}$

37. $\ell = \dfrac{V}{wh}$ **39.** $r = \dfrac{C}{2\pi}$ **41.** $t = \dfrac{V}{g}$ **43.** $t = \dfrac{I}{Pr}$ **45.** $m = \dfrac{F-b}{x}$ **47.** $w = \dfrac{P - 2\ell}{2}$ **49.** $a = \dfrac{2S - n\ell}{n}$

51. $d = \dfrac{2(S - na)}{n(n-1)}$ **53.** $d = \dfrac{L-a}{m-1}$ **55.** $t = \dfrac{mpv - km}{k}$

PROBLEM SET 4.4   page 169

1. $x + 6$ **3.** $10x - 7$ **5.** $5x - 8$ **7.** $2x + 9$ **9.** $x^2 + (x+2)^2$ **11.** $3x + 5$ **13.** $20x + 7$
15. $6x + 4$ **17.** $4x + 24$ **19.** $25(q + 2) + 10(2d + 5)$ **21.** $15n + 40d + 125q$

23. d dimes; $d - 12$ nickels; $2d - 12$ quarters   **25.** $0.3x - 7$ **27.** $0.3x + 5$ **29.** $0.8t + 8$ **31.** $\dfrac{x}{0.289}$

33. $\dfrac{5x+2}{0.48}$ **35.** $\dfrac{p-16}{2}$ **37.** $x(x-8)$ **39.** $1.06x$

PROBLEM SET 4.5   page 183

1. 27   **3.** 32; 34   **5.** 36   **7.** 12 and 17   **9.** 14 and 11   **11.** 40   **13.** 11 quarters; 15 dimes; 45 nickels
15. 43 nickels; 27 dimes   **17.** 30 quarters; 50 dimes   **19.** \$120   **21.** \$200   **23.** \$8,058   **25.** \$16
27. \$31   **29.** \$12,500   **31.** \$28,000 at 13.2%; \$34,000 at 13.7%   **33.** \$4,444.44
35. Length = 64 meters; width = 32 meters   **37.** Length = 60 meters; width = 15 meters   **39.** 10 feet
41. 10 mph; 48 minutes   **43.** 12 hours   **45.** 15 mph   **47.** 6 milliliters of 10% acid; 4 milliliters of 15% acid
49. 200,000 gallons of 9% alcohol; 100,000 gallons of 12% alcohol
51. 75 pounds of the \$5.80 per pound coffee; 25 pounds of the \$6.00 per pound coffee   **53.** 45 hours
55. 4 hours   **57.** $\frac{4}{5}$ hour

PROBLEM SET 4.6   page 195

1.    **3.**    **5.**

7. 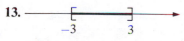   **9.** 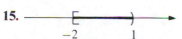   **11.**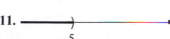

13. 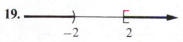   **15.**    **17.**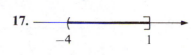

19. 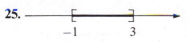   **21.** 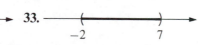   **23.**

25. 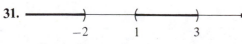   **27.** 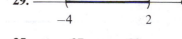   **29.**

31.   **33.**   **35.** <   **37.** <   **39.** >

41. <   **43.** >   **45.** <   **47.** >   **49.** Addition property   **51.** Transitive property
53. Multiplication property (ii)   **55.** Addition property   **57.** Multiplication property (ii)

PROBLEM SET 4.7   page 203

1. $x < 6$;   **3.** $x \geq 1$;
5. $x \leq -5$;   **7.** $x \geq 3$;

9. $x > \frac{3}{2}$;

11. $x > -3$;

13. $x \leq -\frac{2}{7}$;

15. $x < -2$;

17. $x > \frac{1}{5}$;

19. $x \leq \frac{2}{11}$;

21. $x \geq 3$;

23. $x < 1$;

25. $x < -2$;

27. $x \leq -\frac{4}{3}$;

29. $x \leq -\frac{2}{3}$;

31. $x < -1$;

33. $x < 3$;

35. $x \leq 1$;

37. $x \leq \frac{7}{3}$;

39. $x < \frac{7}{5}$;

41. $x \leq -2$;

43. $x \leq 8$;

45. $x \geq -3$;

47. $x \geq \frac{9}{2}$;

49. $x \geq 12$;

51. $x \geq \frac{19}{12}$;

53. $x \geq \frac{7}{5}$;

55. $-\frac{3}{2} \leq x \leq 0$;

57. $-\frac{4}{3} \leq x \leq 2$;

59. $-\frac{4}{3} < x < \frac{7}{3}$;

61. $2 < x < 6$;

63. $3 \le x \le 4$;

65. $-8 < x \le 1$;

67. $x \le -\frac{4}{5}$ or $x \ge 2$;

69. $x < 1$ or $x > 11$;

71. $x < -4$ or $x > -3$;

73. $x < \frac{3}{5}$ or $x > 1$;

75. $x < \frac{3}{4}$ or $x > \frac{7}{4}$;

77. $4x - 5 < 13$ **79.** Between \$297.50 and \$415 for one worker; between \$362.50 and \$480 for the other
81. $\frac{170}{9} \le C \le \frac{260}{9}$ **83.** \$35,000

PROBLEM SET 4.8 page 209

1. 5   **3.** 10   **5.** 7   **7.** 8   **9.** 5.2   **11.** 14   **13.** $-2, 2$ **15.** $-\frac{5}{2}, \frac{5}{2}$ **17.** $-3, 3$ **19.** $-3, 3$
21. $-2, 2$ **23.** $-\frac{4}{3}, \frac{4}{3}$ **25.** $-5, 5$ **27.** $-1, 1$ **29.** No solution   **31.** $-1, 5$ **33.** $-1, 11$ **35.** $\frac{1}{3}, \frac{8}{3}$
37. $-\frac{2}{7}, \frac{6}{7}$ **39.** No solution   **41.** $-\frac{2}{3}$ **43.** $-6, 10$ **45.** $-\frac{25}{4}, \frac{75}{4}$ **47.** $-3, 1$ **49.** $-\frac{9}{5}, \frac{3}{5}$
51. $-11, -3$ **53.** $-\frac{2}{5}, \frac{6}{5}$ **55.** $-26, 10$ **57.** $0, \frac{1}{3}$ **59.** $-\frac{1}{2}$ **61.** 0   **63.** -1 **65.** All real numbers
67. All real numbers   **69.** $-4, -3, -2, -1, 0, 1, 2, 3$

PROBLEM SET 4.9   page 214

1. $-1 < x < 1$;

3. $x < -5$ or $x > 5$;

5. $-5 < x < 5$;

7. $-4 \le x \le 4$;

9. $-6 \le x \le 6$;

11. $x < -3$ or $x > 3$;

13. $x \le -3$ or $x \ge 3$;

15. $x \le -4$ or $x \ge 4$;

17. $-2 < x < 4$;

19. $0 \le x \le 2$;

21. $x < -1$ or $x > 9$;

23. $x \le -1$ or $x \ge 7$;

25. $-1 < x < \frac{11}{3}$;

27. $x \le 2$ or $x \ge 3$;

29. $x \le -\frac{5}{4}$ or $x \ge \frac{1}{2}$;

31. $-1 \le x \le 2$;

33. $x \le -\frac{5}{2}$ or $x \ge 3$;

35. $\frac{1}{3} \le x \le 1$;

37. $x \le -1$ or $x \ge 5$;

39. $x < -\frac{9}{5}$ or $x > \frac{3}{5}$;

41. $-\frac{11}{2} \le x \le \frac{13}{2}$;

43. $-\frac{1}{7} \le x \le \frac{3}{7}$;

45. $-2 \le x \le 1$;

47. $x \le -\frac{4}{5}$ or $x \ge 2$;

49. All real numbers;

51. No solution;

53. All real numbers;

55. All real numbers;

57. No solution;

59. $|x| \le 5$

REVIEW PROBLEM SET   page 215

1. 2   **3.** 13   **5.** -3 **7.** -7 **9.** 4   **11.** -7 **13.** -4 **15.** $\frac{7}{4}$ **17.** 4   **19.** $-\frac{2}{7}$ **21.** $\frac{20}{9}$

23. 3   **25.** 6   **27.** 80   **29.** 3   **31.** 8   **33.** No solution   **35.** $\frac{37}{99}$ **37.** $\frac{11}{225}$ **39.** $-2a$ **41.** $-\dfrac{2a}{b}$

43. $\dfrac{b}{a-c}$ **45.** $\dfrac{12a - a^3}{6b}$ **47.** $\dfrac{2b}{x}$ **49.** $3a - 1$ **51.** $\dfrac{a^2 + 3a - ac}{a - c}$ **53.** $m = \dfrac{E}{c^2}$ **55.** $h = \dfrac{S - 2\pi r^2}{2\pi r}$

57. $m = G_i r^3$ **59.** $h = \dfrac{v^2 - gR}{g}$ **61.** $2x + 5$ **63.** $2x + 7$ **65.** Joe, \$15; Dawn, \$30; Mike, \$34

67. Width = 42 meters; length = 168 meters   **69.** \$18,750 and \$6,250   **71.** $1\frac{2}{3}$ hours   **73.** 6.9 miles

75.

77.

79.

81.

83.

85. $x > 2$;

87. $x > 6$;

89. $x < -4$;

91. $x \leq -10$;

93. $x \geq 8$;

95. $x \geq \frac{1}{18}$;

97. Its weight must be less than 52.5 ounces   **99.** $-5, 5$ **101.** $-33, 29$ **103.** No solution   **105.** $-4, \frac{5}{2}$

107. $-2, \frac{36}{11}$ **109.** $-4, \frac{2}{3}$

111. $-11 < x < 11$;

113. $x \leq -4$ or $x \geq 4$;

115. $-83 \leq x \leq 3$;

117. $x \leq -4$ or $x \geq \frac{1}{2}$;

119. $-14 < x < 16$;

121. $-\frac{14}{3} \leq x \leq \frac{8}{3}$;

CHAPTER 4 TEST   page 218

1. (a) $\frac{8}{3}$; (b) -1; (c) -14; (d) No solution   **2.** (a) $-\frac{b}{c}$; (b) $\frac{21a + 4}{5}$ **3.** (a) $h = \frac{S - 2A}{p}$; (b) $R = \frac{v^2 - gh}{g}$

4. Son's age is $x + 6$; Mary's age is $4x + 6$.   **5.** Joe sold 24, Jim sold 12, and Mary sold 14.

6. (a)   (b)   (c)

(d)   **7.** (a) $x \geq 4$ (b) $y > -\frac{5}{2}$ (c) $x \geq \frac{4}{13}$ **8.** (a) $-8, 8$ (b) $-\frac{5}{3}, 3$

9. (a) $-2 < x < 8$;   (b) $x \leq -\frac{1}{2}$ or $x \geq \frac{5}{2}$;

Chapter 5

PROBLEM SET 5.1   page 228

1. 1   **3.** 1   **5.** 1   **7.** Undefined   **9.** $\frac{1}{25}$ **11.** $\frac{1}{49}$ **13.** $-\frac{1}{10}$ **15.** $\frac{27}{64}$ **17.** $\frac{1}{p^5}$ **19.** x **21.** $\frac{1}{u^3 v^6}$

23. $\frac{4b^2}{a^7}$ **25.** 1,000   **27.** $-\frac{1}{24}$ **29.** $\frac{1}{6}$ **31.** $\frac{x^2}{7y^5}$ **33.** 290   **35.** $y - x$ **37.** 4   **39.** 49   **41.** $\frac{x^4 - 1}{x^2}$

43. $2a^2b^2$ **45.** $\dfrac{2x-1}{x}$ **47.** $\dfrac{1+x}{1-3x}$ **49.** 49 **51.** $\frac{1}{81}$ **53.** $\dfrac{1}{x^5}$ **55.** $\frac{1}{64}$ **57.** $\frac{1}{256}$ **59.** $\dfrac{1}{81x^4}$ **61.** $\dfrac{25}{p^8}$

63. $\dfrac{x^2}{25}$ **65.** $\dfrac{q^4}{p^4}$ **67.** 64 **69.** $\dfrac{1}{x^8}$ **71.** $\dfrac{1}{y^3}$ **73.** $a+b$ **75.** x^5 **77.** $\frac{1}{9}$ **79.** $\dfrac{x^2}{y^{10}}$ **81.** $\dfrac{c^3}{a^9}$

83. $\dfrac{u^{28}z^4}{v^{20}}$ **85.** $\dfrac{1}{a^4b^4}$ **87.** 1.77273×10^{11} **89.** 1.67316×10^4 **91.** (a) \$197.86; (b) \$1,122.96

93. (a) \$316.01; (b) \$3,168.33 **95.** (a) \$436.38; (b) \$11,953.20 **97.** (a) \$351.03; (b) \$3,624.79

99. 4.6872×10^{-1} mile **101.** 3×10^2 hours

PROBLEM SET 5.2 page 239

1. 4 **3.** 21 **5.** -3 **7.** Undefined **9.** $13^{1/2}$ **11.** $(3x^2)^{1/5}$ **13.** $(a^3b^2)^{1/4}$ **15.** $(-3a^2b^4)^{1/5}$

17. $\sqrt[6]{243}$ **19.** $\sqrt{343y^3}$ **21.** $\sqrt[7]{1,024x^5y^5}$ **23.** 6 **25.** $-\frac{1}{4}$ **27.** $-\frac{3}{2}$ **29.** $\frac{3}{4}$ **31.** 4 **33.** 16

35. -8 **37.** 27 **39.** Undefined **41.** 9 **43.** $\frac{1}{16}$ **45.** $\frac{1}{27}$ **47.** $-\frac{1}{32}$ **49.** 2 **51.** x **53.** 25

55. $\dfrac{1}{x^2}$ **57.** $16p^{12}$ **59.** $128y^3$ **61.** $\dfrac{y}{5}$ **63.** $5^{17/21}$ **65.** $x^{1/2}$ **67.** $\dfrac{8d}{c}$ **69.** $\dfrac{2,304a^6}{b^5}$ **71.** $\frac{2}{125}$

73. $\dfrac{2y^3}{5x^2}$ **75.** q^6 **77.** $-\dfrac{pq^3}{3}$ **79.** $4x-x^2$ **81.** $\dfrac{4ab-1}{b}$ **83.** $\dfrac{25x^3+10x^{3/2}+1}{x^3}$ **85.** $x+y$

87. $\dfrac{15-11x^{1/4}-14x^{1/2}}{x^{1/2}}$ **89.** $\dfrac{4+x}{x^{1/2}}$ **91.** $\dfrac{x-2}{2(x-1)^{3/2}}$ **93.** $\dfrac{1}{(x^2+1)^{1/2}}$

95. (a) 1.592; (b) 0.1752; (c) 4.743; (d) 11.46; (e) 33.77; (f) 4.711 **97.** 522.4 square meters **99.** 361

PROBLEM SET 5.3 page 248

1. $3\sqrt{3}$ **3.** $-3\sqrt[3]{2}$ **5.** $12\sqrt{2}$ **7.** $4\sqrt{3x}$ **9.** $3\sqrt[3]{6c^2}$ **11.** $2y\sqrt{5y}$ **13.** $2x\sqrt[3]{x}$ **15.** $-2y^3\sqrt[3]{2y}$

17. $7p\sqrt{2p}$ **19.** $\dfrac{\sqrt{7}}{2}$ **21.** $-\dfrac{\sqrt[3]{5}}{2}$ **23.** $-\dfrac{\sqrt[3]{x^2}}{4}$ **25.** $\dfrac{\sqrt{3}}{2|w|}$ **27.** $-\dfrac{\sqrt[3]{a^2}}{2}$ **29.** $\dfrac{\sqrt{3}}{5|x|}$ **31.** $\dfrac{\sqrt{17}}{y^2}$ **33.** $\dfrac{\sqrt{3}}{3x^2}$

35. $\dfrac{3\sqrt[4]{6}}{5|y|}$ **37.** $-y^3$ **39.** $-2x^2$ **41.** $(a+b)^2$ **43.** $|x|$ **45.** $-2y^4z^2$ **47.** $2|a|b^2\sqrt[4]{4}$ **49.** 5 **51.** $-p$

53. x^2 **55.** y **57.** m **59.** $2|(3x+1)y^3|\sqrt[4]{2}$ **61.** $6a^2b^5$ **63.** $2xy\sqrt[3]{y^2}$ **65.** $x^2\sqrt[8]{x^3y}$ **67.** u^5

69. $ab\sqrt[4]{a}$ **71.** $\dfrac{5\sqrt[3]{p^2q}}{2}$ **73.** $\sqrt[6]{x^5}$ **75.** $y^{15}\sqrt{y^7}$ **77.** $b^{20}\sqrt[3]{a^{14}b^3}$ **79.** $4\sqrt{3}$ **81.** $|x+4|$ **83.** $3,000$

85. 10.08 **87.** 12.37 **89.** 7.65761×10^7

PROBLEM SET 5.4 page 252

1. $8\sqrt{7}$ **3.** $8\sqrt{5}$ **5.** $7\sqrt[3]{4}$ **7.** $5\sqrt{2}$ **9.** $3\sqrt{2}$ **11.** $3\sqrt{3}$ **13.** $6\sqrt{5}$ **15.** $13\sqrt[3]{3}$ **17.** $2\sqrt{3x}$

19. $(6p+3)\sqrt{p}$ **21.** $7\sqrt{2p}$ **23.** $7\sqrt{3x}$ **25.** $-m\sqrt[4]{m}$ **27.** $\dfrac{5x\sqrt{2x}}{y}$ **29.** $-\dfrac{33y\sqrt{y}}{7}$

31. $(3x^2-10xy)\sqrt[3]{3x^2y}$ **33.** $(10p^2+7q^2)\sqrt[3]{pq}$ **35.** $10\sqrt{3x}+\sqrt[3]{2x}$ **37.** $14m\sqrt[3]{m^2}$ **39.** $7\sqrt{y-1}$

41. $(3|x|-6|y|+2x-4y)\sqrt{x-2y}$ **43.** (a) 5.8416; (b) 4.2426; (c) no

PROBLEM SET 5.5   page 259

1. 6   **3.** $2\sqrt{6}$ **5.** $6\sqrt{42}$ **7.** $30\sqrt{3}$ **9.** $-231\sqrt{2}$ **11.** $-14\sqrt{tu}$ **13.** $5\sqrt[3]{6}$ **15.** 30

17. $\sqrt{x^2 + 4x + 3}$ **19.** $(a-3)\sqrt{a+3}$ **21.** $2\sqrt{2} - 2$ **23.** $4\sqrt{3} - \sqrt{6}$ **25.** $\sqrt{15x} - \sqrt{30y}$

27. $20 - 3\sqrt{10}$ **29.** $2\sqrt{33} - 44$ **31.** $x - 5 - \sqrt{x^2 - 4x - 5}$ **33.** $4 - \sqrt{6}$ **35.** $2x + 15\sqrt{x} + 7$

37. $34 + 3\sqrt{6}$ **39.** $18x + 2\sqrt{3xy} - 8y$ **41.** $8 - 2\sqrt{15}$ **43.** $4x + 4\sqrt{xy} + y$ **45.** $4x - 4\sqrt{xy} + y$

47. $x + 6\sqrt{x - 1} + 8$ **49.** 11   **51.** 42   **53.** $9x - 121$ **55.** $x - 1$ **57.** $\dfrac{2\sqrt{3}}{3}$ **59.** $\dfrac{8\sqrt{11x}}{77x}$ **61.** $\dfrac{\sqrt{7y}}{7y}$

63. $\dfrac{5\sqrt[3]{49}}{7}$ **65.** $\dfrac{5\sqrt[3]{3x^2}}{3x}$ **67.** $\sqrt[4]{27}$ **69.** $\dfrac{7\sqrt[6]{32}}{2}$ **71.** $\dfrac{5\sqrt{x+2}}{x+2}$ **73.** $\sqrt{5} - \sqrt{2}$ **75.** $\dfrac{5(\sqrt{5}+1)}{2}$

77. $2 - \sqrt{2}$ **79.** $\sqrt{7} + \sqrt{5} - \sqrt{21} - \sqrt{15}$ **81.** $\dfrac{3\sqrt{xy} + 2y}{9x - 4y}$ **83.** $\dfrac{x - 2\sqrt{xy} + y}{x - y}$ **85.** $\dfrac{37\sqrt{10} + 129}{227}$

87. $\dfrac{3\sqrt{2} + 2\sqrt{3} - \sqrt{30}}{12}$ **89.** 3   **91.** $\dfrac{\sqrt[3]{25} + \sqrt[3]{10} + \sqrt[3]{4}}{3}$ **93.** 0   **95.** $\dfrac{11\sqrt{5}}{5}$ **97.** $\dfrac{5\sqrt{15}}{3}$

PROBLEM SET 5.6   page 267

1. $9i$ **3.** $-\frac{5}{2}i$ **5.** $-x^2 i$ **7.** $15y^2 i$ **9.** $5 + 9i$ **11.** $2 - 2x^2 i$ **13.** -6 **15.** -4 **17.** $x = 5, y = 4$

19. $-1 + 9i$ **21.** $-6 + 8i$ **23.** 12   **25.** $1 - i$ **27.** $-16i$ **29.** $1 - 11i$ **31.** $3 - 6i$ **33.** $20 + 16i$

35. $-9 + 6i$ **37.** $14 - 8i$ **39.** $25 - 8i$ **41.** $4 - 19i$ **43.** $27 - 5i$ **45.** $5 + 12i$ **47.** $2i$ **49.** 41

51. 170   **53.** 53   **55.** (a) $2 - 4i$; (b) 20   **57.** (a) $-2i$; (b) 4   **59.** $\frac{4}{25} - \frac{3}{25}i$ **61.** $\frac{35}{41} + \frac{28}{41}i$ **63.** $\frac{7}{13} - \frac{4}{13}i$

65. $\frac{14}{29} - \frac{23}{29}i$ **67.** $\frac{10}{53} + \frac{35}{53}i$ **69.** $-\frac{1}{2} - \frac{1}{2}i$ **71.** $2 - 5i$ **73.** $\frac{19}{29} - \frac{4}{29}i$ **75.** $-\frac{9}{29} - \frac{21}{29}i$ **77.** i **79.** -1

81. i **83.** 1   **85.** i **87.** $-i$ **89.** 0

REVIEW PROBLEM SET   page 268

1. $\frac{1}{9}$ **3.** $\dfrac{1}{y^4}$ **5.** $\dfrac{1}{t^4 s^2}$ **7.** $\dfrac{y^3}{x^2}$ **9.** 1   **11.** $\frac{40}{9}$ **13.** $\dfrac{2x^3 y^2}{y^2 + x^3}$ **15.** 81   **17.** x^7 **19.** $\frac{1}{64}$ **21.** t^{12}

23. $16x^6$ **25.** $\dfrac{1}{8u^6}$ **27.** 25   **29.** $\dfrac{1}{r^3}$ **31.** $4y^5$ **33.** $\dfrac{x^2}{y^{12}}$ **35.** $\dfrac{1}{8u^2 v^7}$ **37.** $\dfrac{s^2 t^6}{r^8}$ **39.** 1.3643×10^2

41. 9   **43.** -6 **45.** Not a real number   **47.** $15^{1/2}$ **49.** $(a + b)^{3/4}$ **51.** $\sqrt[3]{13^2}$ **53.** $\sqrt[7]{(2x + 3y)^2}$

55. -8 **57.** $\frac{1}{8}$ **59.** -8 **61.** $\frac{1}{4}$ **63.** 7   **65.** $x^{1/4}$ **67.** $\frac{1}{8}$ **69.** $\dfrac{1}{x^{1/2}}$ **71.** $-2t^2$ **73.** $4m^2$ **75.** $\dfrac{v^6}{4}$

77. $\dfrac{c^{75}}{d^{12}}$ **79.** $-\dfrac{8z^9}{x^3 y^6}$ **81.** $a^{1/21}$ **83.** $r^8 y^{16/3}$ **85.** $\dfrac{1}{x^7 y^2}$ **87.** $2x - 3x^2$ **89.** $\dfrac{1 + 6z^{2/3} + 9z^{4/3}}{z^{4/3}}$

91. $\dfrac{3 + x}{x^{2/3}}$ **93.** $\dfrac{x^2 - 3}{(x^2 + 2)^{1/2}}$ **95.** $3(y - 1)^{1/3}[4(y - 1) + \frac{16}{3}]$ **97.** 133.3   **99.** 22.36   **101.** $5\sqrt{5}$

103. $2t\sqrt[4]{2t}$ **105.** $-2xy^3\sqrt[3]{3xy^2}$ **107.** $\frac{2}{5}$ **109.** $-\dfrac{t^2}{2}$ **111.** $-\dfrac{\sqrt[3]{5}}{c^2 d^3}$ **113.** $5u^3$ **115.** $2x^2$ **117.** v

119. $4u^2 v^3$ **121.** $\sqrt{x}$ **123.** $\sqrt[12]{z^7}$ **125.** $u\sqrt[6]{72u}$ **129.** $8\sqrt{2}$ **131.** $6\sqrt{x}$ **133.** $10\sqrt{2}$ **135.** $3\sqrt{3}$

137. $10\sqrt{2z}$ **139.** $5\sqrt{7u}$ **141.** $\sqrt{6} + \sqrt{15}$ **143.** $3\sqrt{2} - 2\sqrt{6}$ **145.** $12 + 4\sqrt{5}$ **147.** 6   **149.** $3t - 5$

151. $3\sqrt{2}$ **153.** $2x - \sqrt{xy} - y$ **155.** $\dfrac{4\sqrt{3}}{3}$ **157.** $\dfrac{5\sqrt{2}}{8}$ **159.** $\dfrac{2\sqrt{3ts}}{3s}$ **161.** $\dfrac{25 - 5\sqrt{11}}{14}$

163. $\dfrac{10\sqrt{x} + 20}{x - 4}$ **165.** $13 - 2\sqrt{42}$ **167.** $\dfrac{u^2 + u\sqrt{v} - 2v}{u^2 - 4v}$ **169.** $\dfrac{7\sqrt{3}}{9}$ **171.** $12i$ **173.** $3 + ti$

175. $-1 + 16i$ **177.** $-4 + 4i$ **179.** $-27 + 10i$ **181.** $23 - 10i$ **183.** $12 + 16i$ **185.** $46 + i$

187. $3 - 5i$ **189.** $-4 + 7i$ **191.** $\frac{14}{17} + \frac{5}{17}i$ **193.** $\frac{15}{13} + \frac{23}{13}i$ **195.** $-\frac{11}{26} - \frac{3}{26}i$ **197.** i **199.** $-i$

201. -1

CHAPTER 5 TEST   page 272

1. (a) $\dfrac{1}{z^3}$; (b) $\dfrac{1}{p}$; (c) $\dfrac{y^4}{x^3}$ **2.** (a) $\dfrac{1}{x^4}$; (b) $\dfrac{y^6}{4x^4}$; (c) $\dfrac{1}{xy}$ **3.** (a) $\dfrac{w^8}{z^4}$; (b) $\dfrac{1}{x^7}$ **4.** 1.48213×10^1

5. (a) 2; (b) -5; (c) Not a real number   **6.** $(x + 2)^{3/5}$ **7.** $\sqrt[7]{(3w - z)^4}$ **8.** (a) 8; (b) 9; (c) $\frac{1}{16}$

9. (a) $y^{5/3}$; (b) $\dfrac{1}{x^{1/10}}$; (c) $\dfrac{2x^2}{y^{1/3}}$ **10.** (a) $\dfrac{2y^3 - 4}{y}$; (b) $\dfrac{xy^{2/3} - 1}{y^{2/3}}$

11. (a) $5x^2y\sqrt{3xy}$; (b) $-3uv^4\sqrt[3]{3uv^2}$; (c) $\dfrac{4w\sqrt{2w}}{5|x|y^2}$; (d) $4v^4\sqrt{u}$ **12.** $x\sqrt[15]{x^7}$

13. (a) $(4x + 2)\sqrt{3x}$; (b) $3\sqrt{5} + 3\sqrt{2}$; (c) $7x - 3y$ **14.** (a) $\dfrac{\sqrt{21}}{3}$; (b) $\dfrac{3x - 5\sqrt{x} - 2}{x - 4}$

15. (a) $3 + 3i$; (b) $-5 + 2i$; (c) $-15 + 3i$; (d) $12 + i$; (e) $\frac{1}{13} + \frac{5}{13}i$

Chapter 6

PROBLEM SET 6.1   page 279

1. $1, 2$ **3.** $2, 4$ **5.** $-4, 5$ **7.** $-7, 5$ **9.** $-3, 4$ **11.** $-1, \frac{5}{3}$ **13.** $-\frac{1}{2}, \frac{2}{5}$ **15.** $-\frac{4}{3}, \frac{2}{3}$ **17.** $\frac{1}{2}, \frac{3}{5}$

19. $\frac{5}{4}, 4$ **21.** $-\frac{5}{2}, \frac{4}{3}$ **23.** $-\frac{2}{5}, \frac{7}{2}$ **25.** $0, 7$ **27.** $-7, 0$ **29.** $\frac{1}{7}$ **31.** $-\frac{1}{2}, 1$ **33.** $-5, \frac{4}{3}$ **35.** $-1, \frac{3}{5}$

37. $-\frac{3}{2}, 11$ **39.** $-7, 4$ **41.** $-\frac{4}{3}, \frac{3}{2}$ **43.** $\frac{3}{7}, 4$ **45.** $-1, 7$ **47.** $-\frac{5}{2}, \frac{7}{2}$ **49.** $-3a, 5a$

51. $-p - q, p - q$ **53.** $-\dfrac{2b}{3}, \dfrac{b}{2}$ **55.** $2d, 2d + 1$

PROBLEM SET 6.2   page 289

1. $-8, 8$ **3.** $-3, 3$ **5.** $-\sqrt{15}, \sqrt{15}$ **7.** $-2, 4$ **9.** $-\frac{7}{2}, \frac{5}{2}$ **11.** $\frac{1}{2}, \frac{7}{6}$ **13.** $-\frac{1}{4}, \frac{7}{4}$ **15.** $-9i, 9i$

17. $6 - 5i, 6 + 5i$ **19.** $\frac{2}{3} - \frac{7}{3}i, \frac{2}{3} + \frac{7}{3}i$ **21.** $(x + 3)^2$ **23.** $(y - 5)^2$ **25.** $(m + 10)^2$ **27.** $(x + \frac{7}{2})^2$

29. $(c + \frac{17}{2})^2$ **31.** $-3, 1$ **33.** $6 - \sqrt{53}, 6 + \sqrt{53}$ **35.** $2 - \sqrt{5}, 2 + \sqrt{5}$ **37.** $\dfrac{5 - \sqrt{30}}{5}, \dfrac{5 + \sqrt{30}}{5}$

39. $-\frac{2}{5}, \frac{7}{5}$ **41.** $\dfrac{4}{5} - \dfrac{\sqrt{69}}{5}i, \dfrac{4}{5} + \dfrac{\sqrt{69}}{5}i$ **43.** $-\dfrac{7}{8} - \dfrac{\sqrt{31}}{8}i, -\dfrac{7}{8} + \dfrac{\sqrt{31}}{8}i$ **45.** $\frac{2}{3}, 2$ **47.** $\dfrac{1 - \sqrt{2}}{3}, \dfrac{1 + \sqrt{2}}{3}$

49. $\frac{1}{4}, \frac{5}{4}$ **51.** $\dfrac{-b - \sqrt{b^2 - 12ab}}{2a}, \dfrac{-b + \sqrt{b^2 - 12ab}}{2a}$ **53.** $\dfrac{v - \sqrt{v^2 - 4gs}}{2g}, \dfrac{v + \sqrt{v^2 - 4gs}}{2g}$

61. $4\sqrt{3}$ feet per second   **63.** 19.21%.

PROBLEM SET 6.3   page 295

1. 1, 4   **3.** $-\frac{1}{3}, \frac{1}{2}$ **5.** $-\frac{1}{2}, 3$ **7.** $-\frac{1}{2}, \frac{5}{3}$ **9.** $\dfrac{4 - \sqrt{34}}{6}, \dfrac{4 + \sqrt{34}}{6}$ **11.** $\dfrac{5 - \sqrt{17}}{4}, \dfrac{5 + \sqrt{17}}{4}$

13. $\dfrac{1}{5} - \dfrac{\sqrt{34}}{5}i, \dfrac{1}{5} + \dfrac{\sqrt{34}}{5}i$ **15.** $\dfrac{4}{7} - \dfrac{\sqrt{5}}{7}i, \dfrac{4}{7} + \dfrac{\sqrt{5}}{7}i$ **17.** $-\frac{4}{5}, \frac{2}{3}$ **19.** $\dfrac{1}{6} - \dfrac{\sqrt{83}}{6}i, \dfrac{1}{6} + \dfrac{\sqrt{83}}{6}i$

21. $\dfrac{1 - 2\sqrt{5}}{2}, \dfrac{1 + 2\sqrt{5}}{2}$ **23.** $\dfrac{5 - \sqrt{21}}{2}, \dfrac{5 + \sqrt{21}}{2}$ **25.** $3 - \sqrt{13}, 3 + \sqrt{13}$ **27.** $-1.45, 1.65$

29. $-1.06, 3.66$ **31.** $-3.91, 1.80$ **33.** $\dfrac{-mn + m\sqrt{n^2 + 4n}}{2n}, \dfrac{-mn - m\sqrt{n^2 + 4n}}{2n}$

35. $\dfrac{-RC - \sqrt{R^2C^2 - 4LC}}{2LC}, \dfrac{-RC + \sqrt{R^2C^2 - 4LC}}{2LC}$ **37.** $D = 64$; 2 unequal real roots

39. $D = 0$; 1 real root   **41.** $D = -11$; 2 unequal complex roots   **43.** $D = 1$; 2 unequal real roots

45. $D = 0$; 1 real root   **47.** $k > -\frac{4}{3}$ **49.** $k > \frac{49}{24}$

PROBLEM SET 6.4   page 301

1. 5, 6   **3.** 12, 13   **5.** 6   **7.** 6 feet and 10 feet   **9.** 30 feet and 44 feet   **11.** 30 feet   **13.** 5 feet

15. 3, 4, 5   **17.** 5 meters; 12 meters   **19.** 5 mph; 12 mph   **21.** 1 second and 7 seconds

23. 5 seconds and 25 seconds   **25.** 120 mph   **27.** 12 hours   **29.** $540   **31.** 5 kilometers per hour

PROBLEM SET 6.5   page 307

1. $-3, -2, 2, 3$ **3.** $-2, 2, -i, i$ **5.** $-5, -2, 2, 5$ **7.** $-\frac{1}{2}, \frac{1}{4}$ **9.** $-\frac{2}{5}, 3$ **11.** $-\dfrac{\sqrt{5}}{5}, -\frac{1}{2}, \frac{1}{2}, \dfrac{\sqrt{5}}{5}$

13. $-\frac{1}{2}, -\frac{1}{4}, \frac{1}{4}, \frac{1}{2}$ **15.** $-1, 0, 1$ **17.** $-3, -1, 1$ **19.** $-5, -1, 3$ **21.** $-3, 1, -1 - \sqrt{3}i, -1 + \sqrt{3}i$

23. $-1, \frac{1}{2}, \dfrac{-1 - \sqrt{39}i}{4}, \dfrac{-1 + \sqrt{39}i}{4}$ **25.** $-2, -1, \frac{1}{3}, \frac{2}{3}$ **27.** $-4, 3, \dfrac{-1 - \sqrt{13}}{2}, \dfrac{-1 + \sqrt{13}}{2}$ **29.** $-\frac{1}{4}$

31. $-1, -\frac{1}{2}, \dfrac{-1 - i}{2}, \dfrac{-1 + i}{2}$

PROBLEM SET 6.6   page 312

1. 25   **3.** 3   **5.** $\frac{11}{2}$ **7.** 5   **9.** $\frac{5}{2}$ **11.** $\frac{7}{2}$ **13.** 1   **15.** 30   **17.** 4   **19.** 8   **21.** No solution   **23.** 1

25. $\frac{261}{7}$ **27.** 9   **29.** 4   **31.** 4   **33.** 0, 3   **35.** 27   **37.** 7   **39.** 5   **41.** $\frac{1}{4}$ **43.** $-\frac{1}{2}, \frac{3}{2}$ **45.** $V = \pi r^2 h$

47. $T = 273\left[\left(\dfrac{C}{331}\right)^2 - 1\right]$

PROBLEM SET 6.7   page 317

1. -243 **3.** -128 **5.** ± 64 **7.** ± 128 **9.** 0 **11.** 5 **13.** $\frac{8}{3}$; 2 **15.** $-\frac{127}{2}$; $\frac{129}{2}$ **17.** $\frac{513}{1,024}$ **19.** $\frac{7}{3}$
21. -64; 8 **23.** 64; -27 **25.** 1; 4 **27.** -3 **29.** -19; 61 **31.** $-\frac{7}{2}$; 3

PROBLEM SET 6.8   page 327

1. $-2 < x < 1$;      **3.** $1 \leq x \leq 2$;

5. $x < -3$ or $x > 1$;   **7.** $x \leq -2$ or $x \geq 6$;

9. $-4 \leq x \leq 5$;   **11.** $-1 \leq x \leq \frac{1}{2}$;

13. $x < -1$ or $x > \frac{5}{3}$;   **15.** $x \leq -\frac{1}{4}$ or $x \geq 7$;

17. All real numbers;

19. All real numbers except $-\frac{1}{3}$;

21. No solution   **23.** $1 < x < 4$;

25. $x < -1$ or $x > 2$;   **27.** $-\frac{2}{5} < x \leq 4$;

29. $x \leq \frac{1}{5}$ or $x > \frac{2}{3}$;   **31.** $x \leq 1$ or $x > 3$;

33. $x < 3$;   **35.** $x < 0$ or $x > \frac{1}{2}$;

37. $x \leq -1$ or $x \geq 10$ **39.** $3 - \sqrt{2} < t < 3 + \sqrt{2}$

REVIEW PROBLEM SET   page 328

1. $-1, 15$ **3.** $-4, 3$ **5.** $0, 2$ **7.** $-4, 2$ **9.** $-\frac{4}{9}, 1$ **11.** $-\frac{4}{3}, \frac{5}{2}$ **13.** $-\frac{2}{5}, \frac{3}{4}$ **15.** $-3b, 2b$
17. $-12, 12$ **19.** $-4i, 4i$ **21.** $-\frac{1}{5}, \frac{7}{5}$ **23.** $-5, \frac{17}{3}$ **25.** $(u - 4)^2$ **27.** $(x + \frac{5}{2})^2$ **29.** $3 \pm \sqrt{2}$

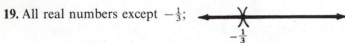

31. $4 \pm \sqrt{7}$ **33.** $\frac{1}{3} \pm \frac{1}{3}i$ **35.** $\frac{3}{2} \pm i$ **37.** $-\frac{3}{2}, \frac{1}{2}$ **39.** $2c, -3c$ **41.** $-5 \pm \sqrt{38}$ **43.** $-\frac{1}{2} \pm \frac{\sqrt{11}}{2}i$

45. $\frac{1 \pm \sqrt{17}}{8}$ **47.** $\frac{-5 \pm \sqrt{161}}{4}$ **49.** $\frac{-3 \pm \sqrt{65}}{4}$ **51.** $-0.67, 5.22$ **53.** Two real and unequal solutions

55. Two complex solutions   **57.** One real solution   **59.** $-2, -1, 1, 2$ **61.** $-1, -\frac{1}{4}, \frac{1}{4}, 1$

63. $-2i, 2i, -\frac{\sqrt{3}}{3}, \frac{\sqrt{3}}{3}$ **65.** $-3, -2, 1, 2$ **67.** $-3, 1, -1 \pm \sqrt{2}$ **69.** 17   **71.** -1 **73.** No solution

75. 22   **77.** $\frac{7}{4}$ **79.** 4   **81.** 0   **83.** 4   **85.** -8 **87.** 16   **89.** $29, -25$ **91.** $-\frac{5}{8}$ **93.** $-27, 1$

95. $-5 < x < 3$;   **97.** $t \le -2$ or $t \ge 5$;

99. $x < -3$ or $x > 2$;   **101.** $\frac{1}{3} \le y < \frac{7}{5}$;

103. 6, 7, 8   **105.** 10 feet   **107.** 89.04 seconds   **109.** 3 mph; 4 mph

CHAPTER 6 TEST   page 330

1. (a) 0, 3; (b) $-3, 5$; (c) $-1, 2$ **2.** (a) $-6, 6$; (b) $\frac{2}{3} \pm i$ **3.** (a) $2 \pm \sqrt{2}$; (b) $\frac{-6 \pm \sqrt{33}}{3}$

4. (a) $\frac{5}{6} \pm \frac{\sqrt{23}}{6}i$; (b) $\frac{1 \pm \sqrt{7}}{2}$ **5.** (a) Two real solutions; (b) Two complex solutions   **6.** $\pm 2, \pm i$

7. (a) 13; (b) 0   **8.** (a) $-\frac{26}{81}$; (b) $-125, 1$

9. (a) $x < -1$ or $x > \frac{1}{2}$;   (b) $-\frac{3}{2} \le x < 5$;

10. 8 feet, 12 feet

Chapter 7

PROBLEM SET 7.1   page 336

1. (a) I; (b) II; (c) III; (d) IV; (e) y axis; (f) x axis; (g) x axis; (h) y axis

3. (a) (2, 3); (b) $(-4, 2)$; (c) $(-3, 1)$; (d) $(5, -4)$ **5.** (10, 6)   **7.** $(-2, 2)$; $(-2, -2)$; $(2, -2)$

9. $(-2, -6)$; $(-1, -\frac{9}{2})$; $(0, -3)$; $(1, -\frac{3}{2})$; $(2, 0)$ **11.** $(-2, 8)$; $(-1, 5)$; $(0, 2)$; $(1, -1)$; $(2, -4)$

13. $(-2, 1)$; $(-1, \frac{3}{2})$; $(0, 2)$; $(1, \frac{5}{2})$; $(2, 3)$ **15.** $(-2, \frac{7}{3})$; $(-1, \frac{5}{3})$; $(0, 1)$; $(1, \frac{1}{3})$; $(2, -\frac{1}{3})$

17. $(-2, -\frac{1}{2})$; $(-1, \frac{3}{2})$; $(0, \frac{7}{2})$; $(1, \frac{11}{2})$; $(2, \frac{15}{2})$ **19.** $(-2, -\frac{25}{3})$; $(-1, -\frac{20}{3})$; $(0, -5)$; $(1, -\frac{10}{3})$; $(2, -\frac{5}{3})$

21. $(-2, \frac{21}{5})$; $(-1, \frac{18}{5})$; $(0, 3)$; $(1, \frac{12}{5})$; $(2, \frac{9}{5})$ **23.** (0, 3); (3, 0); $(-1, 4)$; (1, 2)   **25.** (0, 0); (0, 0); (1, 1); $(-1, -1)$

27. (0, 2); (1, 2); $(-1, 2)$; (2, 2)   **29.** $(-5, 0)$; $(-5, -1)$; $(-5, -3)$; $(-5, 4)$ **31.** $y = 2x + 3$

33. $y + 3x = 4$ **35.** $y = x$ **37.** $x = -3$ **39.** IV   **41.** III

PROBLEM SET 7.2 page 346

1. 10 **3.** $\sqrt{17}$ **5.** $\sqrt{29}$ **7.** 5 **9.** 4 **11.** 5 **13.** 4 **15.** $\dfrac{\sqrt{481}}{6}$ **17.** Yes **19.** Yes **21.** No

27. $-2, 3$ **31.** $(-1, 7)$ **33.** $(1, \frac{1}{2})$ **35.** $(10, -4)$ **37.** 17.71 **39.** 28.39

41. Radius $= 2$; center at $(-1, 2)$ **43.** Radius $= 4$; center at $(3, 4)$

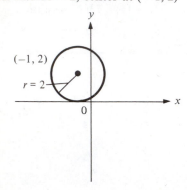

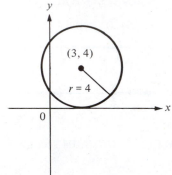

45. $r = \dfrac{\sqrt{26}}{2}$; center at $(-3, 2)$ **47.** $r = \dfrac{5\sqrt{10}}{2}$; center at $\left(-\dfrac{7}{2}, \dfrac{5}{2}\right)$

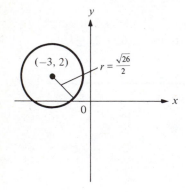

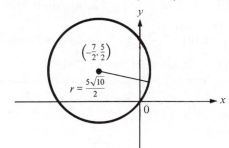

PROBLEM SET 7.3 page 354

1.

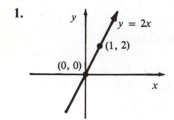

3.

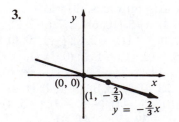

5.

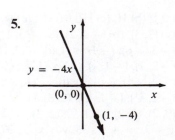

7.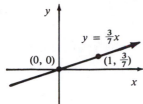

$y = \frac{3}{7}x$

$(0, 0)$ $(1, \frac{3}{7})$

9.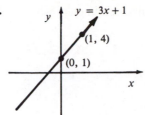

$y = 3x + 1$

$(1, 4)$

$(0, 1)$

11.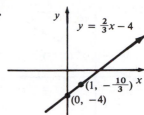

$y = \frac{2}{3}x - 4$

$(1, -\frac{10}{3})$

$(0, -4)$

13.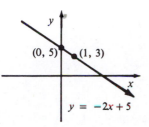

$(0, 5)$ $(1, 3)$

$y = -2x + 5$

15.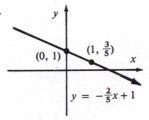

$(0, 1)$ $(1, \frac{3}{5})$

$y = -\frac{2}{5}x + 1$

17.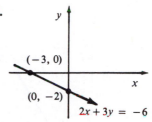

$(-3, 0)$

$(0, -2)$

$2x + 3y = -6$

19.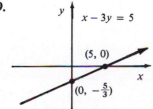

$x - 3y = 5$

$(5, 0)$

$(0, -\frac{5}{3})$

21.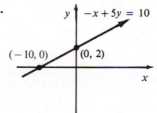

$-x + 5y = 10$

$(-10, 0)$ $(0, 2)$

23.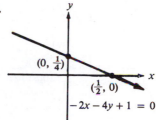

$(0, \frac{1}{4})$

$(\frac{1}{2}, 0)$

$-2x - 4y + 1 = 0$

25.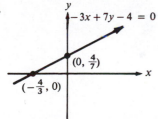

$-3x + 7y - 4 = 0$

$(0, \frac{4}{7})$

$(-\frac{4}{3}, 0)$

27.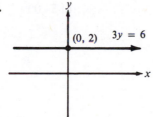

$(0, 2)$ $3y = 6$

29.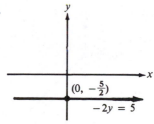

$(0, -\frac{5}{2})$

$-2y = 5$

31.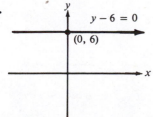

$y - 6 = 0$

$(0, 6)$

33.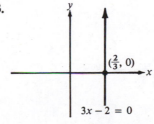

$(\frac{2}{3}, 0)$

$3x - 2 = 0$

35.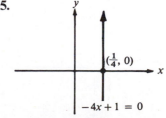

$(\frac{1}{4}, 0)$

$-4x + 1 = 0$

37.

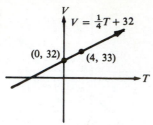

$V = 33\frac{1}{4}$ cubic centimeters when $T = 5°C$

39. (a)

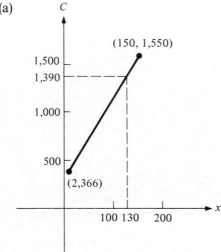

(b) $1,390

41. (a)

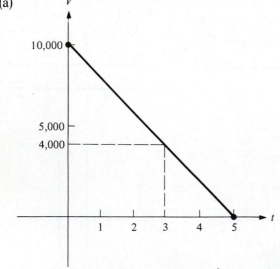

(b) $4,000

PROBLEM SET 7.4   page 362

1. -1 **3.** $\frac{2}{3}$ **5.** $-\frac{1}{2}$ **7.** 0 **9.** Undefined

11.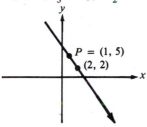

13.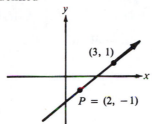

15.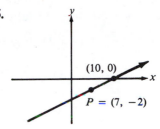

17.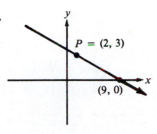

19.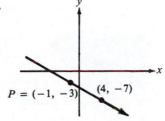

21. Parallel   **23.** Neither   **25.** Perpendicular   **27.** Neither   **29.** (a) $\frac{3}{2}$; (b) -1 **31.** (a) -10; (b) 0
33. (a) $-\frac{13}{2}$; (b) $\frac{13}{3}$ **35.** $\overline{AD}$ and $\overline{BC}$ have slope $-\frac{3}{2}$; $\overline{AC}$ and $\overline{BD}$ have slope 3
37. $\overline{AB}$ has slope $\frac{2}{5}$; $\overline{BC}$ has slope $-\frac{5}{2}$ **39.** $\overline{AB}$ has slope 1; $\overline{BC}$ has slope -1 **41.** Not collinear
43. Collinear   **45.** $\overline{AD}$ and $\overline{BC}$ have slope $-\frac{4}{3}$

PROBLEM SET 7.5   page 371

1. $y - 2 = 5(x + 1)$ **3.** $y - 3 = -3(x - 7)$ **5.** $y + 1 = -\frac{3}{7}(x - 5)$ **7.** $y = \frac{3}{8}x$ **9.** $y = -5$
11. $y = \frac{1}{2}x + \frac{7}{2}$ **13.** $y = -3x + 5$ **15.** $y = -\frac{3}{7}x$ **17.** $x = -3$
19. $y = \frac{2}{3}x - \frac{1}{3}$
$\quad m = \frac{2}{3}$
$\quad x$ intercept $= \frac{1}{2}$
$\quad y$ intercept $= -\frac{1}{3}$

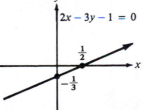

21. $y = -2x + 5$
$\quad m = -2$
$\quad x$ intercept $= \frac{5}{2}$
$\quad y$ intercept $= 5$

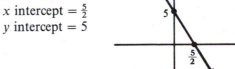

23. $y = 4x + 5$
$\quad m = 4$
$\quad x$ intercept $= -\frac{5}{4}$
$\quad y$ intercept $= 5$

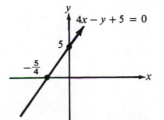

25. $y = 2x$
$\quad m = 2$
$\quad x$ intercept $= 0$
$\quad y$ intercept $= 0$

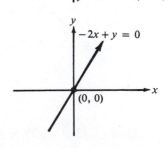

27. (a) $y + 1 = -\frac{1}{2}(x - 3)$; (b) $y = -\frac{1}{2}x + \frac{1}{2}$; (c) $x + 2y - 1 = 0$ **29.** (a) $y - 5 = 10(x + \frac{1}{2})$; (b) $y = 10x + 10$;
(c) $10x - y + 10 = 0$ **31.** (a) $y - 5 = \frac{3}{2}(x - 1)$; (b) $y = \frac{3}{2}x + \frac{7}{2}$; (c) $3x - 2y + 7 = 0$
33. (a) $y - 0 = \frac{5}{3}(x - 0)$; (b) $y = \frac{5}{3}x$; (c) $5x - 3y = 0$ **35.** (a) $y - 3 = \frac{1}{2}(x - 0)$; (b) $y = \frac{1}{2}x + 3$;
(c) $x - 2y + 6 = 0$ **37.** (a) $y - 0 = -\frac{2}{3}(x + 1)$; (b) $y = -\frac{2}{3}x - \frac{2}{3}$; (c) $2x + 3y + 2 = 0$

39. (a) $y - 6 = 0 \,(x + 1)$; (b) $y = 6$; (c) $y - 6 = 0$ **41.** (a) $-\frac{5}{2}$; (b) $\frac{7}{3}$ **43.** $\dfrac{x}{5} + \dfrac{y}{6} = 1$ **45.** $\dfrac{x}{-3} + \dfrac{y}{-1} = 1$

47. $\dfrac{x}{1} + \dfrac{y}{-\frac{5}{2}} = 1$ **49.** (a) $N = \frac{5}{2}V + 50$; (b) 16 feet per second

PROBLEM SET 7.6   page 376

1.

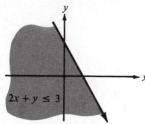

$y \le 2x + 5$

3.

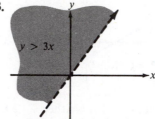

$y > 3x$

5.

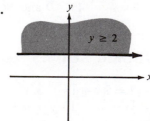

$y > -2x + 3$

7.

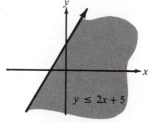

$2x + y \le 3$

9.

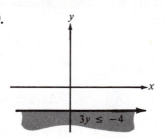

$3y \le -4$

11.

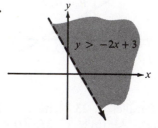

$y \ge 2$

13.

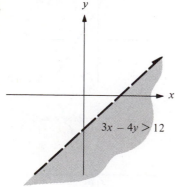

$3x - 4y > 12$

15.

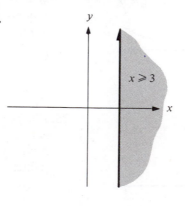

$x \ge 3$

17.

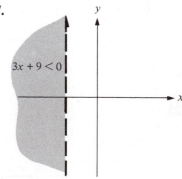

$3x + 9 < 0$

19. f **21.** a **23.** d **25.** $x + y \ge 3$ **27.** $y < \frac{1}{2}x + 1$ **29.** $2x + y \ge 4$

REVIEW PROBLEM SET   page 378

1. I   **3.** II   **5.** y axis   **7.** IV   **9.** $(-2, 4)$; $(-1, \frac{7}{2})$; $(0, 3)$; $(1, \frac{5}{2})$; $(2, 2)$ **11.** 13   **13.** 5
15. $|\overline{AB}| = |\overline{AC}| = \sqrt{53}$ **17.** $(3, 4)$
19. Radius $= 6$; center at $(3, -2)$ **21.** Radius $= 3$; center at $(-2, 3)$

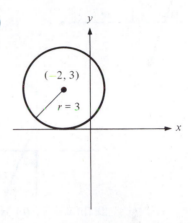

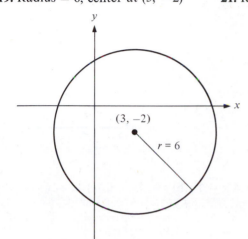

23.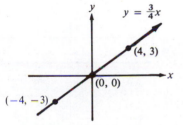

25.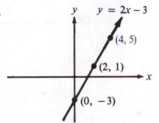

27.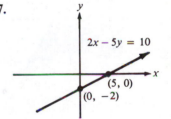

29. $4x + 12 = 0$ 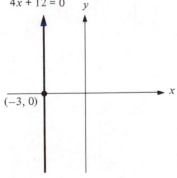   **31.** 1   **33.** $-\frac{1}{2}$ **35.**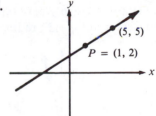

37.

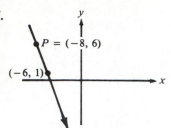

39. Parallel **41.** Perpendicular **43.** $y - 1 = 3(x - 1)$ **45.** $y = -2$
47. $y = 2x + 4$ **49.** $x = 3$

51. $y = -2x + 4$
$m = -2$
x intercept $= 2$
y intercept $= 4$

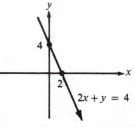

53. $y = \frac{3}{4}x + 3$
$m = \frac{3}{4}$
x intercept $= -4$
y intercept $= 3$

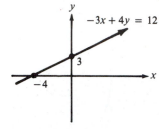

55. (a) $y - 3 = \frac{3}{2}(x - 2)$; (b) $y = \frac{3}{2}x$; (c) $3x - 2y = 0$ **57.** (a) $y + 2 = 4(x - 1)$; (b) $y = 4x - 6$; (c) $4x - y - 6 = 0$

59.

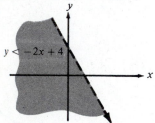

61.

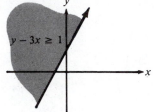

63.

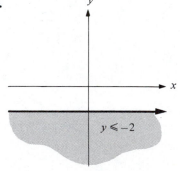

CHAPTER 7 TEST page 380

1. (a) II; (b) Positive y axis; (c) IV; (d) III; (e) I; (f) Negative x axis **2.** 13 **3.** (1, 7)
4. (a) Center at $(3, -2)$; radius $= 4$; (b) Center at $(2, -3)$; radius $= 5$
5. (a)

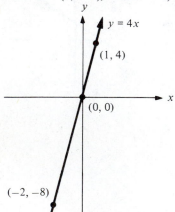

(b)

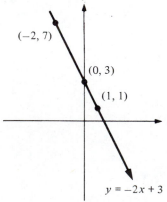

6. (a)

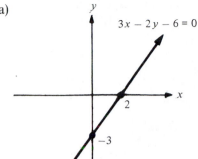

$3x - 2y - 6 = 0$

(b)

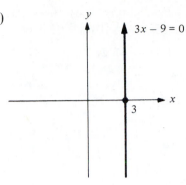

$3x - 9 = 0$

(c)

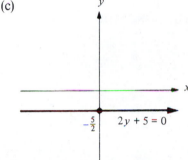

$2y + 5 = 0$

7. $\frac{5}{2}$ **8.** (a) $y - 4 = \frac{2}{3}(x + 3)$; (b) $y = 2x + 1$; (c) $x = -4$; (d) $y = \frac{2}{3}x - 5$ **9.** (a) (i) $y + 1 = -2(x - 3)$; (ii) $y = -2x + 5$; (iii) $2x + y - 5 = 0$; (b) (i) $y - 4 = \frac{5}{3}(x + 2)$; (ii) $y = \frac{5}{3}x + \frac{22}{3}$; (iii) $5x - 3y + 22 = 0$
10. (a)

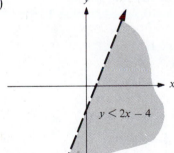

$y < 2x - 4$

(b)

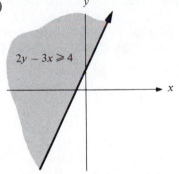

$2y - 3x \geqslant 4$

Chapter 8

PROBLEM SET 8.1   page 390

1. Independent; $(\frac{1}{3}, 0)$ **3.** Inconsistent   **5.** Independent; $(2, 1)$ **7.** $(4, 3)$ **9.** $(1, 5)$ **11.** $(\frac{7}{3}, -\frac{4}{3})$
13. $(159, -186)$ **15.** $(\frac{7}{11}, -\frac{10}{11})$ **17.** $(3, 1)$ **19.** $(1, 1)$ **21.** $(3, 2)$ **23.** $(-5, 3)$ **25.** $(2, -1)$

27. $(-\frac{3}{2}, \frac{27}{2})$ **29.** $(\frac{2}{5}, \frac{21}{5})$ **31.** $(\frac{22}{25}, \frac{21}{25})$ **33.** $(10, 24)$ **35.** $\left(\dfrac{3a+b}{10}, \dfrac{a-3b}{10}\right)$ **37.** $(1, 2)$ **39.** $(21, 3.6)$
41. $(\frac{1}{13}, \frac{1}{17})$ **43.** $(\frac{1}{4}, -\frac{1}{5})$ **45.** $(-1, \frac{1}{3})$ **47.** 2

PROBLEM SET 8.2   page 396

1. $(2, 3, -1)$ **3.** $(4, 3, 2)$ **5.** $(2, -1, 3)$ **7.** $(1, 1, 1)$ **9.** $(-5, \frac{5}{3}, \frac{28}{3})$ **11.** $(2, 2, 1)$ **13.** $(5, 2, 1)$
15. $(\frac{94}{35}, \frac{1}{7}, \frac{1}{5})$ **17.** $(1, -3, -2)$ **19.** $(6, -3, 12)$ **21.** $(3, 4, 6)$ **23.** $(2, 3, 5)$

PROBLEM SET 8.3   page 400

1. 26   **3.** 24   **5.** $-\frac{13}{2}$ **7.** 0   **9.** 1   **11.** 116   **13.** 1   **15.** 18   **17.** -438 **19.** 4   **21.** -1
23. 2 or 8   **25.** $\frac{23}{2}$ **27.** $x < -2$ or $x > 2$

PROBLEM SET 8.4   page 405

1. $(\frac{1}{3}, \frac{2}{3})$ **3.** $(0, 0)$ **5.** $(\frac{85}{4}, \frac{35}{4})$ **7.** $(-11, 7)$ **9.** $(5, -6)$ **11.** $(1, 2, 3)$ **13.** $(\frac{19}{4}, -\frac{3}{4}, -\frac{29}{4})$
15. $(-5, -\frac{14}{3}, -\frac{16}{3})$ **17.** $(2, 1, -1)$ **19.** $(-4, 5, 4)$

PROBLEM SET 8.5   page 410

1. 5, 7   **3.** 3 and 12   **5.** 4,925 prey; 725 predators   **7.** Dawn is 14; Joe is 17   **9.** 6 dimes; 12 quarters
11. 17 dimes; 28 quarters   **13.** 63 fives; 15 tens   **15.** fixed charge is $30; hourly rate is $20
17. Tom takes 12 hours; John takes 6 hours   **19.** $40; $20   **21.** $25,000 at 10.5%; $15,000 at 13.5%
23. $2,250 at 8.5%; $1,750 at 7.9%   **25.** $20,000 at 18%; $10,000 at 19.5%
27. 210 dimes; 60 quarters; 24 half-dollars   **29.** 45 type I; 20 type II; 15 type III

PROBLEM SET 8.6   page 415

1.

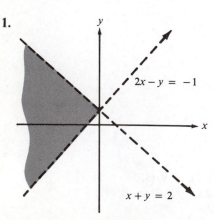

3.

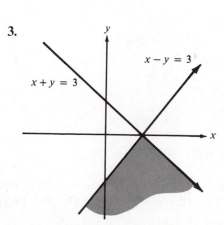

5.

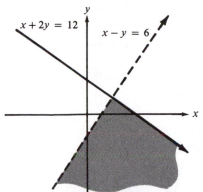

$x + 2y = 12$

$x - y = 6$

7.

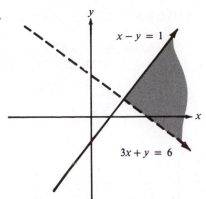

$x - y = 1$

$3x + y = 6$

9.

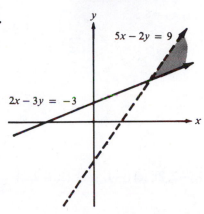

$5x - 2y = 9$

$2x - 3y = -3$

11.

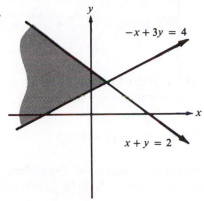

$-x + 3y = 4$

$x + y = 2$

13.

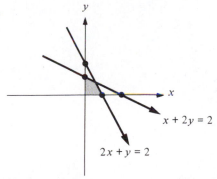

$x + 2y = 2$

$2x + y = 2$

15.

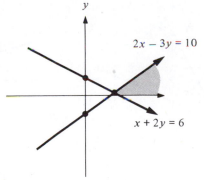

$2x - 3y = 10$

$x + 2y = 6$

17. $\begin{cases} x \geq 0, \ y \geq 0 \\ x + y \leq 3{,}000 \\ 56x + 86y \leq 200{,}000 \end{cases}$

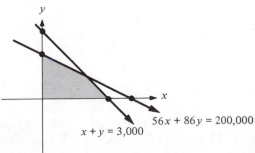

$56x + 86y = 200{,}000$

$x + y = 3{,}000$

REVIEW PROBLEM SET   page 416

1. Independent   **3.** Inconsistent   **5.** $(3, 1)$ **7.** $(1, 2, -2)$ **9.** $(2, -1)$ **11.** $(-1, 3, 2)$ **13.** $(\frac{1}{3}, \frac{1}{2})$
15. 11   **17.** 26   **19.** $-3, 1$ **21.** $(2, -1)$ **23.** $(-1, 3, 2)$
25.

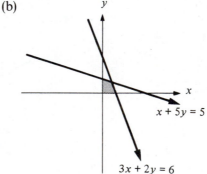

$-2x + y = 4$

$2x + 4y = 3$

27.

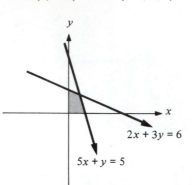

$2x + 3y = 6$

$5x + y = 5$

29. $7, -5$ **31.** 37 at 30¢; 63 at 22¢   **33.** 11 nickels; 21 dimes
35. \$10,000 at 14%; \$5,000 at 15%; \$15,000 at 12%

CHAPTER 8 TEST   page 417

1. (a) Independent; (b) Inconsistent; (c) Dependent   **2.** (a) $(-1, -5)$; (b) $(1, -2, 2)$; (c) $(-1, -5)$; $(1, -2, 2)$
3. (a) -3; (b) 11   **4.** $-7, 1$ **5.** (a) $(1, \frac{1}{2})$; (b) $(1, 2, 3)$
6. (a)

$-2x + y = 6$

$3x - 5y = 15$

(b)

$x + 5y = 5$

$3x + 2y = 6$

7. 13 nickels; 7 dimes

Chapter 9

PROBLEM SET 9.1   page 424

1. 2   **3.** 2   **5.** 8   **7.** -1 **9.** $\frac{9}{4}$ **11.** $-3, 1$ **13.** $3 \pm \sqrt{6}$ **15.** $3 = \log_5 125$ **17.** $5 = \log_{10} 100,000$
19. $-2 = \log_4 \frac{1}{16}$ **21.** $-2 = \log_6 \frac{1}{36}$ **23.** $\frac{1}{2} = \log_9 3$ **25.** $-\frac{3}{2} = \log_{100} 0.001$ **27.** $0 = \log_7 1$

29. $3 = \log_x a$ **31.** $9^2 = 81$ **33.** $27^{2/3} = 9$ **35.** $10^{-3} = 0.001$ **37.** $(\frac{1}{3})^{-2} = 9$ **39.** $10^{0.64} = 4.35$
41. $(\sqrt{16})^{1/2} = 2$ **43.** $x^0 = 1$ **45.** 6   **47.** $\frac{1}{2}$ **49.** $-\frac{1}{2}$ **51.** 2   **53.** -3 **55.** $\frac{5}{2}$ **57.** -4 **59.** 6
61. 0   **63.** -3 **65.** 9   **67.** 5   **69.** 25   **71.** 3   **73.** 243   **75.** 0   **77.** 5.6   **79.** 4   **81.** 2   **83.** 5

PROBLEM SET 9.2   page 430

1. $\log_4 5 + \log_4 y$ **3.** $\log_3 u + \log_3 v$ **5.** $\log_5 x - \log_5 3$ **7.** $\log_2 7 - \log_2 15$ **9.** $5 \log_7 3$
11. $5 \log_3 c$ **13.** $\frac{1}{2} \log_4 w$ **15.** $4 \log_7 3 + 2 \log_7 5$ **17.** $4 \log_{11} x + 2 \log_{11} y$ **19.** $\frac{1}{5} \log_2 x + \frac{1}{5} \log_2 y$
21. $2 \log_5 a - 4 \log_5 b$ **23.** $3 \log_7 x + \frac{1}{4} \log_7 y - 3 \log_7 z$ **25.** $4 \log_4 u + 5 \log_4 v - \frac{3}{4} \log_4 z$
27. $\frac{1}{7} \log_3 y - \frac{1}{7} \log_3 (y + 7)$ **29.** $\log_5 \frac{4}{3}$ **31.** $\log_7 \frac{1}{6}$ **33.** $\log_3 \dfrac{5z}{y}$ **35.** $\log_3 \frac{2}{25}$ **37.** $\log_2 x^3 y^7$
39. $\log_a \dfrac{y}{3}$ **41.** $\log_4 \dfrac{\sqrt{a}}{b^3 z^4}$ **43.** $\log_7 (x^2 - 1)$ **45.** $\log_e (y + 5)$ **47.** 4   **49.** $\frac{1}{5}$ **51.** 0.7781
53. 1.2552   **55.** 0.1761   **57.** 0.6990   **59.** 1.9084   **61.** -0.3010 **63.** $\frac{1}{2}$ **65.** 18   **67.** 5   **69.** $\frac{6}{5}$
71. 103   **73.** Let $\log_b b^r = t$, so that $b^t = b^r$, or $t = r$. Therefore, $\log_b b^r = r$

PROBLEM SET 9.3   page 439

1. 3.782×10^3; 3   **3.** 3.81×10^{-3}; -3 **5.** 3.75×10^5; 5   **7.** 1.321×10^{-4}; -4 **9.** 2.71312×10^{-4}; -4
11. 2.1×10^2; 2   **13.** 3.14×10^{-5}; -5 **15.** 1.13×10^4; 4   **17.** 5.41×10^{-6}; -6
19. 3.127×10^{-4}; -4 **21.** 2.5011   **23.** 4.7275   **25.** 1.2330   **27.** 0.8331   **29.** 0.0719   **31.** -0.5017
33. -1.1469 **35.** -3.7496 **37.** -5.1549 **39.** 3.1889   **41.** 1.9007   **43.** 0.7253   **45.** -0.2422
47. -1.2723 **49.** 2.59   **51.** 19.7   **53.** 553   **55.** 3,560   **57.** 0.543   **59.** 0.00793   **61.** 0.0235
63. 0.000325   **65.** 0.696   **67.** 0.00712   **69.** 1.397   **71.** 34.88   **73.** 0.01682   **75.** 0.00003448

PROBLEM SET 9.4   page 444

1. 1.518   **3.** 0.7912   **5.** 2.781   **7.** -2.872 **9.** -2.448 **11.** -4.379 **13.** 84.74   **15.** 1.244
17. 29.77   **19.** 25,500   **21.** 0.0328   **23.** 0.001426   **25.** 8.899   **27.** 2.297   **29.** -0.6270
31. 8.166   **33.** 4.113   **35.** 0.8607   **37.** 0.04505   **39.** 41.02   **41.** 0.05388   **43.** 0.04002   **45.** 15.30
47. 10.20   **49.** 1.637   **51.** 2.469   **53.** 0.7357   **55.** 1.519   **57.** 5.758   **59.** 1.240   **61.** 2.807
63. -1.850 **65.** 0.5493   **67.** 1.396   **69.** 3.229   **71.** 0.1989   **73.** 7.8   **75.** 2.51×10^{-6} **77.** 1.2091
79. 1.3383

PROBLEM SET 9.5   page 449

1. \$10,100.28   **3.** \$20,229.19   **5.** 11%   **7.** 8.243%   **9.** \$6,661.68   **11.** \$17,131.04   **13.** 2,983,649
15. \$2,013.75   **17.** 0.505   **19.** 0.0806   **21.** 13.253%   **23.** 23.834%   **25.** (a) 40; (b) 7.2 weeks

REVIEW PROBLEM SET   page 451

1. $\frac{3}{2}$ **3.** -3 **5.** $\frac{5}{3}$ **7.** 4   **9.** $5 = \log_2 32$ **11.** $\frac{2}{3} = \log_8 4$ **13.** $-\frac{2}{3} = \log_{27}(\frac{1}{9})$ **15.** $n = \log_z w$
17. $2^3 = 8$ **19.** $10^2 = 100$ **21.** $17^0 = 1$ **23.** $10^{-1} = \frac{1}{10}$ **25.** 2   **27.** 0   **29.** $-\frac{3}{2}$ **31.** $-\frac{7}{2}$
33. -4 **35.** $\frac{7}{4}$

37. 2   **39.** $\frac{17}{7}$ **41.** 2   **43.** $\log_6 7 + \log_6 u$ **45.** $6\log_2 3 + 14$ **47.** $\log_4 x + 5\log_4 y$

49. $2\log_a w - 4\log_a z$ **51.** $\log_2(\frac{2}{9})$ **53.** $\log_5(\frac{1}{6})$ **55.** $\log_a \dfrac{x^5}{y^3}$ **57.** $\frac{3}{2}$ **59.** $\frac{9}{4}$ **61.** 15

63. $\dfrac{\sqrt{26}}{2}$ **65.** 4.68×10^7; 7   **67.** 1.2×10^{-6}; -6 **69.** 5.6×10^3; 3   **71.** 1.92×10^{-4}; -4

73. 2.9274   **75.** 1.8762   **77.** 0.5030   **79.** -0.3251 **81.** -2.2316 **83.** 5.58   **85.** 89.8

87. 1,728   **89.** 0.0283   **91.** 0.0003912   **93.** 1.168   **95.** -3.502 **97.** 3,102   **99.** 0.004822

101. 3.814   **103.** 9.356   **105.** 0.1627   **107.** 2.211   **109.** 2.322   **111.** 0.8510   **113.** 1.315

115. $11,565.30   **117.** 14.49%   **119.** $3,941.04   **121.** 664,363

CHAPTER 9 TEST   page 453

1. (a) $\frac{1}{2}$; (b) $\frac{11}{5}$ **2.** (a) $4 = \log_3 81$; (b) $c = \log_z b$ **3.** (a) $6^3 = 216$; (b) $a^t = B$ **4.** (a) 6; (b) $\frac{3}{2}$; (c) -3

5. (a) $3\log_5 2 + \log_5 x$; (b) $\log_3 z - \log_3 7$; (c) $3\log_4 x + 4\log_4 y$ **6.** (a) $\log_7 \frac{2}{5}$; (b) $\log_2 \dfrac{x^3}{y^2}$

7. (a) $\frac{7}{2}$; (b) 3   **8.** (a) 2.8663; (b) -2.0496; (c) 1.2463   **9.** (a) 353; (b) 61.18; (c) 0.00212

10. (a) 1.445; (b) 0.0001516; (c) 2.635; (d) 0.4868   **11.** (a) -0.2167; (b) 0.3005   **12.** $29,718.95

Chapter 10

PROBLEM SET 10.1   page 460

1. 4   **3.** -5 **5.** 1   **7.** $3a + 3z + 1$ **9.** 169   **11.** 4   **13.** 0   **15.** 0   **17.** 3   **19.** 5   **21.** 5

23. 2   **25.** $|-a - 2|$ **27.** All real numbers   **29.** All real numbers except 0   **31.** $x > 1$ **33.** $x \le 2$

35. All real numbers   **37.**   **39.**

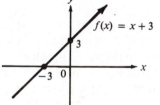

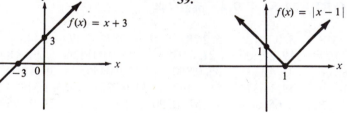

41. 3   **43.** -4 **45.** $5x + 4$ **47.** $7x - 2$ **49.** $6x^2 + 7x - 5$ **51.** $\dfrac{2x - 1}{3x + 5}$; $x \ne -\frac{5}{3}$ **53.** $6x + 9$

55. $f(50) = 3{,}000$; $f(70) = 2{,}600$ **57.** Function   **59.** Function   **61.** Not a function

PROBLEM SET 10.2   page 466

1. $f(x) = 2x$; $f(x + 2) = 2x + 4$; $f(2) + f(3) = 10$; $\dfrac{f(x + h) - f(x)}{h} = 2$ **3.** $y = \frac{1}{2}x^3$ **5.** $y = 3x^3$

7. $y = \frac{14}{1331}x^3$ **9.** 4   **11.** $T = \dfrac{0.01x}{y}$ **13.** $y = \dfrac{192}{x^3}$ **15.** 3:2; 27:8   **17.** $S = 6x^2$; 864 square inches

19. N is tripled

PROBLEM SET 10.3   page 474

1. Domain = all real numbers; range = all real numbers; slope = -3
3. Domain = all real numbers; range = all real numbers; slope = $-\frac{3}{4}$
5. Domain = all real numbers; range = $\{-1\}$; slope = 0
7. Domain = all real numbers; range = all real numbers; slope = 2
9. Domain = all real numbers; range = all real numbers; slope = 4
11. $f(x) = 2x + 1$ **13.** $f(x) = -\frac{4}{3}x + 4$ **15.** $f(x) = 3$ **17.** $f(x) = mx$ **19.** $f(x) = \frac{1}{5}x + \frac{14}{5}$
21. y intercept = 0; x intercept = 0; vertex = $(0, 0)$; domain = all real numbers; range = $y \geq 0$

23. y intercept = -3; x intercepts = $\pm\dfrac{\sqrt{6}}{2}$; vertex = $(0, -3)$; domain = all real numbers; range = $y \geq -3$

25. y intercept = -1; x intercept = -1; vertex = $(-1, 0)$; domain = all real numbers; range = $y \leq 0$
27. y intercept = 6; x intercepts = $-3, -2$; vertex = $(-\frac{5}{2}, -\frac{1}{4})$; domain = all real numbers; range = $y \geq -\frac{1}{4}$
29. y intercept = 0; x intercepts = $0, \frac{3}{2}$; vertex = $(\frac{3}{4}, -\frac{9}{8})$; domain = all real numbers; range = $y \geq -\frac{9}{8}$
31. y intercept = 3; x intercepts = $-3, -1$; vertex = $(-2, -1)$; domain = all real numbers; range = $y \geq -1$
33.   **35.** (a) 48 feet; (b) 2 seconds

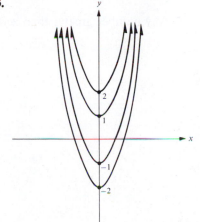

PROBLEM SET 10.4   page 481

1. domain = all real numbers;
   range = positive real numbers

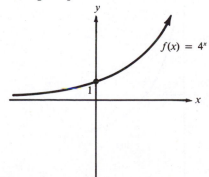

3. domain = all real numbers;
   range = positive real numbers

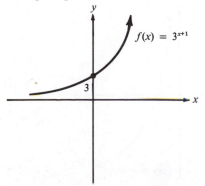

5. domain = all real numbers;
range = negative real numbers

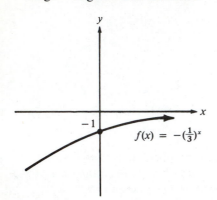

7. domain = all real numbers;
range = positive real numbers

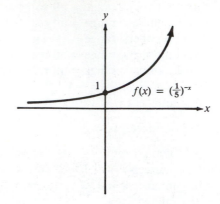

9. domain = all real numbers;
range = positive real numbers

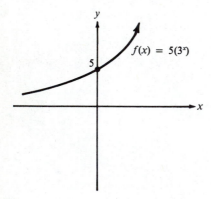

11. 3   **13.** 4   **15.** Any number greater than zero

19. domain = positive real numbers;
range = all real numbers

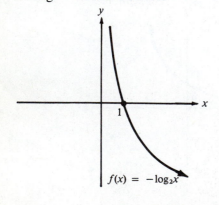

21. domain = positive real numbers;
range = all real numbers

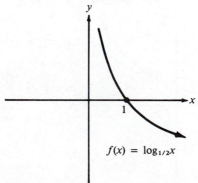

23. domain = positive real numbers;
range = all real numbers

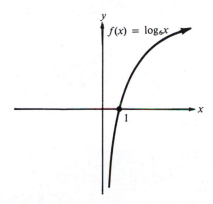

25. 2   **27.** 4   **29.** 9   **33.** 304   **35.** 64.5

PROBLEM SET 10.5   page 490

1.

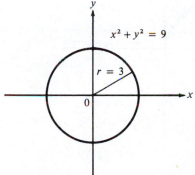

3.

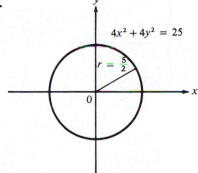

5.

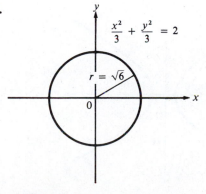

7.

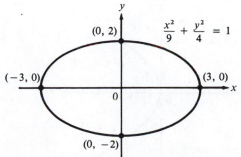

9.

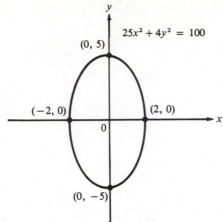

$25x^2 + 4y^2 = 100$

$(0, 5)$

$(-2, 0)$ $(2, 0)$

0

$(0, -5)$

11.

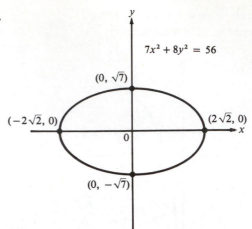

$7x^2 + 8y^2 = 56$

$(0, \sqrt{7})$

$(-2\sqrt{2}, 0)$ $(2\sqrt{2}, 0)$

0

$(0, -\sqrt{7})$

13.

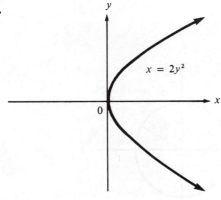

$x = 2y^2$

0

15.

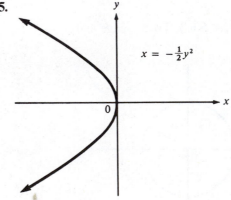

$x = -\frac{1}{2}y^2$

0

17.

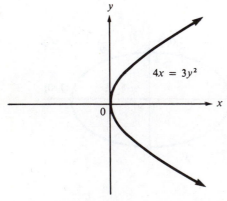

$4x = 3y^2$

0

19.

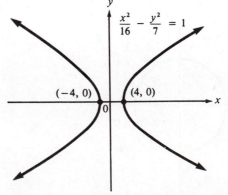

$\frac{x^2}{16} - \frac{y^2}{7} = 1$

$(-4, 0)$ $(4, 0)$

0

21.

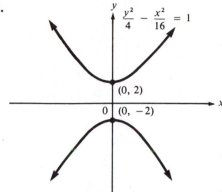

$\dfrac{y^2}{4} - \dfrac{x^2}{16} = 1$

$(0, 2)$

$(0, -2)$

23.

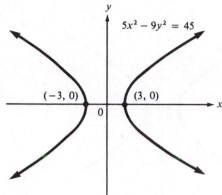

$5x^2 - 9y^2 = 45$

$(-3, 0)$ $(3, 0)$

25.

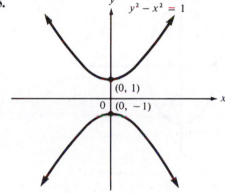

$y^2 - x^2 = 1$

$(0, 1)$

$(0, -1)$

27.

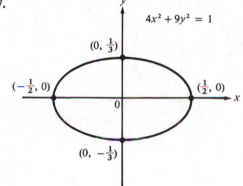

$4x^2 + 9y^2 = 1$

$(0, \frac{1}{3})$

$(-\frac{1}{2}, 0)$ $(\frac{1}{2}, 0)$

$(0, -\frac{1}{3})$

29.

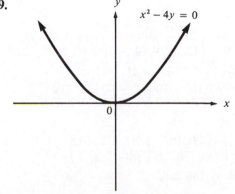

$x^2 - 4y = 0$

31.

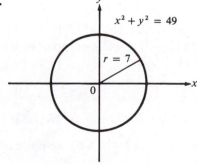

$x^2 + y^2 = 49$

$r = 7$

33.

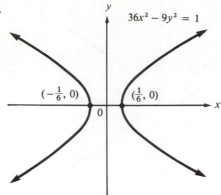

$36x^2 - 9y^2 = 1$

$(-\tfrac{1}{6}, 0)$ $(\tfrac{1}{6}, 0)$

35.

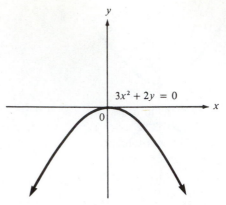

$3x^2 + 2y = 0$

37.

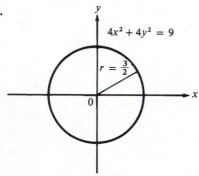

$4x^2 + 4y^2 = 9$

$r = \tfrac{3}{2}$

39.

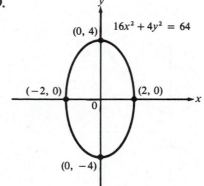

$(0, 4)$ $16x^2 + 4y^2 = 64$

$(-2, 0)$ $(2, 0)$

$(0, -4)$

41. $\dfrac{x^2}{240^2} + \dfrac{y^2}{140^2} = 1$

PROBLEM SET 10.6   page 495

1. $(-1, -2); (2, 1)$ **3.** $(2, 4); (-0.8, -4.4)$

5. $(-1 + 2i, 2 - 3i); (-1 - 2i, 2 + 3i); (-6, \sqrt{6}i); (-6, -\sqrt{6}i)$; no point of intersection   **7.** $(\tfrac{7}{2}, \tfrac{5}{2}); (\tfrac{11}{4}, \tfrac{5}{4})$

9. $(2 + 3i, 1 - 2i); (2 - 3i, 1 + 2i)$; no point of intersection   **11.** $(5, 4)$ **13.** $(4, 2); (4, -2)$

15. $(3, 2); (-3, 2); (3, -2); (-3, -2)$ **17.** $(-2\sqrt{6}, 1); (-2\sqrt{6}, -1); (2\sqrt{6}, 1); (2\sqrt{6}, -1)$

19. $(-2, 2\sqrt{3}); (-2, -2\sqrt{3}); (2, 2\sqrt{3}); (2, -2\sqrt{3})$ **21.** $(4, 2); (-4, 2); \left(\dfrac{4\sqrt{15}}{3}, -\dfrac{10}{3}\right); \left(\dfrac{-4\sqrt{15}}{3}, -\dfrac{10}{3}\right)$

23. $(3, 2); (-3, 2); (-3, -2); (3, -2)$ **25.** $(2i, 3); (-2i, 3); (-2, 1); (2, 1)$

27. $(3i, 5); (3i, -5); (-3i, 5); (-3i, -5)$; no point of intersection   **29.** $\left(\dfrac{41}{10}, \dfrac{3\sqrt{91}}{10}\right); \left(\dfrac{41}{10}, -\dfrac{3\sqrt{91}}{10}\right)$

31. 8 centimeters by 12 centimeters   **33.** \$6,800; 2.5%   **35.** 150 feet by 180 feet

REVIEW PROBLEM SET   page 496

1. 9   **3.** 0   **5.** 3   **7.** 25   **9.** $2u + 5$ **11.** 4   **13.** All real numbers

15. All real numbers except -2 and 1   **17.** $x \leq \frac{3}{2}$ **19.** Function   **21.** Not a function

23. $f(x) = \frac{2}{3}x$ **25.** $f(x) = 4\sqrt{x}$ **27.** $f(x) = \dfrac{20}{x}$ **29.** 48 pounds

31. Domain = all real numbers; range = all real numbers; slope = 2

33. Domain = all real numbers; range = all real numbers; slope = $\frac{3}{2}$

35. y intercept = -4; x intercepts = $-\frac{1}{2}, \frac{4}{3}$; vertex = $(\frac{5}{12}, -\frac{121}{24})$; domain = all real numbers; range = $y \geq -\frac{121}{24}$

37. y intercept = 9; x intercept = -3; vertex = $(-3, 0)$; domain = all real numbers; range = $y \geq 0$

39. y intercept = -3; x intercepts = $-\frac{3}{4}, -\frac{1}{2}$; vertex = $(-\frac{5}{8}, \frac{1}{8})$; domain = all real numbers; range = $y \leq \frac{1}{8}$

41. domain = all real numbers;
range = positive real numbers

43. domain = all real numbers;
range = positive real numbers

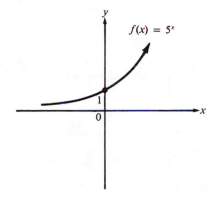

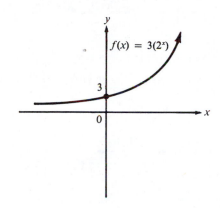

45. domain = all real numbers;
range = positive real numbers

47. domain = positive real numbers;
range = all real numbers

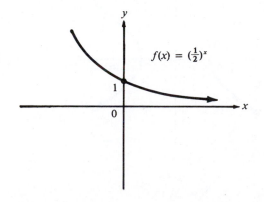

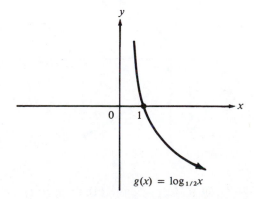

49. domain $= x > -2$;
range = all real numbers

51. 1   **53.** $\frac{1}{2}$ **55.** $\frac{5}{4}$ **57.** $\frac{1}{20}$

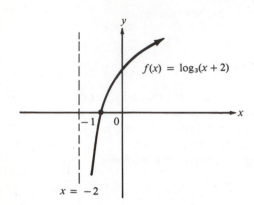

59. circle

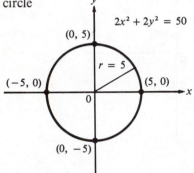

61. ellipse

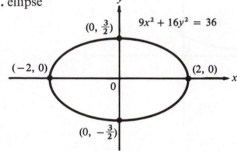

63. parabola

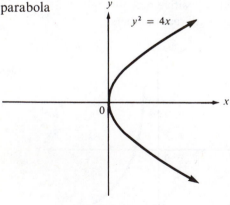

65. hyperbola

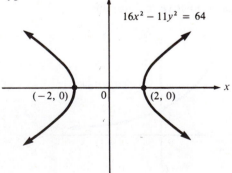

67. $(3, -4)$ **69.** $(6, 0)$; $(-5, -\sqrt{11})$; $(-5, \sqrt{11})$ **71.** $(-5, -2)$; $(-5, 2)$; $(5, -2)$; $(5, 2)$
73. 16 feet by 30 feet   **75.** 4 centimeters by 5 centimeters

CHAPTER 10 TEST   page 498

1. (a) -10; (b) $2\sqrt{2}$; (c) 10; (d) $3y$ **2.** (a) Set of all real numbers except 2;

(b) set of all real numbers x such that $x \le \frac{5}{3}$ **3.** (a) $y = 5x$; (b) $y = \dfrac{8}{x}$

4. (a)   slope $= -3$
      domain $= \mathbb{R}$ (a)
      range $= \mathbb{R}$

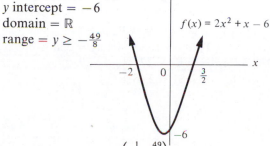

$f(x) = -3x + 4$

(b)   slope $= \frac{1}{3}$
    domain $= \mathbb{R}$ (b)
    range $= \mathbb{R}$

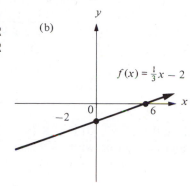

$f(x) = \frac{1}{3}x - 2$

5. (a) vertex $= \left(-\frac{1}{4}, -\frac{49}{8}\right)$
    x intercepts $= -2, \frac{3}{2}$
    y intercept $= -6$
    domain $= \mathbb{R}$
    range $= y \ge -\frac{49}{8}$

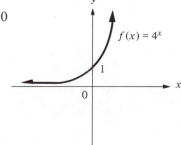

$f(x) = 2x^2 + x - 6$

$\left(-\frac{1}{4}, -\frac{49}{8}\right)$

(b) vertex $= \left(-\frac{7}{6}, \frac{25}{12}\right)$
    x intercepts $= -2, -\frac{1}{3}$
    y intercept $= -2$
    domain $= \mathbb{R}$
    range $= y \le \frac{25}{12}$

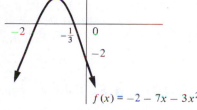

$\left(-\frac{7}{6}, \frac{25}{12}\right)$

$f(x) = -2 - 7x - 3x^2$

6. (a) domain $= \mathbb{R}$
    range $= y > 0$

$f(x) = 4^x$

(b) domain $= \mathbb{R}$
    range $= y > 0$

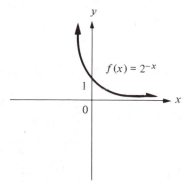

$f(x) = 2^{-x}$

(c) domain $= x > 0$
range $= \mathbb{R}$

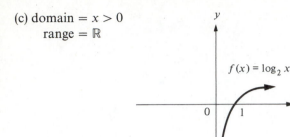

$f(x) = \log_2 x$

(d) domain $= x > -1$
range $= \mathbb{R}$

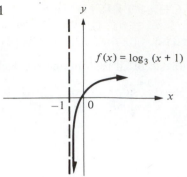

$f(x) = \log_3 (x + 1)$

7. (a) $\frac{1}{25}$; (b) 1; (c) 2; (d) $\frac{4}{3}$ **8.** (a) (3, 2), (3, −2), (−3, 2), (−3, −2); (b) (2, 1)
9. Length $= 10$ feet; width $= 5$ feet

Chapter 11

PROBLEM SET 11.1   page 504

1. $\frac{3}{2}$; 4; $\frac{15}{2}$; 12; $\frac{35}{2}$ **3.** -1; -1; 0; 2; 5   **5.** 2; 4; 2; 4; 2   **7.** A.P.; $d = 3$ **9.** A.P.; $d = 5$
11. A.P.; $d = -13$ **13.** A.P.; $d = 1.2$ **15.** $a_{10} = 50$; $a_{15} = 85$ **17.** $a_6 = 16a + 4b$; $a_9 = 25a - 8b$
19. G.P.; $r = 3$ **21.** G.P.; $r = -2$ **23.** G.P.; $r = \frac{2}{3}$ **25.** G.P.; $r = -\frac{2}{3}$ **27.** $\frac{1}{128}$ **29.** 2
31. $(1 + a)^{n-1}$ **33.** $a_6 = 192$; $a_{10} = 3,072$ **35.** 48   **37.** Sixth   **39.** \$1,262.48

PROBLEM SET 11.2   page 511

1. 15   **3.** 660   **5.** 15   **7.** 15   **9.** $\frac{163}{60}$ **11.** 500   **13.** $\frac{5}{6}$ **15.** 145   **17.** -28 **19.** $63b - 63$
21. $a_{10} = 33$; $S_{10} = 195$ **23.** $d = 13\frac{5}{51}$; $a_{18} = 239\frac{2}{3}$ **25.** $n = 32$; $d = \frac{21}{31}$ **27.** $\frac{189}{2}$ **29.** 5,460
31. $10(1 + a + a^2 + a^3 + a^4 + a^5)$ **33.** -4 or 3   **35.** -2 or 2   **37.** $a_9 = \frac{1}{128}$; $S_8 = \frac{85}{64}$
39. $n = 10$; $S_{10} = \frac{1,023}{16}$

PROBLEM SET 11.3   page 517

1. $\frac{1}{30}$ **3.** $\frac{1}{9}$ **5.** 2   **7.** 0   **9.** $(n + 1)n(n - 1)(n - 2)$ **11.** $x^5 + 10x^4 + 40x^3 + 80x^2 + 80x + 32$
13. $x^6 + 12x^4y^2 + 48x^2y^4 + 64y^6$ **15.** $a^{18} - 6a^{14} + 15a^{10} - 20a^6 + 15a^2 - 6a^{-2} + a^{-6}$
17. $32 + 80\left(\dfrac{x}{y}\right) + 80\left(\dfrac{x}{y}\right)^2 + 40\left(\dfrac{x}{y}\right)^3 + 10\left(\dfrac{x}{y}\right)^4 + \left(\dfrac{x}{y}\right)^5$
19. $256z^8 + 1,024z^7x + 1,792z^6x^2 + 1,792z^5x^3 + 1,120z^4x^4 + 448z^3x^5 + 112z^2x^6 + 16zx^7 + x^8$
21. $y^{10} - 10y^8x + 40y^6x^2 - 80y^4x^3 + 80y^2x^4 - 32x^5$ **23.** $x^{20} - 20x^{18}a + 180x^{16}a^2 - 960x^{14}a^3$
25. $\left(\dfrac{x}{2}\right)^{7/2} + 14\left(\dfrac{x}{2}\right)^3 y + 84\left(\dfrac{x}{2}\right)^{5/2} y^2 + 280\left(\dfrac{x}{2}\right)^2 y^3$ **27.** $x^{16} + 16x^{15}y + 120x^{14}y^2 + 560x^{13}y^3 + 1,820x^{12}y^4$

29. $a^{11} - 22a^{10}b^2 + 220a^9b^4 - 1,320a^8b^6 + 5,280a^7b^8$ **31.** $x^7 - 14x^6y + 84x^5y^2 - 280x^4y^3 + 560x^3y^4$
33. $a^{27} - 9a^{26} + 36a^{25} - 84a^{24} + 126a^{23}$ **35.** $\frac{455}{4,096}x^{24}a^3$ **37.** $\frac{224}{243}x^6a^{12}$ **39.** $\frac{14}{9}a^5x^{12}$

REVIEW PROBLEM SET   page 518

1. 4, 2, 4, 2 **3.** 2; $\frac{7}{2}$; 4; $\frac{17}{4}$ **5.** 1; $\frac{18}{5}$; $\frac{39}{5}$; $\frac{68}{5}$ **7.** $a_9 = 44$; $s_9 = 216$ **9.** $a_{11} = 12$; $s_{11} = 297$
11. $a_{24} = 4$; $s_{24} = 50$ **13.** 3 **15.** 3 or -2 **17.** $a_8 = 49,152$; $s_8 = 65,535$ **19.** $a_6 = -\frac{1}{3}$; $s_6 = 60\frac{2}{3}$
21. $a_{18} = -768\sqrt{2}$; $s_{18} = 1,533(1 - \sqrt{2})$ **23.** -2 or 24 **25.** $\frac{1}{7}$ or 10 **27.** 95 **29.** 20 **31.** 160
33. $x^4 + 8x^3y + 24x^2y^2 + 32xy^3 + 16y^4$ **35.** $1 + 5x + 10x^2 + 10x^3 + 5x^4 + x^5$
37. $1 - 12x + 60x^2 - 160x^3 + 240x^4 - 192x^5 + 64x^6$ **39.** $81x^4 + 108x^3y + 54x^2y^2 + 12xy^3 + y^4$
41. $243x^5 + 405x^{9/2} + 270x^4 + 90x^{7/2} + 15x^3 + x^{5/2}$ **43.** $8x^3 + \dfrac{12x^2}{y} + \dfrac{6x}{y^2} + \dfrac{1}{y^3}$ **45.** $210x^6y^4$
47. $13,440x^6y^4$ **49.** $1,082,565x^8y^3$

CHAPTER 11 TEST   page 519

1. $\frac{1}{2}$; $\frac{2}{3}$; $\frac{3}{4}$; $\frac{4}{5}$; $\frac{5}{6}$ **2.** $a_{12} = -19$; $S_{12} = -96$ **3.** $\frac{7}{4}$ **4.** $a_7 = \frac{1}{4}$; $S_7 = \frac{43}{4}$ **5.** $-\frac{7}{2}$ **6.** 82
7. $32x^5 + 80x^4y + 80x^3y^2 + 40x^2y^3 + 10xy^4 + y^5$ **8.** $-2,268x^6y^3$

Chapter 12

PROBLEM SET 12.1   page 525

3. (a) $\angle ACB$; (b) $\angle ACD$; (c) $\angle BCD$ **5.** 90° **7.** $\angle 1 = 70°$; $\angle 2 = 70°$; $\angle 3 = 40°$; $\angle 4 = 140°$; $\angle 5 = 110°$
9. (a) Vertical angles; (b) complementary angles; (c) adjacent angles; (d) supplementary angles;
(e) complementary angles **11.** (a) 70°; (b) 70°; (c) 40°; (d) 70°

PROBLEM SET 12.2   page 534

1. Congruent **3.** Congruent **5.** ASA **7.** (a) $\sqrt{84.39}$; (b) $\frac{131}{32}$; (c) 35°; (d) $\dfrac{5\sqrt{84.39}}{16}$
9. (a) $\frac{20}{3}$; (b) 4; (c) 12; (d) 6 **11.** $\angle C = 35°$; $\angle D = 48°$; $\angle E = 97°$ **13.** Similar **15.** Similar **17.** 3

PROBLEM SET 12.3   page 541

1. 15 square inches **3.** 25.42 square centimeters **5.** 40.88 square feet **7.** 64 square centimeters
9. 39.69 square feet **11.** 94.09 square yards **13.** 17 square feet **15.** 62.01 square meters
17. 26 square inches **19.** 9 square feet **21.** 18.615 square meters **23.** 18.69 square inches
25. 40 square feet **27.** 24.51 square inches **29.** $40\frac{3}{8}$ square feet **31.** 78.54 square feet
33. 30.19 square meters **35.** 60.82 square inches

PROBLEM SET 12.4   page 545

1. 16 inches   **3.** 20.6 centimeters   **5.** 25.8 feet   **7.** 32 centimeters   **9.** 25.2 feet   **11.** 38.8 yards
13. 16.8 feet   **15.** 34 meters   **17.** $20\frac{2}{3}$ inches   **19.** 12 inches   **21.** 17.7 meters   **23.** $10\frac{3}{4}$ inches
25. 24 inches   **27.** 50.4 meters   **29.** $1\frac{1}{3}$ feet   **31.** 43.98 feet   **33.** 32.67 inches   **35.** 103.04 centimeters

PROBLEM SET 12.5   page 550

1. $V = 60$ cubic inches; $S = 94$ square inches   **3.** $V = 80\frac{8}{9}$ cubic yards; $S = 114\frac{7}{9}$ square yards
5. $V = 919.275$ cubic feet; $S = 578.14$ square feet   **7.** $V = 250\pi$ cubic inches; $LS = 100\pi$ square inches;
$S = 150\pi$ square inches   **9.** $V = 189.953\pi$ cubic meters; $LS = 92.66\pi$ square meters;
$S = 126.28\pi$ square meters   **11.** $V = 87.856\pi$ cubic feet; $LS = 51.68\pi$ square feet; $S = 74.8\pi$ square feet
13. $V = 400$ cubic inches; $LS = 260$ square inches   **15.** $V = 92.38$ cubic feet; $LS = 123.6$ square feet
17. $V = 743.4$ cubic inches; $LS = 366$ square inches   **19.** $V = 16\pi$ cubic inches; $LS = 20\pi$ square inches
21. $V = 194.672\pi$ cubic feet; $LS = 105.8\pi$ square feet   **23.** $V = 8.192\pi$ cubic centimeters;
$LS = 12.8\pi$ square centimeters   **25.** $V = \dfrac{256\pi}{3}$ cubic feet; $S = 64\pi$ square feet

27. $V = 234.155\pi$ cubic meters; $S = 125.44\pi$ square meters

29. $V = \dfrac{4{,}000\pi}{3}$ cubic inches; $S = 400\pi$ square inches

REVIEW PROBLEM SET   page 551

1. (a) $\overline{AB}, \overline{AC}, \overline{AD}, \overline{BC}, \overline{CD}$; (b) $\overline{AC}, \overline{BC}, \overline{DC}$; (c) B **3.** 2   **5.** (a) $133°$; (b) $47°$; (c) $133°$
7. (a) $120°, 30°$; (b) $145°, 55°$; (c) $178°, 88°$; (d) $180°, 90°$ **9.** (a) $53°$; (b) $53°$; (c) $37°$ **11.** $15°$

13. Congruent; SAS   **15.** Not congruent   **17.** 3   **19.** $280 + \dfrac{49\pi}{2}$ square feet   **21.** 2π square units

23. $\overline{AD} = 12$; 60 square units   **25.** 81 square inches   **27.** 27 square inches   **29.** 27 square inches
31. 500 square centimeters   **33.** $\frac{6,561}{16}$ square meters   **35.** $10x^2$ square units   **37.** 480 square meters
39. $V = 84$ cubic centimeters; $S = 136$ square centimeters   **41.** $V = 45\pi$ cubic inches; $S = 48\pi$ square inches
43. $V = \dfrac{500\pi}{3}$ cubic inches; $S = 100\pi$ square inches

CHAPTER 12 TEST   page 557

1. (a) Two; (b) one   **2.** $135°$ **3.** $\angle a = \angle b = 40°$ **4.** $\angle A = \angle 3$ and $\angle C = \angle 2$ **5.** Yes   **6.** 15

7. 25 square centimeters   **8.** 20 inches   **9.** 16 inches   **10.** 4 centimeters   **11.** $\dfrac{2401\pi}{4}$ square meters

12. $36 + \dfrac{9\pi}{2}$ square units   **13.** $V = 288\pi$ cubic centimeters; $S = 144\pi$ square centimeters

14. $V = \dfrac{20\pi}{3}$ cubic inches; $S = 2\sqrt{29}\pi + 4\pi$ square inches   **15.** $V = 96$ cubic meters; $S = 136$ square meters

16. $V = 360\pi$ cubic centimeters; $S = 192\pi$ square centimeters

Index

Geometry

Assume A = area, C = circumference, V = volume, S = surface area, r = radius, h = altitude, l = length, w = width, b (or a) = length of a base, and s = length of a side.

1	Square	$A = s^2$; $P = 4s$
2	Rectangle	$A = lw$; $P = 2l + 2w$
3	Parallelogram	$A = bh$
4	Triangle	$A = \frac{1}{2}bh$
5	Circle	$A = \pi r^2$; $C = 2\pi r$
6	Trapezoid	$A = \frac{1}{2}(a + b)h$
7	Cube	$S = 6s^2$; $V = s^3$
8	Rectangular Box	$S = 2(lw + wh + lh)$; $V = lwh$
9	Cylinder	$S = 2\pi rh$; $V = \pi r^2 h$
10	Sphere	$S = 4\pi r^2$; $V = \frac{4}{3}\pi r^3$
11	Cone	$S = \pi r \sqrt{r^2 + h^2}$; $V = \frac{1}{3}\pi r^2 h$

English–Metric Conversions

Length:

1 inch = 2.540 centimeters

1 foot = 30.48 centimeters

1 yard = 0.9144 meter

1 mile = 1.609 kilometers

Volume:

1 pint = 0.4732 liter

1 quart = 0.9464 liter

1 gallon = 3.785 liters

Weight:

1 ounce = 28.35 grams

1 pound = 453.6 grams

1 pound = 0.4536 kilogram